BIOS *Instant Notes*

# Human Physiology

**BIOS INSTANT NOTES**

*Series Editor: B.D. Hames, School of Biochemistry and Molecular Biology, University of Leeds, Leeds, UK*

**Biology**
Animal Biology, Second Edition
Biochemistry, Third Edition
Bioinformatics
Chemistry for Biologists, Second Edition
Developmental Biology
Ecology, Second Edition
Genetics, Third Edition
Human Physiology
Immunology, Second Edition
Mathematics & Statistics for Life Scientists
Medical Microbiology
Microbiology, Third Edition
Molecular Biology, Third Edition
Neuroscience, Second Edition
Plant Biology, Second Edition
Sport & Exercise Biomechanics
Sport & Exercise Physiology

**Chemistry**
*Consulting Editor: Howard Stanbury*
Analytical Chemistry
Inorganic Chemistry, Second Edition
Medicinal Chemistry
Organic Chemistry, Second Edition
Physical Chemistry

**Psychology**
*Sub-series Editor: Hugh Wagner, Dept of Psychology, University of Central Lancashire, Preston, UK*
Cognitive Psychology
Physiological Psychology
Psychology
Sport & Exercise Psychology

BIOS *Instant Notes*

# Human Physiology


Daniel P McLaughlin

Associate Professor in Physiology, School of Biomedical Sciences, University of Nottingham, Nottingham, UK

Jonathan A Stamford

Associate Lecturer in Biological Psychology, The Open University (South East), East Grinstead, UK

David A White

Associate Professor and Reader in Medical Biochemistry, School of Biomedical Sciences, University of Nottingham, Nottingham, UK

With a contribution by Terence Bennett, Professor of Physiology, School of Biomedical Sciences, University of Nottingham, Nottingham, UK



Taylor & Francis
Taylor & Francis Group

First published 2007

Published by:

**Taylor & Francis Group**

In US: 711 Third Avenue, New York,
NY 10017, USA

In UK: 2 Park Square, Milton Park
Abingdon, OX14 4RN



First issued in hardback 2018

ISBN 13: 978-1-138-37310-5 (hbk)
ISBN 13: 978-0-415-35546-9 (pbk)



A catalog record for this book is available from the British Library.

**Library of Congress Cataloging-in-Publication Data**

McLaughlin, Daniel P.
Instant notes in human physiology / Daniel P. McLaughlin, Jonathan A. Stamford, David A. White ; with a contribution by Terence Bennett.
p. ; cm. -- (BIOS instant notes)
Includes bibliographical references and index.
ISBN-13: 978-0-415-35546-9 (alk. paper)
ISBN-10: 0-415-35546-X (alk. paper)
1. Human physiology. 2. Human physiology -- Handbooks, manuals, etc. I. Stamford, J. A. II. White, David A., Ph. D. III. Title. IV. Title: Human physiology. V. Series.
[DNLM: 1. Physiological Processes. 2. Physiology. QT 104 M4785i 2007]
QP34.5.I574 2007
612 -- dc22 2006025140

Editor: Elizabeth Owen
Editorial Assistant: Kirsty Lyons
Production Editor: Georgina Lucas
Typeset by: Phoenix Photosetting, Chatham, Kent, UK

T&F informa

Taylor & Francis Group, an informa business

Visit our web site at http://www.garlandscience.com

# CONTENTS

# ABBREVIATIONS

| | |
|---|---|
| ABP | androgen-binding protein |
| ABPI | ankle:brachial pressure index |
| ACE | angiotensin-converting enzyme |
| acetyl-CoA | acetyl coenzyme A |
| ACh | acetylcholine |
| AChE | acetylcholinesterase |
| AChR | acetylcholine receptor |
| ACTH | adrenocorticotropic hormone, corticotropin |
| ADH | antidiuretic hormone (vasopressin) |
| ADP | adenosine diphosphate |
| AMP | adenosine monophosphate |
| AMPA | α-amino-3-hydroxyl-5-methyl-4-isoxazole-propionic acid |
| ANP | atrial natriuretic peptide |
| ANS | autonomic nervous system |
| AP | action potential |
| AT | angiotensin |
| ATP | adenosine triphosphate |
| AV node | atrioventricular node |
| BP | blood pressure |
| $Ca^{2+}$ | calcium ion |
| cAMP | cyclic adenosine monophosphate |
| CBG | cortisol-binding globulin, transcortin |
| CCK | cholecystokinin |
| CF | cystic fibrosis |
| CFTR | cystic fibrosis transmembrane conductance regulator |
| cGMP | cyclic guanosine monophosphate |
| ChAT | choline acetyl transferase |
| CIA | central inspiratory activity |
| $Cl^-$ | chloride ion |
| CNS | central nervous system |
| CO | cardiac output |
| CoA | coenzyme A |
| $CO_2$ | carbon dioxide |
| COMT | catechol-*O*-methyltransferase |
| COPD | chronic obstructive pulmonary disease |
| COX1 | cyclo-oxygenase 1 |
| COX2 | cyclo-oxygenase 2 |
| CSF | cerebrospinal fluid |
| CRH | corticotropin-releasing hormone |
| DAG | diacylglycerol |
| DBH | dopamine-β-hydroxylase |
| DBP | diastolic blood pressure |
| DCT | distal convoluted tubule |
| 2-DG | 2-deoxyglucose |
| DHA | dehydroepiandrosterone |
| DIT | di-iodotyrosine |
| DNA | deoxyribonucleic acid |
| DOPA | 3,4-dihydroxyphenylalanine |
| DOPAC | 3,4-dihydroxyphenylacetic acid |
| DPG | 2,3-diphosphoglycerate |
| DPPC | dipalmitoylphosphatidyl-choline |
| DRG | dorsal root ganglion/dorsal respiratory group |
| ECF | extracellular fluid |
| ECG | electrocardiogram |
| EDRF | endothelium-derived relaxant factor |
| $E_m$ | membrane potential |
| EPO | erythropoietin |
| EPP | endplate potential |
| EPSP | excitatory postsynaptic potential |
| ER | endoplasmic reticulum |
| ESR | erythrocyte sedimentation rate |
| $FADH_2$ | flavine adenine dinucleotide, reduced form |
| $FEV_1$ | forced expiratory volume in 1 s |
| fMRI | functional magnetic resonance imaging |
| FSH | follicle-stimulating hormone |
| FVC | forced vital capacity |
| GABA | γ-aminobutyric acid |
| GDP | guanosine diphosphate |
| GERD | gastro-esophageal reflux disease |
| GFR | glomerular filtration rate |

| | |
|---|---|
| GH | growth hormone, somatotropin |
| GHRH | growth hormone-releasing hormone |
| GI | gastrointestinal |
| GLP-1 | glucagon-like peptide 1 |
| GLP-2 | glucagon-like peptide 2 |
| GLUT | glucose transporter |
| GNG | gluconeogenesis |
| GnRH | gonadotropin-releasing hormone |
| GRE | glucocorticoid response element |
| GTN | glyceryl trinitrate |
| GTO | Golgi tendon organ |
| GTP | guanosine triphosphate |
| $H^+$ | hydrogen ion, proton |
| $H_2CO_3$ | carbonic acid |
| Hb | hemoglobin |
| HbA | hemoglobin, adult |
| HbF | hemoglobin, fetal |
| HCG | human chorionic gonadotropin |
| HCl | hydrochloric acid |
| $HCO_3^-$ | bicarbonate ion |
| HDL | high-density lipoprotein |
| HHb | hemoglobin, reduced form |
| 5-HIAA | 5-hydroxyindoleacetic acid |
| HPL | human placental lactogen |
| HPV | hypoxia-induced pulmonary vasoconstriction |
| HR | heart rate |
| ICM | inner cell mass |
| IGF-1 | insulin-like growth factor-1 |
| IL-1 | interleukin-1 |
| IL-6 | interleukin-6 |
| $IP_3$ | inositol trisphosphate |
| IPSP | inhibitory postsynaptic potential |
| $K^+$ | potassium ion |
| $K_a$ | dissociation constant of an acid |
| LAD | left anterior descending (coronary artery) |
| L-DOPA | L-3,4-dihydroxyphenylalanine |
| LDL | low-density lipoprotein |
| LES | lower esophageal sphincter |
| LGN | lateral geniculate nucleus |
| LH | luteinizing hormone |
| LMN | lower motor neuron |
| LVST | lateral vestibulospinal tract |
| MABP | mean arterial blood pressure |
| mAChR | muscarinic acetylcholine receptor |
| MALT | mucosa-associated lymphoid tissue |
| MAO | monoamine oxidase |
| MDRD | Modification of Diet in Renal Disease |
| mEPP | miniature endplate potential |
| MGN | medial geniculate nucleus |
| MIT | mono-iodotyrosine |
| MLCK | myosin light chain kinase |
| MLF | medial longitudinal fasciculus |
| mRNA | messenger ribonucleic acid |
| α-MSH | α-melanocyte-stimulating hormone |
| MVST | medial vestibulospinal tract |
| $Na^+/K^+$ ATPase | ATP-dependent sodium/potassium pump |
| $Na^+$ | sodium ion |
| nAChRs | nictonic acetylcholine receptors |
| NADH | nicotinamide adenine dinucleotide, reduced form |
| NADPH | nicotinamide adenine dinucleotide phosphate, reduced form |
| NAPQI | N-acetyl-*p*-benzo-quinone imine |
| NE | norepinephrine |
| NFP | net filtration pressure |
| $NH_3$ | ammonia |
| $NH_4^+$ | ammonium |
| NK | natural killer (cell) |
| NMDA | *N*-methyl-D-aspartate |
| NMJ | neuromuscular junction |
| NO | nitric oxide |
| $O_2$ | oxygen |
| $ONOO^-$ | peroxynitrite anion |
| $Pa\text{CO}_2$ | carbon dioxide tension of arterial blood |
| PAH | para-aminohippuric acid |
| $P\text{AO}_2$ | oxygen tension of alveolar air |
| $Pa\text{O}_2$ | oxygen tension of arterial blood |
| $P_{BS}$ | Bowman's space hydrostatic pressure |

| | |
|---|---|
| $P_{cap}$ | capillary hydrostatic pressure |
| PCT | proximal convoluted tubule |
| PEF | peak expiratory flow |
| PepT1 | proton-coupled oligopeptide transporter-1 |
| PET | positron emission tomography |
| $PGE_2$ | prostaglandin $E_2$ |
| $PGG_2$ | prostaglandin $G_2$ |
| $PGH_2$ | prostaglandin $H_2$ |
| $PGI_2$ | prostacyclin |
| pH | $-\log_{10} [H^+]$ |
| $P_i$ | inorganic phosphate |
| PIF | peak inspiratory flow |
| $PIP_2$ | phosphatidylinositol bisphosphate |
| PKA | protein kinase A |
| $pK_a$ | $-\log_{10} [K_a]$ |
| PKC | protein kinase C |
| PLC | phospholipase C |
| PMN | polymorphonuclear leukocyte |
| PNMT | phenylethanolamine-*N*-methyltransferase |
| PNS | peripheral nervous system |
| $PO_2$ | oxygen tension |
| POMC | pro-opiomelanocortin |
| PPP | pentose phosphate pathway |
| PRH | prolactin-releasing hormone |
| $P_t$ | tissue hydrostatic pressure |
| PTH | parathyroid hormone |
| PTHrP | parathyroid hormone-related protein |
| $\dot{Q}$ | cardiac output |
| RCA | right coronary artery |
| RER | rough endoplasmic reticulum |
| RNA | ribonucleic acid |
| ROS | reactive oxygen species |
| rRNA | ribosomal ribonucleic acid |
| $rT_3$ | reverse triiodothyronine |
| RPF | renal plasma flow |
| RQ | respiratory quotient |
| RV | residual volume |
| SA | sinoatrial (node) |
| SBP | systolic blood pressure |
| SGLT1 | sodium-glucose ligand transporter-1 |
| SHBG | sex-hormone-binding globulin |
| SR | sarcoplasmic reticulum |
| SRIF | somatotropin release-inhibiting factor, somatostatin |
| SV | stroke volume |
| $T_3$ | triiodothyronine |
| $T_4$ | thyroxine |
| TBG | thyroid-binding globulin |
| TCA | tricarboxylic acid |
| TG | thyroglobulin |
| $T_m$ | transport maximum |
| TNF-α | tumor necrosis factor-α |
| TPR | total peripheral resistance |
| TRH | thyrotropin-releasing hormone |
| tRNA | transfer ribonucleic acid |
| TSH | thyroid-stimulating hormone, thyrotropin |
| $TxA_2$ | thromboxane $A_2$ |
| UDP | uridine diphosphate |
| UES | upper esophageal sphincter |
| UMN | upper motor neuron |
| UP | ultrafiltration pressure |
| UV | ultraviolet |
| $\dot{V}$ | ventilation |
| $\dot{V}/\dot{Q}$ | ventilation–perfusion ratio |
| VC | vital capacity |
| VIP | vasoactive intestinal peptide |
| $1{,}25(OH)_2$ vitD | 1,25-dihydroxy vitamin D |
| 25(OH) vitD | 25-hydroxy vitamin D |
| VLDL | very low-density lipoprotein |
| VRG | ventral respiratory group |
| VS | voltage-sensitive (ion channel) |
| VSCC | voltage-sensitive calcium channel |
| $\pi_{BS}$ | Bowman's space colloid osmotic pressure |
| $\pi_{cap}$ | blood colloid osmotic pressure |
| $\pi_t$ | tissue colloid osmotic pressure |

# PREFACE

The majority of students who study human physiology as part of their degree programs throughout the world are not studying the subject in isolation. Coupled with the burgeoning body of information that is being uncovered every year, this means that students are often presented with information overload. The purpose of this text is to give a full overview of the important aspects of human physiology in the *Instant Notes* format. Information is presented in 13 Sections and 85 Topics, but the sheer scope of human physiology inevitably means that the book is non-exhaustive in its detail. However, we have given pointers to useful 'Further reading' at the end of the book to help students who feel the need to go into more depth in their understanding of particular elements of human physiology (we hope this will be all of you). The individual Topics in the book should be self-explanatory, although a list of 'Related topics' is given at the start of each. If you choose to read the book in its entirety, you will find that our aim has been to approach the study of human physiology from a fundamental starting point: that our life purpose, in common with all other species, is to reproduce. However, to find a mate and reproduce successfully, we must move around in our environment and respond to the challenges with which we are presented. Non-reproductive bodily functions such as circulation, respiration, locomotion, and waste disposal simply facilitate this process.

Sections A–C deal with fundamental physiological concepts, important cell biology and basic information on body structure that are essential learning if one hopes to understand most of human physiology. Sections D and E deal with the cardiovascular and respiratory systems, the human body's mechanism for distributing nutrients and oxygen, and for collecting and partly discarding waste products. Sections F, G, and H (brain, spinal cord, and peripheral nerve) describe the neural mechanisms by which many of our bodily processes are controlled. The nervous system, together with the musculature (Section I) are particularly important in allowing us to move around in our environment. Inevitably, all of this movement and potential perturbation of physiological equilibrium (homeostasis) means that we must obtain nutrients to run these processes, hence this is why Section J covers gastrointestinal physiology. Section K describes the hormonal control of many aspects of body function, the endocrine system being very important in control of circulating nutrients and in reproduction (Section L). Finally, waste products must be eliminated by the kidneys (Section M), which just happen to also be major homeostatic organs. Living is a challenge. The fact that we hardly notice this is testament to our exquisite evolution.

*Danny McLaughlin*

# A1 INTRODUCTION TO PHYSIOLOGY

## Key Notes

**What is physiology?**

Physiology is an academic discipline that attempts to explain the function of body systems in chemical and physical terms. Recent advances in a number of other academic disciplines such as cell biology and molecular biology are important in understanding modern physiological concepts.

**Essential principles**

Humans consist of between $10^{13}$ and $10^{14}$ cells, bathed in a fluid rich in electrolytes and macromolecules such as proteins. All cellular functions can be described in chemical terms, so it is important to have solid understanding of fundamental chemistry in order to understand cellular physiology. The structure of tissues and organs is important in determining their function, so a reasonable amount of histological and anatomical knowledge is essential in understanding the physiology of all body systems.

## What is physiology?

Nowadays, human physiology is rarely taught as an academic discipline in its own right, and one is often asked by students what physiology actually is. This question is entirely valid; where does one draw a line that distinguishes physiology from biochemistry? Or physiology from cell biology? The answer to both these questions is that such lines of demarcation do not exist. Nor should they, in today's study of the biomedical sciences. Human physiology is now, quite rightly, taught as part of more integrated degree courses. This fact may cause pain to some of us who were inspired and delighted in equal measure by our lecturers and tutors at university, career physiologists to a man/woman, but this is a fact of life. The definition of the academic discipline of 'physiology' given to me – *the study of body function that seeks to describe biological phenomena in chemical and physical terms* – seemed vast, even then (and that was as recently as the 1980s).

In the interim, the enormous advances that have been made in genetics, molecular biology and receptor classification (to give only a few examples) have made this even more wide ranging. For example, one cannot expect to understand how a biological molecule exerts its panoply of effects *in vivo* if one does not have a solid grasp of the reasons why the receptors through which it acts are so different, or why some cells do different things in response to the same stimulus.

## Essential principles

It is also very important for students of physiology to understand the basic principles of chemistry, especially as related to the physicochemical properties of ions, carbohydrates, proteins and lipids; after all, human organisms consist of relatively few cells bathed in fluids that are solutions/suspensions of such chemicals. A number of topics in this text deal explicitly with such concepts, but a small amount of revision of fundamental chemistry would not go amiss before starting.

A degree of understanding of cell biology and tissue structure is also an essential prerequisite for the study of physiology. Nowadays, 'physiology' combines the more traditional approach to body systems with cellular biochemistry, and this is essential if one hopes to have a solid understanding of the subject. Where appropriate, one must try to gain knowledge of the structure and function of cells and tissues in order to understand why cells designed to do roughly the same things (e.g. contract) can have very different characteristics (e.g. skeletal, cardiac and smooth muscle). Thus a modicum of cell biological, histological and anatomical knowledge is also required before one can fully comprehend the physiology of some cells/tissues. Where appropriate, we have cited useful texts in these areas that should help support these aims.

# A2 HOMEOSTASIS AND INTEGRATION OF BODY SYSTEMS

## Key Notes

**The *milieu interieur***

In order to live, we do things that might affect the composition of extracellular fluid in which our cells are bathed. Physiological control systems seek to maintain the composition of extracellular fluid, also known as the *milieu interieur*. Together these processes maintain homeostasis.

**Feedback loops**

Physiological control systems normally function by means of feedback loops; these can be negative feedback or positive feedback loops. Negative feedback loops, where the end result counteracts some perturbation in a measured parameter, are more common in physiological processes.

**Components of a control system**

Physiological control systems are similar to those in engineering. The common components of a physiological control system are: a sensor (receptor), an input to the control system (often a neural input to the brain), and an output (either neural or hormonal in nature) to effector tissues.

**Integration**

In an effort to maintain homeostasis (maintenance of the *milieu interieur*), our bodies often utilize several body systems. For example, in order to maintain blood pH at its 'normal' level of 7.4, we use both the respiratory system and the renal system to alter the level of various chemicals that have an effect on pH. Integrated responses of body systems to a change in the *milieu interieur* allow us to adapt rapidly to our environment.

**Related topics**

The cardiac cycle, blood pressure and its maintenance (D6)
Chemoreceptors and respiratory acid–base balance (E6)
Structure and function of the autonomic nervous system (H5)
Hormone action (K3)
Birth and lactation (L6)
Acid–base balance (M4)
Hormonal control of fluid balance (M5)

## The *milieu interieur*

In the 18th century, the French scientist Claude Bernard first described the concept of the ***'milieu interieur'***, the internal environment of the body. This concept led directly to the study of the maintenance of the extracellular environment (**homeostasis**), since this influences the function of all cells in the body. It is important to note that not all organisms are capable of homeostasis. For example, unicellular organisms may be able to secrete some substance that provides an environment that is more conducive to cell growth, or they may have a limited ability to move themselves around by way of primitive flagellae, but on the whole there is very little that such organisms can do to modify their extracellular environment.

However, as one moves up the phylogenetic tree, the ability of organisms to control the composition of their extracellular fluid increases until, in the case of humans, there is a remarkable capacity to adjust cell processes in order to maintain a more or less constant extracellular environment. Why should this be so? In order to carry out our daily lives, we constantly place ourselves in the path of challenges to our *milieu interieur*. For example, the simple act of running to catch a bus means that our muscles require an increased blood supply and oxygen delivery in order to complete the task. If we did not increase our ventilation and cardiac output in order to match demand, there is a possibility that our blood might become dangerously hypoxic (low on oxygen) and hypercapnic (high in carbon dioxide ($CO_2$)).

**Feedback loops**

The other, more difficult question is: how is it possible? Our responses to such simple challenges are feats of integrated control engineering, the mimicry of which would test the capabilities of the most talented scientists on earth; and yet the individual components are simple. Almost all of our actions are enabled by two basic control processes: negative and positive feedback loops. In **negative feedback**, the more common of the two, an increase in the level of one factor is corrected for by another action that serves to decrease the level of the factor. One example is body temperature, where a fall in core temperature results in an increase in shivering, non-shivering thermogenesis, piloerection (raising of hairs on the skin) and an increase in metabolic rate, allowing us to generate more heat and raise body temperature back to its 'ideal' mean of 37°C. This is akin to a fall in temperature in your hallway at home that results in the central heating system kicking in to increase the temperature near the thermostat.

In **positive feedback**, an increase in the level of one factor leads to another action that serves to increase the level of the factor further still. Thus, the role of positive feedback mechanisms is not to stabilize or regulate a physiological parameter. In most cases, this could be detrimental, but it is used only rarely in physiological systems. A good example is the interplay between the hormones **estrogen** and **luteinizing hormone** (LH) in preparing for ovulation, where estrogen acts to release LH, which in turn leads to increased production of estrogen by the ovaries. The purpose of this positive feedback is to accelerate the development of ovarian follicles in preparation for **ovulation**.

**Components of a control system**

Physiological control systems have the same components as control systems in engineering (*Fig. 1*):

- a sensor, with a set point and an output to:
- a control unit with a link to:
- an effector system, which has variable levels of activation, and which may have two antagonistic arms.

Variability of parameters around a set point is an inherent flaw in such systems and the same is true in physiology. Our core temperature does not remain at 37°C, but fluctuates a little around this mark, slight increases and decreases being compensated for by negative feedback mechanisms in either direction (again, just like a central heating system switching on and off). What is

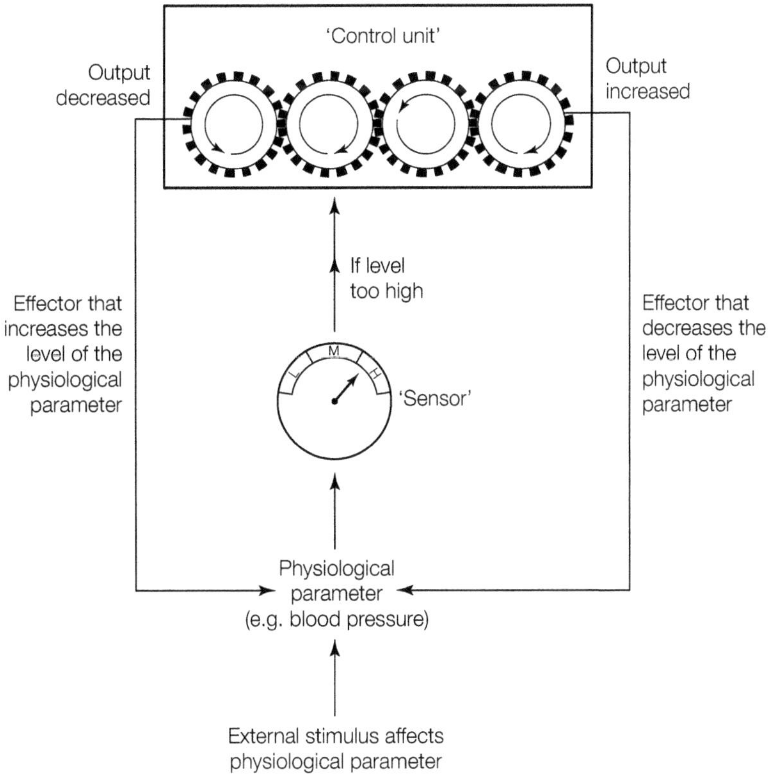


*Fig. 1. Components of a physiological control system. In the example presented (blood pressure), an increase in the physiological parameter being monitored results in a reduction in the output of an effector (in this case the sympathetic branch of the autonomic nervous system) that would increase the level of the physiological parameter, and an increase in the output of an effector (in this case the parasympathetic branch of the autonomic nervous system) that would decrease the level of the physiological parameter.*

important is the degree of error around the set point that will trigger the negative feedback mechanisms. One example of where such mechanisms can go awry is in the control of blood pressure. **Blood pressure** is sensed by receptors (**baroreceptors**) in the walls of major blood vessels, which send information back to the central nervous system so that neuronal output to the heart and blood vessels can be controlled. An increase in blood pressure above the 'normal' mean of 90–100 mmHg would normally result in reflex slowing of heart rate and dilatation of blood vessels to lower blood pressure; but in people with high blood pressure (hypertension), the sensitivity of the baroreceptors to increases in blood pressure is diminished, which means that larger increases in blood pressure are required to initiate the reflex negative feedback loop. This, in turn, leads to prolonged periods of hypertension and further desensitization of the baroreceptors, a vicious pathological cycle that causes end-organ damage, some of which may result in further hypertension.

All feedback loops involve the action of nerves or hormones, or both, as in the examples given above. These are the 'brakes' and 'accelerators' of physiological systems, and normal function is usually impossible without them. Pathological conditions that lead to abnormal nerve function or endocrine deficiency or hyperactivity are debilitating. You will be able to read more about these aspects of physiology in later sections (see Sections H and K). For now, it is important that you understand the concept of homeostasis, i.e. the maintenance of the *milieu interieur*, the state and composition of the extracellular fluid.

## Integration

As we explore human physiology in this book, it will become apparent that no one body system acts in isolation. The brain, for example, is the regulator of all body functions at some level, whether this regulation is achieved through neural (nervous) or neurohumoral (nervous and hormonal) influences. Similarly, since none of the cells in the body can exist without a supply of oxygen-rich blood, the heart and blood vessels (the cardiovascular system) have a central role in the function of all organs. Problems with the circulatory system rapidly lead to pathology of one form or another. Neuronal output from the brain plays a large part in modulation of the cardiovascular system and the brain is the part of the body most dependent upon an oxygen-rich blood supply, so it should be immediately apparent that these two systems are intimately linked.

Let us throw another two body systems into the mix, namely the respiratory and renal (kidney) systems. Achieving adequate oxygen delivery to the tissues is impossible unless ventilation is matched to perfusion of the lungs by blood from the right side of the heart, and return of blood from the lungs to the left side of the heart is matched by a concomitant output to the systemic circulation. Where do the kidneys come in? Carbon dioxide ($CO_2$) produced by cellular metabolism is normally excreted by blowing this off at the lungs, but if this system is compromised by poor ventilatory function, dangerously low blood pH (high $H^+$ ion concentration – **acidosis**) can result. The kidneys can compensate for this by excreting excess acid in the form of **ammonium** ions ($NH4^+$) if sufficient **ammonia** ($NH_3$) is available in kidney epithelial cells. The level of detail and backup in such systems is exquisite, reflecting millions of years of evolutionary selection.

What all of this means is that in reading through this text, you will need to be aware that when physiological processes are described, the knock-on effect may extend much further than you imagine. Understanding the physiological response of one type of cell is not enough – try to think of the bigger picture!

# B1 CELL ORGANELLES

## Key Notes

**The plasmalemma**

Virtually all cells in the body have a full complement of cell organelles. Organelles are cell constituents that have a particular role to play in the function of the cell, akin to one activity of a community or city. The plasmalemma (cell membrane) is a modified phospholipid bilayer that acts as the cell boundary; some substances can pass through the plasmalemma easily, whereas it is almost impermeable to others.

**Cytoplasm and nucleus**

The cytoplasm is the intracellular fluid in which many activities of the cell take place. The activities of the cell are controlled by the nucleus, which contains the genomic deoxyribonucleic acid (DNA) that encodes for almost all of the proteins that the cell synthesizes. Proteins are essential for life.

**Protein production and transport**

Genomic DNA is transcribed into messenger ribonucleic acid (mRNA) in the nucleus. This mRNA is exported to the cytoplasm where it is translated into proteins on free ribosomes (for cytoplasmic proteins) or on rough endoplasmic reticulum (in the case of proteins destined for cell membranes or for export to other cells). Proteins are transported along intracellular 'scaffolding' known as the cytoskeleton.

**Energy requirements**

Numerous different metabolic processes supply an energy substrate for the work that the cell is doing. This energy substrate is normally adenosine triphosphate (ATP), and the chemical energy is stored in the phosphate bond between adenosine diphosphate (ADP) and inorganic phosphate in the ATP molecule. The mitochondria are the sites of oxidative phosphorylation, the most efficient method that the cell has for making ATP.

**Waste disposal**

All cellular activities produce waste. In some cases, this waste is chemical and is eventually excreted by the kidneys or in the feces. Other cell waste is structural and is degraded in an organized fashion by endosomes and lysosomes.

**Related topics**

Gene transcription and protein translation (B2)
Metabolic processes (B3)
Membrane potential (B4)
Utilization of $O_2$ and production of $CO_2$ in tissues (E5)

## The plasmalemma

The cell is like a city. What it does, when viewed at the molecular level, seems impossibly complex. In the same way, getting to know the function of everyone in your home town is impossible, but we can all find the phone number of the local Chinese take-away or an emergency plumber when the need arises; we use the *Yellow Pages*. It's not necessary to know the function of every protein in the

*Table 1. Functions of the major cellular organelles*

| Organelle | Major function(s) |
|---|---|
| Plasmalemma (cell membrane) | Regulation of movement of substances into and out of the cell |
| Nucleus | Site of genomic DNA; transcription of DNA into RNAs takes place here |
| Ribosomes | Site of translation of mRNA into protein |
| Endoplasmic reticulum (ER) | Intracellular membrane system – rough ER has ribosomes that translate mRNAs into proteins destined for membranes in the cell or for transport out of the cell |
| Golgi apparatus | Intracellular membrane system that is the site of much post-translational modification of proteins into their final forms |
| Cytoskeleton | Intracellular protein filaments used to transport substances around the cell |
| Mitochondrion | Site of mitochondrial DNA. Krebs' cycle and oxidative phosphorylation take place here (much ATP production) |

cell, but understanding the major things that the cell can do is a great help. *Table 1* shows the functions of the most important cellular organelles

The **plasmalemma** is basically a double layer of a specific type of fat molecule, known as a **phospholipid**. These molecules arrange themselves in such a way that their water-miscible parts stick out into the extracellular fluid and into the intracellular fluid (**cytoplasm**), and their lipid parts come together to form a fluid membrane. Like oil and water, they don't mix easily. This means that any molecule that isn't readily soluble in lipid (fat) finds it difficult to gain entry to the cell. *Figure 1* shows a schematic of the various activities of the cell.

## Cytoplasm and nucleus

The cytoplasm is the atmosphere in this city-cell. Since it's an aqueous environment, it contains dissolved proteins, ions, and the essentials of life, like amino acids and glucose. However, like our atmosphere, it also contains some substances that cannot build up as they may become toxic.

The controller of the city-cell, or its mayor if you will, is the **nucleus**. Like city hall, the nucleus houses the local legislature, which passes laws about what can and cannot be done within the city boundary, and signs off on planning applications. In the context of the city-cell, this role is performed by the **genomic DNA**, which contains the 'blueprint for life', deciding which genes should be expressed when (and a good many other things besides, the like of which we don't yet fully understand). Genes run the show. They're the building control officers, the public health inspectors, and the like.

## Protein production and transport

Like all cities, things run on exchange of goods. These commodities are **proteins**. Genes give the go-ahead and proteins can be synthesized. The small factories that manufacture the proteins are the **ribosomes**, encoded by stretches of DNA, the ribosomes themselves being synthesized right in the heart of the nucleus, the **nucleolus**. However, the ribosomes need to get their product into the marketplace and that is where the **endoplasmic reticulum** and **Golgi apparatus** come in. The endoplasmic reticulum ships the products to the Golgi apparatus in bulk, devoid of the nice shiny packaging that attracts consumers, and the Golgi

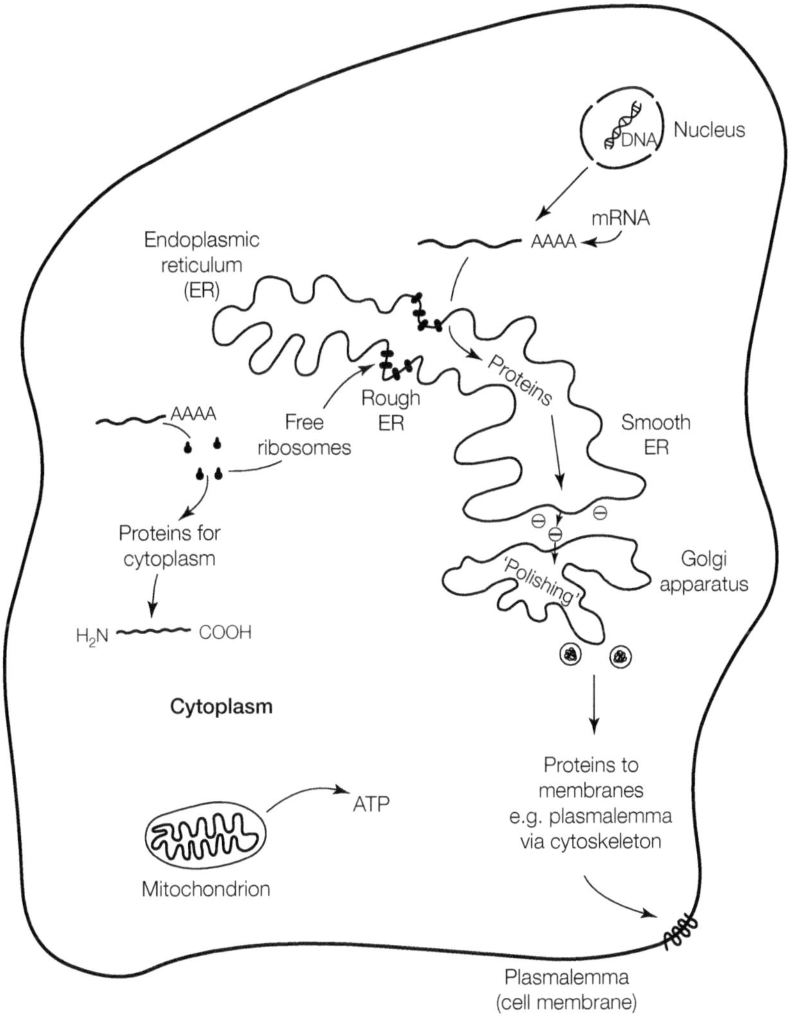


*Fig. 1. Activities of the cell. Messenger ribonucleic acid (mRNA) is polyadenylated ($AAA_n$) at the 3′-end.*

apparatus polishes things up before allowing the proteins to be used by the consumers in the cytoplasm and elsewhere.

It is important to realize that transport of the various substances in the cell does not rely solely on diffusion or Brownian motion. There are many intracellular structural proteins that help to give the cell a more rigid structure (like internal scaffolding), but which also provide structural connections between the various parts of the cell along which proteins can be transported. This intracellular protein network is known as the **cytoskeleton**, and is like the road network of the city-cell.

## Energy requirements

An important question is how all of this runs without power (or money). The molecule **adenosine triphosphate** (ATP) is the energy currency of the cell (see Topic B3). Splitting the high-energy phosphate bond to convert ATP to adenosine diphosphate (ADP) and **inorganic phosphate** ($P_i$) releases energy that can be used by cellular processes. ATP is produced in the cell by splitting a molecule of **glucose** into two three-carbon **carbohydrate** molecules (**glycolysis**, producing only a few molecules of ATP), then 'burning' these short-chain carbon molecules in the presence of oxygen in the **mitochondrion** (**oxidative phosphorylation**, producing more than 30 molecules of ATP). As such, the mitochondria can be considered to be the power stations of the city-cell, generating an easily-used energy source from carbon-based fuel. More details of these processes are given in Topic B3.

**Waste disposal**

Finally, all this work inevitably means that some things get broken. Cell components need to be mopped up and repaired on a regular basis. Also, rogue elements and invaders can threaten the delicate equilibrium of the city-cell. Getting rid of rubbish, and destroying invaders is the job of **endosomes** and **lysosomes**. Cellular material, and in some cases whole foreign cells, can be enclosed in a lipid sphere (vesicle) known as an endosome. When lysosomes containing destructive enzymes are fused with these, the material in the vesicle (now an **endolysosome**) can be degraded for recycling.

# B2 GENE TRANSCRIPTION AND PROTEIN TRANSLATION

## Key Notes

**The genome**

The nucleus of cells contains genomic DNA, 50% of which is inherited from each parent. Genomic DNA is organized into 46 chromosomes: 22 pairs of autosomal chromosomes and 1 pair of sex-chromosomes that determine sex. This complement of 46 chromosomes is known as the human genome and contains between 30 000 and 40 000 genes.

**Structure of DNA**

Genomic DNA is composed of two complementary strands, arranged in a double helix. The strands of DNA are composed of bases, sugars (deoxyribose) and phosphate. The bases in DNA are adenine, guanine, cytosine and thymine; adenine pairs with thymine and guanine pairs with cytosine on the complementary strands.

**DNA transcription**

In DNA transcription, one of the strands of genomic DNA is copied using the enzyme RNA polymerase II to make mRNA. RNA is composed of bases, sugars (ribose) and phosphate. The bases in RNA are adenine, guanine, cytosine and uracil. The mRNA that is synthesized in DNA transcription passes into the cytoplasm, where it is used to make a protein on ribosomes. Not all genes are transcribed at all times.

**mRNA translation**

mRNA is translated into protein on ribosomes (either free ribosomes in the cytoplasm, or sitting on rough endoplasmic reticulum (RER)). The amino acid sequence of the protein to be synthesized is determined by the base sequence of the mRNA, which in turn is dependent on the base sequence of the genomic DNA from which it was derived. The mRNA sequence is read three bases (a codon) at a time. Transfer RNA (tRNA) molecules with specific three-base anticodon sequences, and coupled to specific amino acids, are used in the process of mRNA translation to make a protein with a predetermined sequence. Proteins undergo post-translational modification in the cell.

**Related topics**

Cell organelles (B1)
The panoply of transmitters (F4)
Hormone action (K3)

## The genome

Everything that the cells of the body do is controlled by having a set of plans, more commonly referred to as the human **genome**. Genomic deoxyribonucleic acid (**genomic DNA**), found in the **nucleus** of the cell, contains elements known as **genes** that code for the **proteins** that our cells synthesize in order to carry out their various functions. Different cells express different genes, depending on what the cell must do, so the process is tightly controlled and regulated.

The genomic DNA contained in the nucleus of the cell can be gathered together into 23 pairs of **chromosomes**, 22 pairs that do not determine sex and 1 pair that does (the sex chromosomes, two X chromosomes in females and one X and one Y chromosome in males). Numbering of the chromosomes in the human cell is based roughly on their relative size; chromosome 1 is the largest, and chromosome 22 the smallest. However, this is unrelated to the number of genes on each chromosome. A current estimate of the number of genes in the human genome runs to between 30 000 and 40 000, but simply because a chromosome is larger doesn't mean it will definitely have more genes than its neighbor. Also, only around 2% of the human genome has protein-coding function, and so much of the genome is composed of stretches of DNA with, as yet, cryptic function.

## Structure of DNA

So how is genomic DNA used to synthesize proteins? This topic is dealt with in more detail in *Instant Notes in Genetics* and *Instant Notes in Molecular Biology*, but a brief description is appropriate here. First, one must consider the chemical structure of DNA. DNA is composed of two polymer strings arranged opposite one another. The monomers of DNA are small molecules called **bases**, joined together into a polymer by a sugar (**deoxyribose**) and phosphate backbone. The bases in DNA are **adenine** (A), **guanine** (G), **cytosine** (C) and **thymine** (T). Two of these (A and G) are **purines** and two (C and T) are **pyrimidines**. In 1953, James Watson and Francis Crick discovered that these four bases paired up with each other in a very specific fashion to form two complementary strands; A and T pair together with two hydrogen bonds, and G and C pair together with three hydrogen bonds. Thus, if one strand of the DNA molecule has a specific base sequence, the complementary strand will also have a specific base sequence, as shown in *Fig. 1*.

```
DNA
'Sense' strand       5'ATGTGCTATCGATACCGCTCGCTAAGATAAGATCCGATCCGCTTCGGCT3'
                       |||||||||||||||||||||||||||||||||||||||||||||||||
'Antisense' strand   3'TACACGATAGCTATGGCGAGCGATTCTATTCTAGGCTAGGCGAAGCCGA5'

mRNA                 5'AUGUGCUAUCGAUACCGCUCGCUAAGAUAAGAUCCGAUCCGCUUCGGCU3'
```

*Fig. 1. The relationship between the sequence of bases in the two strands of genomic DNA and the sequence of bases in the RNA transcribed.*

## DNA transcription

This process occurs in the nucleus of the cell and is dealt with in more detail in *Instant Notes in Molecular Biology*, but again a brief description is necessary here. When genomic DNA is copied prior to cell division, a copy of both these strands is made, to be used in the daughter cell, by specific enzymes known as **DNA polymerases**. However, as part of normal cell function, other enzymes known as ribonucleic acid polymerases (**RNA polymerases**) are able to read the antisense strand of the genomic DNA contained in the genes and produce the various RNAs that the cell needs, such as ribosomal RNA (**rRNA**), transfer RNA (**tRNA**) and messenger RNA (**mRNA**). All of these RNAs have a role to play in protein synthesis in the cell (see below). The mRNAs, which code for proteins, are complementary strands with the 'same' sequence as the sense strand of genomic DNA (Figure 1). However, RNAs differ from DNA in two regards: RNA bases are A, G, C, and U (**uracil**), and the sugar molecules making up part of the backbone of the polymer are **ribose**, not deoxyribose, hence its name. When mRNA has been synthesized, it is processed and released into the **cytoplasm**, where it is used to make protein polymers from **amino acids**. It is important to note that the process of **gene transcription** is highly regulated. Only some genes are transcribed at any one time; intercellular signaling molecules such as hormones can affect gene transcription in either direction.

**mRNA translation**

This process takes place in the cytoplasm of the cell, on free **ribosomes**, and at the specialized rough endoplasmic reticulum (ER with associated ribosomes, RER). Generally speaking, mRNAs that code for cytoplasmic proteins are translated on free ribosomes, whereas those destined for membranes in the cell or for export to other cells are translated at the rough ER. Mammalian ribosomes are assemblies of four rRNAs and approximately 75 proteins.

A more detailed description of this process is given in *Instant Notes in Molecular Biology*. The base sequence of the mRNA is a message to the cell to make a protein with a particular amino acid sequence. Runs of three bases in mRNA known as **codons** (of which there are 64 possible combinations) tell the cell which amino acids to use. Amino acids that are to be used for protein synthesis (**translation**) are attached to tRNA. tRNA is folded in such a fashion that anticodons complementary to the codons of mRNA are available for tRNA–mRNA interaction at the ribosomes. Thus, the mRNA sequence from *Fig. 1* is converted to a short protein with the sequence methionine-cysteine-tyrosine-proline-tyrosine-arginine-serine-leucine-arginine, as shown in *Fig. 2*.

| | |
|---|---|
| mRNA | 5'AUG-UGC-UAU-CGA-UAC-CGC-UCG-CUA-AGA-UAA-GAT......3' |
| Protein | Met-Cys-Tyr-Pro-Tyr-Arg-Ser-Leu-Arg-Stop |

*Fig. 2. The relationship between the sequence of bases in mRNA and the sequence of amino acids in the protein or peptide produced.*

All proteins begin (at least in the early stages of translation) with the amino acid methionine, since the codon AUG is the '**start codon**' and this one codes for methionine. For most other amino acids, there are a few possible codons (e.g. in *Fig. 2*, both UAU and UAC encode tyrosine). There are also three '**stop codons**' (UAA, UAG, and UGA), which are used to terminate the translation of mRNA to protein. In the case of proteins translated at the rough ER, the first protein product, it will be processed and packaged by the **Golgi apparatus**. Specific amino acid sequences in proteins 'tell' the cell machinery what must be done with the protein and where in the cell it should be sent in order to do its 'job', by reading the 'signal peptide', usually in the first part of the amino acid sequence of the protein. Additionally, protein phosphorylation and/or other modifications may be needed for the protein to have its maximal activity in the cell.

Almost everything that is done by human cells relies on proteins, whether they are used as structural components of the cell, to carry oxygen, or as the biological catalysts known as **enzymes**. Controlled gene transcription and protein translation is therefore an essential part of the work of all cells.

# B3 METABOLIC PROCESSES

## Key Notes

**Introduction**

Metabolism describes the chemical reactions of both anabolic and catabolic pathways in the cell. These reactions are usually catalyzed by enzymes, proteins which modify the rate of the particular reaction, and many metabolic pathways are controlled by enzymes which catalyze early steps in the pathway, thereby limiting the build-up of large amounts of pathway intermediates. Anabolism defines the building processes responsible for the biosynthesis of molecules, particularly macromolecules used for energy storage, while catabolism defines destructive processes whereby macromolecules are degraded, usually to provide energy. The energy for metabolic processes is derived from oxidation of fuel molecules: glucose, fatty acids and amino acids. Oxidations are exothermic reactions, and the cell harnesses some of the energy released on oxidation as chemical bond energy in ATP.

**Catabolism of carbohydrate**

Glucose is stored as glycogen predominantly in liver and skeletal muscle. The absence of the key enzyme, glucose-6-phosphatase, precludes the muscle from releasing free glucose into the circulation for use by other tissues. Muscle glycogen serves as a glucose store solely for its own use. Glycogen is hydrolyzed (glycogenolysis) in both muscle and liver to glucose-1-phosphate by glycogen phosphorylase. Glucose-1-phosphate is isomerized to glucose-6-phosphate which serves as a substrate for two pathways: glycolysis and the pentose phosphate pathway. Under aerobic conditions, glycolysis converts glucose-6-phosphate to pyruvate, accompanied by the production of ATP via substrate-level phosphorylation at two steps in the pathway (4ATP/glucose-6-phosphate). When oxygen supply is limited, pyruvate is reduced to lactate to enable glycolysis to continue. Lactate can serve as a substrate for gluconeogenesis. When oxygen supply is not limiting, however, pyruvate enters the mitochondrion and is oxidized to carbon dioxide and water, and then some of the energy released from this oxidation is used to drive the synthesis of ATP by the process of oxidative phosphorylation. Metabolism of glucose via glycolysis also generates intermediates for other metabolic pathways.

The major end-products of the pentose phosphate pathway are the reduced pyridine nucleotide NADPH (reduced nicotinamide adenine dinucleotide phosphate) and five carbon sugars. The pentoses derived from the pentose phosphate pathway are components of nucleosides and hence DNA and various cofactors including $NAD^+$, FAD (flavine adenine dinucleotide) and coenzyme A.

**Gluconeogenesis**

Despite being a synthetic process (the synthesis of glucose from non-carbohydrate precursors), gluconeogenesis (GNG) occurs when the body is in a catabolic state, and serves to provide glucose to glucose-dependent tissues in circumstances when supply of glucose is limited. GNG also occurs during periods of peripheral insulin resistance such as the hypermetabolic phase of trauma and uncontrolled diabetes mellitus type II. The minimal structural requirement for a GNG substrate is a three-carbon molecule, and the major substrates are lactate from glycolysis, glycerol from metabolism of triglyceride, and glucogenic amino acids (mainly from hydrolysis of skeletal muscle proteins). GNG is under hormonal control, being inhibited by anabolic hormones (insulin) and stimulated by catabolic hormones (glucagon, cortisol)

**Catabolism of fat**

Fat is stored mainly in adipose tissue as triglyceride (a molecule composed of free fatty acids and glycerol). Hydrolysis of triglyceride (lipolysis) in adipose tissue by a hormone-sensitive lipase releases free fatty acids into the circulation, where they are transported to tissues by albumin. All tissues apart from brain, erythrocytes and nervous tissue use fatty acids as a major energy source between meals and during fasting. Fatty acids enter the tissue and are activated to their coenzyme A (CoA) derivative in the cytosol before being transported across the mitochondrial membranes. In the matrix of the mitochondrion, they are converted back to their CoA derivative before metabolism by β-oxidation, which produces acetyl-CoA. Acetyl-CoA enters the citric acid cycle and is oxidized to $CO_2$ and water. Some of the energy released in these processes is used to drive the synthesis of ATP by oxidative phosphorylation. The other product of lipolysis is glycerol, and it has three carbon atoms so can act as a substrate for GNG. Unsaturated fatty acids are converted to intermediates of the β-oxidation pathway before metabolism and odd-numbered fatty acids yield acetyl-CoA and a molecule of propionyl-CoA in the final round of oxidation. The process of fatty acid oxidation in tissues is driven by the circulating concentration of free fatty acids, which in turn reflects the catabolic state of the body, adipose tissue lipolysis being stimulated by catabolic hormones, particularly glucagon and cortisol.

**Ketone body synthesis**

Ketone bodies, acetylacetate and 3-hydroxybutyrate are formed in the liver from acetyl-CoA derived from β-oxidation of fatty acids when the rate of acetyl-CoA production exceeds its rate of oxidation in the citric acid cycle. Thus, in highly catabolic states, when insulin concentrations are low, such as prolonged fasting and uncontrolled diabetes type I, large amounts of fatty acids from lipolysis of adipose tissue triglyceride are transported to the liver and undergo β-oxidation in mitochondria. Three molecules of acetyl-CoA are required for the synthesis of 3-hydroxy-3-methylglutaryl CoA (HMGCoA), the key intermediate, which loses a molecule of acetyl-CoA to form acetoacetate. Acetoacetate is reduced to 3-hydroxybutyrate in a reaction requiring NADH (reduced nicotinamide adenine dinucleotide), and results in the regeneration of NAD required for continued β-oxidation. 3-hydroxybutyrate is the major ketone body released into the circulation, and has to be re-oxidized back to acetoacetate before being used as a metabolic fuel by the heart and, critically, by the brain during starvation when glucose supply is low.

**Catabolism of protein and amino acids**

Proteins are hydrolyzed to amino acids by the action of proteases, and in muscle this is stimulated by cortisol. Glycogenic amino acids are those which, on removal of their α-amino group, yield a carbon skeleton which can be used for glucose synthesis, while those whose carbon skeleton can be used for ketone body production are ketogenic. The nitrogen is usually removed by transamination, a process requiring vitamin $B_6$ and is eventually excreted via the kidneys as urea. Dietary protein is hydrolyzed to amino acids as well as di- and tripeptides in the gut, by proteases secreted from the pancreas.

**Storage of glucose**

Glucose is stored as glycogen mainly in liver and skeletal muscle. This process, glycogenesis, which is promoted by insulin involves the activation of glucose to uridine diphosphate (UDP)-glucose and addition of one glucose molecule at a time to the growing glycogen molecule, a reaction catalyzed by glycogen synthase. The total amount of glucose stored as glycogen is greater in muscle than in liver, although the relative concentration is greater in the latter.

**Lipogenesis**

The pathway of lipogenesis is very active in lactating mammary gland, liver and adipose tissue. All of the carbons for fatty acid synthesis are derived from acetyl-CoA, and the fatty acid chains are elongated two carbons at a time. The donor of the two carbons is 3-carbon malonyl-CoA with the loss of $CO_2$. Formation of malonyl-CoA from carboxylation of acetyl-CoA by acetyl-CoA carboxylase represents the rate-limiting step of fatty acid biosynthesis. Triglyceride formed by esterification of three fatty acids to glycerol, is stored mainly in adipose tissue, and triglyceride synthesis occurs mainly in liver, enterocytes, adipose tissue and lactating mammary gland. In general, lipogenesis, triglyceride synthesis and fat storage are promoted by insulin, i.e. the well-fed state.

**Protein synthesis**

Protein synthesis occurs continually as part of the turnover process of body proteins. Uptake of amino acids into liver and muscle and subsequent protein synthesis are promoted by insulin. Details of mRNA translation to produce nascent proteins in the cell are covered in Topic B2.

**Related topics**

Nutritional requirements of cardiac muscle (D3)
Utilization of $O_2$ and production of $CO_2$ in tissues (E5)
Absorption of nutrients (J6)
Pancreatic hormones (K9)

## Introduction

The body is a dynamic system in which all molecules are in a continual state of turnover, regulated by the processes of synthesis (**anabolism**), and degradation (**catabolism**). **Homeostasis** represents the state of physiological equilibrium arising as a balance between these two opposing metabolic processes, and disturbance of this balance can often lead to disease. Metabolism describes the chemical reactions of both anabolic and catabolic pathways in the cell. These reactions are usually catalyzed by enzymes, proteins which modify the rate of the particular reaction, and many metabolic pathways are controlled by **enzymes** that catalyze early steps in the pathway, thereby limiting the build-up of large amounts of pathway intermediates. Such rate-limiting steps are often the sites chosen for drug therapy.

**Anabolism** defines the building processes responsible for the biosynthesis of molecules, particularly macromolecules used for energy storage, while

**catabolism** defines destructive processes whereby macromolecules are degraded, usually to provide energy. Both anabolism and catabolism can be controlled independently, but the concept of homeostasis implies coordination between them, and this is maintained through the actions of a number of hormones which coordinate and control them to avoid the wasteful degradation of newly synthesized macromolecules. Tissue responses to hormonal stimuli (*Fig. 1*) can be acute, lasting only for a few minutes (e.g. glycogen breakdown in muscle in response to epinephrine during exercise), or chronic, of the order of a few hours or days (lipid degradation in adipose tissue in response to raised cortisol). Often acute control involves a modification of enzyme activity through allosteric modulators or covalent modification, for example, enzyme phosphorylation; chronic control on the other hand often results from a change in the amount of a key enzyme brought about by a change in its rate of synthesis or degradation. These two situations are not mutually exclusive, and a tissue may alter its metabolism chronically after initial changes following an acute response.

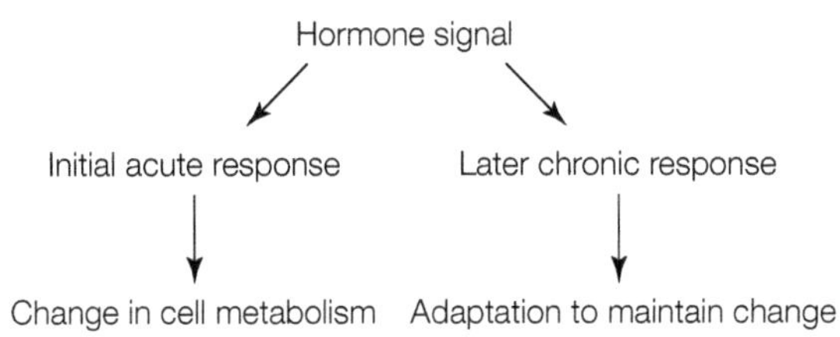


*Fig. 1. Effects of hormone signals.*

The following pages will present an overview of metabolic processes to help in the understanding of later sections in this book. Details of individual reactions are omitted, and the reader is recommended to access *Instant Notes in Biochemistry* for a more comprehensive coverage of metabolic pathways.

The macronutrients of the diet, **carbohydrate** (mainly starch), **protein** and **fat** (triglyceride) supply the building blocks for macromolecular synthesis, and provide fuel for the release of energy for the essential processes of mechanical work such as muscle contraction, macromolecular synthesis and the maintenance of transmembrane ionic gradients (active transport). The diet also contains micronutrients such as trace metals (e.g. copper, selenium, magnesium, nickel) and **vitamins**, both fat-soluble (vitamins A, D, E, and K) and water-soluble (thiamine, riboflavin, nicotinamide, pyridoxine, pantothenic acid, biotin, cobalamin, folic acid, and ascorbic acid) many of which are essential cofactors in enzyme-catalyzed reactions.

The energy for metabolic processes is derived from oxidation of fuel molecules, glucose, fatty acids and amino acids. Oxidations are exothermic reactions and the cell harnesses some of the energy released on oxidation as chemical bond energy in **ATP**, the metabolic currency of energy (*Fig. 2*).

ATP can be generated either through **substrate-level phosphorylation** during glycolysis of glucose in the cytosol (see later) or via the **tricarboxylic acid (TCA) cycle** and **oxidative phosphorylation** in **mitochondria**. The ATP yield from oxidation of **acetyl coenzyme A (acetyl-CoA)**, a common intermediate in the degradation of carbohydrate, fat and protein (*Fig. 2*), from oxidative phosphorylation, however, is much higher than from glycolysis.

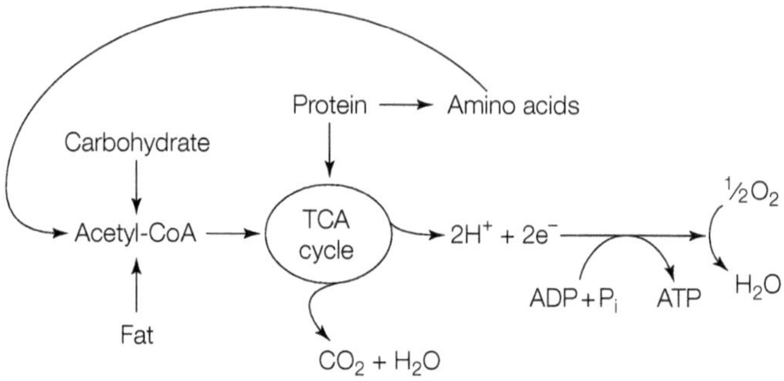


*Fig. 2. Metabolic processes that produce carbon dioxide ($CO_2$), water ($H_2O$) and adenosine triphosphate (ATP). ADP, adenosine diphosphate; $e^-$, electron; $O_2$, oxygen; $P_i$, inorganic phosphate; TCA, tricarboxylic acid.*

Oxidation of acetyl-CoA in the mitochondria is accompanied by the formation of reduced cofactors, NADH and $FADH_2$, which donate their electrons to the **electron transport chain** on the inner membrane of the mitochondria. As the electrons pass down the chain to molecular oxygen, the intermediates of the chain undergo cyclical oxidation and reduction, with energy being released during their oxidation. About 40% of this energy is stored as chemical bond energy of ATP, while the rest is liberated as heat and contributes to the maintenance of body temperature.

As described in Section J, the macromolecules of the diet are hydrolyzed prior to absorption in the intestine. The current section will thus be concerned with intermediary metabolism only.

## Catabolism of carbohydrate

The catabolic state is characterized by an increase in the ratio of catabolic hormones (principally **epinephrine**, **glucagon**, and **cortisol**) to anabolic hormones (principally **insulin**). This results in the breakdown of the stored fuel molecules glycogen, triglyceride, and protein. In the presence or absence of other hormones, some hormones can be net activators of either anabolism or catabolism; one example is growth hormone, which has catabolic effects when circulating levels of insulin are low, but anabolic effects when circulating levels of insulin are high.

The fasting glucose concentration in the blood is about 4–5 mM, and the body has a daily requirement of about 160 g of glucose to maintain this. The tissues that have an absolute requirement for glucose as a metabolic fuel are the brain, nervous tissue, erythrocytes, and to some extent the kidney; the brain is unable to use fatty acids as an energy source *per se*, and mature erythrocytes have no mitochondria in which to oxidize fatty acids. The catabolism of carbohydrate involves a two-stage process, the initial stage of which involves the release of glucose units from its stores followed by metabolism of these released glucose units. Carbohydrate is stored in the body as glycogen, a polymer of glucose, mainly in liver and skeletal muscle. The breakdown of glycogen to glucose-1-phosphate, **glycogenolysis** (*Fig. 3*), involves the release of individual glucose molecules from the stored polymer by the action of **glycogen phosphorylase**. Raised plasma epinephrine, or decreased plasma insulin, cause a rapid hydrolysis of glycogen to glucose units, glucose 1-phosphate, in liver and skeletal muscle, while raised plasma glucagon stimulates glycogenolysis in the liver

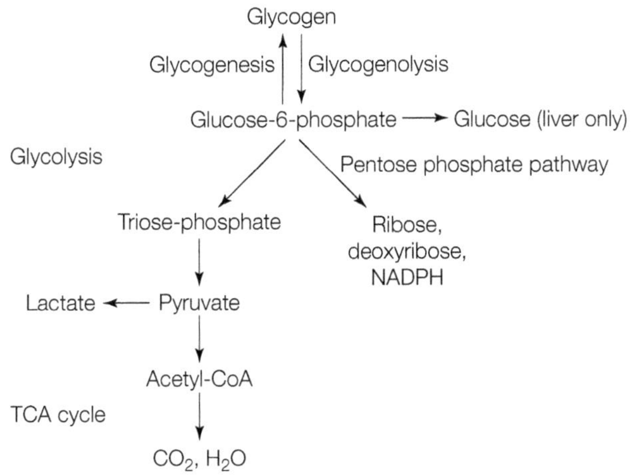


*Fig. 3. Catabolic pathways for carbohydrates. $CO_2$, carbon dioxide; $H_2O$, water; NADPH, reduced nicotinamide adenine dinucleotide phosphate.*

only. The actions of both epinephrine in the muscle, and glucagon in the liver are mediated via **cyclic adenosine monophosphate (cAMP)** and phosphorylation of glycogen phosphorylase (see Topic K3).

Glucose-1-phosphate, the product of phosphorylase action on glycogen, isomerizes to glucose-6-phosphate which, in the liver, has three possible fates:

- *hydrolysis to glucose by glucose 6-phosphatase*: the capacity of the body to maintain blood glucose at 4–5 mM depends on it being able to provide glucose from body stores in the short term, and to synthesize glucose from other molecules in the longer term. The liver has the enzyme glucose-6-phosphatase which allows it to provide glucose from hepatic glycogen for glucose-dependent tissues via the circulation. In the absence of a dietary input of glucose, the body has sufficient hepatic glycogen to supply glucose for just one day. Although skeletal muscle also stores glycogen, it lacks glucose-6-phosphatase, and thus cannot provide glucose for other tissues. Muscle glycogen is used solely by the muscle itself
- *metabolism via the glycolytic pathway*: in the glycolytic pathway, six-carbon glucose is split into two three-carbon molecules which are converted eventually to **pyruvate**. Small amounts of ATP are generated in glycolysis, and a cell like the mature erythrocyte, which lacks mitochondria, is absolutely dependent on this pathway for its energy supply. When oxygen supply to the cell is not limited, further metabolism of pyruvate occurs when it enters the mitochondrion and undergoes oxidative decarboxylation to acetyl-CoA. The acetyl-CoA is then oxidized to $CO_2$ and water in the TCA cycle, with the formation of many more molecules of ATP via oxidative phosphorylation. Under conditions where oxygen supply to a tissue may be limiting, however, for example, prolonged high-activity exercise in skeletal muscle, and in the ischemic heart where oxygen tension is low, substrate-level phosphorylation becomes the major route of ATP synthesis. In this situation, pyruvate is converted to **lactate** by the action of **lactate dehydrogenase**, instead of being converted to acetyl-CoA. As mentioned above, this situation obtains permanently in the mature erythrocyte, which has no mitochondria with which to

oxidize acetyl-CoA, and erythrocytes contribute most of the lactate in plasma in the resting state

- *metabolism via the pentose phosphate pathway*: an alternative metabolic pathway for glucose-6-phosphate is the **pentose phosphate pathway (PPP)** sometimes known as the hexose monophosphate shunt. The major products of this pathway are five-carbon sugars, such as ribose and deoxyribose required for nucleic acid synthesis, and reducing equivalents in the form of reduced **nicotinamide adenine dinucleotide phosphate (NADPH)**. NADPH is required for hepatic drug metabolism and for the reductive biosynthesis of macromolecules, including cholesterol and fatty acids. PPP activity is thus high in liver, lactating mammary gland, erythrocytes, and activated macrophages where NADPH provides electrons for the production of **reactive oxygen species** required for the degradation of 'foreign' macromolecules. In the erythrocyte, formation of NADPH via PPP is essential to maintain glutathione in its reduced form as part of the battle against oxidative stress in these cells (*Fig. 4*).

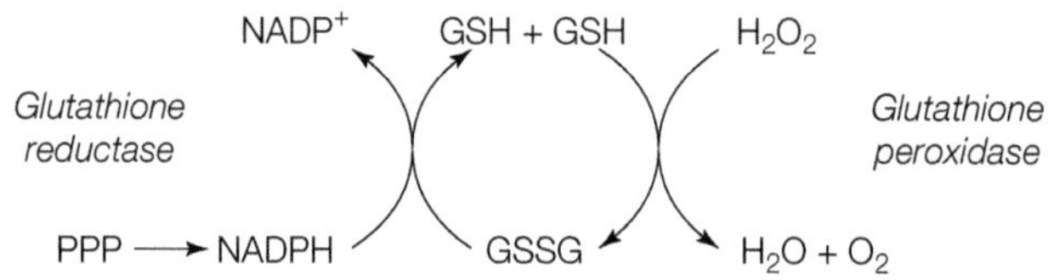


*Fig. 4. The pentose phosphate pathway. GSH, reduced glutathione (glutamylcysteinylglycine); GSSG, oxidized glutathione; $H_2O_2$, hydrogen peroxide; PPP, pentose phosphate pathway.*

## Gluconeogenesis

**Gluconeogenesis (GNG)** describes the synthesis of glucose from non-carbohydrate precursors such as lactate, amino acids, and glycerol. The minimum requirement for a molecule to serve as a precursor for glucose synthesis is that it must contain three or more carbon atoms. Humans cannot make glucose from fatty acids, because the pathway for the metabolism of fatty acids ends with acetyl-CoA, which is a two-carbon molecule. GNG occurs in the liver, kidney, and to some extent in the small intestine. It serves as a mechanism for resynthesizing glucose from lactate released from skeletal muscle during acute **anaerobic exercise** (Cori cycle), and from glycerol released from adipose tissue during **lipolysis**. Furthermore, when hepatic glycogen levels become depleted during fasting, the body responds by synthesizing glucose from amino acids (particularly alanine and glutamine) released from skeletal muscle protein. This represents an attempt to maintain blood glucose concentration at around 4 mM in the absence of dietary glucose input.

During more long-term fasting, the brain's demand for glucose is reduced by its adaptation to the use of **ketone bodies** rather than glucose as its preferred energy substrate (see 'Ketone body synthesis' below). Gluconeogenic activity is also raised in poorly controlled **diabetes mellitus** and in the hypermetabolic phase of the trauma response. Although the initial reaction of gluconeogenesis occurs in the mitochondrion, much of the pathway, apart from three energetically unfavorable steps, is simply a reversal of the glycolytic pathway. The use of amino acids as substrates for gluconeogenesis requires the removal of the amino group and conversion of the carbon skeleton to pyruvate or a member of the TCA cycle.

Most tissues have the ability to convert glucose to glucose-6-phosphate, and metabolize it via glycolysis, and when glucose is abundant, just after feeding, it becomes the major energy substrate.

**Catabolism of fat**

Fat is stored in **adipose tissue** as **triglyceride**, a molecule in which a fatty acid is esterified to each of the three hydroxyl groups of glycerol. Fatty acids can be released from this store to be used as a source of energy by many tissues. In the resting state between meals, fatty acids serve as the energy supply for virtually all tissues apart from those mentioned above which have an obligatory requirement for glucose (brain, red blood cells, nervous tissue, and kidney). Fatty acids are released from adipose tissue in response to the action of catabolic hormones, such as epinephrine and glucagon, which signal via their membrane receptors to activate an intracellular **hormone-sensitive lipase**. This lipase hydrolyzes triglyceride to free fatty acids, and glycerol and the free fatty acids are transported in the circulation bound to albumin to be used in peripheral tissues such as heart, skeletal muscle, and liver. Albumin-bound free fatty acids serve as the major respiratory substrate for the majority of tissues in the period between meals and the fasting state. The uptake of free fatty acids into tissues, and their subsequent metabolism is proportional to their plasma concentration. The glycerol released is water-soluble and returns to the liver where it is phosphorylated to glycerol-3-phosphate to provide a building block for GNG.

Fatty acids enter the tissues and bind to fatty acid-binding proteins. They then become activated to their CoA derivative, fatty acyl-CoA, under the action of the enzyme, **thiokinase**.

$$\text{fatty acid} + \text{ATP} \rightarrow \text{fatty acyl-CoA} + \text{pyrophosphate}$$

Oxidation of fatty acids occurs in the matrix of the mitochondrion by the process of β-oxidation. However, the mitochondrial inner membrane presents a barrier to fatty acyl-CoA due to the large and polar nature of the coenzyme A, and although fatty acyl-CoA is the substrate for **β-oxidation**, fatty acids are transported through the inner membrane linked to **carnitine**, a small hydrophobic molecule. The fatty acyl-carnitine derivative is formed in the outer mitochondrial membrane under the action of carnitine palmitoyl transferase I and an acyl carnitine transporter in the inner mitochondrial membrane facilitates the transfer of the fatty acid into the mitochondrial matrix. Here it is converted back to fatty acyl-CoA by the enzyme carnitine palmitoyl transferase II using the pool of matrix coenzyme A. As mentioned above, the fatty acyl-CoA in the mitochondrial matrix is a substrate for β-oxidation, a process in which the fatty acid is degraded two carbons at a time, releasing acetyl-CoA for oxidation in the TCA cycle. Most of the fatty acids in man are composed of an even number of carbon atoms, usually $C_{16}$–$C_{20}$, and are completely oxidized to acetyl-CoA. Thus, a molecule of palmitic acid with 16 carbons will yield 8 molecules of acetyl-CoA, while **arachidonic acid** with 20 carbons will yield 10 molecules of acetyl-CoA. In addition to one molecule of acetyl-CoA, each round of β-oxidation yields one molecule of $FADH_2$ and one of NADH. It has been described above how the oxidation of acetyl-CoA in the TCA cycle is also accompanied by the production of the reduced cofactors, $FADH_2$, and NADH. All of these reduced cofactors donate their electrons to the **electron transport chain**, and some of the energy released by the redox reactions of the electron transport chain during electron transport to oxygen is converted to chemical energy as the terminal phosphate bond of ATP. Thus, the total yield of ATP from fatty acid oxidation far exceeds that from oxidation of carbohydrate, mole for mole. In a tissue such as muscle, the rate of β-oxidation is dependent on the supply of fatty acids to the tissue, and this in turn reflects the rate of triglyceride breakdown in adipose tissue.

The process described above is for even-numbered saturated fatty acids. Unsaturated fatty acids undergo metabolism to convert them to intermediates of the β-oxidation pathway prior to oxidation as described for saturated fatty acids. Any odd-numbered fatty acids, such as those derived from dietary microorganisms or plants, are metabolized via β-oxidation to a three-carbon fatty acid, propionyl-CoA, and this undergoes metabolism eventually to four-carbon butyryl-CoA via a process requiring vitamin $B_{12}$ and then acetyl-CoA.

## Ketone body synthesis

In situations where lipolysis of adipose tissue triglyceride is greatly increased, such as prolonged starvation and untreated diabetes mellitus type II, the supply of fatty acids to the liver is also increased and the rate of β-oxidation is such that acetyl-CoA production exceeds that at which acetyl-CoA can be oxidized in the TCA cycle. As described above, β-oxidation generates NADH, and this will slow down the TCA cycle, as $NAD^+$ is required for three reactions of the cycle. In response, the liver synthesizes **ketone bodies**, acetoacetate and 3-hydroxybutyrate, from this excess acetyl-CoA at a rate faster than acetyl-CoA is produced. In an attempt to regenerate $NAD^+$ in the mitochondrion, about 75% of the acetoacetate (3-ketobutyrate) produced is reduced to 3-hydroxybutyrate such that the ratio of 3-hydroxybutyrate:acetoacetate in the circulation is 3:1.

Ketone bodies are cleared rapidly from the circulation being used by extrahepatic tissues such as heart, skeletal muscle and kidney (*Fig. 5a*). The brain will also adapt to the use of ketone bodies as an alternative fuel to glucose when the plasma concentration rises sufficiently to allow their rapid uptake, for example during starvation. The use of ketone bodies as tissue fuel involves the conversion of acetoacetate to acetoacetyl-CoA by a 3-ketoacyl-CoA transferase which catalyzes the transfer of CoA from succinyl-CoA to acetoacetate. Succinyl-CoA is an intermediate of the TCA cycle. The acetoacetyl-CoA produced can then undergo the final steps of β-oxidation to acetyl-CoA, which can be oxidized in the tissues. Liver lacks the 3-ketoacyl-CoA transferase and thus cannot use ketone bodies as a metabolic fuel.

An alternative metabolic pathway (*Fig. 5b*) exists in the liver when excessive amounts of fatty acid are released from adipose tissue; this involves the reformation of triglyceride and export of the triglyceride from the liver as part of **very low-density lipoprotein (VLDL)**. Thus, in situations where there is a peripheral resistance to insulin, e.g. untreated diabetes mellitus type II or severe trauma, hepatic synthesis becomes significant and, somewhat anomalously, the plasma concentrations of both free fatty acids and VLDL are raised.

## Catabolism of protein and amino acids

The catabolism of protein and amino acids represents an essential component of the turnover and daily renewal of body proteins. On average, an adult human degrades about 300–400 g of protein each day, and in homeostasis this will be replaced by newly synthesized protein. **Nitrogen balance** is defined as the difference between nitrogen entering the body via the diet, mostly as protein, and nitrogen excreted via the urine, mainly as **urea**. When lean body mass (muscle mass) remains constant during adulthood, the nitrogen balance is zero. In excessive catabolic states such as after major trauma, surgery, or infection,

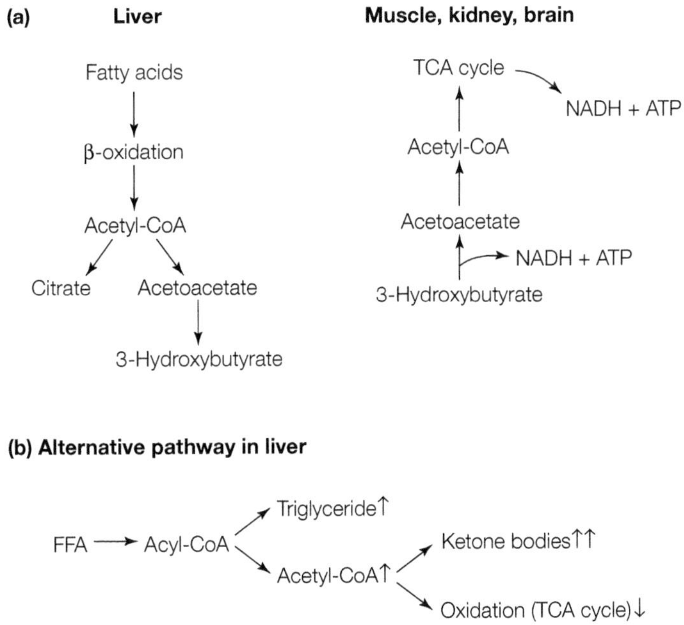


*Fig. 5. Metabolic pathways for formation of ketone bodies. ATP, adenosine triphosphate; FFA, free fatty acids; NADH, reduced nicotinamide adenine dinucleotide; TCA, tricarboxylic acid.*

nitrogen balance can become negative as the body uses the carbon skeletons of amino acids, via GNG, as a metabolic fuel. However, even in extreme cases, amino acids will not provide more than 20% of the metabolic fuel requirements. Conversely, during growth in children and pregnancy, nitrogen balance is positive as the body builds new muscle mass.

The breakdown of protein, particularly in muscle, is stimulated by raised plasma cortisol concentration. Exactly how this happens is unclear, but it is antagonized by an increase in plasma insulin and the net rate of protein breakdown at any time represents a balance between the actions of anabolic and catabolic hormones (*Fig. 6a*).

The hydrolysis of proteins liberates free amino acids, and the initial reaction in the catabolism of these molecules involves the removal of the unwanted

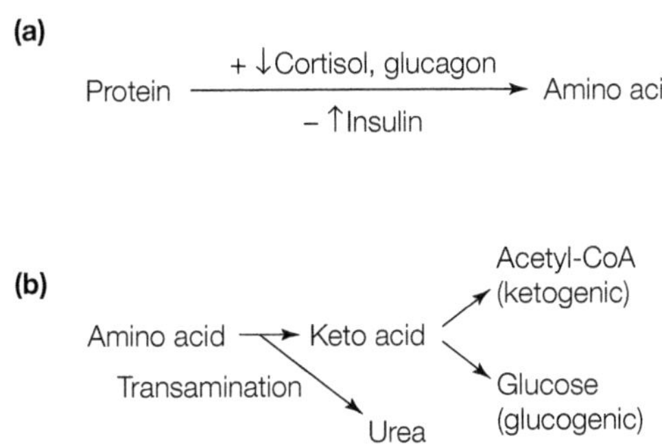


*Fig. 6. Catabolic pathways for protein (a) and amino acids (b).*

α-amino group, usually by pyridoxal phosphate (vitamin $B_6$)-dependent transamination to a keto-acid (*Fig. 6b*). This yields a keto-acid containing the carbons which can be used for gluconeogenesis or ketogenesis. The α-amino group nitrogen is eventually excreted in the urine as part of urea.

The actions of both cortisol in liver and muscle, and glucagon in liver only increase the amount of hydrolytic enzymes in these tissues. Glucagon also stimulates the uptake of amino acids, and particularly alanine, into liver. This is of particular importance since muscle protein is the major source of amino acids during times of physiological stress when the body seeks substrates for GNG. The liver is the only tissue capable of urea synthesis, and is also the major tissue involved in GNG; thus a molecule such as alanine serves to transport three-carbon molecules from muscle to the liver for GNG as well as the waste nitrogen for excretion as urea. Most amino acids, after removal of the nitrogen, are converted to TCA cycle or glycolytic intermediates, and can be used for the synthesis of glucose (glucogenic), while others (e.g. leucine) can be converted to acetyl-CoA (ketogenic). Interestingly, *in vivo*, most of the amino acids released by hydrolysis of muscle proteins are metabolized to alanine and glutamine prior to release into the circulation, and the plasma concentration of these amino acids rises much more than that of others after trauma or during starvation.

## Storage of glucose

The anabolic state is characterized by a relative rise in the plasma ratio of insulin to the major catabolic hormones glucagon, cortisol and catecholamines. Such a situation arises just after a meal, especially one high in carbohydrate content. A rise in plasma glucose and amino acids derived from the diet, along with the release of gastrointestinal peptides stimulates the secretion of insulin from **pancreatic β-cells**, and insulin promotes the storage of glucose as glycogen in liver, and muscle and fatty acids as triglyceride in adipose tissue. It also stimulates the uptake of amino acids into muscle and muscle protein synthesis.

Glucose is stored as glycogen, primarily in liver and muscle. The total amount of glycogen stored in muscle is greater than in the liver, but it is present at a higher concentration in the latter. As described above, the liver alone is capable of supplying glucose from glycogen directly to blood and hence to other tissues. Its glycogen content will thus fluctuate throughout the day as it releases glucose to the blood between meals and during an overnight fast. After digestion and absorption of dietary carbohydrate, the concentration of glucose in the hepatic portal vein may be in excess of 10 mM, as glucose moves from the gut to the liver. Raised plasma insulin will inhibit glucagon secretion by the pancreas, thereby inhibiting hepatic glycogen breakdown, and stimulate the synthesis of glycogen in the liver through activation of glycogen synthase. Raised insulin and glucose levels will also promote the synthesis of hepatic glucokinase and ensure the net flux of glucose into liver glycogen. Insulin has no effect on the uptake of glucose by the liver; the rate of uptake reflects the glucose concentration difference between blood and the cytosol of the hepatocytes. However, insulin does promote the uptake of glucose into muscle and adipose tissue, and the synthesis of glycogen in muscle.

## Lipogenesis

The storage of fatty acids as triglycerides represents the major mechanism for conserving the excess energy of dietary fat, carbohydrate, and amino acids. The biosynthesis of fatty acids involves the addition of two carbons to the growing fatty acid chain such that all mammalian fatty acids are even numbered. The donor of the two carbons is a three-carbon molecule, malonyl-CoA, and the synthesis of this molecule, catalyzed by acetyl-CoA carboxylase, is rate limiting in the fatty acid biosynthetic pathway. Acetyl-CoA carboxylase is controlled acutely by insulin (stimulatory) and glucagon (inhibitory). The preferred substrate precursors for fatty acid synthesis are glucose and lactate, and raised insulin levels (e.g. after a meal), will ensure the provision of acetyl-CoA by stimulating the glycolytic pathway.

The rate of synthesis of triglyceride in both liver and adipose tissue is dependent on the supply of fatty acyl-CoA. In liver this reflects the *de novo* synthesis of fatty acids while in adipose tissue fatty acids are derived from hydrolysis of the triglyceride component of VLDLs, synthesized in the liver, or **chylomicrons**, VLDL particles synthesized in the epithelial cells of the small intestine. Insulin stimulates the activity of adipose tissue lipoprotein lipase which carries out this hydrolysis. In western societies, where approximately 35% of dietary energy is provided by fat, little *de novo* synthesis of fatty acids occurs, and stored triglyceride is derived from dietary fat.

Dietary fat is transported from the gut in the form of chylomicrons. These lipoproteins are secreted into the lymphatic system and drain into the circulation via the thoracic duct. In this way, dietary fatty acids can be offloaded to adipose tissue before the chylomicron remnants, containing fat-soluble vitamins and cholesterol, are taken up by the liver.

VLDLs transport any newly synthesized triglyceride to adipose tissue or muscle, depending on the prevailing hormonal conditions. Under anabolic conditions (raised insulin) VLDL-triglyceride is hydrolyzed by adipose tissue lipoprotein lipase, to provide fatty acids for storage as triglyceride in adipose tissue. Under stress conditions (catabolic) however, where **glucocorticoids** are raised, the VLDL is metabolized preferentially by skeletal muscle lipoprotein lipase to provide fatty acids for β-oxidation and energy production for the tissue.

## Protein synthesis

An average adult of 70 kg synthesizes approximately 400 g of protein a day, of which only 70 g comes from dietary amino acids. The majority of the amino acids required for protein synthesis come from the daily turnover of body proteins (*Fig. 7a*).

Amino acids derived from hydrolysis of dietary protein in the gut reach the liver via the **hepatic portal vein** and are used in the synthesis of new protein or catabolized. A rise in the plasma concentration of free fatty acids, as is seen after a meal, stimulates the release of glucagon by **pancreatic α-cells**, in addition to the glucose-stimulated release of insulin from the pancreatic β-cells. The effect of glucagon on the liver is twofold. Firstly, it promotes the uptake of amino acids into the liver to serve protein synthesis, and secondly, it allows sufficient GNG to occur to prevent an insulin-stimulated hypoglycemia. Thus even in the anabolic state, some conversion of amino acids to glucose takes place. Besides its effect on glucose uptake into peripheral tissue, insulin also stimulates the uptake of amino acids, particularly branched-chain amino acids, into muscle, where they are incorporated into protein or transaminated and catabolized. This involves intracellular signaling via a phosphorylation cascade, which eventually modifies gene transcription. The acute actions of insulin are to stimulate amino acid transport into skeletal muscle to supply substrate for amino acyl-tRNA synthesis and to increase polysome number and **translation of mRNA** into protein in both muscle and liver (*Fig. 7b*).

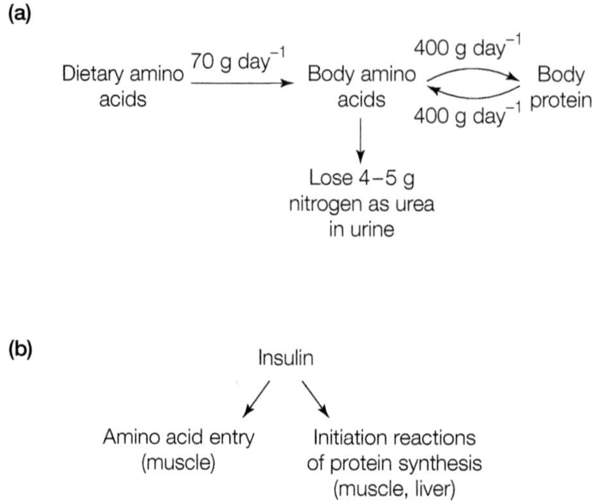


*Fig. 7. (a) Protein and amino acid balance. (b) Effects of insulin on amino acid uptake and protein synthesis in muscle and liver.*

# B4 Membrane potential

## Key Notes

**Separation of ions**

The plasmalemma is a modified phosphilipid bilayer and is only selectively permeable to charged chemical species, even those as small as inorganic ions. Separation of ions on either side of the plasmalemma results in the interior of most cells being slightly negatively charged with respect to the exterior of the cell.

**Intracellular and extracellular ion concentrations**

The only way in which ions can cross the plasmalemma is through ion channels or ion transporters. Active pumping of $Na^+$ ions in exchange for $K^+$ ions, for example, results in a greater concentration of $Na^+$ ions outside the cell and a greater concentration of $K^+$ ions inside the cell. $Na^+$ and $K^+$ ions can diffuse through specific ion channels, down their concentration gradients, but only if these channels are open. In most cells at rest, the permeability of the plasmalemma to $K^+$ ions is around 100-fold greater than that for $Na^+$ ions, because more $K^+$ ion channels are open at rest.

**Equilibrium potentials**

Separation of ions at different concentrations on either side of the plasmalemma results in a tendency for ions to diffuse down their concentration gradients. $Na^+$ ions are in electrochemical equilibrium with $Cl^-$ ions in the extracellular fluid, and $K^+$ ions are in equilibrium with large organic anions in the intracellular fluid. Diffusion of ions sets up equilibrium potentials between ions that are in equilibrium, and as membrane permeability to a particular ion increases, membrane potential will tend towards the equilibrium potential for that ion.

**The importance of permeability**

As a result of equilibrium potentials, the inside surface of the cell becomes slightly negative when $K^+$ permeability is high. The outside surface of the cell becomes slightly negative when $Na^+$ permeability is high (this is the same as relative intracellular positivity). The relatively high resting permeability of the plasmalemma to $K^+$ ions means that most cells have a slight intracellular negative charge at rest. Fixed negative charges associated with membrane proteins also have a role to play in determining this membrane potential.

**Changes in permeability**

Membrane potential changes as the permeability of the plasmalemma to various ions is altered through opening and closing of ion channels. Such changes in membrane potential are important in the function of excitable tissues such as nerves, muscle, and the heart.

**Related topics**

Electrical activity of the heart – the ECG (D4)
Action potentials (F2)
Integration of nerve cell function (F7)
The neuromuscular junction (I2)
Excitation–contraction coupling (I3)
Smooth muscle function and properties (I6)

## Separation of ions

All cells in the human body have a separation of charge on either side of the **plasmalemma**, so that the inside of the cell is slightly negative with respect to the **extracellular fluid**. If one uses a sharp glass microelectrode filled with a saline solution, introduced inside a cell, together with another reference electrode in the extracellular fluid (set to 0 mV), the potential difference between the two electrodes will range from around –70 mV to –40 mV, depending on the cell type being studied. This potential difference across the plasmalemma is known as the **membrane potential**, denoted $E_m$. Nerve cells have an $E_m$ of around –70 mV and red blood cells (**erythrocytes**) have an $E_m$ of around –40 mV.

Part of the reason why the plasmalemma has a negative $E_m$ is the intrinsic chemical properties of the membrane itself. As a **phospholipid** bilayer, it is almost impermeable to chemical substances that are not readily lipophilic (soluble in lipid). Ions, whether negatively or positively charged, are one such substance. They are readily water soluble (hydrophilic), and therefore find it difficult to cross the plasmalemma. However, it is less apparent why there should be a net negative 'charge' inside the cell compared to the extracellular fluid.

## Intracellular and extracellular ion concentrations

*Table 1* shows the concentration of ions on either side of the plasmalemma in one specialized type of cell, the nerve cell or **neuron**:

*Table 1. Intracellular and extracellular concentrations of various ions in a nerve cell*

| Ion | Extracellular fluid concentration (mM) | Axoplasm concentration (mM) |
|---|---|---|
| $K^+$ | 2.5 | 115 |
| $Na^+$ | 145 | 14 |
| $Cl^-$ | 90 | 6 |
| Proteins$^-$ | Negligible | 100 |

$Na^+$ and $Cl^-$ ions are present at higher concentrations in the extracellular fluid, and $K^+$ ions and large negatively charged ions (anions) like proteins are present at higher concentration inside the cell. These differential concentrations of $Na^+$ and $K^+$ ions in the intracellular and extracellular compartments are set up by the action of specific ion pump and ion channel proteins embedded in the plasmalemma. The most important of these is the **$Na^+/K^+$ ATPase** or $Na^+/K^+$ pump. This pump is at work in most cells almost all of the time, and serves to pump $Na^+$ out of the cell and $K^+$ into the cell, a process that is dependent on a supply of ATP from cellular respiration. Three $Na^+$ ions are exchanged for two $K^+$ ions in each cycle of the pump, and so it is slightly electrogenic, making the extracellular environment slightly more electropositive than the intracellular space. In most cells, this pump is fed by $K^+$ ions that can leak slowly out of the cell through a $K^+$-specific ion channel embedded in the membrane, but membrane permeability to other ions is low because other ion channels (those for $Na^+$, for example) are usually closed at rest. At any rate, the permeability of the plasmalemma to $K^+$ is about 100-fold higher than for any other ion at rest.

## Equilibrium potentials

Each of the individual ions has an **equilibrium potential**, termed $E_{ion}$. This $E_{ion}$ is the potential difference that would be set up close to the plasmalemma if the membrane were freely permeable to only that particular ion. $Na^+$ ions are in electrical equilibrium with (predominantly) $Cl^-$ ions in the extracellular fluid, whereas $K^+$ ions are in equilibrium with negatively charged proteins, phosphate and sulfate ions in the intracellular environment. None of these ions can cross the cell membrane except through specific **ion channels**. Thus, if $Na^+$ ion channels in the membrane were to open, some $Na^+$ ions would attempt to enter the cell down their concentration gradient via specific $Na^+$ ion channels, and the $Cl^-$ ions would be left behind in the extracellular fluid. The small potential difference set up by this potential movement of $Na^+$ ions away from their equilibrated anions ($Cl^-$) would tend to make the immediate extracellular surface of the cell membrane slightly more electronegative than it was before the $Na^+$ ion channels opened. This extracellular electronegativity can also be viewed as intracellular electropositivity, when sampled with an intracellular electrode and compared to a reference electrode in the extracellular space.

The equation that describes this relationship between the concentrations of an ion on either side of the cell membrane and $E_{ion}$ is known as the **Nernst equation**. At core body temperature (37°C, i.e. 310 Kelvin), this equation takes the form:

$$\mathbf{E_{ion} = 61/z \log_{10}([ion]_o/[ion]_i)}$$

where $z$ = the valency of the ion (e.g. +1 for $Na^+$ and –1 for $Cl^-$), and $[ion]_i$ and $[ion]_o$ represent the respective molar concentrations of the ion inside and outside the cell.

Thus, as membrane permeability to a particular ion rises by opening of more of the specific channels for that ion, $E_m$ will move closer to the $E_{ion}$ for that particular ion. So if, for example, the plasmalemma were to become more permeable to $Na^+$, then $E_m$ would tend to move closer to $E_{Na}$.

Thus, one can calculate neuronal $E_{ion}$ for a number of ions:

$$E_{Na} = +62 \text{ mV}$$
$$E_K = -101 \text{ mV}$$
$$E_{Cl} = -72 \text{ mV}$$

**The importance of permeability**

If one knows $[ion]_i$ and $[ion]_o$ and the relative permeability of the plasmalemma ($P_{ion}$) to each of the major ions at any point in time, one can use the Goldman equation to calculate a theoretical $E_m$. At core body temperature, and correcting for the valency of the various ions, this equation takes the form:

$$E_m(\text{mV}) = -61\log_{10}\left(\frac{P_K[K^+]_o + P_{Na}[Na^+]_o + P_{Cl}[Cl^-]_i}{P_K[K^+]_i + P_{Na}[Na^+]_i + P_{Cl}[Cl^-]_o}\right)$$

where $P_{ion}$ is the relative permeability of the membrane to the ion in question.

Using the data from *Table 1* and the Goldman equation, it is apparent that under resting conditions, where $P_{Na}$ and $P_{Cl} = 0.01$, and $P_K = 1.0$ (100-fold higher), neurons will have a theoretical $E_m$ somewhere around –90 mV, and that the major determining factor in the Goldman equation is the higher permeability of the plasmalemma to $K^+$ at this time.

$$E_m = -61\log_{10}\left(\frac{0.025 + 0.00145 + 0.00006}{0.115 + 0.00014 + 0.009}\right) = -89.1 \text{ mV}$$

*Resting* $E_m$ is therefore said to be dependent mainly upon the permeability of the plasmalemma to $K^+$ ions. That is not to say, however, that $K^+$ ions flow out of the cell readily. As described above for $Na^+$ and $Cl^-$ ions, they are attracted by the electrostatic force of the protein, phosphate and sulfate anions, to which the plasmalemma is entirely impermeable. Other ions also have a part to play in the determination of the *actual* membrane potential (the equation above shows only the major ions involved), which is why $E_m$ in neurons is closer to 70 mV.

Another factor that influences $E_m$ is the presence of **fixed negative charges** on the extracellular and intracellular surfaces of the cell membrane, usually associated with membrane proteins, which are net negatively charged at physiological pH. The influence of these fixed negative charges on the extracellular surface is normally minimized by binding of extracellular $Ca^{2+}$ ions to these proteins, but when serum $Ca^{2+}$ concentrations fall, these fixed negative charges have a much larger influence and can affect $E_m$ (e.g. see Topics K7 and K8).

## Changes in permeability

Selective, controlled alterations in the permeability of the plasmalemma to various ions by opening and closing specific ion channels (such as $Na^+$ channels) can produce predictable changes in $E_m$. For example, if equal numbers of $Na^+$ channels and $K^+$ channels are open at the same time, then $E_m$ will fall somewhere between $E_K$ and $E_{Na}$. As even more $Na^+$ channels open, then $E_m$ will tend to approach $E_{Na}$. If performed quickly and accurately enough, many such alterations can be done in the space of even one second.

It is essential that you know and understand this. Don't worry about the equations. Concentrate on the principles. A solid understanding of these concepts will help you understand nerve impulses, skeletal muscle contraction, and the pumping action of the heart.

# B5 Phenotypic diversity of cells

## Key Notes

**Undifferentiated versus differentiated**

All cells in the human body are ultimately derived from a single zygote (ovum fertilized by a sperm cell), which undergoes many rounds of cell division to form a ball of embryonic cells. The mechanisms by which the range of different types of differentiated cells such as nerve cells arises from these undifferentiated cells are not, as yet, fully understood.

**Size and shape**

Differentiated human cells have a particular function. This function is often best served by the cell having a particular size and shape. Two examples of cells (derived from the same zygote) that have very different size and shape are red blood cells, and motor neurons; red blood cells are small and ideally designed for oxygen transport, whereas motor neurons have long processes to enable them to provide communication between the spinal cord and peripheral muscles that may be as much as a metre distant from the spinal cord.

**Nucleation and number of organelles**

Another form of differentiation that has important implications for function and gives clues to cell lineage is the number of various cellular organelles. For example, cells that are very metabolically active have numerous mitochondria, and those that must contract (muscle cells) have large amounts of cytoskeletal proteins to allow this to happen.

**Chemistry**

Cells that seemingly have the same phenotype from a structural perspective, can have a very different chemical phenotype. For example, endocrine cells in the anterior portion of the pituitary gland look similar, but individual cells release different peptide hormones. Similarly, neurons that look similar can have entirely different functions by releasing different neurotransmitters with diverse effects on effector cells. This type of chemical phenotypic diversity is enabled by the ability of the cell to change the expression (transcription) of various genes.

**Related topics**

Cell organelles (B1)
Gene transcription and protein translation (B2)
Neurons and glia I (C4)
Bone and muscle (C5)
Blood and immunity (C6)
Alveolar exchange and gas transport (E4)
The panoply of transmitters (F4)
Skeletal muscle function and properties (I1)
Pituitary hormones (K5)
From ovum to fetus (L4)

## Undifferentiated versus differentiated

In the human body, there are between $10^{13}$ and $10^{14}$ cells, all of which develop from a single fertilized **ovum**, the **zygote**. The mechanisms for development of this zygote into an embryo, fetus, baby, child, and adult are better understood now than ever before, but still only a tiny amount is known about how it is actually controlled. The cells of the embryo, and a tiny proportion of cells in adults, are **stem cells** that can develop into fully differentiated cells with a particular form and function in response to an appropriate stimulus. This process is highly controlled at the genetic level, and when it goes wrong can lead to cancerous change and malignancy. It is not within the remit of this text to describe the mechanisms controlling development (see *Instant Notes in Developmental Biology*), but it is important that you gain an understanding of the diversity of cell types in the human body so as to better understand their function.

Differentiated cells in the human body vary in many different ways and, where appropriate, this will be discussed in the relevant topics of this text. There follow some examples of differentiation being related to function.

## Size and shape

The function of particular cell types is best served by having a particular size and shape (*Fig. 1*). Two extreme examples are red blood cells (**erythrocytes**), and motor **neurons** (nerve cells). Erythrocytes are extremely small, some of the smallest cells in the body. Their major function is to transport oxygen around the body, bound to intracellular **hemoglobin**, to the cells that need it. This means that they have to be able to enter any capillary in the body. They are biconcave discs, resembling doughnuts with the hole filled in, have no nucleus, and have a diameter of around 7 μm. Their biconcave shape keeps the distance for diffusion of oxygen low and allows it to rapidly associate with hemoglobin in the lungs and to dissociate with hemoglobin in respiring tissues. Their size, although small, is larger than the diameter of the smallest capillaries that they need to pass through. This means that they need to fold up to enter the smallest capillaries, an act that is facilitated by their shape and their relative paucity of intracellular organelles, most notably the nucleus.

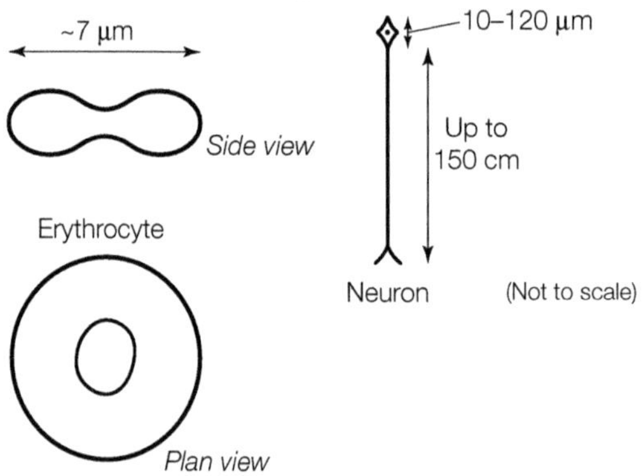


*Fig. 1. Cells differ in terms of shape and size (e.g. erythrocytes and neurons).*

The lower **motor neurons** that control the contraction of **skeletal muscle** (voluntary muscle), on the other hand, are extremely large. The cell body of these cells resides in the **spinal cord** and they have processes that may be up to a metre long (running from the lower part of the spinal cord to the muscles that make up the calf, for example). This means that during development, neurotropic factors have to 'attract' the growth cone of the nerve fiber to the muscle fibers that it will innervate, but also that proteins that are made in or near the cell body must be able to find their way to the very end of the cell. This is why such neurons have a well-developed **cytoskeleton**, in order to transport substances from the cell body to the nerve terminal.

**Nucleation and numbers of organelles**

As has been described previously, erythrocytes have no nucleus and few organelles, but there are some cells in the body that have developed to the other extreme (*Fig. 2*). Skeletal muscle fibers, for example, have several nuclei; if they are cut in cross-section, three or four nuclei can often be seen at the periphery of the cell. In this case, being multinucleate is a function of size. Muscle fibers may have to be as large as 20 cm in length, and they develop by fusion of several cells into one, each one leaving its nucleus behind in the mature muscle fiber. These cells also have many **mitochondria**, but not as many as cardiac muscle fibers, where around 35% of cell volume is taken up by these organelles. In this case, the large number of mitochondria is necessary to meet the energy requirements of the cardiac muscle fiber, which will contract and relax around 100 000 times every day, and which must use aerobic metabolism exclusively. If a cell has more than the average number of a given organelle, this is usually no coincidence; they are there for good functional reasons.

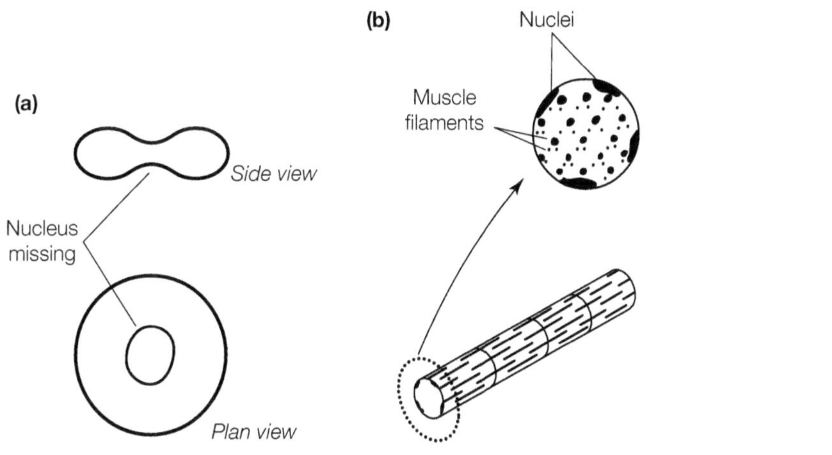


*Fig. 2. Cells differ in terms of the number of certain organelles (e.g. erythrocytes (a) and skeletal muscle fibers (b)).*

**Chemistry**

Even if cells don't look different, they can differ in more subtle ways, such as the chemical processes that they use in order to carry out their role in the body (*Fig. 3*). Two good examples are **endocrine** cells and neurons. There are many endocrine organs in the body (e.g. pituitary gland, hypothalamus, adrenal gland, pancreas, thyroid gland) but even within these tissues, several different endocrine cells may exist. For example, in the anterior part of the **pituitary gland**, different cells synthesize and secrete around six different hormones. In the cortex of the **adrenal gland**, cells in different layers release three classes of **steroid hormones** (mineralocorticoids, glucocorticoids, and androgens). This means that specific endocrine cells must express different enzymes in order to

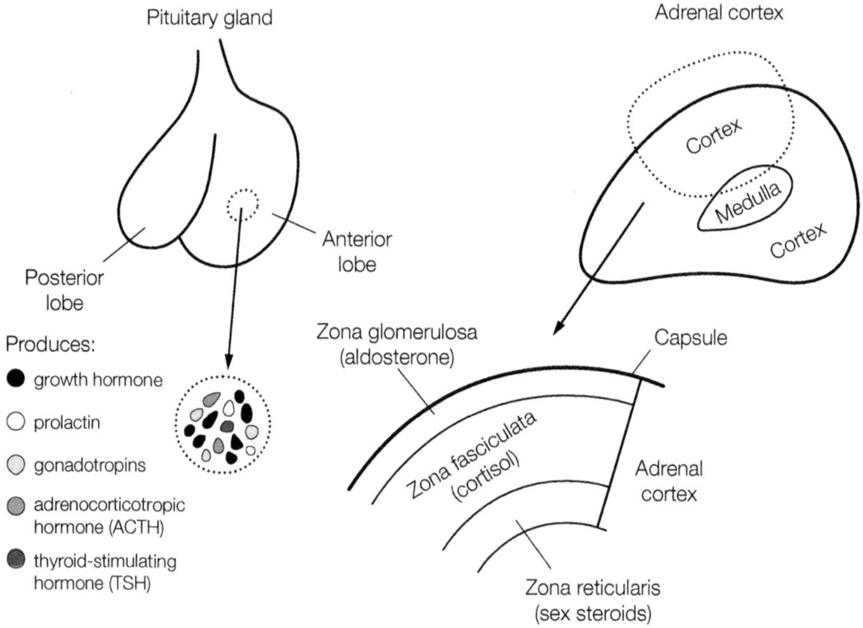


*Fig. 3. Cells differ in terms of their biochemical processes (e.g. cells of the anterior pituitary and cells of the adrenal cortex).*

synthesize these different hormones from their chemical precursors. In the anterior pituitary, the cells are mixed in together, so the only way of telling them apart is by detection of the enzymes or hormones themselves. Nonetheless, in terms of function, the various cells are as different as skeletal muscle and cardiac muscle.

In the same way, different neurons in the brain, spinal cord, and peripheral nervous system use different chemical messengers (**transmitters**) to exert their effects. Two neurons in the brain may look similar, but have different effects on the neurons they make contact with, simply by using different transmitter substances. Imagine two neurons, both of which release amino acid transmitters. One neuron may release **glutamate** to excite its target neuron, and the other may release **glycine** to inhibit its target. In fact, they may both have the same target neuron and compete with each other in an integrated fashion to affect activity of that cell. By having different neurochemistry, the two neurons can have different effects on brain function.

# C1 BASIC STRUCTURE OF THE BODY

## Key Notes

| | |
|---|---|
| **General considerations** | It is important to gain a small amount of understanding of general anatomy in order to learn about physiology. Some background information on body structure is given in this topic. |
| **External surfaces** | There are several external surfaces in the human body that cannot be seen from the outside. These include the gut tube, the lungs and the urogenital system. All of these external surfaces are lined with epithelial tissue that protects us from colonization by pathogens, but which also allows us to interact with our environment by absorbing nutrients and excreting waste products. |
| **Body cavities** | There are several body cavities that are formed during fetal development. These cavities (pleural, pericardial, peritoneal and cranial) are lined by membranes that have the role of protecting the organs inside the cavities. The membranes do this by cushioning the organs, or by preventing them being damaged by friction as they move inside the cavity. |
| **The meninges** | The meninges are the membranous sheets that protect the central nervous system. There are three meningeal layers: dura mater (outermost), arachnoid mater (middle), and pia mater (closest to the brain and spinal cord, adherent to the tissue). Cerebrospinal fluid flows in the space between the arachnoid mater and pia mater, serving to cushion the brain from movement inside the skull and, to some degree, for nutrition. |
| **Segmentation** | In common with all other vertebrates, we are segmented organisms. Our segmented nature is based upon the segmentation of the embryo that takes place *in utero* and is retained in the pattern of innervation of our bodies. |
| **Limbs** | Our limbs are the mechanism that we use to move around in our environment. The limbs develop *in utero* as outgrowths of the body wall, and are innervated by spinal nerves from the lower cervical and upper thoracic segments of the spinal cord (upper limb) and lower lumbar and upper sacral segments of the spinal cord (lower limb). The pattern of innervation of the limbs by spinal nerves is less obviously segmented than that of the neck and torso. |

**Related topics**

Epithelia and connective tissue (C2)
Layout and structure of the respiratory tract (E1)
Spinal cord structure (H1)
Spinal nerves and plexuses (H2)
From ovum to fetus (L4)

## General considerations

When attempting to understand human physiology, a small degree of anatomical knowledge is required. A more detailed understanding of the morphology of different cell types and tissues is needed. These details are presented in later topics in this section, but let us start with a brief overview of the gross anatomical structure of the human body. As will become apparent, the body is made up of an external surface, through which we interact with our environment, supported by services to provide oxygen, blood, nutrients, and instructions in order that we may move around. Let us consider each of these in turn.

## External surfaces

Aside from the skin, which is the obvious external surface of the body, there are several other regions that are still external. These include the airways, the gut tube (from mouth to anus and associated structures such as the gall bladder, and pancreas), the urinary system, and the reproductive system. If one were to shrink to a tiny size, one would be able to gain access to the air sacs in the lungs, the bile ducts, the filtration sites in the kidney, the gonads, all by entering the body from the outside world via the correct orifice. The same opportunity for access is available for bacteria and viruses, as shown by the incidence of diarrhea, chest infections and urinary tract infections in the general population. All of the external surfaces are lined with a covering layer known as **epithelium** (more of which in Topic C2) that has the role of either:

- protecting the interior of our body from substances in the outside world, or retaining fluid and electrolytes
- allowing us to absorb substances
- allowing us to excrete substances into the outside world.

Epithelial function is crucially important in maintaining health and wellbeing.

## Body cavities

Within the human body, there are several cavities that are formed during development of the **embryo**. These are the pleural, pericardial, peritoneal and cranial cavities, which house the lungs, heart, intestines, and the brain and spinal cord, respectively. Each cavity is lined with membranous sheets of cells, all of which have the purpose of preventing damage to the organs within the cavities when they move. This is enabled by the presence of a thin layer of watery fluid between the membranes lining these cavities (for the **pleural cavity**, see *Fig. 1*). This is especially important in the pleural and **pericardial cavities**, when one considers that we inhale and exhale up to 15 000 times a day and that the heart

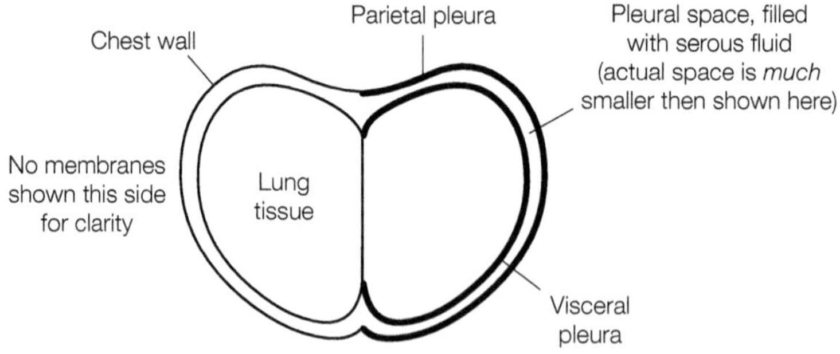


*Fig. 1. Schematic diagram of the pleural cavity, as might be viewed when the thorax is cut in cross-section.*

beats on average 100 000 times a day. The intestines must be free to move in order to mash up food and propel it from the oral to the anal end of the gut. The pleural, pericardial, and **peritoneal cavities** are lined by two layers of membranes, separated by this thin layer of fluid. In most circumstances, the visceral layer adheres very tightly to the surface of the organ(s), and the parietal layer is outside of this, closer to the body wall. The thin layer of fluid acts as a partial vacuum and permits friction-free movement of the organ against the body wall.

## The meninges

The situation with the brain and spinal cord is more complex, since the degree of movement is only very slight, but the potential for harm is enormous. In this case, there are three membrane layers, known as the **meninges** (singular: **meninx**). These three membranes are called (from inside to outside) **pia mater**, **arachnoid mater**, and **dura mater** (*Fig. 2*). Between the arachnoid mater and pia mater (the latter of which adheres very closely to the surface of the brain and spinal cord), there is a specialized form of extracellular fluid known as **cerebrospinal fluid** (CSF), produced by secretory cells in the **choroid plexus** in the brain. One of the purposes of the CSF is to prevent damage to the brain and spinal cord when they move inside the skull and vertebral column, so it can be thought of as similar in this regard to the fluid in the pleural, pericardial and peritoneal cavities.

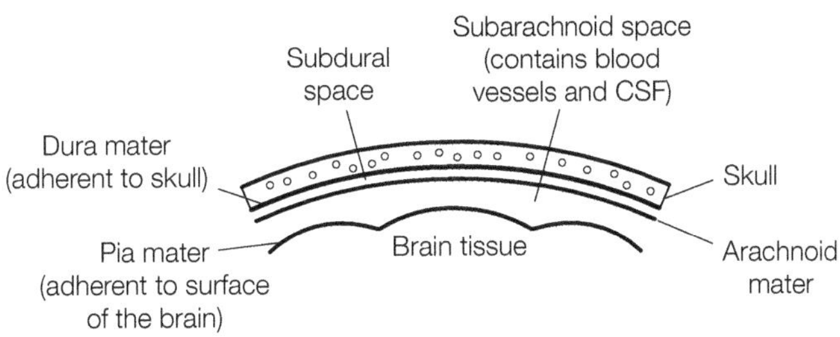


*Fig. 2. Schematic diagram of the meningeal membranes covering the brain, as might be viewed when the head is cut in cross-section.*

## Segmentation

As with all vertebrates (animals with a backbone), we are segmented, based upon the various segments of the **spinal cord** that developed from embryonic structures. A repeating pattern of nerves arising from the spinal cord (known as the spinal nerves) innervates skin, muscle, and all other structures in the body. From the top of the neck to the bottom of the human vertebral column, there are 7 cervical (C1–C7), 12 thoracic (T1–T12), 5 lumbar (L1–L5) and 5 sacral (S1–S5) vertebrae. The spinal nerves emerge from the vertebral column in small spaces between these vertebrae, and go on to innervate skin, muscle, joints, and internal organs. All spinal nerves have a mixed function; they contain nerve fibers that conduct information away from the brain toward the body (motor fibers), and nerve fibers that conduct information from the body to the brain (sensory fibers).

## Limbs

The limbs are simply the mechanism that we have evolved to transport us (and our genes in the form of eggs and sperm) through the outside world. Developmentally, they are outgrowths of the body wall that drag their nerve and blood supplies with them during development. This is obvious when one examines a map of which nerves supply which sections of skin on the human body (a **dermatome** map – see *Fig. 3*). The upper limb is supplied by nerves arising between the fifth cervical and first thoracic segment of the spinal cord; the pattern from the head to the small of the back is interrupted. The same is true of the lower limb. The skin there is supplied by nerves arising between the second lumbar and second sacral segment of the spinal cord, and the pattern continues in the 'saddle' area with nerves arising from the third to the fifth sacral segments.

The limbs receive information about intended movement that is sent from the brain, and convert this into muscle contractions and **flexion** or **extension** of joints. They also act as one of the mechanisms by which we interact with and receive information about the outside world, in the form of sensations like touch, temperature, pain, etc. They are highly specialized for the various forms of movement that humans undertake; no other animal on earth has the range of movement that we do in our limbs, especially in the hands, mechanisms that allow us to carry out extremely dextrous tasks.

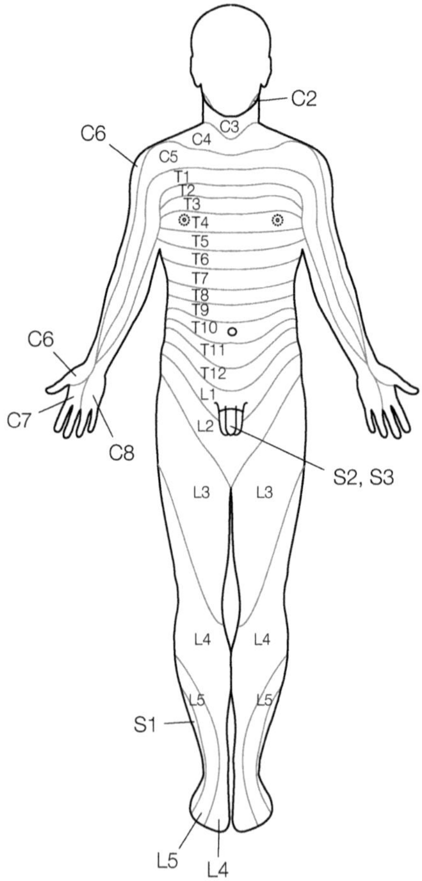


*Fig. 3. The human dermatome map, showing the anterior (front) surface only. C2, T1, L1, S1, etc. refer to areas innervated by second cervical spinal nerve, first thoracic spinal nerve, first lumbar spinal nerve, first sacral spinal nerve, etc.*

# C2 Epithelia and connective tissue

## Key Notes

| | |
|---|---|
| **Epithelial versus non-epithelial** | All body tissues can be classed as either epithelial or non-epithelial. Most epithelial tissue lines the external surfaces of the body, although some cells of epithelial origin (those in the thyroid gland, for example) are no longer on the external surface. Another name for non-epithelial tissue is connective tissue. |
| **General structure of surface epithelium** | All epithelia have the same general structure. Epithelial cells project out into the environment from the external surfaces of the body. There may be a single layer of cells or several layers; new cells are produced by mitotic division of cells in the basal layer of the epithelium. Directly below the epithelial cells is the basement membrane or basal lamina, and below that the connective tissue. Blood vessels and nerve fibers do not generally cross the basement membrane. |
| **Epithelial form and function** | There are several different types of specialized epithelia, designed for different roles. For example, an epithelium that is one cell thick and flat (squamous) is specialized for exchange, and one that is multilayered (stratified) with the upper layers being flattened is specialized to resist friction and to be watertight. |
| **Specialized connective tissue** | The non-epithelial tissue in the body can be loosely organized to form loose connective tissue (e.g. fibroblasts secreting collagen) or highly organized (e.g. skeletal muscle). Such specialization is very important in determining the functional characteristics of the connective tissue. |
| **Related topics** | Basic structure of the body (C1)<br>Skin, hair, and nails (C3)<br>Bone and muscle (C5)<br>Layout and function of the vasculature and lymphatics (D5)<br>Alveolar exchange and gas transport (E4) |

## Epithelial versus non-epithelial

**Epithelial** cells are derived from those cells that, in the fertilized embryo, were on the outside surface. In most cases, epithelial cells are on the outside surface of the body, but there are a few notable exceptions such as the cells that make up the thyroid gland and the cells of the nervous system. These two examples, along with others, become internalized during embryonic development, yet are still classed as epithelial in origin. **Mesothelial** cells, which make up the membranes lining the body cavities (see Topic C1) and endothelial cells (see Topic D5), which line the inside surface of blood and lymphatic vessels, are not epithelial.

Thus all of the cells of the body can be categorized according to a very simple scheme; a cell is either epithelial or non-epithelial. As indicated in Topic C1, epithelial cells are the mechanism by which we make contact with the outside world, or interact with the outside world on a chemical basis. All non-epithelial cells make up **connective tissues**, indicating that connective tissues simply 'connect' epithelial cells together and help move the epithelia around in the outside world. Thus all cells make up either epithelial tissue or connective (non-epithelial) tissue.

## General structure of surface epithelium

**Epithelium** that covers the outside surface of the body has a very consistent general structure (*Fig. 1*). The layer of epithelial cells can be either simple (one cell thick) or stratified (more than one cell thick) and always rests on a so-called '**basement membrane**'. In simple epithelium, all cells make contact with the basement membrane, while in stratified epithelium, only those cells closest to the basement membrane (in the basal layer) do. There is a particular type of simple epithelium that is known as '**pseudostratified**' but this is present only in the respiratory tract and appears to be stratified simply because of the arrangement of the cells. Dead cells are replaced by division of neighboring cells in the case of simple epithelia, or by division of cells in the **basal layer** (closest to the basement membrane) in the case of **stratified** epithelia. The basement membrane is not like other membranes in the body, but consists of interwoven sheets of proteins such as **collagen** and **laminin**. Cells below the basement membrane form connective tissue of various types.

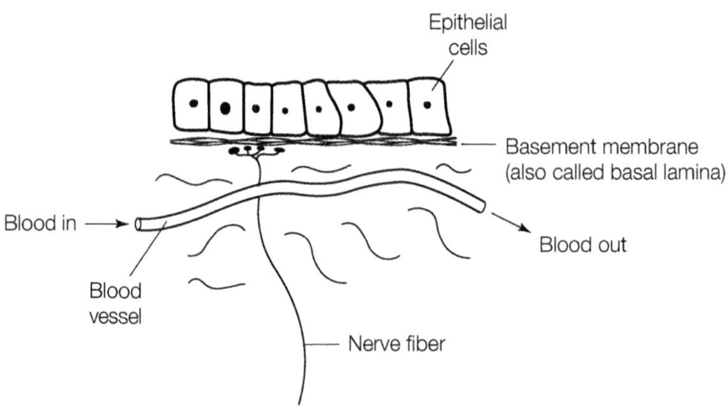


*Fig. 1. General arrangement of epithelia in the body. Epithelial cells rest on a basement membrane, which is not penetrated by nerves, blood vessels, or lymphatics. All tissue below the basement membrane is connective tissue.*

The basement membrane also has a barrier function, in that blood vessels do not penetrate it. This is important in limiting the spread of cancerous epithelial cells either locally or to other tissues in the body via blood vessels. The fact that blood vessels do not penetrate the basement membrane also means that the nutritional requirements of epithelial cells have to be met by diffusion from below the basement membrane. Again, this has relevance to cancer, in that when cancerous cells are increasing in number (proliferating), their nutritional requirements obviously increase. Sometimes this increased need for nutrition is met by an increase in the number of blood vessels supplying the area (angiogenesis). Cancerous epithelial cells form **carcinoma** (a name derived from the 'crab-like' appearance of the new blood vessels supplying the tumor), and cancerous connective tissue cells form **sarcoma**.

**Epithelial form and function**

As indicated above, there are two basic types of epithelium: simple and stratified. However, within these two general types, there are a multitude of different variations. Some examples are given in *Fig. 2*. Epithelia that are specialized for exchange are always simple in form; in fact the cells may even be **squamous** (flattened). Epithelia that have a barrier function or which must resist friction are always stratified, and may even have a layer of tough protein such as **keratin** on their outermost surface as waterproofing or to help resist wear and tear. The skin is covered in this **keratinized** stratified epithelium.

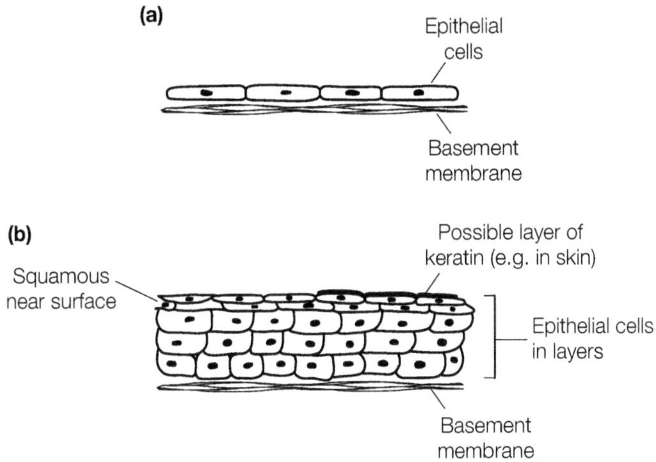


*Fig. 2. Examples of epithelia: (a) simple squamous epithelium; (b) stratified squamous epithelium.*

There are many further specializations that may be observed, and all are related to the function of the epithelium (*Fig. 3*):

- some cells are specialized to produce **mucus** (e.g. **goblet cells** in the airways)
- some cells form complex glandular structures specialized for secretion (e.g. **sweat glands**)
- some cells have hair-like projections called **cilia** on their surface that help to move the secretions (e.g. in the trachea or in the female reproductive tract)
- some cells have finger-like projections called **microvilli** on their surface to increase surface area for absorption (e.g. in the intestine or in certain parts of the kidney tubule)
- the layers of epithelial cells may change shape depending on the degree of distension of the organ that they help line (e.g. the **transitional epithelium** of the urinary bladder).

**Specialized connective tissue**

The cells that compose connective tissue are also specialized for their function, but it is beyond the remit of this topic to describe them all in detail here. Two examples are shown in *Fig. 4*.

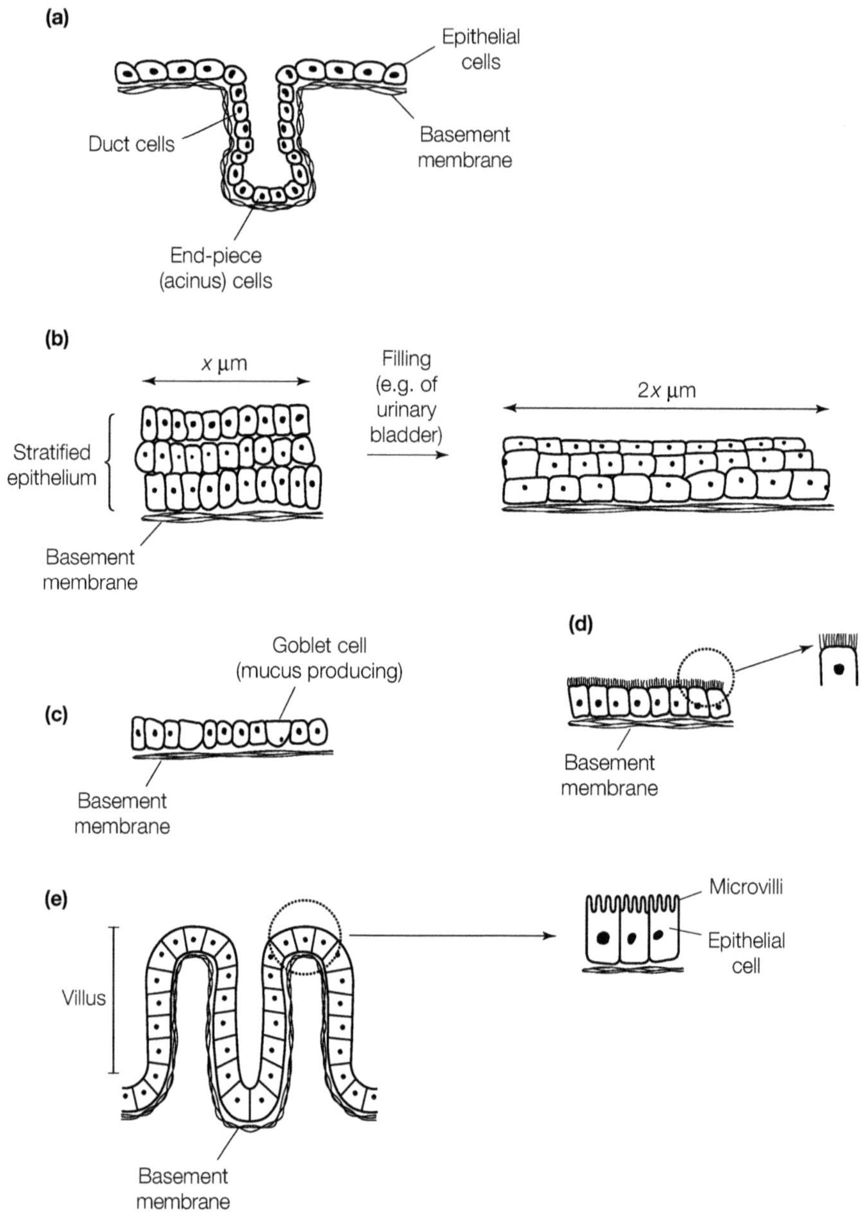


*Fig. 3. Examples of specializations in epithelia: (a) glandular epithelium; (b) transitional epithelium; (c) goblet cells; (d) ciliated epithelium; (e) microvilli.*

**Skeletal muscle cells** (also known as skeletal muscle fibers; *Fig 4(a)*) may have to be up to 40 mm long, which is enormous in terms of other cells in the body (red blood cells are more than 1000 times smaller at around 7 µm). However, in order that a skeletal muscle can act on two bones that form a joint in order to move that joint, each of the cells making up the muscle must be relatively long. This is achieved during development, by fusion of a number of cells

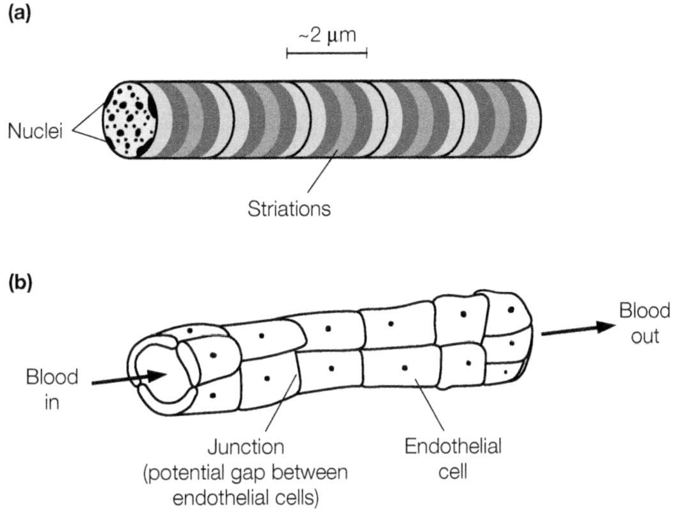


*Fig. 4. Structural features of two types of connective tissue: (a) skeletal muscle fibers; (b) capillary endothelial cells.*

to form one skeletal muscle fiber, and thus skeletal muscle fibers have many nuclei. These nuclei are also squeezed out to the peripheral margins of the cell in order to make way for the bundles of contractile proteins that allow the cell to shorten in response to a nerve impulse and act on the joint. You can read more about this in Topic C5.

The **endothelial cells** (*Fig 4(b)*) that line blood vessels are extremely flattened. This is essential so that when blood is flowing through the vessel, it moves in a relatively friction-free fashion. If endothelial cells become damaged and are not repaired, or if they have fatty substances stuck to them, then turbulent blood flow can lead to endothelial damage and eventually to rupture of the blood vessel wall. **Capillaries** (the vessels where exchange of substances between the blood and the tissues take place) are made up of only endothelial cells. The degree of leakiness of these cells and their intercellular contacts with each other also varies, depending on where the capillaries are in the body. Most capillaries are slightly leaky, helping to facilitate fluid movement between the blood and the interstitial space between cells in the tissues. However, capillaries in the brain are much 'tighter' than average, because fluid movement into brain tissue could have devastating consequences if it were to go wrong (the brain is contained inside an essentially rigid box), and capillaries in the liver are so leaky that proteins from the blood 'wash over' the liver cells (**hepatocytes**). This is ideal since hepatocytes synthesize many essential blood components and also act as a store for substances like **glucose** and **cholesterol** that are needed by all cells in the body. You can read more about the function of the various types of capillaries in Topic D5.

# C3 Skin, hair, and nails

## Key Notes

| | |
|---|---|
| **Skin layers** | The skin, which is our largest organ, is arranged into two main layers: the epidermis (epithelial) and the dermis (connective tissue). The epidermis takes the form of keratinized stratified squamous epithelium. Other features of the skin are hair follicles, sweat glands and subcutaneous fat. Numerous nerve endings innervate the skin to convey information about touch, temperature and pain to the central nervous system. |
| **Skin cells** | There are several different types of epithelial cells in the skin. Keratinocytes are cells that die when they reach the outer layers of the epidermis, leaving behind a layer of keratin to improve friction resistance and waterproofing of the skin. Melanocytes exist in the basal layer of the epidermis and produce melanin, a pigment that they donate to the keratinocytes. Activity of melanocytes is correlated with skin color; the more active the melanocytes, the darker the skin color. |
| **Regional variation** | Although the skin has the same basic structure in all areas of the body, some regional variations exist. The epidermis is thicker in areas where friction is greater. There are more nerve endings in the skin on the fingertips and around the mouth, areas where we receive a lot of tactile information about our environment. |
| **Structure of hair and nails** | Hair and nails are composed of the protein keratin. Epithelial cells in the hair follicle and in the nail bed die when they grow too far away from their nutrient blood supply below the basal lamina. The keratin that the cells contain is left behind as hair and nail. Hair color varies from one person to another in the same way that skin color varies; if melanocytes in the hair follicle are more active, the hair will be darker. |

**Related topics** Epithelia and connective tissue (C2) Structure and function of peripheral nerve (H3)

### Skin layers

In terms of weight, the skin is the largest single organ in our body and is the most obvious way in which we interact with the outside world. It is specialized to keep water out, to preserve heat and fluid, and to inform us of various types of tactile and noxious stimuli. The skin is arranged into two main layers (*Fig. 1*). These are the **epidermis**, which is the highly organized layer of epithelial cells resting on a basement membrane, and the **dermis**, which is the layer of connective tissue directly underlying the basement membrane. Below these two layers lies the **hypodermis** with its associated subcutaneous fat. When the skin is cut in cross-section, it may appear as if certain elements of epithelial origin are in the dermis, but this is simply an illusion caused by being forced to look at something in two dimensions instead of three. The secretory cells that constitute

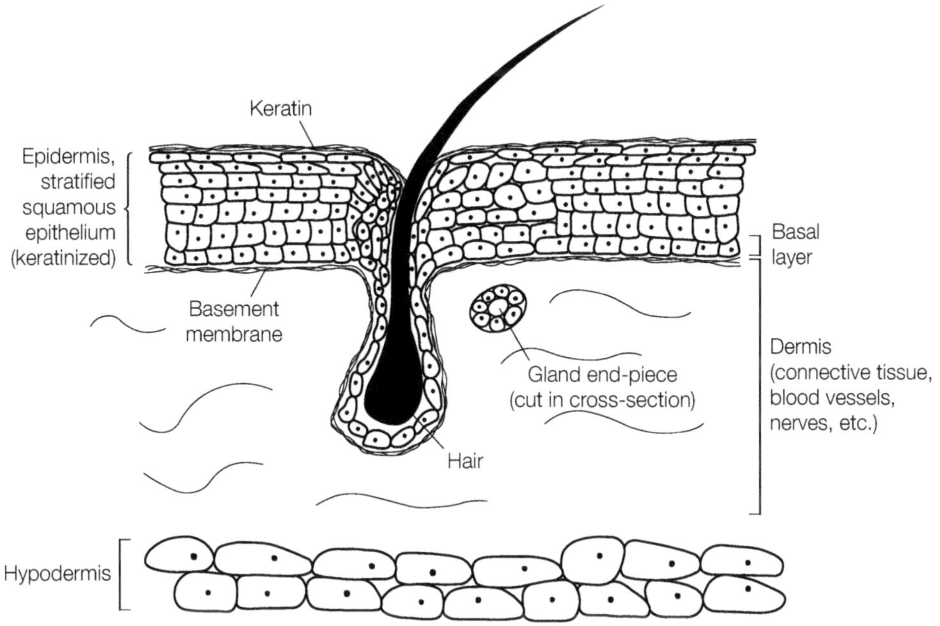


*Fig. 1. Cross-sectional structure of the skin.*

**sweat glands** are still epithelial, and are part of the epidermis (since they are actually above the **basement membrane**), as are **hair follicles** and their associated **sebaceous glands** producing oily **sebum**. However, in common with other epithelia, blood vessels and nerves do not cross the basement membrane in the skin, and are thus confined to the dermis. Within the dermis, there are specialized nerve endings that act as sensory receptors for touch, vibration, pressure, heat, cold and pain. Our skin is our biggest sensory organ.

**Skin cells**

The epithelial cells that make up the epidermis fall into several different categories, such as **squamous cells**, **basal cells**, and **melanocytes**. Basal cells are the most active cells in the epidermis in terms of cell division (**mitosis**), and it is from these cells that the epidermis can be regenerated after injury. Squamous cells exist near the surface of the skin and help to resist friction by virtue of their flattened shape and also by producing keratin. **Keratin** also helps to make the skin waterproof. Melanocytes exist near the basal layer and synthesize **melanin**, which helps to protect the skin from ultraviolet (UV) light. Tanning occurs when melanocytes increase production of melanin as a defense against UV rays, and the melanin is taken up by other epidermal cells. Variations in the skin color of the various races is related to activity of the melanocytes in their skin; ethnic Africans have more active melanocytes than do ethnic Scandinavians for example, but the number of melanocytes in the skin is roughly similar.

**Regional variation**

All of the skin covering the body has this basic structure, but regional differences exist. It is not surprising that the epidermis should be substantially thicker on the sole of the foot or on the fingertips than on the face, for example. Where skin is subjected to greater levels of wear and tear, it will inevitably be thicker.

The hands of carpenters are much rougher and more calloused than those of teachers. Variations in other features also exist. The skin on the palm of the hand is hairless, but that on the back of the hand or arm is hairy. There are more sweat glands in the skin of the armpit (axilla) and groin than in the skin on the chest. The density of specialized nerve endings for touch is far greater on the lips and fingertips than it is on the back. Each of these features is relevant to the function of that part of the body. For example, it makes perfect sense that the density of **touch receptors** is greater on the lips and fingertips than on the back, since we are much more likely to encounter our surroundings via the first two routes. We are naturally inquisitive and pick things up, and as infants we often put objects in our mouths in an effort to find out more about them. You can read more about these sensory receptors in Topic H5.

## Structure of hair and nails

Hair and nails (and horns and hooves in some other vertebrates) are specialized forms of the protein keratin that is produced by the cells of the epidermis. In the case of the **hair follicle**, new cells formed by mitosis push older cells upwards. These older cells die because they are now further away from their nutrient blood supply which exists in the dermis, and the keratin (and melanin content) in them remains as hair of varying color. The cells of the hair follicles have some of the fastest rates of division of any cells in the body. That is why hair loss is a common side-effect of **chemotherapy** for **cancer**, since the chemotherapy drugs kill cells that are actively dividing. This just happens to include not only cancerous cells, but those of the hair follicles as well. Our fingernails and toenails are also composed of keratin and take between 6 months (in the case of healthy fingernails) and 1 year (in the case of healthy toenails) to grow out completely.

# C4 NEURONS AND GLIA I

## Key Notes

**Basics of neural embryology**

Neurons and glia develop from neural ectoderm. The neural tube (which later develops into the brain and spinal cord) is formed from the neural plate early in gestation. Neural crest cells go on to become the neurons and glial cells of the peripheral nervous system and the enteric nervous system in the gut, as well as neuroendocrine cells of several different organs.

**Common nervous system stem cells**

Neurons and glial cells develop from common precursor cells (stem cells). These stem cells form the neural tube, by which time their fate is to some extent determined; they are no longer totipotent, merely pluripotent.

**Types of neurons**

The nervous system is composed of a multitude of different types of neurons. These differ in terms of shape, neurochemistry and function. Neurons need not necessarily be excitatory; some have inhibitory influences on their target cells.

**Types of glial cells**

There are three major forms of glial cells. Oligodendrocytes (in the central nervous system (CNS)), and Schwann cells (in the peripheral nervous system) are the cells that myelinate neurons. Astrocytes have an important role in forming the blood–brain barrier, and in providing trophic support for neurons. Microglia are small resident phagocytic cells in the CNS.

**Related topics**

Neurons and glia II (F1)
Action potentials (F2)
Synaptic transmission (F3)
The panoply of transmitters (F4)
Parts of the brain (G1)
Spinal cord structure (H1)
Introduction to the gastrointestinal tract (J1)
From ovum to fetus (L4)

## Basics of neural embryology

All of the cells in the human body develop from a single fertilized egg, which is known as the **zygote**. This is a single cell with the proper complement of chromosomes, half provided from the mother's egg and half from one of the sperm cells of the father. This cell, and some of the cells that it can divide to produce, has the ability to develop into any human cell type. It is therefore said to be **totipotent**. As the cells keep dividing, a ball of cells known as the **morula** develops at around 96 hours after fertilization.

As described in Topic C2, the cells that make up the nervous system are of ectodermal origin (see Topic L4). The cells that will make up the nervous system migrate (*Fig. 1*) to the inside of the **embryo** from the surface (**neural ectoderm**) within the first five weeks after fertilization because they are attracted by chemical signals being produced by other cells. More details of this process (called

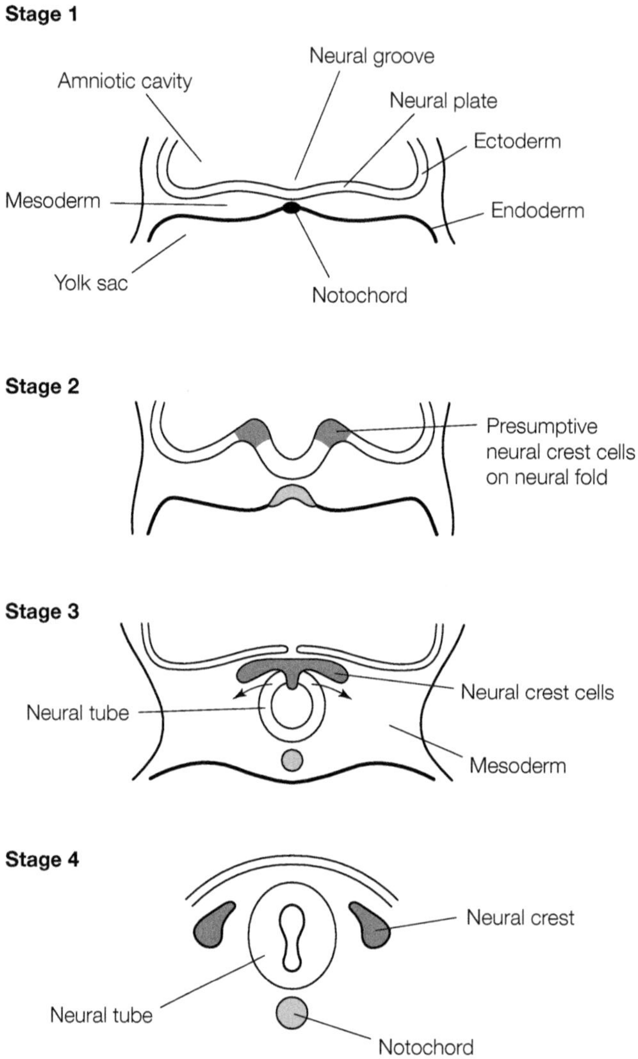


*Fig. 1. The process of migration of neural ectoderm to form the neural tube in the human embryo from day 20 to day 24. Closure of the rostral and caudal ends of the neural tube is not complete until day 28.*

**neurulation**) are given in *Instant Notes in Developmental Biology*, Section J. The **neural tube** is formed from the **neural plate** (part of the ectodermal layer of the embryo), along virtually the entire length of the embryo. It is important that the neural tube closes properly, or a condition known as **spina bifida** can occur, leading to neurological symptoms such as paralysis. The cells of the neural tube then go on to form the nerve cells and glial (supporting) cells of the brain and spinal cord, the **central nervous system** (CNS). The most critical period for development of the CNS is in the first 12 weeks after fertilization. We know this because the likelihood of the **fetus** having CNS defects if exposed to an insult such as rubella infection (German measles) is much higher if the insult occurs in the first 12 weeks of development, rather than later.

At around the time of neurulation, a set of cells known as **neural crest** (distinct from the neural tube) also develops. It is from the cells of the neural crest that the nerve cells and glial cells of the **peripheral nervous system** (PNS) develop, together with **melanocytes** and the cells of the **enteric nervous system** in the gut. Collections of connective tissue cells known as **somites** also form, and these go on to become the cartilage and bone of the backbone (**vertebral column**).

## Common nervous system stem cells

Within the nervous system, there exist only two populations of cells; nerve cells (**neurons**) and supporting cells (**glia**). However, there are a multitude of different neurons that have different characteristics (phenotypes), and the same is true of glia. Each of these different cells is produced from a common type of stem cell; those which made up the neural ectoderm during embryonic development. Thus all the cells of the nervous system are epithelial in origin, and cancers of both neurons (**neuromas**) and glia (**gliomas**) are **carcinomas**.

Cells of the mesoderm cannot become neurons and glia, but can develop into **muscle**. Cells of the ectoderm cannot become muscle, but can develop into neurons and glia. This tells us that by the time the neural tube starts to form, the cells in the various layers of the embryo are already committed to becoming a specific set of cell types. They are no longer totipotent, merely pluripotent. In fact, this is true at a much earlier stage. If the fertilized human zygote (combination of egg and sperm) is split at any stage up to the third cell division (the eight-cell stage), then any of the cells can go on to make up a complete embryo. In fact this is what happens when one zygote develops into identical twins. However, by the time that the cells have completed their fourth division (the 16-cell stage), this ability has been lost; the removal of a single cell at that stage will result in the embryo having a part missing.

## Types of neurons

As stated above, neurons are extremely phenotypically diverse (you can read more about this in the various topics in Section F). If we were able to look down a microscope at all of the neurons in the human nervous system, we would be able to distinguish many different types by shape alone, but their diverse nature goes further than that. Some neurons have a motor function in the nervous system; they influence the activity of muscle cells, or affect the secretory capacity of glands. Others are sensory; they conduct information to the CNS about what is going on in the outside world. Some neurons are under voluntary control, meaning that we can influence their level of activation; others are not under voluntary control (e.g. those of the **autonomic** branch of the PNS), meaning that we are unable to influence their activity. Some neurons, when activated, stimulate the other neurons that they make contact with and are known as **excitatory neurons**. Others, when activated, inhibit activity in the other neurons with which they make contact, and are known as **inhibitory neurons**. The types of chemical (**neurotransmitters**) that neurons use to interact with each other are also extremely diverse and include peptides, amino acids, amines, purines and a host of other chemical species. In recent years, it has even become apparent that a gas (**nitric oxide, NO**) can be utilized as a transmitter between some neurons. Within each class there are many possibilities; for example, aminergic neurons (those that use amines as their neurotransmitter) can use **norepinephrine**, **epinephrine**, **dopamine** or **serotonin**, each of which act through specific **receptors** on the neurons that are being influenced.

## Types of glial cells

Glial cells are also very diverse. Their roles are to support neuronal function and to act as defense mechanisms in the nervous system. They fall into three main categories, as outlined in *Table 1*.

*Table 1. Main types of glial cells in the nervous system*

| Type of glial cell | Main function(s) |
|---|---|
| Oligodendrocytes | Myelination of several neurons (CNS) |
| Schwann cells | Myelination of a single neuron (PNS) |
| Astrocytes | Structural and metabolic support |
| Microglial cells | Resident phagocytes |

**Oligodendrocytes** (in the CNS) and **Schwann cells** (in the PNS) are the cells that help to insulate the long axon processes of neurons by wrapping their specialized cell membranes around the axon. This process is known as **myelination** and increases the speed with which nerve impulses can be conducted along the neuron. The process of myelination is slightly different in the CNS and PNS, in that oligodendrocytes myelinate several neighboring axons, but Schwann cells myelinate only one axon. Non-myelinated axons are still associated with oligodendrocytes and Schwann cells, but they are not covered in myelin.

**Astrocytes** (so called because of their star-like shape when stained in histological preparations) are by far the most numerous glial cells in the CNS. Their predominant role is to help form the **blood–brain barrier** between brain capillaries and brain tissue itself. The physical aspects of this barrier are, as already mentioned in Topic C2, made up of 'tight' connections between capillary **endothelial cells** in the brain. The processes of astrocytes make contact with the endothelial cells, and in this way the astrocytes are able to influence both the 'leakiness' of the capillaries and regulate the types of substances that can enter brain tissue. Excessive amounts of fluid would be detrimental to brain function, but brain cells need **glucose, amino acids** and other substances in order to function properly. Movement of these substances between brain tissue and the blood is regulated by the astrocytes.

**Microglia** are, as their name suggests, extremely small glial cells. The brain is often said to be an immunologically privileged area, in that few, if any, white blood cells are found there. This is thought to be because of the dangers that inflammatory processes would pose to brain function (since the organ is contained inside the skull, which is in effect a rigid box). Microglia are the brain's resident immune cell, similar to **macrophages** in peripheral tissues (see Topic C6) and their role is to engulf and digest foreign material in the brain, or pieces of dead or dying neurons. In this way, immunological inflammation is limited, other than in exceptional circumstances. In circumstances where large numbers of neurons have died (as a result of a stroke, or spinal injury, or in **Alzheimer's disease**, for example), astrocytes can also migrate to the area of injury and form scar tissue which is functionally useless, but which might help preserve structure.

# C5 Bone and muscle

## Key Notes

**Functions of bone**

Bone consists of cells involved in its synthesis and degradation, and an extracellular matrix consisting predominantly of type I collagen (osteoid) which is mineralized by calcium hydroxyapatite (hydroxylated calcium phosphate). This forms a strong, rigid skeleton allowing it to support the tissues of the body as well as protect some of them from external insult. Muscles attached to bone allow movement. Bone is a reservoir of calcium, and the movement of calcium into and out of bone is important for whole-body calcium homeostasis. The outer shell of most bones consists of dense, compact bone (cortical bone), while the more central part is less dense with a honeycomb network of narrow trabeculae (cancellous or lamellar bone).

**Cells of bone**

Osteoblasts are bone-forming cells which synthesize collagen for production of osteoid, and help in the deposition of calcium and phosphate for mineralization of osteoid. They also secrete factors in response to parathyroid hormone which stimulate the resorptive action of osteoclasts. Osteocytes are osteoblasts that have been left behind in the bone matrix during mineralization. They are in contact with each other within the bone matrix through a canalicular network, and play an important part in the maintenance of day-to-day plasma calcium concentration by moving calcium between bone extracellular fluid and the circulation. Osteoclasts, which resorb bone, are large multinucleate cells derived from fusion of pluripotent hematopoietic mononuclear cells. They sit in depressions on the bone surface, making contact via a ruffled border, and under local stimuli secrete acid to dissolve the hydroxyapatite and lysosomal proteases, particularly collagenases, to hydrolyze the organic matrix.

**Bone in calcium and phosphate homeostasis**

While the calcium content of bone remains approximately constant in adults until middle age, there is a daily movement of about 500 mg into and out of bone each day. This is controlled by the action of both osteocytes and osteoblasts, with the latter responding to parathyroid hormone.

**Bone structure and remodeling**

The skeleton is continually subject to gravitational stresses which cause microfractures in the bone, and the concerted action of osteoclasts, macrophages and osteoblasts repair these fractures such that bone mineral density remains constant. Osteoclasts dissolve the bone around the fracture, and macrophages smooth out the surface in preparation for the laying down of new osteoid and mineralization by the osteoblasts. This process is known as remodeling.

**The role of muscle**

Muscle tissue takes three forms: skeletal, cardiac, and smooth. All three types of muscle are contractile, but their features and roles are different. Skeletal muscles are most commonly used to flex or extend joints like the knee or elbow. Cardiac muscle has the function of pumping blood around the body. Smooth muscle is an important structural feature of blood vessels, the gut, the urogenital system, the airways and some glandular tissue.

**Comparative structure of muscle cells**

Skeletal muscle and cardiac muscle are forms of striated muscle; the cells appear 'striped' when viewed with a microscope. Smooth muscle cells lack this striated appearance. Skeletal muscle cells are long (up to 20 mm), whereas smooth muscle cells are much smaller (≤0.5 mm). Cardiac muscle cells are extensively branched. Skeletal muscle cells require some nervous input in order to contract, whereas cardiac muscle cells and many smooth muscle cells are spontaneously active.

**Related topics**

Structure of cardiac muscle (D2)
Layout and function of the vasculature and lymphatics (D5)
The locomotor system (G2)
Skeletal muscle function and properties (I1)
Smooth muscle function and properties (I6)

**Functions of bone**

Bone is a specialized form of **connective tissue** consisting of an organic matrix (30%) of predominantly **type I collagen**, onto which calcium and phosphate, in the form of **calcium hydroxyapatite** crystals, deposit as the inorganic matrix (70%). This structure provides rigidity and strength, but also a certain amount of elasticity to allow the bone to respond to mechanical stresses such as gravity. Thus, bone is a dynamic system that can be remodeled by the action of **osteoclasts** and **osteoblasts** (see below) in reaction to **gravitational stress** and the demands of **calcium homeostasis**.

Bone has several functions:

- it provides a rigid, protective structure for most soft tissues of the body, e.g. the skull protecting the brain
- it serves as a store of minerals, especially calcium, magnesium, phosphorus (as phosphate) and, to some extent, sodium
- structural, providing the skeleton on which to hang other tissues
- it protects the bone marrow, the major site of synthesis of blood cells.

Although originally considered to be a somewhat inert tissue, bone is metabolically active, undergoing turnover to repair **microfractures** induced by gravitational stress. It also has a nerve supply, is vascularized and has an active hematopoietic function. The bones of the limbs (tibia, fibula, femur, humerus, etc.) are known as long bones, while bones such as the skull, mandible and scapula are known as flat bones. Two different types of bone can be described, the structures of which reflect their function. **Cortical, lamellar or compact bone** forms the shafts of long bones and much of the exterior surfaces of flat bones. It is characterized by little metabolic activity and has few cells associated with it. Accounting for about 80% of the skeleton, it provides protection and support, and is resistant to the stresses imposed by weight bearing and movement. **Trabecular, cancellous, or spongy bone**, on the other hand, is found in the extremities and narrow cavities of long bones and vertebrae and in the center of flat bones. It is characterized by an irregular network of thin columns of bone creating spaces that may contain the marrow and the osteoclasts. It is thus much more metabolically active than cortical bone, and better suited to rapid turnover.

## Cells of bone

Osteoblasts are immature cells derived from mesenchymal precursor cells and are the major bone-forming cells. They synthesize and secrete the organic extracellular matrix of bone (**osteoid**), and control the mineralization of the osteoid to form new bone. Osteoblasts are rich in **mitochondria** and have a well-developed **endoplasmic reticulum**, characteristic of cells involved in high rates of protein synthesis and secretion. The major protein synthesized and secreted by the cells is type I collagen, and an indication of collagen synthetic activity can be seen by the measurement, in the urine, of procollagen telopeptides released during processing of the nascent collagen chains. Osteoblasts also secrete a specific isoform of **alkaline phosphatase**, the activity of which in blood is indicative of osteoblastic activity. The cells also possess receptors for **parathyroid hormone (PTH)** and **1,25-dihydroxyvitamin D** (see Topic K7).

**Osteocytes** are osteoblasts which have become entrapped within the bone matrix during bone growth. They lie in extracellular fluid within the **canalicular spaces**, and communicate with one another via pseudopod (finger-like) extensions throughout the matrix. In this situation, they probably have no major biosynthetic activity but are primarily responsible for maintenance of the **bone matrix**. They also contribute to the acute movement of calcium to and from the bone surface during the day-to-day maintenance of calcium homeostasis.

Osteoclasts are large multinucleated cells of the monocyte–macrophage lineage and are responsible for the resorptive processes associated with bone remodeling and also for bone loss in diseases such as **osteoporosis**. Thus, they secrete acid to dissolve the inorganic matrix, and enzymes to hydrolyze the organic matrix – proteases, particularly collagenase for collagen, and **hyaluronidase** to break down proteoglycans. Interestingly, although osteoclasts are responsible for releasing calcium from bone, they do not possess receptors for PTH. It appears that the response of the osteoclast to PTH is indirect and comes via the osteoblast.

## Bone in calcium and phosphate homeostasis

Calcium moves in and out of bone in response to hormonal stimuli and extracellular fluid (ECF) calcium concentration to maintain calcium homeostasis. The mechanical integrity of bone is also maintained through a continuous cycle of bone remodeling involving **osteoclastic resorption** and **osteoblastic formation** to repair microfractures induced by gravitational stresses and insult. In this case, resorption and formation are tightly coupled within discrete regions of the bone, without affecting calcium homeostasis in the ECF.

Osteocytes embedded in the bone matrix itself (see above) are in contact with an intact layer of osteoblasts which line the bone surface. This layer of osteoblasts separates bone ECF from bulk ECF. The calcium concentration of bulk ECF is much greater than that of bone ECF, and this concentration gradient promotes the movement of calcium into bone. Movement of calcium out of bone, however, is under the control of PTH and is probably a role of the layer of osteoblasts, since these cells have PTH receptors on their surface.

## Bone structure and remodeling

Besides the cells responsible for its turnover, bone consists of an organic, connective tissue matrix of predominantly collagen fibers (90–95%), onto which is deposited the inorganic component, calcium hydroxyapatite. The non-mineralized component, known as osteoid, is composed of bundles of type I collagen, triple helical proteins of three α-chains (two $\alpha_1$ and one $\alpha_2$) wrapped around each other, rather like a three-stranded rope. Other minor, non-collagenous proteins such as alkaline phosphatase, **osteocalcin** and **osteopontin** are also present. The extracellular matrix collagen is synthesized by osteoblasts in a precursor form, **tropocollagen**, and exported from the cell. The tropocollagen is assembled into insoluble fibrils extracellularly. Stability of the organic matrix derives from the unique cross-linking of the collagen fibrils involving hydroxylysine and hydroxyproline residues in the collagen peptides. Enzymatic hydroxylation of such lysine and proline residues occurs post-translationally (i.e. after synthesis of the peptide) in a reaction requiring ascorbic acid (vitamin C). Deficiency of vitamin C, scurvy, leads to the reduced secretion of collagen molecules which have a decreased ability to cross-link, and a defective collagen matrix. This may present as defective ossification and bone growth in infants. Mutations in the collagen gene can also lead to pathological conditions; for example, mutation in the type I collagen gene can give rise to **osteogenesis imperfecta** ('brittle-bone disease').

As mentioned earlier, the inorganic, or mineral component of bone consists of calcium and phosphate in the form of calcium hydroxyapatite, $Ca_{10}(PO_4)_6(OH)_2$. This salt constitutes about 70% of the bone mass and gives the bone its rigidity.

**Bone mass** changes over a person's lifetime (*Fig. 1*). Peak bone mass is normally reached between the age of 20–30 years and remains constant (the **consolidation period**) in adults until approximately 40 years of age. From this time onwards, age-related factors contribute to bone loss to a greater or lesser extent. For example, there is an accelerated rate of bone loss in women during the menopause, associated with the **estrogen withdrawal** that typifies that period of a woman's life.

During the time when bone mass is maximal and constant, the consolidation period, bone undergoes slow and orderly turnover, or remodeling, in response to mechanical stress, hormones and probably also cytokines (*Fig. 2*). This complex process occurs in basic multicellular units and involves communication between all three types of bone cell, to first resorb bone and then rebuild it.

Mechanical stress in the bone is sensed by the osteocytes which signal to the osteoblasts on the bone surface via their long cytoplasmic projections. Cytokines

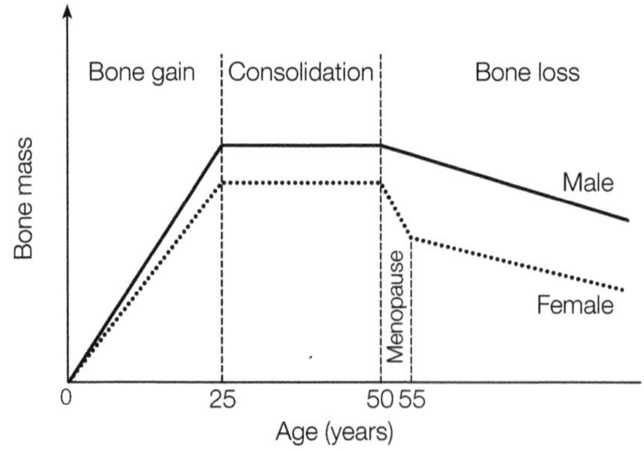


*Fig. 1. Changes in bone mass during life. Note the accelerated rate of loss of bone mass in females during the menopause.*

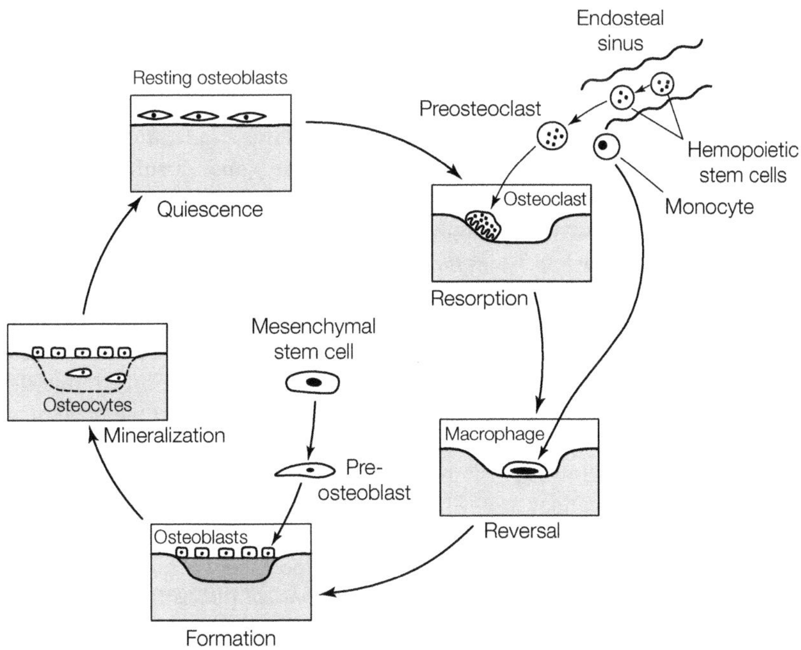


*Fig. 2. Bone remodeling. See text for details.*

such as **tumor necrosis factor-α, interleukin-1** and **interleukin-6**, and **insulin-like growth factor-1** released by the osteoblasts recruit multinucleate osteoclasts from the circulating monocyte precursors to the bone surface, and these cells begin to excavate a cavity in the mature bone. Osteoclasts attach themselves to the bone surface via a ruffled border, releasing acid to dissolve the inorganic matrix, and lysosomal proteases (collagenases, cathepsins) to hydrolyze the organic matrix exposed as a consequence. This erosion of bone is localized to the area surrounded by the ruffled border. **Macrophages** are then recruited to the site of resorption and 'tidy up' the surface in preparation for formation of new bone. This involves removal of dead cells, any residual collagen or non-collagen proteins, and formation of a cement line to which the new osteoid binds. The process of resorption releases cytokines such as transforming growth factor-β from the bone, which were deposited in the bone matrix during bone formation. Their release stimulates the differentiation of precursor stem cells into osteoblasts, and the formation of new osteoid at the cement line on the previous resorption site. This new osteoid provides a nidus for mineralization. In this way, bone resorption is coupled to formation. Mineralization initially involves the formation by the osteoblast of vesicles containing alkaline phosphatase and small crystals of hydroxyapatite. These vesicles concentrate calcium and phosphate, allowing the crystals to grow and finally rupture and deposit clusters and nodules of hydroxyapatite onto the collagen matrix. The function of the alkaline phosphatase is to provide phosphate from hydrolysis of organic phosphates in the cell. In the final stages of mineralization, the needle-like crystals of hydroxyapatite align themselves along the collagen fibrils to form the calcified collagen of mature bone.

The remodeling process described above is essential for the day-to-day maintenance of mature bone, and does not lead to a change in bone mass. However, from the age of about 40 years, there is a net loss of bone mineral, particularly in women during and immediately after the menopause. Here, the coupling seen in remodeling no longer obtains, and although bone is mineralized it is of poorer quality and integrity, and insufficient to replace that resorbed. Eventually, the decreased bone mass and deficient micro-architecture leads to increased fragility and risk of fracture, the condition known as osteoporosis. The risk of fracture, particularly of the wrist and neck of femur, increases exponentially with age.

## The role of muscle

All **muscle** in the body is derived from embryonic **mesoderm** (the middle layer of cells in the **embryo**). When we hear the word 'muscle' we immediately think of the obvious mass of skeletal muscle that we have in our body, the purpose of which is to move joints so that we can move around in our environment. However, there are parts of our body that are composed of muscle that we cannot see, and which do not have this locomotor function. The heart is composed predominantly of muscle, yet it does not move any joints, and the walls of the gut, airways, blood vessels, urogenital tract and even some glands have muscle as an integral part of their structure and function. In each case, the muscle moves something, even if it is not a joint (*Fig. 3*). In the case of the heart, the rhythmic pumping action of the muscle is the mechanism by which blood is moved around our body. In the case of the gut, the muscle in the wall is the means by which we mix food and propel it from the oral to the anal end of the gastrointestinal system. Activation of muscle in blood vessels (the vasculature) narrows the tube and limits blood flow, or raises blood pressure in the system as a whole. Activation of specialized muscle cells in some glands can help to eject the glandular secretion onto the surface of the body, as is the case in breast tissue during **lactation**.

Muscle cells in different parts of the body do very different things, so one must expect there to be variation in structure that reflects this functional diversity. More information is given in Topic D2 and Section I, but a brief description is given here.

## Comparative structure of muscle cells

There are three types of muscle cell (**myocyte** or muscle fiber) in the body; **skeletal muscle** fibers, **cardiac muscle** fibers, and **smooth muscle** fibers. Of these, the majority of skeletal muscle fibers exist in muscles that act on bones to move joints; cardiac muscle fibers make up the heart; and smooth muscle fibers are an integral part of the gut, airways, urogenital tract, vasculature and some glands. The myocytes in each type of muscle are very different, and the main differences in structure and function are shown in Table 1.

The major function of skeletal muscle is to move bones and flex or extend joints, and this process should normally be under voluntary control. This is why skeletal muscle fibers only contract in response to a nervous impulse. The fibers are long (up to 20 mm) and develop from many individual cells. The nuclei of these cells exist within the skeletal muscle fiber, but are pushed to the periphery by the contractile proteins that they must be packed with in order to move large bones like those in the lower limb. As will be described later (Section I), the contractile proteins are highly ordered within the muscle fiber, and so striations (stripes) are obvious when the fibers are viewed down a microscope.

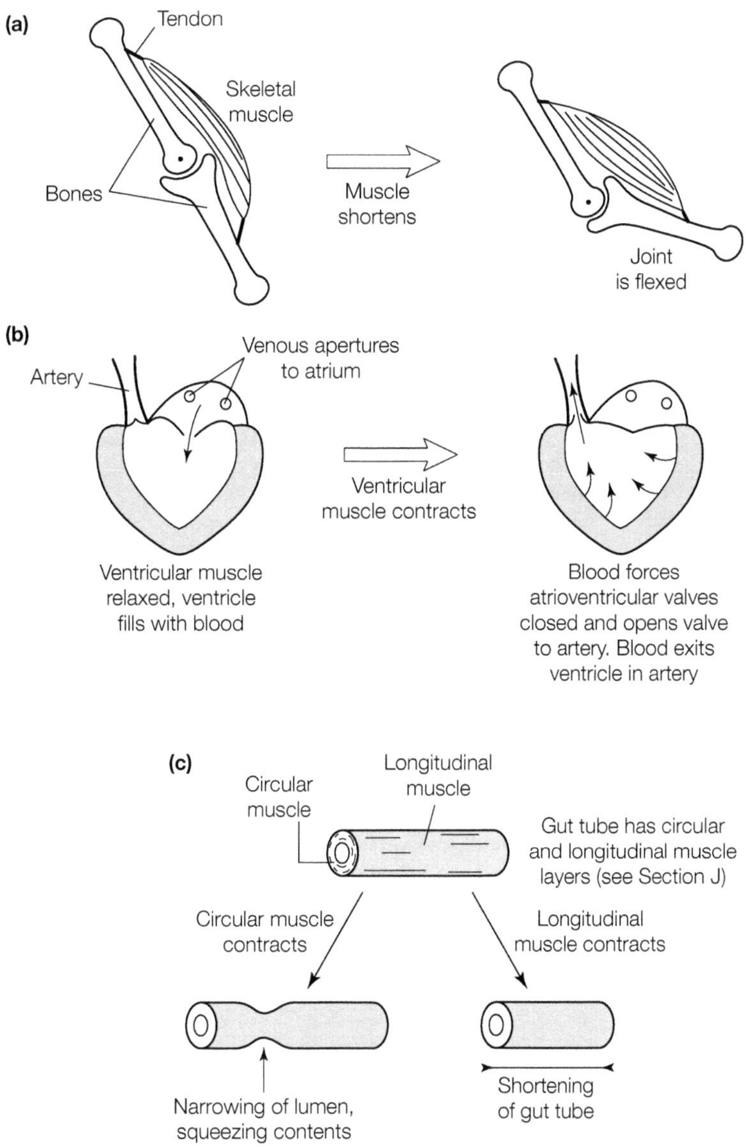


*Fig. 3. Functions of various muscle types: (a) skeletal; (b) cardiac; (c) smooth.*

Cardiac muscle fibers contract rhythmically in the absence of any nervous influence, although nervous activity (and hormone levels) can influence both the frequency of contractions (heart rate) and the force of contractions (**inotropy**). More details of why this is so are given in Topics D2, D3, and D4. Another important feature of cardiac muscle is the extensive branching of the fibers so that they make contact with many other neighboring fibers; this allows the contraction to spread in an organized way so that blood is pumped out of the heart in a coordinated fashion. It is also important that cardiac muscle contraction is not sustained, or the heart would not fill with blood between beats.

*Table 1. Distinguishing features of the different types of muscle in the body*

| Feature | Type of muscle fiber | | |
|---|---|---|---|
| | Skeletal | Cardiac | Smooth |
| Striations | Present | Present | Absent |
| Composition | Multinucleate single cell | Mononucleate single cell | Mononucleate single cell |
| Branching of cell (fiber) | Absent | Present | Absent |
| Nucleus | Many, at periphery of fiber | Single, central | Single, central |
| Fiber length | 1–40 mm | ≤0.08 mm | 0.02–0.5 mm |
| Fiber diameter | 10–40 μm | ~15 μm | 8–10 μm (at thickest part) |
| Myofibrils | Obvious | Obvious | Inconspicuous |
| Contraction | Rapid, powerful twitch | Moderately rapid, unsustained | Slow, rhythmic, sustained |
| Control of contraction | Neurogenic | Myogenic, modulated by nerves | Myogenic, modulated by nerves |
| Distribution | Locomotor muscles, abdominal wall, diaphragm, middle ear muscles | Heart only | Gut, respiratory tract, urogenital tract, glands, ciliary muscles of the eye, piloerector muscles of hairy skin |

Smooth muscle, like cardiac muscle, will contract in the absence of any neural influence. However, this activity is *influenced* by nervous input from the **autonomic nervous system**. Smooth muscle does not appear striated, because the purpose of the contractions of smooth muscle is not to move bones or large amounts of fluid, and so the orientation of the contractile machinery is different from skeletal and cardiac muscle. The force of contractions in smooth muscle is much lower, and the contractions are more sustained than in skeletal or cardiac muscle. This is useful if, for example, the body wishes to divert blood flow away from the gut for a substantial period of time, in order to make a greater proportion of blood flow available to exercising muscle.

# C6 BLOOD AND IMMUNITY

## Key Notes

**Composition of blood**

Blood is primarily composed of serum (and dissolved electrolytes), plasma proteins and blood cells. Plasma is serum and plasma proteins, but not the cellular constituents of blood. The blood is important in oxygen and carbon dioxide transport, immunity, and clotting, and it has a role as a delivery system for blood-borne hormones and nutrients

**Red blood cells**

Red blood cells or erythrocytes are derived from stem cells in the bone marrow, and are the most numerous cells in the blood. They are small, non-nucleated cells that transport oxygen and carbon dioxide in the blood. Most of the oxygen carried in the blood is bound to hemoglobin in erythrocytes. The mechanism of binding of oxygen to the four heme groups at the center of the four globin chains of hemoglobin has important implications for the function of hemoglobin as an oxygen and carbon dioxide transport protein.

**White blood cells and platelets**

There are far fewer white blood cells (leukocytes) and platelets in blood than there are erythrocytes. Leukocytes are important in immune function; with each class of leukocyte important in distinct immune functions. Platelets are actually tiny cell fragments, formed from megakaryocytes. Their major role is to help regulate clotting of the blood.

**Related topics**

Layout and function of the vasculature and lymphatics (D5)
Alveolar exchange and gas transport (E4)
The liver, gall bladder and spleen (J5)
Glomerular filtration and renal plasma flow (M2)

## Composition of blood

The circulating blood volume of a healthy adult male is around 5–6 l and accounts for around 7% of body weight. Although blood looks homogeneous when it is drawn by a doctor, or if we cut ourselves, it is in fact a complex mixture of cells, electrolytes, proteins, vitamins, nutrients and hormones, all of which are essential for life. Blood cells are suspended in a liquid known as **plasma**, the composition of which is given in *Table 1*.

The cells in blood are essential for oxygen delivery, immunity and clotting. The electrolytes, vitamins and nutrients are essential for normal function of all of the cells in the body, because some exchange of these components occurs between the blood and the **extracellular fluid** in capillary beds (see Topic D5). One of the main functions of plasma proteins is to enable **clotting** in response to vessel injury, so that bleeding is arrested (**hemostasis**). **Hormones** that are released into the blood are thereby transported to their sites of action and enable cellular processes to be regulated in the absence of neural influence. This is particularly important in the regulation of cell metabolism and the uptake and release of nutrients from sites of storage such as the liver. To suggest that

*Table 1. Composition of plasma*

| | | Normal plasma concentrations |
|---|---|---|
| Ions | Sodium ($Na^+$) | 135–145 mM |
| | Chloride ($Cl^-$) | 95–105 mM |
| | Potassium ($K^+$) | 3.5–5 mM |
| | Calcium ($Ca^{2+}$) | 2.1–2.6 mM |
| | Magnesium ($Mg^{2+}$) | 0.75–1.0 mM |
| | Iron ($Fe^{3+}$) | 9–30 μM |
| | Hydrogen ($H^+$) | 35–45 nM |
| | Bicarbonate ($HCO_3^-$) | 22–26 mM |
| Proteins | Total | 6–8 g $dl^{-1}$ |
| | Albumin | 3.5–5.5 g $dl^{-1}$ |
| | Globulin | 2.3–3.5 g $dl^{-1}$ |
| Fats | Cholesterol | 150–200 mg $dl^{-1}$ |
| | Phospholipids | 150–220 mg $dl^{-1}$ |
| | Triglycerides | 30–160 mg $dl^{-1}$ |
| Carbohydrates | Glucose | 4–6 mM |
| Vitamins | Vitamin $B_{12}$ | 200–800 pg $ml^{-1}$ |
| | Vitamin A | 0.15–0.6 μg $ml^{-1}$ |
| | Vitamin C | 4–15 μg $ml^{-1}$ |
| Hormones | Cortisol | 5–18 μg $dl^{-1}$ |
| | Aldosterone | 3–10 ng $dl^{-1}$ |
| Others | Creatinine | 0.6–1.2 mg $dl^{-1}$ |
| | Uric acid | 0.18–0.49 mM |
| | $CO_2$ | 23–30 mM |

**oxygen transport** is the only real important role of the blood is to trivialize its importance to so many bodily functions and our survival in general. It is a truly amazing fluid.

## Red blood cells

These cells are also known as **erythrocytes** (*Fig. 1a*). They have no nucleus and have a diameter of around 7 μm and a maximum thickness of around 2.5 μm. There are approximately $5 \times 10^{12}$ such cells in every liter of normal blood so they are by far the most numerous blood cells. They have a lifespan of around 120 days and new cells are formed in bone marrow. One measure of red cell content of the blood is the **hematocrit**, which is normally around 0.45. This means that 45% of blood volume is made up of erythrocytes. Another measure is the **erythrocyte sedimentation rate** (ESR). This is a measure of how quickly the erythrocytes in a blood sample sediment in one hour. Normal values for healthy men are between 2 and 8 mm $h^{-1}$. Abnormal ESRs can be indicative of blood disorders related to erythrocyte form or content.

Their function is to transport oxygen (and to some extent carbon dioxide) around the body, bound to the protein **hemoglobin**, which consists of four protein subunits (usually two α subunits and two β subunits), each of which can carry one molecule of diatomic oxygen ($O_2$) bound to an iron-containing **heme** group in the center of the protein chain. In healthy individuals, there are around 30 pg of hemoglobin per cell, amounting to an astonishing 150 g of hemoglobin per liter of blood. Thus the oxygen transport potential of the blood is enormous; around 200 ml of oxygen per liter of blood.

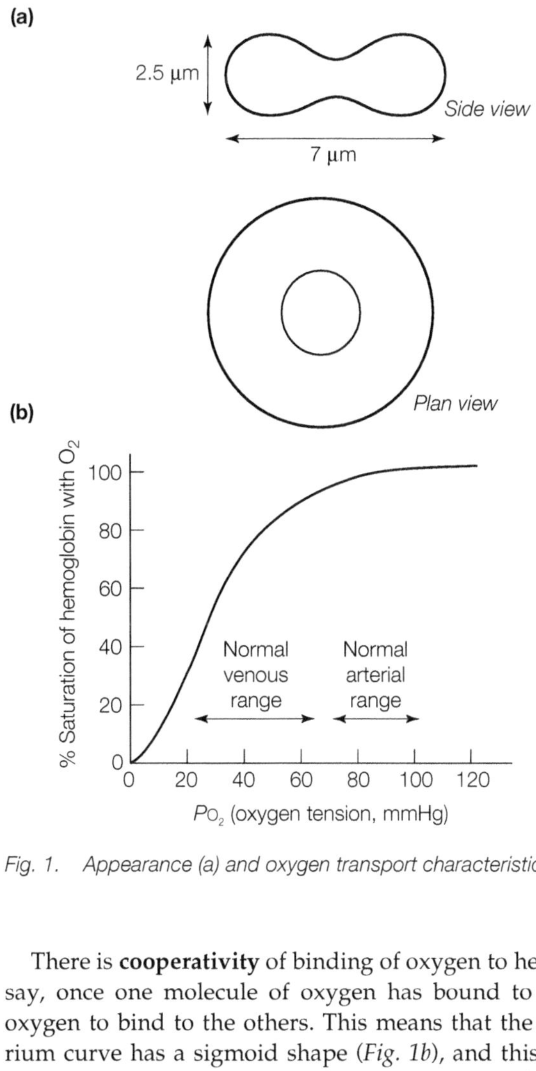


*Fig. 1. Appearance (a) and oxygen transport characteristics (b) of erythrocytes.*

There is **cooperativity** of binding of oxygen to hemoglobin subunits. That is to say, once one molecule of oxygen has bound to one subunit, it is easier for oxygen to bind to the others. This means that the oxygen–hemoglobin equilibrium curve has a sigmoid shape (*Fig. 1b*), and this has implications for oxygen delivery to respiring tissues (see Topics E5 and E6). A number of diseases, known as **hemoglobinopathies**, can affect the ability of the hemoglobin to carry oxygen and/or make it difficult for the hemoglobin molecule to hold its shape in low-oxygen conditions. One such disease is sickle cell anemia, in which low-oxygen conditions in the blood result in misfolding of the hemoglobin, deformation of erythrocytes, blockage of blood vessels, pain, and possibly even death.

Hemoglobin can also carry $CO_2$ from respiring tissues and release it in the lungs, where it can be exhaled. The binding of $CO_2$ to hemoglobin is reversible. **Carbon monoxide (CO)** can also bind to hemoglobin (200 times more avidly than $O_2$), and the resultant low amount of oxygenated hemoglobin in erythrocytes can lead to death.

## White blood cells and platelets

The other cells in blood (white blood cells or **leukocytes**, and **platelets**) are far less numerous than erythrocytes, but are just as important. Without them, we might die from infection or bleed to death. In each liter of blood, there are approximately $8 \times 10^9$ leukocytes, and $3 \times 10^{11}$ platelets. However, not all leukocytes are the same (*Fig. 2*). In fact there are five main types of circulating leukocyte (**neutrophils**, 60%; **lymphocytes**, 30%; **monocytes**, 6%; **eosinophils**, 3%; and **basophils**, 1%), and each of these has a different role to play in the body's defense mechanisms.

Neutrophils are the characteristic cell of acute inflammation and are the first cells to leave the blood and attack invading microbes. Their hallmark multilobed nuclei (hence their other name of **polymorphonuclear leukocytes** or PMNs) make them easy to see on tissue sections. They are phagocytic, meaning that they surround microbes and swallow them, before digesting them with enzymes. Lymphocytes can be classed as either B or T lymphocytes, depending upon whether they originated in the bone marrow or the thymus. The majority of circulating lymphocytes are T lymphocytes and these can be subclassified into helper T cells, suppressor T cells or natural killer (NK) cells, depending upon the types of receptors they express on their surface and their role in immune defense of the host. B lymphocytes mediate immune responses by releasing antibodies that bind to a specific antigenic molecule on the invading microbe and help to destroy it. Monocytes possess a kidney-shaped nucleus and are like neutrophils in that they are phagocytic. However, they can only do this after they have been activated and have left the blood and entered the tissues, from which point they become known as **macrophages**.

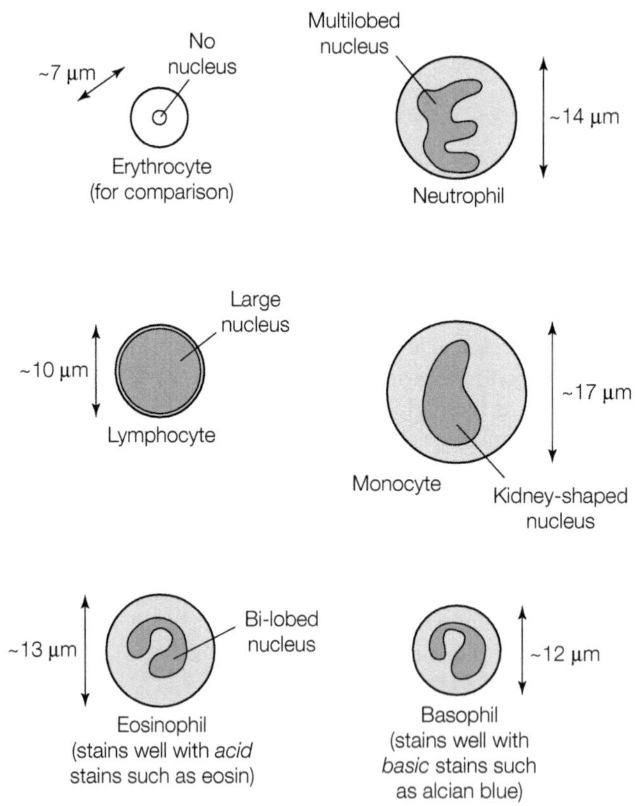


*Fig. 2. Appearance of blood leukocytes (see text for their roles).*

Eosinophils and basophils are the least numerous of all blood cells and are named because they either stain well with the acidic dye eosin (eosinophils) or with basic dyes (basophils). Eosinophils are most often involved in defense against parasites and in allergic reactions. Basophils, on the other hand, release heparin and histamine, and influence blood flow to areas where other leukocytes may be needed in order to fight infection. They are also implicated in allergic reactions and other forms of hypersensitivity.

Platelets are essential for blood **clotting**. Without them, we would find it difficult to stem the flow of blood from a wound. As detailed above, they are more numerous than any other cell in the blood, with the exception of the erythrocytes. However, they are less than half the size of erythrocytes. This is because platelets are formed by fragmentation of a much larger progenitor cell, the **megakaryocyte**. For the same reason, they have no nucleus. They contain granules of important proteins that help in blood clotting.

# D1 OVERVIEW OF THE HEART AND GREAT VESSELS

## Key Notes

**Introduction**

The heart beats more than one thousand million times in an average lifetime, supplying our tissues with blood and, thereby, oxygen. The heart pumps deoxygenated blood to the lungs and then pumps oxygenated blood to the rest of the body. At rest, cardiac output is approximately 5 l $min^{-1}$.

**Cardiac output, blood pressure and peripheral resistance**

Blood flows in a closed system, propelled by the pumping action of the heart. The blood exerts pressure on the vessels in which it flows; this pressure is known as blood pressure. Arterial blood pressure is the product of cardiac output and the resistance to flow in the system (peripheral resistance); as cardiac output and resistance increase or decrease, so does arterial blood pressure.

**Chambers, valves and conduction system**

On each side of the heart, there are two chambers: an upper atrium and lower ventricle. During the cardiac cycle, blood must flow from veins into the atria, from the atria into the ventricles, and from the ventricles into arteries; any backflow in this system causes problems with cardiac function. The heart uses valves between the atria and ventricles and between the ventricles and arteries to prevent backflow. The direction of flow during the cardiac cycle is also maintained by the way in which heart muscle is activated. The atria are activated first, followed by the ventricles. Ventricular contraction is stimulated by a specialized conducting system.

**Importance of the septum**

The septum sits between the right and left sides of the heart and prevents the deoxygenated blood from the right side of the heart mixing with the oxygenated blood on the left side of the heart. In some congenital heart diseases, septal defects are present ('hole in the heart' conditions).

**Coronary blood flow**

The heart has its own blood supply, the coronary circulation. The coronary arteries originate just above the cusps of the aortic valve between the left ventricle and the aorta. The major coronary artery branches are the right coronary artery, the left anterior descending coronary artery, and the circumflex artery. Most coronary blood flow to the (larger) left ventricle takes place during the relaxatory phase of the cardiac cycle, diastole.

**Related topics**

Electrical activity of the heart – the ECG (D4)
The cardiac cycle, blood pressure, and its maintenance (D6)
Investigating cardiovascular function (D7)
Cardiovascular response to exercise (D8)
Hormonal control of fluid balance (M5)

## Introduction

The heart is the organ that has a special place in our minds. We speak of 'getting to the heart of the matter' and of 'heartfelt best wishes'. We even forget that our emotions reside in some part of our CNS and ascribe them to the heart instead. No-one has ever said that they had a broken limbic system after a break-up, and Valentine's Day cards are not plastered with little pictures of the brain. If our heart fails, we do (quite literally). This section will explore the function of the heart and blood vessels, the cardiovascular system.

If it had been developed by a commercial company, the heart would be described as the most amazing pump ever designed. It goes through its cycle of contraction and relaxation 70 times every minute, every minute of the day, every day of our lives. This means that in a lifetime of 70 years, the heart pumps blood to the lungs and around the body more than $2.5 \times 10^9$ times. The world's most expensive pump simply couldn't live up to the demands placed on the heart. Not only that, but the heart is able to vary its rate of contraction and the strength with which it contracts in order to meet the varying needs of our body for oxygenated blood, depending on activity. The heart is, in fact, two pumps in series. The role of the right side of the heart is to pump blood around the lungs, where carbon dioxide can be lost and oxygen obtained, and back to the left side of the heart. The role of the left side of the heart is to pump this oxygenated blood around the rest of the body (providing oxygen for cellular respiration), and back to the right side of the heart (*Fig. 1*). The cardiac septum separates the right and left sides of the heart.

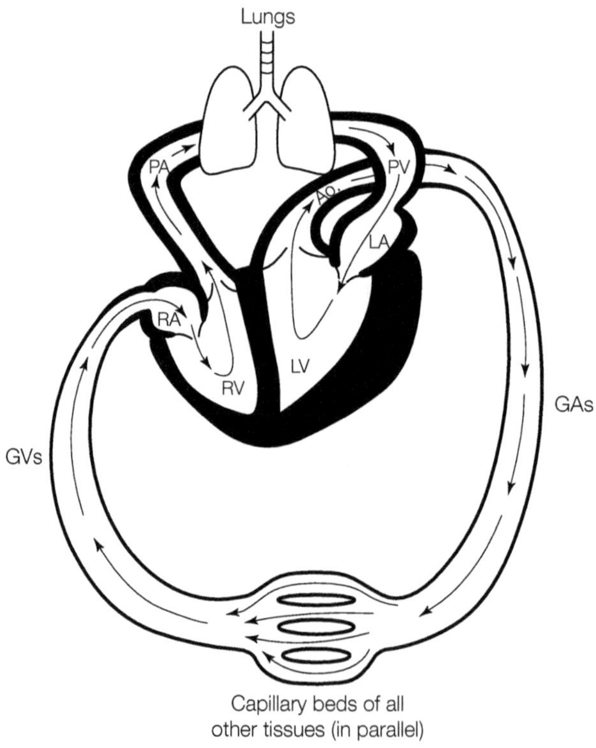


*Fig. 1. The heart as two pumps in series. Note that this drawing is not to scale. Ao, aorta; GAs, great arteries; GVs, great veins; LA, left atrium; LV, left ventricle; PA, pulmonary artery; PV, pulmonary vein; RA, right atrium; RV, right ventricle.*

Normal **cardiac output** at rest in a healthy 70 kg male is around 5 l $min^{-1}$, meaning that with each beat, the heart pumps out around 70 ml of blood (**stroke volume**). Several large blood vessels conduct blood away from, and towards the heart. Deoxygenated venous blood returns to the right side of the heart via two large veins, the superior (upper) **vena cava** and inferior (lower) vena cava, and leaves the right side of the heart and enters the lungs via the **pulmonary arteries**. Oxygenated arterial blood returns to the left side of the heart from the lungs via the **pulmonary veins**, and leaves the left side of the heart to perfuse the rest of the body via the **aorta**. Note that, in contrast to other arteries, the pulmonary artery does not contain oxygenated blood and, in contrast to other veins, the pulmonary veins do not contain deoxygenated blood.

**Cardiac output, blood pressure and peripheral resistance**

When thinking about **hemodynamics** (blood flow), it is useful to start off by examining principles that most people already understand. Consider the two following examples, of which most people have a working knowledge. Current will not flow in an electrical circuit unless the voltage at one end of the circuit is higher than that at the other end (potential difference), hence this is why we use batteries with higher voltages for electronic gadgets that require greater current flow. Using narrower wires in the circuit will reduce current flow (5 amp fuse wire is narrower than 30 amp fuse wire). Water will not flow down a hose attached to a tap unless the tap is turned on, creating higher water pressure at that end than at the nozzle. Compression of the hose by kinking will reduce the flow of water. In both of these examples, the relationship between flow, pressure difference (or potential difference) and resistance to flow can take the same form as **Ohm's law**:

$$V = I \times R \qquad (1)$$

where $V$ is pressure difference, $I$ is flow, and $R$ is resistance to flow

In the same way, blood would not flow through the system of hoses in our body that we call the cardiovascular system unless the left side of the heart generated higher pressure than exists near the right side of the heart. Contraction of the left side of the heart creates higher **blood pressure** in the major arteries than in the right side of the heart, and this difference in pressure is called the '**perfusion pressure**' of the systemic circulation. Mean arterial blood pressure is 90–95 mmHg (millimeters of mercury) and the pressure in the right side of the heart is normally no more than 1–2 mmHg. Thus, perfusion pressure of the systemic circulation is roughly the same size as mean arterial blood pressure. Thus, in Equation (1), one can substitute cardiac output for the term $I$, mean arterial pressure for the term $V$ and **total peripheral resistance** (the resistance to flow across all the vessels of the systemic circulation) for the term $R$, giving another more useful equation:

$$\text{MABP} = \text{CO} \times \text{TPR} \qquad (2)$$

where MABP is mean arterial blood pressure, CO is cardiac output, and TPR is total peripheral resistance.

Also, in the case of the human cardiovascular system, where $n$ vascular beds are arranged in parallel rather than in series:

$$1/\text{TPR} = 1/R_1 + 1/R_2 + 1/R_3 + 1/R_4 + \ldots\ldots 1/R_n \qquad (3)$$

where $R_1$ to $R_n$ are the individual resistances to flow in the various vascular beds.

It is evident from Equation (2) that if CO or TPR change, then MABP will also change, and that if TPR increases and MABP must stay within tight physiological limits, then CO must be decreased to compensate for the increase in TPR. The relevance of this to the function of the heart should be immediately apparent; in order for the heart to maintain adequate perfusion of the body, there must by mechanisms by which information about CO and/or MABP can be 'fed back' to the heart in order that adequate tissue perfusion is maintained. There are numerous receptors for arterial pressure (**baroreceptors**) which sense such changes and signal to the brain to alter its modulatory activity on the heart; these are known as baroreceptor reflexes and will be covered in more detail in Topic D6. Since CO is the mathematical product of heart rate and **stroke volume**, altering either of these parameters can obviously alter CO.

## Chambers, valves and conduction system

Each side of the heart also consists of two chambers, the (upper) **atrium** and (lower) **ventricle**. As their name suggests (*atrium* Latin: hallway), the atria act as chambers that accommodate blood that has returned to the heart, before it can be ejected by the ventricles to the lungs or to the rest of the body. In view of the fact that the atria sit above the ventricles, it is important that blood being ejected by the ventricles finds its way into the pulmonary artery or the aorta and not back into the atrium. It is equally important that blood that is destined for the lungs or systemic circulation does not leak back towards the heart. This unidirectional flow of blood from atria to ventricles and from ventricles to arteries is facilitated by valves which sit between the atria and ventricles and at the outflow tracts of the pulmonary artery and aorta (*Fig. 1*).

The **atrioventricular valves** are known as the **tricuspid valve** (between right atrium and right ventricle) and the **mitral valve** (also known as the **bicuspid valve**, between left atrium and left ventricle). The valves at the outflow tracts of the right and left ventricle are the **pulmonary valve** and **aortic valve**, respectively. They are also known as the **semilunar valves**, but this is an older term. All four valves act as one-way valves that close when blood flows back towards them in the 'wrong' direction. The tricuspid and mitral valves are also prevented from being 'blown out' into the atria when the ventricles contract, by being tethered to the wall of the ventricle by specialized connective tissue strands known as the **chordae tendineae**. During contraction of the ventricles, the chordae tendineae become 'tensed' by contraction of the **papillary muscles** of the ventricles, to which they are attached. This helps keep the valves closed in the face of the high pressure that exists in the ventricles during their contraction.

All of these valves function by essentially passive mechanisms. For example, the mitral valve only opens when pressure in the left atrium exceeds that in the left ventricle and closes only when left ventricular pressure exceeds that in the left atrium. The aortic valve only opens when pressure in the left ventricle exceeds that in the aorta, and closes when aortic pressure exceeds left ventricular pressure and some blood starts to flow back towards the heart. For more details of pressure and volume changes in the heart and great vessels, see Topic D6.

The function of valves can become compromised by infection, heart attack, or connective tissue disease. Valves that do not open as much as they should are said to be stenosed. **Mitral stenosis**, for example, is commonly caused by infections like rheumatic fever. This results in blood 'backing up' into the lungs, because not enough drains from the left atrium to the left ventricle in any one cardiac cycle. This can lead to problems with breathing. **Aortic stenosis** can result in lowered cardiac output, tiredness, lethargy and enlargement of the left ventricle. Backflow of blood from the left ventricle to the left atrium, known as **mitral regurgitation**, can be caused by papillary muscle defects or the rupture of one or more chorda tendinea. Again, this results in blood 'backing up' into the lungs, causing breathing problems. Such problems can be definitively diagnosed by listening to abnormal heart sounds, or by performing cardiac ultrasound (**echocardiography**).

One final feature of the heart that must be mentioned is its specialized conducting system (*Fig. 2*). Changes in electrical activity that start in the **sino-atrial node** (SA node) are conducted through the heart in a highly organized way. This electrical activity spreads through the atria until it reaches the **atrio-ventricular node** (AV node) and is conducted down two branches of the conducting system, the left and right bundle branches, to affect the apex of the heart. From here, electrical activity spreads throughout the ventricles via **Purkinje fibers** to produce a coordinated contraction that ejects blood into the pulmonary artery and aorta. You can read more about this electrical activity in topic D4.

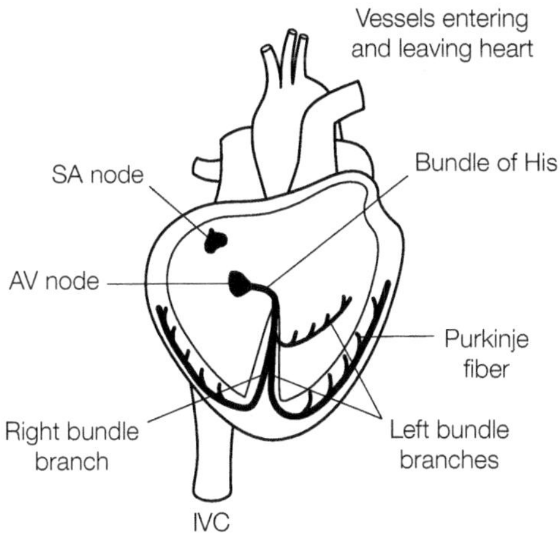


*Fig. 2. The conducting system in the heart. AV, atrio-ventricular; IVC, inferior vena cava; SA, sinoatrial. The positions of the atrial septum and the heart valves have been omitted for clarity.*

## Importance of the septum

The heart and great vessels develop *in utero* between the second and seventh weeks of gestation. A beating heart can be visualized using conventional ultrasound scans on the embryo around 5 weeks, and fetal heartbeats can be detected by **Doppler ultrasound** probes as early as 10 weeks after the mother's last menstrual period. In the embryo, the heart and great vessels develop from lateral **mesoderm**. This begins as a primitive linear tube, which then undergoes a series of complex looping and folding movements to generate the heart with its four chambers. In doing so, the tube must undergo **septation**; the right and left sides of the heart, each with two chambers, must become separated by a wall of cardiac muscle known as the **septum**. This is important because, as mentioned above, the heart must function as two pumps in series. If septation is in some way defective, this can manifest itself as so-called 'hole in the heart' problems, usually in the first few weeks after birth. Such problems usually mean that the baby has problems breathing or may even look blue (**cyanosis**). This is because deoxygenated blood bypasses the lungs and mixes with oxygenated blood, resulting in an inadequate oxygen supply to the body. Again, such problems can be definitively diagnosed using echocardiography.

## Coronary blood flow

The heart also has its own blood supply, flowing in the coronary arteries. There are three coronary arteries: the **right coronary artery** (RCA), the **left anterior descending coronary artery** (LAD), and the **circumflex artery**. The RCA emerges from a coronary sinus in the wall of the aorta near the anterior cusp of the aortic valve, and supplies both atria, most of the right ventricle and some of the most posterior parts of the left ventricle. A branch of this artery also supplies the main pacemaker tissue of the heart, the SA node. The LAD and circumflex coronary arteries are branches of a larger coronary artery which emerges from a **coronary sinus** in the wall of the aorta near the left posterior cusp of the aortic valve. The LAD coronary artery supplies both atria, the anterior part of the left ventricle, and some adjacent parts of the right ventricle. The circumflex artery is so called because it curves around to supply the posterior parts of the left ventricle and anastomoses (joins up with) the branches of the RCA. Note that one cusp of the aortic valve has no coronary artery associated with it. This is why the right posterior cusp is sometimes known as the non-coronary cusp.

The greater proportion of blood flow to the left ventricle occurs during the part of the cardiac cycle known as **diastole** (when the ventricle is relaxing). There are two main reasons for this. Firstly, when muscle of the left ventricle contracts (during **systole**), the vessels supplying it are squeezed shut; in fact some blood may flow back towards the aorta in the coronary arteries supplying the left ventricle. It is only when the ventricular muscle relaxes that flow can resume. Secondly, when the aortic valve is open during systole, the inlets to the coronary arteries are covered over by the valve cusps because of their position at the base of the cusps. It is only when the aortic valve closes that a greater amount of blood is able to gain access to the coronary arteries and perfuse the heart tissue itself.

# D2 STRUCTURE OF CARDIAC MUSCLE

## Key Notes

**Microscopic structure of cardiac muscle**

Cardiac muscle cells are small striated myocytes. They are highly branched and are heavily invested with mitochondria. The sarcoplasmic reticulum (see Topic I3) is not as extensive as in skeletal muscle. The intercalated disks between myocytes are not uniform; gap junctions exist so that electrical signals can pass easily from one cell to its neighbors.

**Specialization related to function**

The striated nature of cardiac muscle means that it can contract in a highly organized manner. Electrical contact and extensive branching of fibers mean that cardiac muscle acts as a functional syncytium. Between 30% and 40% of the volume of cardiac mycocytes is taken up by mitochondria; these mitochondria provide ATP from oxidative phosphorylation to power contraction.

**Innervation of cardiac muscle**

Cardiac muscle is spontaneously active and, unlike skeletal muscle, does not require nervous input in order to contract. However, both branches of the autonomic nervous system (parasympathetic and sympathetic) innervate cardiac muscle and modify its function. The parasympathetic branch of the autonomic nervous system slows heart rate, but has only a minor effect on force of contraction, whereas the sympathetic branch increases both heart rate and the force of contraction of atrial and ventricular myocytes. These effects are mediated by acetylcholine acting on muscarinic receptors in the case of the parasympathetic branch, and norepinephrine acting on β-adrenoceptors in the case of the sympathetic branch.

**Related topics**

Nutritional requirements of cardiac muscle (D3)

Structure and function of the autonomic nervous system (H5)

Autonomic pharmacology (H6)

### Microscopic structure of cardiac muscle

In comparison with skeletal muscle, where the cells are long (1–40 mm) and straight, **cardiac muscle** cells are extremely small (≤80 μm) and are extensively branched. There are also specialized intercellular junctions in cardiac muscle known as **intercalated disks**, which do not exist in skeletal muscle (*Fig. 1a*). In common with skeletal muscle, cardiac muscle is striated. That is to say, the intracellular proteins that make up the contractile machinery are arranged in a highly organized fashion, and this is apparent as 'banding' of the cells. In common with all other types of muscle, shortening of the cell is brought about by different types of **muscle filaments** sliding past each other. The two major proteins that make up these filaments are **actin** and **myosin** (*Fig. 1b*). It is the regular arrangement of these filaments that gives striated muscle (both skeletal and cardiac) its

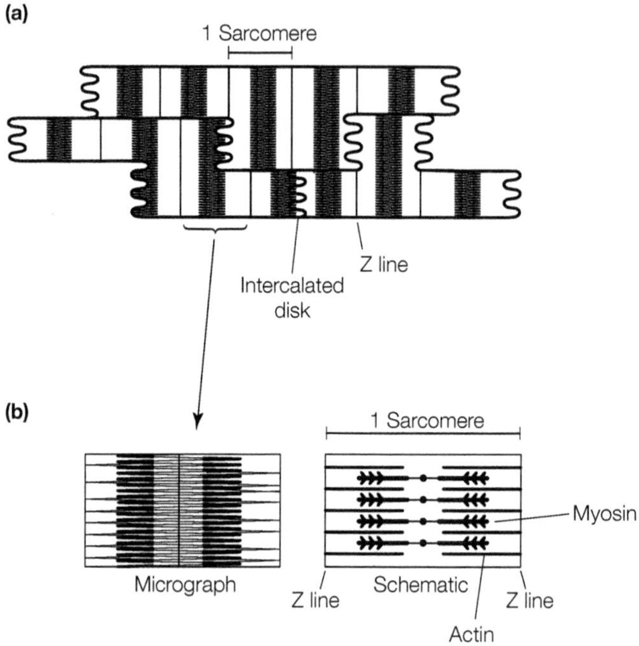


*Fig. 1. Structure of cardiac muscle. (a) Cardiac muscle cells are branched and make connections with their neighbors through intercalated disks. Three cardiac muscle cells are depicted here; (b) striations are caused by the regular orientation of thick (myosin) and thin (actin) muscle filaments.*

characteristic appearance. The lengths of the protein filaments themselves do not change, but in sliding past each other, the cell becomes shorter and inevitably, fatter. The mechanics and bioenergetics of this interaction are broadly similar in all muscle cells and are covered in more detail in Topics I3 and I6.

When viewed with an electron microscope, which has much higher resolving power than a conventional light microscope, other structural features of cardiac myocytes can be seen (*Fig. 2*). The cells have extensive numbers of **mitochondria**, so much so that 30–40% of cell volume is taken up by these organelles. There is a relative paucity of other organelles, particularly the intracellular membrane systems (**sarcoplasmic reticulum**) that are so prominent in skeletal muscle, and which store the calcium ions that are necessary for contraction. Also, when one views an electron micrograph of cardiac muscle, it is apparent that the intercalated disks between cells are not uniform along their length; some portions are much less electron dense than others. These are **gap junctions**, where the two cells are 'electrically connected' to each other; pores between the cells allow ions (and thus charge) to pass between cells. Each cardiac myocyte is extensively branched, and has connections with several neighboring cells through these intercalated disks with gap junctions.

## Specialization related to function

The various structural peculiarities of cardiac muscle cells are not there by chance. Each has a particular role to play in the function of the heart muscle. Let us consider them in turn:

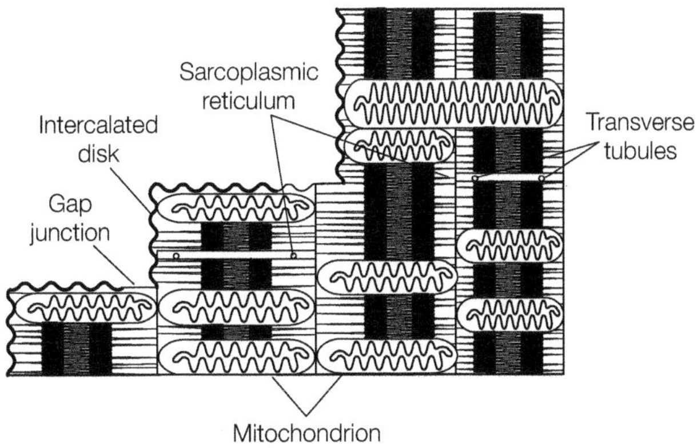


*Fig. 2. Ultrastructure of a cardiac muscle cell.*

- *striations*: in order to be able to contract powerfully enough to move blood around the body, about 20% of the volume of cardiac muscle fibers is taken up by **myofibrils**, the protein contractile elements. It is easier to have these in organized bundles, and this organized form is seen as striations
- *intercalated disks and gap junctions*: electrically leaky contact between cardiac muscle cells means that a wave of electrical activity and contraction can be propagated through the myocardium, without the need for nervous input and complicated activation patterns. Cardiac muscle is said to act as a '**functional syncytium**'
- *extensive branching*: when cardiac muscle contracts, this must be done in a coordinated fashion, so neighboring cells should contract almost synchronously
- *numerous mitochondria*: the heart cannot build up an oxygen debt that it can pay back later by resting. When the products of **anaerobic metabolism** (e.g. lactate) build up in cardiac muscle, they result in chest pain. Thus the heart relies almost entirely upon **aerobic metabolism** to generate the large amounts of **ATP** required for contraction. Each cell needs lots of mitochondria, the sites of **Krebs' cycle** and **oxidative phosphorylation**
- *less sarcoplasmic reticulum than skeletal muscle* (see Topic I3): some of the **calcium ions** that are used to trigger contraction come from the extracellular fluid and pass through calcium channels in the cell membrane, and some are released from the sarcoplasmic reticulum. This has important implications for function that are made more explicit in Topic D4.

## Innervation of cardiac muscle

Skeletal muscle has **motor innervation** that must be activated in order that the muscle can contract. In order to flex our elbow, our brains *must* send a signal to contract to the myocytes in the muscles making up the biceps group. This is not true for cardiac muscle. Cardiac muscle cells are spontaneously active and, if examined in isolation in a laboratory, will contract at a given rate dependent upon the area of the heart from which the muscle cell has been taken. Thus all cardiac myocytes have some '**pacemaker**' activity. The predominant pacemaker area in the heart is the **sinoatrial node** (SA node), the cells of which contract between 60 and 200 times per minute, dependent on nervous and hormonal influences (the intrinsic firing rate of SA node cells is around 100 times per minute). The electrical properties of the cells in the SA node and other parts of the heart are covered in depth in Topic D4.

However, despite the fact that the heart beats according to its spontaneous electrical rhythm, other non-voluntary (autonomic) nerves can affect both heart rate and force of contraction (and thus stroke volume). The autonomic nervous system (ANS) consists of two branches, known as the **sympathetic** and **parasympathetic** branches (see Topic H5). The **vagus nerve** conducts the parasympathetic innervation of the heart, and several cardiac nerves conduct the sympathetic innervation. Circulating **epinephrine** from the adrenal medulla can also be considered as an additional sympathetic input to the heart. Sympathetic innervation of the atria and ventricles is widespread, but parasympathetic innervation of the heart is much more discrete; innervation of the ventricles is particularly sparse. The major role of the sympathetic branch of the ANS can be considered 'fight or flight', and that of the parasympathetic branch can be considered 'rest and digest'. It should not be surprising, then, to hear that the major effects of the sympathetic branch on the heart are to increase **heart rate** (positive **chronotropy**) and increase **stroke volume** (positive **inotropy**), whereas those of the parasympathetic branch are to decrease heart rate (negative chronotropy) and, under certain circumstances, to decrease stroke volume (negative inotropy).

The sympathetic nerves to the heart release **norepinephrine**, which acts on **β-adrenoceptors** to produce an increase in heart rate (see Topic D4) and force of contraction. The vagus nerve releases **acetylcholine (ACh)**, which acts on **muscarinic ACh receptors** to decrease heart rate (see Topic D4) and, where there is a pre-existing level of sympathetic activity, to reduce the force of contraction. Under resting conditions, both branches of the ANS to the heart are active, but the parasympathetic branch predominates slightly. Thus, humans are said to have a resting level of **'vagal tone'**. When **cardiac output** must be increased, this can be done either by shutting down vagal input to the heart and increasing sympathetic input, or both. This will increase heart rate (above 70 beats $min^{-1}$) and stroke volume (above 70 ml $beat^{-1}$) to increase cardiac output (above 4.9 l $min^{-1}$). Heart rate and cardiac output can be increased pharmacologically by the administration of epinephrine (which also acts on β-adrenoceptors) or by the administration of **atropine**, a muscarinic ACh receptor blocker.

# D3 NUTRITIONAL REQUIREMENTS OF CARDIAC MUSCLE

## Key Notes

**The heart as a pump**

The heart must continue beating for our entire lives, irrespective of our activity levels. The heart generates the energy substrate ATP by aerobic methods and therefore depends upon an adequate oxygen supply derived from coronary blood flow.

**Nutritional requirements**

When glucose is abundant, for example, after a meal, the heart will metabolize the sugar preferentially via glycolysis and the citric acid cycle to generate energy for ATP synthesis. Between meals and during short-term fasting, the heart spares glucose and uses fatty acids derived from triglyceride hydrolysis in adipose tissue as its major fuel. During prolonged fasting (in excess of 2–3 days), the heart is able to use ketone bodies produced by the liver as an energy source.

**Effects of cardiac ischemia**

In cardiac ischemia, reduced coronary blood flow can cause cardiac pain. The pain of angina is one form of cardiac pain, caused by build-up of toxic metabolites in cardiac muscle when oxygen supply is inadequate for its needs. In myocardial infarction, severe, prolonged lack of blood supply to a part of the heart can cause cell death due to failure of energy-dependent processes such as maintenance of ion concentrations on either side of the cell membrane by the $Na^+/K^+$ ATPase pump.

**Related topics**

Metabolic processes (B3)
Membrane potential (B4)
Overview of the heart and great vessels (D1)
Cardiovascular response to exercise (D8)
Utilization of $O_2$ and production of $CO_2$ in tissues (E5)

## The heart as a pump

Unlike skeletal muscle, the activity of heart is constant and rhythmic, with the heart beating more than $2 \times 10^9$ times in a lifetime of 70 years. Thus, heart muscle cannot be prone to fatigue or this would rapidly result in death. The heart is a completely aerobic organ and the cells of heart muscle, the cardiac myocytes, are rich in **mitochondria** which occupy almost half of the cytoplasmic volume (see Topic D2). It is dependent on oxygen availability to maintain its function in the steady state, and energy for contraction and relaxation of cardiac myocytes comes from the hydrolysis of **adenosine triphosphate** (ATP). Cardiac blood flow (**coronary blood flow**) is closely matched to cardiac activity (see Topic D8), so that cardiac myocytes need never build up an oxygen debt, unlike skeletal muscle fibers.

## Nutritional requirements

Because of its good blood supply and thus good oxygen supply, heart muscle can oxidize metabolic fuels completely to **carbon dioxide** ($CO_2$) and water.

However, fuel reserves in the muscle are poor: a little **glycogen** and **phosphocreatine**, much less than in skeletal muscle. Thus the heart relies on blood-borne fuel for its energy supply. Post-prandially (in the fed state), **glucose** is the preferred respiratory substrate, being metabolized via the **glycolytic pathway** to **pyruvate**, which is oxidized to **acetyl coenzyme A** by the enzyme pyruvate dehydrogenase in mitochondria (see Topic C3). Acetyl-CoA is further oxidized to $CO_2$ and water in the **citric acid cycle**, releasing energy for the synthesis of ATP. Under these conditions, the **respiratory quotient** of heart muscle (RQ: number of molecules of $CO_2$ produced for each molecule of oxygen consumed by the tissue) is 1.0.

As blood glucose levels fall to their homeostatic concentration of 4–5 mmol $l^{-1}$, free **fatty acids**, released from adipose tissue **triglyceride**, are transported to the heart bound to **albumin** and become the major fuel source for the tissue. The heart extracts fatty acids preferentially from the coronary circulation for oxidative energy production. Fatty acids undergo activation in the cytosol of the cardiac myocytes prior to transport into the mitochondria where they are oxidized to acetyl-CoA via the **β-oxidation** pathway (see Topic B3). Subsequent oxidation of the acetyl-CoA in the citric acid cycle again generates energy for ATP synthesis. Under these conditions the RQ is 0.75. This switch from carbohydrate to fat for metabolic fuel serves to conserve glucose for those tissues that have an absolute requirement for it as an energy source (brain, nervous tissue, erythrocytes, and kidney).

As the body enters the fasting state, **ketone bodies**, acetoacetate and 3-hydroxybutyrate, are synthesized in the liver from acetyl-CoA derived from high rates of β-oxidation in that tissue (see Topic B3). These ketone bodies serve as a respiratory substrate for the heart under fasting conditions. This requires the action of **β-ketoacyl transferase**, an enzyme present in heart but not in liver, to convert acetoacetate to acetoactyl-CoA which is then converted to two molecules of acetyl-CoA. The acetyl-CoA thus produced is oxidized in the citric acid cycle. 3-Hydroxybutyrate must first be oxidized to acetoacetate before it too can follow this path and serve as a metabolic fuel.

## Effects of cardiac ischemia

ATP produced from the various fuels in cardiac muscle is used for three main important cellular actions:

- contraction of myocytes by **cross-bridge cycling** (see Topic I3)
- pumping of **calcium ions** back into intracellular stores to induce relaxation (see Topic I3)
- maintenance of function of membrane **$Na^+/K^+$ ATPase** pumps (see Topic B4).

If coronary blood flow is compromised in some way (by **atheroma** in a branch of a coronary artery, for example), then cardiac activity may not be able to meet the demands being placed on the heart by the body. The chest pain of **angina pectoris** (where there is narrowing of the coronary arteries) is commonly brought on by a level of exercise that increases cardiac work above the level that can be sustained by the compromised blood supply. This is known as 'stable' angina, where coronary blood flow at rest is sufficient, but becomes inadequate during exercise. In 'unstable' angina, transient chest pain can arise at rest. In both cases, pain receptors in the heart are activated by substances leaking out of overworked cardiac myocytes (e.g. $K^+$ ions, lactic acid, adenosine), and the chest pain serves to limit activity and allow the cardiac myocytes to recover.

If blood flow to a portion of the heart is completely occluded, this is known as a **heart attack** or **myocardial infarction** (MI). In this case, the chest pain can occur at rest and is unlikely to abate until the occlusion has been relieved. During an MI, some cells in the heart may die and be replaced by scar tissue; this is why some people who survive an MI can succumb to heart failure months or years later. Cardiac myocytes die as a result of an MI, not because they have insufficient ATP to contract, but because they have insufficient ATP to run the $Na^+/K^+$ and other ATPases and maintain **membrane potential** (see Topic B4). This results in the myocytes accumulating $Ca^{2+}$ and $Na^+$ ions, which can cause the cell to swell as water enters down its osmotic gradient or, in the case of calcium ions, by activating degradative enzymes that destroy the cell by **necrotic** or **apoptotic** mechanisms. Some damage can be done to cardiac myocytes during reperfusion of the tissue, as **reactive oxygen species (ROS)** can be synthesized when ATP levels become depleted and **adenosine diphosphate** (ADP) and **adenosine monophosphate** (AMP) levels rise (*Fig. 1*). The conversion of hypoxanthine to uric acid is catalyzed by **xanthine oxidase** and produces ROS as by-products. **Xanthine dehydrogenase** is cleaved to xanthine oxidase by the action of proteases in the presence of $NAD^+$ (nicotinamide adenine dinucleotide) and calcium ions.

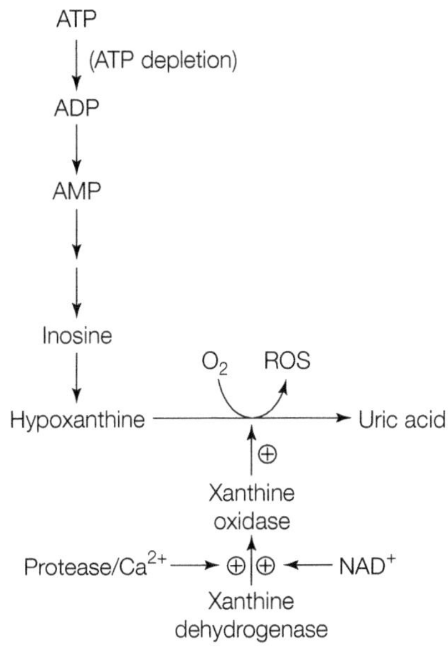


*Fig. 1. Generation of reactive oxygen species (ROS) after reperfusion of previously ischemic tissue. ADP, adenosine diphosphate; AMP, adenosine monophosphate; ATP, adenosine triphosphate; $NAD^+$, nicotinamide adenine dinucleotide; $O_2$, oxygen.*

# D4 Electrical activity of the heart – the ECG

## Key Notes

**Cardiac muscle action potentials**

In common with neurons, skeletal muscle cells, and smooth muscle cells, cardiac muscle fibers are excitable. This means that they have a relatively negative membrane potential at rest and can be induced to undergo action potentials by excitation. Cardiac myocytes from different regions of the heart have different forms of action potential, and cells are electrically coupled to each other by gap junctions. The sinoatrial node is the heart's normal 'pacemaker' region. The time taken for sinoatrial node cell membrane potential to go from 'rest' to 'threshold' and fire an action potential is inversely proportional to heart rate. Atrial and ventricular myocytes display action potentials that last several hundred milliseconds (longer than contractions of the muscle fibers). This means that contractions in cardiac muscle cannot summate, so relaxation takes place between beats to allow filling of the heart.

**Effects of sympathetic and parasympathetic input**

Both branches of the autonomic nervous system have effects on heart rate via the sinoatrial node. The parasympathetic branch releases acetylcholine, which acts on muscarinic receptors to alter the permeability of the cardiac myocyte membrane to both $K^+$ and $Na^+$ ions; this results in slowing of the heart rate. Norepinephrine released from sympathetic neurons (and circulating epinephrine from the adrenal medulla) acts on β-adrenoceptors to alter the permeability of the cardiac myocyte cell membrane to $Na^+$ ions, so that heart rate increases.

**Spread of depolarization and repolarization**

In order for the heart to work efficiently as a pump, the muscle of the atria and ventricles must be activated in a particular sequence. Action potentials of sinoatrial node cells passively spread throughout the atria before the atrioventricular node is activated. Atrioventricular node cells then pass on activation to the conducting system of the heart, activating the septum and the ventricular muscle in a particular sequence to facilitate ejection of blood into the pulmonary artery and aorta. Repolarization of cardiac myocytes is important in ensuring that the heart is able to fill with blood before the next beat, again 'paced' by the sinoatrial node.

**ECG lead systems**

The electrocardiogram (ECG) measures some of the electrical activity associated with depolarization and repolarization of the various parts of the heart. In a clinical setting, 10 electrodes will be attached to the patient, allowing an ECG to be recorded from 12 leads (or electrode pairings). These various leads measure the electrical activity associated with depolarization and repolarization of heart muscle from several different perspectives. The ECG normally presented in textbooks is that recorded using the limb leads (normally lead II (left leg/right arm electrode pairing)).

**Form of the basic ECG**

The basic ECG has three main features: a P wave, a QRS complex, and a T wave. The P wave is a measure of the electrical activity associated with depolarization of the myocytes in the atria. The QRS complex represents the electrical activity associated with the depolarization of the myocytes of the septum and the ventricles (atrial repolarization is masked by this large complex). The T wave represents ventricular repolarization. Other measures such as the P–R interval and QRS complex width can also help in the diagnosis of heart diseases.

**Related topics**

Membrane potential (B4)
Investigating cardiovascular function (D7)
Action potentials (F2)
Structure and function of the autonomic nervous system (H5)
Autonomic pharmacology (H6)
Reabsorption of electrolytes and glucose (M3)

## Cardiac muscle action potentials

As outlined in topic B4, all of the cells of the body are negatively charged on the inside of the cell membrane with respect to the outside and this is known as the **membrane potential** ($E_m$) for that cell. Membrane permeability to ions is highly controlled, in that ions can only move across the cell membrane via specific **ion channels**; the number of these ion channels that are open at any one time determines membrane permeability to that ion. Resting $E_m$ for cardiac muscle cells is between –60 and –90 mV, dependent upon cell type; around –60 mV in cells of the **sinoatrial node** (SA node), and around –90 mV in atrial and ventricular cells. As outlined in topic B4, $E_m$ in a given cell changes as the permeability of the cell membrane changes; when cells become more permeable to sodium or calcium ions, $E_m$ will tend to increase (i.e. become more positive), and when cells become more permeable to potassium ions, $E_m$ will tend to decrease (i.e. become more negative). When $E_m$ increases, this is termed **depolarization**, and when $E_m$ decreases, this is termed **polarization**, **repolarization**, or **hyperpolarization**. As cells become depolarized, the net effect is to make the outside surface of the cell more negatively charged, and as they become repolarized, the net effect is to return the outside surface of the cell to its more positively charged resting state. The attraction between elements of opposite charge that exist on the surface of neighboring cells (or groups of cells) when one is depolarized and the other is polarized results in an electric dipole that can be detected using monitoring equipment.

Under resting conditions, cardiac muscle cells have relatively high permeability to potassium ions and low permeability to sodium and calcium ions. However, the cells of the SA node are unable to maintain their resting $E_m$ at –60 mV; some potassium channels close and non-specific cation channels (so-called **'funny' cation channels**) and calcium channels open, resulting in a slow depolarization that is known as the '**pacemaker potential**' (*Fig. 1a*). This means that as time passes, SA node cells become progressively more depolarized, and because many ion channels are voltage dependent (i.e. they open or close in response to changes in $E_m$), this means that there is a greater likelihood that other ion channels will open, causing what is known as an **action potential** (*Fig. 1a*). In nerve and muscle cells, when $E_m$ reaches a specific level known as '**threshold**', large numbers of voltage-dependent ion channels open briefly (for a few milliseconds to a few

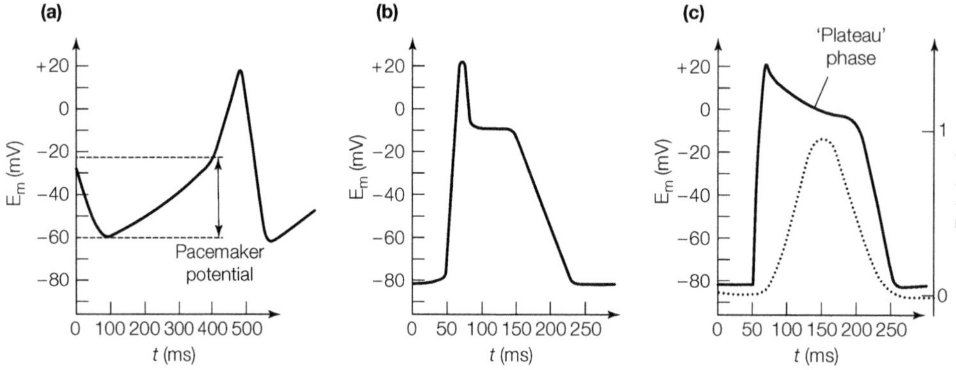


*Fig. 1. Action potentials in cardiac myocytes: (a) SA node cells; (b) atrial cells; (c) ventricular cells. Note the different timescale of the action potential in the SA node in (a) and the duration of ventricular contraction in (c).*

hundred milliseconds), increasing the permeability of the cell membrane to certain ions, and causing depolarization of the cell. In the case of the SA node cell, the majority of voltage-dependent ion channels that open in response to $E_m$ reaching threshold (around –20 mV) are calcium ion channels. This means that the cell becomes further depolarized, with $E_m$ reaching around +20 mV. At this point, voltage-dependent potassium channels open and the cell membrane becomes repolarized, following which the pacemaker potential begins again. This means that SA node cells are spontaneously active; they spend very little time at their resting $E_m$ of around –60 mV. This cycle continues 60–200 times a minute (dependent on nervous or hormonal influences), resulting in 'pacing' of the rest of the cardiac muscle at the rate of the SA node cells. Hence the SA node is the heart's principal **pacemaker** region, although all cardiac muscle cells are spontaneously active to some extent. Cardiac muscle action potentials depend on the concentrations of various ions on either side of the cell membrane, so electrolyte disturbances can cause marked alterations in electrical activity, and should always be investigated in people who have cardiac problems. For example, **hyperkalemia** (higher than normal levels of plasma potassium ions) can result in a resting $E_m$ that is closer to threshold, and abnormal cardiac excitability. Hyperkalemia can also cause problems with repolarization of cardiac myocytes, because the driving force for potassium ion movement out of the cell during the repolarization phase is reduced. **Hypokalemia** can also cause arrhythmias.

As outlined in Topic D2, cardiac muscle cells are electrically coupled to each other, each cell affecting all of its neighbors. This means that when the cells of the SA node undergo an action potential, neighboring cells are affected. Electrical activity originating in the SA node spreads passively through the atria, with atrial cells undergoing an action potential (*Fig. 1b*). Activity then spreads to the **atrioventricular node** (AV node) and the specialized conducting system of the ventricles, the left and right bundle branches of the **bundle of His**, causing ventricular cells to undergo an action potential (*Fig. 1c*). During the action potential, no superimposed depolarization is possible; this can only occur once $E_m$ has returned to its resting value.

The depolarization associated with the action potentials results in entry of calcium ions into the muscle cells and release of calcium ions from intracellular stores. These calcium ions are the catalysts for the complex series of events that results in contraction of the cardiac muscle cell (see Topic I3 for more details). The form of the cardiac muscle action potential also means that the heart cannot be activated too frequently, which might result in a sustained contraction. The ventricles must relax in order for them to fill with blood to be ejected in the next beat. The fact that the cardiac action potential lasts several hundred milliseconds (as opposed to 2–5 ms for nerve fibers and 10–15 ms for skeletal muscle fibers) means that contraction is over before a new action potential can begin (*Fig. 1c*). The heart is forced to undergo a phase of relaxation (**diastole**) before it can contract again (**systole**). At rest, the heart spends twice as long in diastole as in systole, and diastole shortens as heart rate rises.

### Effects of sympathetic and parasympathetic input

As stated above, under normal circumstances the rate of discharge of the cells of the SA node determines heart rate. However, the heart rate can be altered by activity of both of these branches of the **autonomic nervous system**. **Sympathetic** input, either from sympathetic cardiac nerves or circulating **epinephrine**, increases **heart rate** (and also increases force of contraction of cardiac muscle), and **parasympathetic** input from the vagus nerve reduces heart rate. **Norepinephrine** and epinephrine act upon **β-adrenoceptors** to increase the likelihood that 'funny' cation channels will be open during the pacemaker potential in SA node cells. This results in a pacemaker potential slope that is greater than normal, and hence the SA node cells reach 'threshold' and fire an action potential more quickly, causing an increase in heart rate. The major effect of **acetylcholine (ACh)** released from fibers of the **vagus nerve** is to reduce heart rate by acting on **muscarinic ACh receptors**, increasing the likelihood that potassium channels will be open, and decreasing the likelihood that 'funny' cation channels will be open during the pacemaker potential. This results in a pacemaker potential slope that is less steep than normal, and a lower resting $E_m$. Both these effects of ACh combine to increase the time taken for the SA node cells to reach 'threshold', and cause a reduction in heart rate.

### Spread of depolarization and repolarization

Coordinated contraction and relaxation of the heart is dependent upon this cycle of depolarization and repolarization. Under normal circumstances, there is a highly organized sequence of activation of different parts of the heart (*Fig. 2a*). Firstly, the SA node discharges and the electrical activity spreads passively across the atria to the AV node. The AV node is normally the only connection between the atria and the ventricles; they are separated and insulated from each other by rings of fibrous, non-electrically active connective tissue. Thus for excitation of the atria to spread to the ventricles, the electrical activity must pass through the AV node. The tissue of the AV node conducts at only around 10% of the speed of the atrial tissue, so there is always some degree of **AV node block** in conduction of the electrical activity from the atria to the ventricles. This allows for adequate ventricular filling before ventricular contraction takes place. Pathological AV node block is deleterious (e.g. ventricular activation may not occur after every atrial activation, the person may experience their heart 'missing a beat'). Once the AV node tissue has been activated, the electrical activity is passed on to the specialized muscle fibers of the bundle of His and **Purkinje fibers** that pass into ventricular tissue proper. Ventricular activation takes place from the apex of the heart, up towards the base, akin to squeezing a tube of toothpaste from the end rather than near the cap, so that more blood can be ejected from the ventricles per beat. During ventricular activation, the atrial cells are undergoing repolarization.

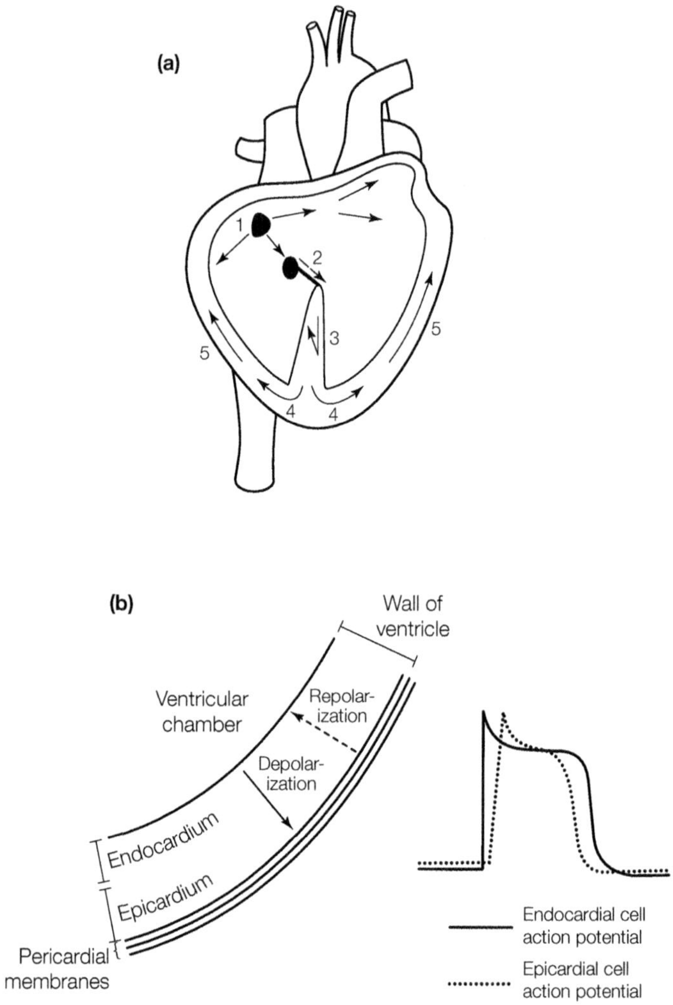


*Fig. 2. The sequence of electrical activation in the heart. (a) The entire heart: 1, SA node depolarization; 2, conduction through the atrioventricular node and the bundle of His; 3, depolarization of the septum; 4/5, ventricular depolarization. The positions of the atrial septum and the heart valves have been omitted for clarity; (b) different layers of ventricular tissue and the relative duration of their action potentials.*

The net depolarization of the ventricular muscle proceeds from the layer closest to the inner surface of the heart (the **endocardium**) to the outer layers of muscle (the **epicardium**). However, because the action potentials of endocardial cells take longer to complete than those of epicardial cells, repolarization of epicardial cells takes place before that of the endocardial cells, and so the net repolarization proceeds from epicardium to endocardium (*Fig. 2b*). Bear in mind that, although it is very tempting to think of cardiac muscle action potentials as depolarization, repolarization also occurs. Thus, for every cell, voltage across the membrane rises and falls for every action potential.

## ECG lead systems

The dipoles that exist because of the electrical activity of the heart can be recorded by placing electrodes in various places on the surface of the body to measure the **electrocardiogram (ECG)**. There are many different systems for recording the ECG. In a hospital setting, the most commonly used system is to attach electrodes to the wrists and ankles (four in total), and a further six electrodes to various positions on the chest. Although there are then 10 electrodes attached to the subject, this allows recording of the **12-lead ECG** (*Table 1*), since in ECG terminology the word 'lead' refers to the combination of a pair of these electrodes, where the electrical potential at one is compared to the electrical potential at the other. This allows measurement of **potential difference**, or **voltage** in a given lead (pair of electrodes). This topic will not consider **chest leads** any further (since clinical scenarios in which the chest leads are more important are outside the remit of this text), but will describe the ECG as detected by the **limb leads** (leads I, II, II, aVR, aVL and aVF).

*Table 1. Electrode positions for the various leads of the ECG*

| Electrode | Placement | Lead | Electrode combination |
|---|---|---|---|
| RA | Right wrist | I | LA to RA |
| LA | Left wrist | II | LL to RA |
| LL | Left ankle | III | LL to LA |
| RL (earth) | Right leg | aVR | RA to 'average' of LA and LL |
| | | aVL | LA to 'average' of RA and LL |
| | | aVF | LL to 'average' of RA and LA |
| $V_1$ | Right edge of sternum, 4th intercostal space | $V_1$ | $V_1$ to 'average' of RA, LA and LL |
| $V_2$ | Left edge of sternum, 4th intercostal space | $V_2$ | $V_2$ to 'average' of RA, LA and LL |
| $V_3$ | Mid-way between $V_2$ and $V_4$ | $V_3$ | $V_3$ to 'average' of RA, LA and LL |
| $V_4$ | 5th intercostal space, mid-clavicular line | $V_4$ | $V_4$ to 'average' of RA, LA and LL |
| $V_5$ | 5th intercostal space, anterior axillary line | $V_5$ | $V_5$ to 'average' of RA, LA and LL |
| $V_6$ | 5th intercostal space, mid-axillary line | $V_6$ | $V_6$ to 'average' of RA, LA and LL |

Lead I describes the potential difference between the LA electrode and the RA electrode and essentially measures potential changes from the left-hand side to the right-hand side of the chest. Lead II describes the potential difference between the LL electrode and the RA electrode, and essentially measures potential changes from the left leg, up and across the body to the right arm. Lead III describes the potential difference between the LL electrode and the LA electrode, and essentially measures potential changes from the left leg, up the axis of the body to the left arm. This allows us to draw **Einthoven's triangle** (*Fig. 3a*), in which the heart in the chest can be viewed from six different angles, using leads I, II, III, aVR, aVL and aVF. This means that the pattern of electrical activity in the heart will be different depending upon which lead one uses to measure the ECG, so it is not surprising that the ECG recorded by each lead looks different. The heart sits in the chest in an orientation more closely related to lead II than to leads I or III, and it is with reference to lead II that the description of the various ECG waves to different types of electrical activity seen in the heart are related here. The mean electrical axis of the heart is also parallel to lead II in most instances.

**Form of the basic ECG**

On an ECG, an upward deflection represents the electrical activity associated with a wave of depolarization that is moving towards the positive electrode of

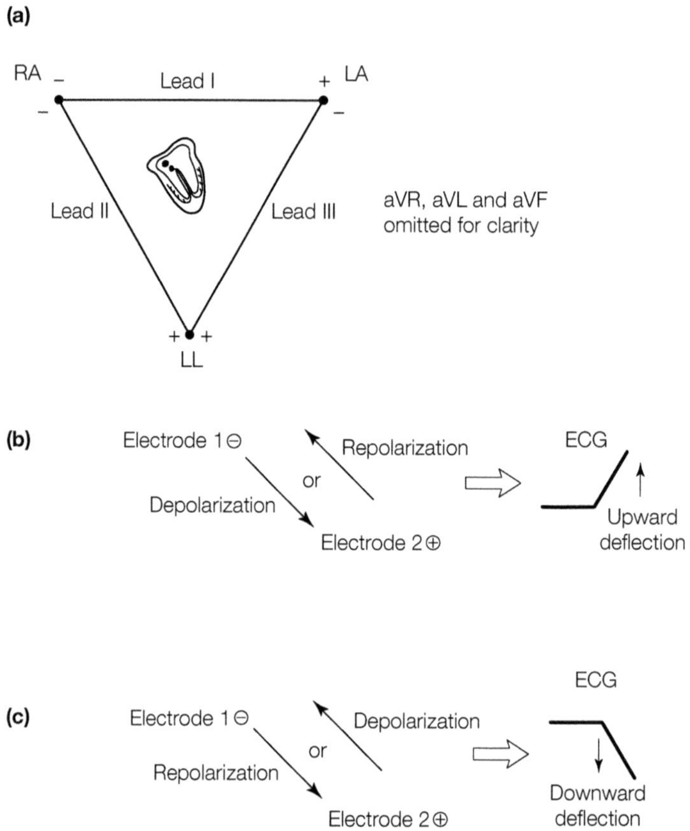


*Fig. 3. Depolarization and repolarization of cardiac muscle and their effect on the electrocardiogram (ECG): (a) Einthoven's triangle (aVR, aVL and aVF leads have been omitted for clarity – see Table 1); (b) causes of upwards deflection in the ECG; (c) causes of downwards deflection in the ECG.*

the lead pairing, or a wave of repolarization that is moving away from the positive electrode of the lead pairing (*Fig. 3b*). Conversely, the electrical activity associated with a wave of depolarization that is moving away from, or a wave of repolarization that is moving towards the positive electrode of the lead pairing results in a downward deflection of the ECG (*Fig. 3c*). It is therefore relatively easy to describe the salient features of an ECG recorded in lead II (LL to RA) in relation to electrical activity of the heart in a given cardiac cycle.

As the atria depolarize after discharge of the SA node, the wave of depolarization of the cardiac myocytes spreads from the top right of the atria to the bottom left. This results in a positive deflection of the ECG trace, known as the **P wave** (*Fig. 4*). There then follows a short delay (consistent with the slow activation of the AV node and conduction of the cardiac muscle action potentials down the bundle of His) before the depolarization of the ventricles is displayed as the **QRS complex** (*Fig. 4*). The major component of this complex in lead II is positive because the major component of the depolarization wave in the ventricles moves towards the positive electrode (LL) of the lead II pairing. During the QRS complex, the atria repolarize, but the electrical signal associated with atrial repolarization is so small in comparison with that being produced by the larger ventricular muscle mass, that it is masked by the QRS complex. There then follows the normally isoelectric **S–T segment** of the ECG; the period when all ventricular myocyte action potentials are in their plateau phase (*Fig. 1c*). This is followed by a **T wave** (*Fig. 4*), which represents the electrical signal associated with repolarization of the ventricles. As mentioned above, ventricular repolarization proceeds in the opposite direction to that of depolarization, and thus the net wave of repolarization is moving away from the positive electrode (LL) of the lead II pairing, which results in the upward deflection of the T wave.

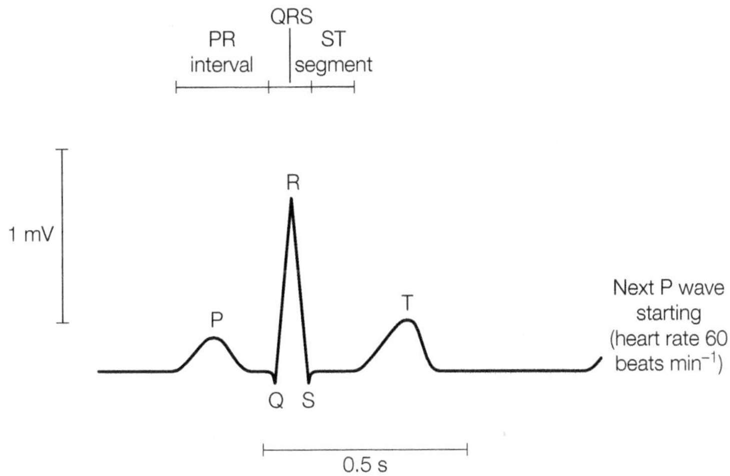


*Fig. 4. The form of the ECG as viewed using lead II of the standard ECG electrode system.*

It is important to not only examine, but also to *measure* the components of the ECG, as these measurements can be of diagnostic importance. The **P–R interval** should normally be between 0.12 and 0.2 s; longer P–R intervals indicate some degree of block of conduction between the atria and the ventricles. A QRS complex should follow every P wave, indicating that the heartbeat is being 'paced' by the SA node. **QRS complex width** should be below 0.12 s. Broad QRS complexes often occur when ventricular depolarization is occurring, due to spontaneous discharge from some cells in the ventricles themselves, rather than being paced from the SA node and conduction system. QRS complexes that are taller than normal can be indicative of ventricular hypertrophy.

# D5 Layout and function of the vasculature and lymphatics

## Key Notes

**Circulatory routes**

The heart is actually arranged as two pumps in series. The right side of the heart pumps blood to the pulmonary circulation; it picks up oxygen and gets rid of carbon dioxide. The left side of the heart pumps oxygenated blood around the rest of the body and back to the right side of the heart. From the left ventricle, blood can take a number of routes: to the heart itself, into the abdomen and lower limbs, or into the head, the trunk and the upper limbs. Some blood going from the left ventricle to the lungs in the bronchial circulation returns to the left side of the heart via the pulmonary vein. Some blood returning from the intestines to the right atrium passes through the liver via the hepatic portal vein before being drained into the hepatic vein and inferior vena cava.

**The structure of blood vessels**

All blood vessels have the same basic three-layered structure; an inner tunica intima, middle tunica media and outer tunica adventitia. The contribution of these three layers to the vessel wall and the types of connective tissue in the layers varies depending upon the function of the vessel. For example large elastic arteries have a lot of elastic protein fibres in their tunica media, whereas capillaries have only the endothelial cells and basement membrane of the tunica intima.

**Arteries and arterioles**

Arteries and the smaller arterioles supply oxygenated blood to the tissues. Some arteries have a lot of elastic tissue in their walls to withstand the variations in pressure caused by the pumping action of the heart. Most arteries and arterioles have a substantial amount of smooth muscle in the tunica media; in the case of arterioles, changes in the degree of constriction of this smooth muscle can alter resistance to flow in a vascular bed.

**Capillaries**

Capillaries are thin-walled blood vessels composed of endothelial cells and their associated basement membrane. They are designed to facilitate transport of substances in and out of the tissues. Fluid can normally pass through spaces between the endothelial cells to produce interstitial fluid; the forces that govern the net flow of fluid in a capillary bed are known as Starling forces. Edema can be caused by disturbances in the factors affecting these Starling forces.

**Venules and veins**

Blood returns to the heart from the tissues via venules and veins. These are thin-walled, low-pressure, high-capacitance vessels, some of which have one-way valves; at rest around 70% of circulatory volume is contained in the venous side of the circulation. Blood flow back to the heart is facilitated by the pumping action of muscles in the legs and by fluctuations in both thoracic and abdominal pressure as we breathe in and out. The valves in some veins prevent backflow of blood away from the heart.

**Lymphatic vessels**

Body tissues are also perfused by lymphatic vessels. Excess tissue fluid drains into lymphatic capillaries and eventually drains back into the venous side of the circulation. There are lymph nodes near the points of coalescence of lymphatic vessels; lymphatic fluid is checked for infection here.

**Related topics**

Overview of the heart and great vessels (D1)
Alveolar exchange and gas transport (E4)
The liver, gall bladder and spleen (J5)
Glomerular filtration and renal plasma flow (M2)

## Circulatory routes

As stated in Topic D1, the circulation is in fact two circulatory pathways in series; the right side of the heart pumps blood around the **pulmonary circulation** and the left side of the heart pumps blood around the **systemic circulation**. The route that blood takes around these circuits is as follows: **ventricle**, **artery**, **arteriole**, **capillary**, **venule**, **vein**, **atrium**. The purpose of the pulmonary circulation is to pick up oxygen and get rid of carbon dioxide in the lungs, and the purpose of the systemic circulation is to deliver oxygen and pick up carbon dioxide from respiring tissues. The general scheme of distribution of blood to the various organs in the body is shown in *Fig. 1*.

Blood leaving the left side of the heart can take three initial routes; the **coronary arteries** to heart tissue itself, the **descending aorta** to the abdomen, pelvis and legs, and the subclavian and **carotid arteries** to the upper limbs, trunk and head (including brain). Coronary arteries perfuse the heart tissue itself, and the venous drainage of the heart returns to the right atrium via the **great cardiac vein**, for the most part. The **abdominal aorta** supplies the liver, spleen, stomach, and intestines, as well as the kidneys and gonads. Lower down in the abdomen, the abdominal aorta splits into left and right **common iliac arteries** that supply the pelvic organs and the legs. The subclavian arteries and carotid arteries and their various branches supply the brain, head and neck, spinal cord, trunk, and upper limbs. Venous drainage of the various regions comes together in three places; the **hepatic portal system**, the **inferior vena cava**, and the **superior vena cava**. Venous blood draining the intestines makes up the hepatic portal system, which flows through the liver to deliver nutrients to liver cells, where they can be stored. The **hepatic artery** also supplies the liver, but venous drainage of the liver itself is via the inferior vena cava. Thus the abdominal and pelvic organs and the legs drain into the inferior vena cava, and the head, neck, arms, and trunk drain into the superior vena cava. The inferior and superior venae cavae come together at the right atrium to deliver blood to the right side of the heart for distribution to the pulmonary circulation.

## The structure of blood vessels

With the exception of the smallest blood vessels running through tissues, the capillaries, all of the blood vessels of the body have the same basic structure (*Fig. 2*). They are composed of three layers, named from the luminal surface of

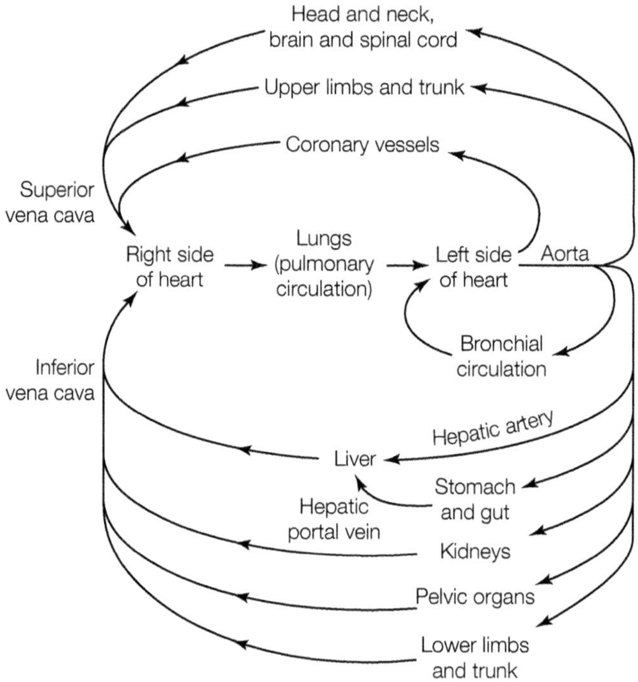


*Fig. 1. Schematic model of circulatory routes in the body.*

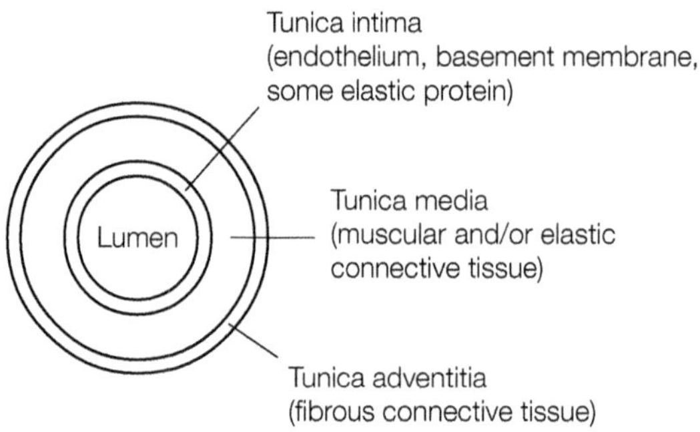


*Fig. 2. Schematic diagram of the composition of the walls of blood vessels.*

the blood vessel to the outside surface. These layers are known as the **tunica intima**, the **tunica media** and the **tunica adventitia**, and each layer contains similar structures in all blood vessels. The tunica intima is composed of **endothelial cells** that line the inside surface of the vessel, their associated **basement membrane**, a thin layer of connective tissue, and an elastic lamina (known as the **internal elastic lamina**). The intima is usually the thinnest layer of the blood vessel wall. The tunica media is composed of **smooth muscle**, a layer of connective tissue and elastic tissue (the **external elastic lamina**). Some elastic tissue is distributed throughout the media, but the majority exists in this outermost layer of the media. The media is usually the thickest of the three layers. The tunica adventitia consists of a tough layer of connective tissue (loose elastic fibers and **collagen**) that connects the blood vessel to its surroundings. The actual composition of the various layers varies depending upon the function of the blood vessel. For example, the media of large arteries contains a lot of elastic tissue because those vessels (e.g. the aorta) have to be able to cope with high pressure distention during ventricular contraction, but spring back when the ventricles relax. Smaller arteries and arterioles have proportionately more smooth muscle in the media in order to be able to regulate blood flow through capillary beds, and some veins have valves – folds of intimal tissue that help keep blood flowing towards the heart.

## Arteries and arterioles

As mentioned above, the largest arteries of the body must withstand the pulsatile changes in intraluminal pressure that come about due to the pumping action of the heart. If they did not contain substantial amounts of elastic tissue, they might easily tear. In view of their potential to expand and recoil, these arteries are known as **'windkessel' vessels**; when the ventricles relax the elastic properties of these vessels mean that blood flow is maintained when **blood pressure** falls from **systolic** levels of around 120 mmHg to **diastolic** levels of around 80 mmHg. As these arteries become less compliant and more stiff with age, this can cause elevated blood pressure (**hypertension**).

Smaller arteries and arterioles have large amounts of smooth muscle in their walls, and the smooth muscle is subject to nervous influence, circulating hormones and locally produced vasoactive chemicals. All capillaries in the body are fed from small arteries and arterioles, and so the levels of constriction of the smooth muscle in the walls of these blood vessels determine how much blood flows through a given tissue. The diameter of these vessels also has a marked effect on **peripheral resistance**, because resistance to blood flow is inversely proportional to the radius of the vessel to the fourth power (i.e. $R$ is proportional to $1/r^4$). So, if $r$ decreases by as little as 10%, $R$ increases by 50%. This means that (because mean **arterial blood pressure** is the mathematical product of **cardiac output** and peripheral resistance – see Topic D1), mean arterial blood pressure can be markedly elevated or reduced by small changes in the contractile state of the smooth muscle in the walls of muscular arteries and arterioles.

## Capillaries

Capillaries consist only of endothelial cells on their basement membrane. As such, there is tremendous potential for exchange across these very thin structures, and for fluid exchange through the gaps between the cells. Thus, capillaries are ideally suited to their role as the vessels that supply nutrients to respiring tissues and pick up toxic metabolites to be excreted from the body. Most capillaries in the body are slightly leaky. This means that fluid filtration can take place between the blood and tissue extracellular fluid. The forces that determine how much fluid will be exchanged along a capillary, and in which direction the net flow of fluid will take place are known as **Starling's forces** (*Fig. 3*).

**Capillary hydrostatic pressure ($P_{cap}$)** and **tissue hydrostatic pressure ($P_t$)** oppose each other; $P_{cap}$ normally predominates at the arterial end of a capillary bed, and $P_t$ normally predominates at the venous end. **Blood colloid osmotic pressure ($\pi_{cap}$)** and **tissue colloid osmotic pressure ($\pi_t$)** (the component of osmotic pressure contributed by substances that are suspended rather than dissolved in a solvent) also oppose each other; because blood contains more

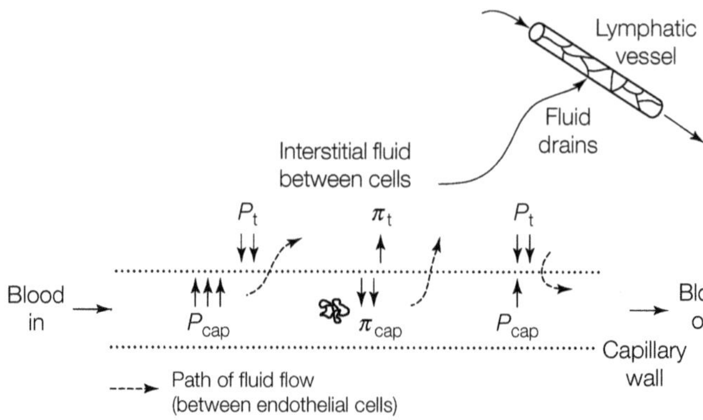


*Fig. 3. Model to describe fluid movement from the blood into the tissues and vice versa. $P_{cap}$, capillary hydrostatic pressure; $P_t$, tissue hydrostatic pressure; $\pi_{cap}$, blood colloid osmotic pressure; $\pi_t$, interstitial fluid colloid osmotic pressure. Mean $P_{cap}$ ~ 25 mmHg; mean $\pi_{cap}$ ~ 25 mmHg; mean $\pi_t$ ~ 4 mmHg; mean $P_t$ ~ 2 mmHg.*

protein than tissue fluid, $\pi_{cap}$ tends to predominate all the way along the length of a capillary bed. The equation that describes **net fluid filtration pressure** (NFP) takes the following form:

$$\text{NFP} = (P_{cap} + \pi_t) - (P_t + \pi_{cap})$$

NFP in most capillary beds is normally around 1–2 mmHg, although this figure varies depending upon which vascular bed one is talking about. Thus only a small amount of net fluid loss from the blood takes place in capillary beds. The resultant small amount of fluid accumulation in the tissues is normally drained away by **lymphatic drainage** of the tissue. However, under conditions where fluid movement out of the blood is high (e.g. with venous congestion or abnormal capillary leakiness), or where lymphatic drainage is compromised (e.g. when pressed on by cancerous tissue), the tissues can begin to swell, a condition known as **edema**.

In some regions of the body, capillaries need to be either exceptionally tight, or exceptionally leaky. In the brain, for example, the blood–brain barrier is tight to prevent swelling of brain tissue inside the skull, and to control which substances enter this crucial organ. In the liver, the capillaries are known as **sinusoids**, with gaps large enough for small blood cells and proteins to get through; this is important since the liver is one of the sites (along with the spleen) of degradation of some old and damaged erythrocytes. The liver is also the site of biosynthesis and secretion of many blood proteins. In the kidney, specialized capillaries known as **glomerular capillaries** (see Topic M2) consist of endothelial cells that are fenestrated. These **fenestrated endothelial cells** allow filtration of the **plasma** in the first stage of urine production.

## Venules and veins

Venules and veins are thin-walled vessels with proportionately much less smooth muscle and elastic tissue in their walls. They are, in effect, capacitance vessels; around 70% of blood volume is in the venous side of the circulation when we are at rest and **venous return** increases upon exercise. Blood pressure in the venous side of the circulation is low; average **venous pressure** is around 10 mmHg, as opposed to around 90 mmHg in the arterial system. Veins in the legs have valves to prevent backflow of blood away from the heart and venous return to the heart is helped by three main mechanisms.

1. As muscles in the legs work, they squeeze the veins, potentially pushing blood in two directions (both towards and away from the heart). However, the valves in leg veins prevent backflow.
2. As we breathe in, **thoracic pressure** decreases (see Topic E3), which reduces compression of the venae cavae in the chest, promoting venous return.
3. Also, on breathing in, **abdominal pressure** rises, pushing blood in abdominal veins back towards the heart (since thoracic pressure is lower at the same point).

Mechanism 1 above is known as the '**muscle pump**', and mechanisms 2 and 3 together are termed the '**respiratory pump**' (see Topic D8).

It is often erroneously assumed that all veins have valves; they do not. One example is the veins of the **vertebral column**, which connect to the veins draining pelvic organs. This is one route of spread of malignancy; **prostate cancer** can spread to the vertebral column if the patient has to strain to pass urine and increases abdominal pressure whilst doing this. Cancerous cells can break off from the tumor and find their way into the vertebrae.

## Lymphatic vessels

**Lymphatic vessels** are technically part of the cardiovascular system, but are often overlooked in textbooks. The function of lymphatic vessels (as outlined above) is to provide a route for drainage of excess tissue fluid back into the circulation. They are similar to thin-walled blood vessels, as one might expect from vessels that operate in a low-pressure system. Some have valves, and lymph flow is facilitated by arterial pulsations. The lymphatic drainage of several organs comes together at certain points in the body, and there are collections of **lymphoid tissue** there (e.g. in the groin, the axilla, the neck, and the abdomen). The purpose of these **lymph nodes** is to act as sentries guarding against infection; when infection is encountered by resident **lymphocytes**, an immune response is mounted that results in aggregation of immune cells in the lymph nodes. This is what is typically known as 'swollen glands'. The lymphatic system is also a route of spread of malignancy, and staging the progression of a cancer in a patient often depends upon whether local or distant lymph nodes have become involved. Lymphatic fluid eventually drains into the **brachiocephalic veins**, where it mixes with the blood.

# D6 THE CARDIAC CYCLE, BLOOD PRESSURE, AND ITS MAINTENANCE

## Key Notes

**Phases of the cardiac cycle**

The heart must contract then relax in each cardiac cycle. These phases of the cardiac cycle are systole (ventricular contraction) and diastole (ventricular relaxation). Systole and diastole each have several different subphases, such as isovolumetric contraction (in the case of systole) and isovolumetric relaxation (in the case of diastole). The heart sounds that one can hear via a stethoscope or simply by placing an ear against someone's chest are known as the first ($S_1$, 'lub') and second ($S_2$, 'dup') heart sounds, and are caused by closure of heart valves.

**Pressure–volume loops**

The relationship between pressure in the ventricle and the volume of blood in the ventricle during the cardiac cycle can be plotted as a pressure–volume loop. Such graphical representations of ventricular function can indicate how the heart is affected by changes in end-diastolic volume or arterial blood pressure.

**Blood pressure**

Blood flows in a closed system of blood vessels, exerting pressure on the walls. The pressure exerted varies during the cardiac cycle, such that systolic blood pressure is the peak pressure exerted by blood against the arterial wall during systole, and diastolic blood pressure is the lowest pressure exerted against the arterial wall, during diastole. At rest, systole lasts twice as long as diastole, so diastolic blood pressure has a greater influence on mean arterial blood pressure. Mean arterial blood pressure is the arithmetic product of cardiac output and resistance to blood flow in the peripheral vasculature (total peripheral resistance); hence as cardiac output or total peripheral resistance change (e.g. increase), mean arterial blood pressure will change (increase in this case). Total peripheral resistance can be affected by changes in blood viscosity, vessel length and radius of the vessels being perfused, but moment-by-moment changes in vessel radius are the biggest contributory factors to changes in resistance.

**Control of blood pressure – neural**

Specialized nerve endings (baroreceptors) in the walls of the aortic arch and the carotid sinus serve as the sensors of arterial blood pressure. If blood pressure increases, these neurons send more frequent nerve impulses back to the cardiovascular center in the brainstem via the vagus nerve and glossopharyngeal nerve, respectively. Neural output from the cardiovascular center, via the sympathetic and parasympathetic branches of the autonomic nervous system, influences both cardiac function and vascular tone, to effect changes in blood pressure, in the opposite direction to that sensed by the baroreceptors. Such responses are so rapid that they are said to form the baroreceptor reflex. Some baroreceptors also exist in the heart and in pulmonary blood vessels.

**Control of blood pressure – hormonal**

In addition to neural mechanisms for control of blood pressure hormonal factors are also important, although responses mediated by hormones take longer to have an effect. Thus, hormonal mechanisms for control of blood pressure operate over the medium-to-long-term timeframe. The most important hormone systems in control of blood pressure are epinephrine, vasopressin, atrial natriuretic peptide, and the renin–angiotensin–aldosterone system. These systems act through a combination of effects on the heart, the blood vessels and the kidneys.

**Related topics**

Homeostasis and integration of body systems (A2)
Investigating cardiovascular function (D7)
Structure and function of the autonomic nervous system (H5)
Innervation of smooth muscle (I7)
Hormonal control of fluid balance (M5)

## Phases of the cardiac cycle

The heart beats 70 times per minute at rest, performing as a pair of pumps in series with each beat. Between beats, the **ventricles** must fill with blood, so ventricular muscle must relax during this time. This phase of the **cardiac cycle** is known as **diastole**. The phase of the cardiac cycle during which the ventricles are contracting and blood is being ejected into the **pulmonary artery** and **aorta** is known as **systole**. When one listens to the heart, one hears the heartbeat as two discrete sounds ('lub' and 'dup'). These sounds are known as the first and second heart sounds (**$S_1$** and **$S_2$**). Systole and diastole can be subdivided into other phases, the names of which serve as a description of the events that are taking place in the heart in each phase. For simplicity, only the details of the left atrium, mitral valve, left ventricle, and aortic valve are described below. Events involving structures on the right side of the circulation are essentially the same, although the pressures involved are much lower.

*Systole*

- **$S_1$ ('lub')** occurs early in systole, but is caused by vibration of the mitral valve when it closes at the end of diastole (see below).
- **Isovolumetric contraction**: pressure in the left ventricle is rising due to contraction, but has not yet risen above that in the **aorta** at the end of diastole, so the **aortic valve** is shut and volume in the ventricle remains the same. This phase begins at the peak of the R wave of the **ECG**.
- **Rapid ejection**: pressure in the ventricle exceeds that in the aorta, and the aortic valve opens, leading to ejection of blood into the aorta. The **mitral valve** may be pushed slightly into the left atrium, but is normally prevented from being pushed into the atrium fully by contraction of the **papillary muscles**. Thus no blood flow between the ventricle and atrium normally takes place during systole. If the papillary muscles or **chordae tendineae** are damaged, or if the valve is incompetent (i.e. it does not close fully), some blood may be forced into the atrium from the ventricle (see Topic D7).
- **Reduced ejection**: pressure in the left ventricle falls so that flow of blood into the aorta slows. This coincides with the **T wave** on the ECG. The aortic valve closes at the end of this phase.

*Diastole*

- **$S_2$ ('dup')** is a heart sound that occurs during diastole, but is caused by vibration of the aortic valve as it closes at the end of systole.
- **Isovolumetric relaxation**: during this subphase and the ejection subphases (above), the left atrium fills with blood on top of the closed mitral valve, and left **atrial pressure** increases gradually. Blood flow out of the ventricle has stopped as the aortic valve is also closed.
- **Rapid ventricular filling**: when atrial pressure rises above **ventricular pressure**, the mitral valve opens and blood enters the ventricle. Stenosis (stiffening and inadequate opening) of the mitral valve may result in inadequate ventricular filling and backup of blood into the left atrium (see Topic D7).
- **Reduced ventricular filling (diastasis)**: atrial and ventricular pressures rise very slowly and filling continues more slowly.

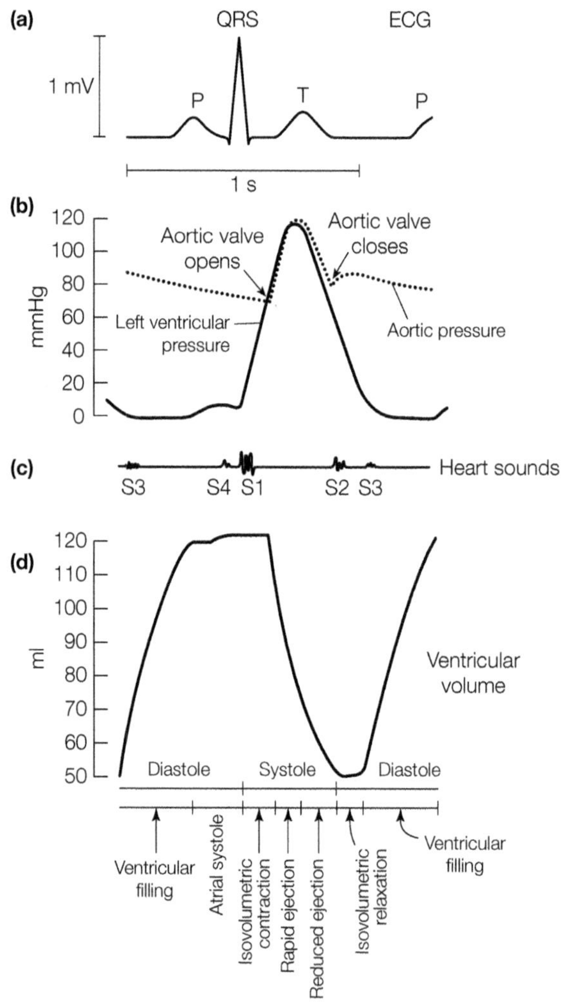


*Fig. 1. Relationship of the electrocardiogram (ECG) to heart sounds and the phases of the cardiac cycle. (a) ECG; (b) pressure profiles from the left ventricle and the aorta; (c) heart sounds; (d) volume of the left ventricle.*

- **Atrial systole**: depolarization of the atria (**P wave** of the ECG) results in contraction of atrial muscle and a small amount of 'topping off' of the ventricle with blood. When the ventricle starts to contract, ventricular pressure rises. When ventricular pressure rises above atrial pressure, the mitral valve closes (see above).

*Figure 1* shows the relationship between the ECG, phases of the cardiac cycle, heart sounds and ventricular function.

## Pressure–volume loops

As described above, pressures and volumes in the ventricle fluctuate as the cardiac cycle proceeds. If one takes these data and plots a graph of ventricular pressure against ventricular volume, one finds that the curve is a loop (*Fig. 2*). Such graphs of ventricular function are known as **pressure–volume loops**, and can yield important information about the function of the ventricle and about **arterial blood pressure**. On a pressure–volume loop such as that shown in *Fig. 2*, the mitral valve opens at point 1 and closes at point 2, and the aortic valve opens at point 3 and closes at point 4. Note, then, that systole is the phase between points 2 and 4, and diastole is the phase between points 4 and 2.

It is also apparent from *Fig. 2* that the isovolumetric contraction phase takes place between points 2 and 3 (no change in volume, but a large increase in pressure), and that isovolumetric relaxation takes place between point 4 and point 1 (no change in volume, but a large decrease in pressure). Filling of the ventricle and ejection of blood from the ventricle takes place between points 1 and 2 and between points 3 and 4, respectively. Another important point to note is that opening of the aortic valve and ejection of blood from the ventricle (point 3) only happens when ventricular pressure rises above around 80 mmHg, and closure of the aortic valve only happens when ventricular pressure falls below around 120 mmHg. This suggests that **aortic pressure** is normally between 80 and 120 mmHg during all phases of the cardiac cycle. The aortic valve will not open until ventricular pressure has risen above the diastolic pressure in the aorta (normally around 80 mmHg). This means that if **diastolic pressure** in the aorta is raised, then the ventricle has to do more work in order to eject a given

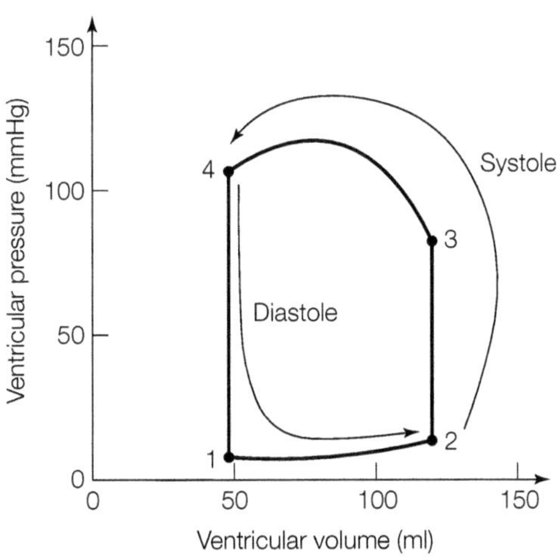


*Fig. 2. The left ventricular pressure–volume loop.*

volume of blood. The ventricle is said to be working against increased **afterload**, and it is a common finding that people with long-term elevation of blood pressure have **left ventricular hypertrophy** (an increase in size of individual cells) because the ventricular muscle has to work harder to maintain **cardiac output**. Note also that the ventricle is not normally empty at the end of systole. Around 50 ml of blood remains in the ventricle, meaning that with a **stroke volume** of 70 ml, the **ejection fraction** is 58% (70/[50 + 70] = 0.58). If **end-diastolic volume** increases, then so does **end-diastolic fiber length**. This is said to be a measure of the degree of **preload**, the initial stretch of ventricular muscle fibers before they begin to contract, and has important implications for function (see topic D8).

## Blood pressure

The blood in the cardiovascular system is being pumped around in a series of closed vessels. As the heart contracts and relaxes, the amount of pressure that is exerted upon the walls of the blood vessels also varies. Topic D5 gave a description of how the large arteries are very elastic in order to accommodate these large changes in pressure. When **blood pressure** is measured (see Topic D7), two figures are quoted (e.g. 120/80 mmHg). The first of these figures is **systolic blood pressure (SBP)**; the maximal pressure exerted on the arterial wall during systole. The second figure is **diastolic blood pressure DBP** – the lowest pressure exerted on the arterial wall during diastole (*Fig. 3a*). Such large variations in blood pressure over the cardiac cycle are only apparent in the arteries; by the time blood is flowing into the smallest arterioles and the capillaries, the blood pressure profile is smooth (*Fig. 3b*). Thus, from the perspective of tissue

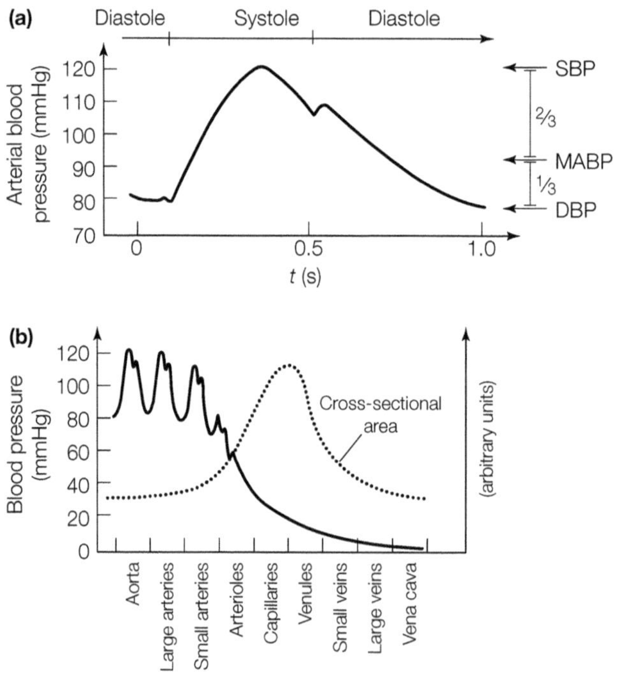


*Fig. 3. Blood pressure profile in (a) the aorta; and (b) throughout the systemic circulation. MABP, mean arterial blood pressure; DBP, diastolic blood pressure; SBP, systolic blood pressure.*

perfusion, mean **arterial blood pressure** is important. One can calculate mean arterial blood pressure (MABP) from SBP and DBP, using a simple equation:

$$MABP = DBP + (SBP - DBP)/3$$

Thus, using the figures above (120/80 mmHg), MABP is 93 mmHg. The reason why MABP is not simply an average of SBP and DBP (if this were the case MABP would be 100 mmHg) is that, at rest, the heart spends twice as much time in diastole as it does in systole. Thus, DBP makes a greater contribution to MABP (*Fig. 3a*), and the figure for MABP is actually the point on the blood pressure curve that, if one were to draw a horizontal line across the curve, would give two equal areas above and below that line.

MABP is also related to cardiac output (CO) and **total peripheral resistance** (TPR) as outlined in topic D1:

$$MABP = CO \times TPR$$

It is apparent from this equation that any factor(s) that increase either CO or TPR will increase MABP, and any factor(s) that decrease either CO or TPR will lower MABP. CO is the mathematical product of **heart rate** and **stroke volume**. Thus, changes in either heart rate or stroke volume have the potential to alter MABP. The factors affecting resistance to flow (*R*) can be described by the following equation:

$$R = 8\eta L/\pi r^4$$

where $\eta$ is blood viscosity, *L* is length of blood vessels, *r* is blood vessel radius, and $\pi$ is constant.

Under normal second-by-second circumstances in a given vascular bed, $\eta$ and *L* do not change, but *r* can be influenced by changing the degree of constriction of the smooth muscle in the wall of the blood vessel. In fact, reducing *r* by only 10% results in a 52% increase in *R*. Reducing *r* by 20% results in a 144% increase in *R*. Hence, **vessel radius** (and the degree of constriction of smooth muscle in the walls of blood vessels) has the largest effect on TPR. This is one of the methods that the body uses to alter MABP; generalized **vasoconstriction** can increase TPR and thereby raise MABP. Capillaries make the greatest contribution to the total cross-sectional area of the cardiovascular system (*Fig. 3*), and so one might think that capillary constriction or dilatation might be important in altering TPR. However, because capillaries have no smooth muscle in their walls, little can be done to alter their cross-sectional area. On the other hand, the arterioles, which feed capillary beds, are very muscular, and constriction or dilatation of these vessels results in large changes in TPR. For this reason, arterioles are commonly known as the **resistance vessels** of the cardiovascular system, and are said to make the greatest contribution to TPR.

## Control of blood pressure – neural

Blood pressure is monitored by **baroreceptors** in the walls of blood vessels in various parts of the body. One of the major baroreceptor sites is in the **carotid body**, where distention of the wall of the carotid artery results in increased stimulation of specialized nerve endings and increased firing of those neurons. Other baroreceptors exist in the arch of the aorta. These neurons send processes via the **glossopharyngeal nerve** (cranial nerve IX) and the **vagus nerve** (cranial nerve X) to various cardiovascular monitoring centers in the **brainstem**. Here, the neurons synapse with other neurons that control the firing rate of **parasympathetic** (vagal) and **sympathetic** nerve fibers to the heart, and sympathetic nerve fibers (vasomotor nerves) to the vasculature. In general, activation of baroreceptors by increased MABP leads to reduced sympathetic outflow from the cardiovascular center, and increased vagal outflow to the heart. A reduction in sympathetic outflow would reduce heart rate and stroke volume (see Topics D2 and D3) and decrease **vasomotor tone**, thereby reducing CO, TPR and MABP. An increase in parasympathetic outflow would reduce heart rate, thereby reducing MABP. Thus, this is a **negative feedback** loop that functions to maintain MABP within relatively tight limits. One problem with baroreceptor control of blood pressure is that such receptors, if exposed to high levels of MABP for significant periods of time (1–2 days), reset their basal firing rate to suit the 'new' levels of basal MABP. This can mean that sustained increases in MABP (as in **hypertension**) cannot be compensated for by normal mechanisms.

**Cardiopulmonary baroreceptors** also exist in the atria, ventricles, and pulmonary vessels. Their main role is to monitor central **blood volume**. Decreased stretch of the atria, ventricles and pulmonary vessels results in neurally induced increases in heart rate, stroke volume and vasoconstriction, via cardiovascular centers in the brainstem.

## Control of blood pressure – hormonal

Baroreceptor mechanisms in the body also serve to maintain blood pressure over the medium to long term. Such mechanisms are generally hormonal in nature and are therefore less rapid than the neural pathways described above. In view of the diverse nature of the potential hormonal responses to a change in blood pressure over the medium to long term, it would be foolish to state that the consequences of a particular alteration in blood pressure would have a predictable effect on the levels of a given set of circulating hormones, but it is important to be able to predict what effect the various hormones themselves might have on different parameters that have a role to play in determining blood pressure.

In general, the four major hormone systems that are at play are adrenal medullary **epinephrine**, the **renin–angiotensin–aldosterone system**, **vasopressin**, and **atrial natriuretic peptide** (see *Table 1* for a summary of their actions).

Epinephrine is released from the **adrenal medulla** in response to a fall in MABP, as part of the sympathetic response. The adrenal medulla is actually a collection of innervated nerve cell bodies that have not developed axonal processes, but discharge their transmitter (80% epinephrine, 20% norepinephrine) into the bloodstream. Thus, medullary epinephrine is a hormone, but is released on activation of the sympathetic branch of the autonomic nervous system. Epinephrine acts on **α- and β-adrenoceptors** in a wide range of body tissues to exert a physiological effect; in the case of the heart and blood vessels, epinephrine stimulates cardiac β-adrenoceptors and vascular α-adrenoceptors (which cause vasoconstriction) to raise CO and TPR, respectively. This is also part of the general 'fight or flight' response to a threatening stimulus. Although populations of both α- and β-adrenoceptors (that mediate vasoconstriction and vasodilatation, respectively) have been identified on vascular smooth muscle of coronary arterioles, the functional role of these receptors has yet to be definitively proven. It should be mentioned, however, that a population of $\beta_2$-adrenoceptors (activated by circulating epinephrine) exists on skeletal muscle arterioles, and that these receptors are important in mediating at least some of the vasodilatation observed in that vascular bed during exercise.

The renin–angiotensin–aldosterone system is complex and will be covered in more detail in section M, when the renal system is explored. **Renin** circulates in the blood and converts **angiotensinogen** (a peptide produced by the liver) to **angiotensin I**. Angiotensin I is, in turn, converted to **angiotensin II** by the action of **angiotensin-converting enzyme (ACE)**, which is present on **vascular endothelial cells**. Angiotensin II acts on specific receptors (predominantly **AT-1 receptors**) to do two things. Firstly, angiotensin II constricts blood vessels and raises TPR. Secondly, angiotensin II releases the steroid hormone **aldosterone** from the **adrenal cortex**. Aldosterone acts on the kidney to increase sodium reabsorption and help keep blood volume up (water follows sodium under osmotic drive in the kidney), thereby increasing stroke volume and CO. However, because aldosterone, like other steroid hormones, takes more than an hour to have any effect, maintenance of blood pressure by this route is slow. Thirdly, angiotensin II stimulates drinking, which in turn increases blood volume. Fourthly, angiotensin II promotes the secretion of vasopressin (see below). **ACE inhibitors** are often used to treat hypertension, since if one can prevent the conversion of angiotensin I to angiotensin II, the activation of multiple pathways leading to raised MABP can be inhibited.

Vasopressin is a hormone released from the **pituitary gland** in response to a fall in blood pressure (again detected by aortic and carotid baroreceptors) or an increase in extracellular fluid osmolality. As its name suggests, it can cause vasoconstriction. However, vasopressin is also known as **antidiuretic hormone (ADH)**, because it acts to promote water reabsorption in the kidneys. This serves to reduce extracellular fluid osmolality, increase blood volume, and reduce urine output. Blood volume has to fall by more than 10% in order for significant amounts of vasopressin to be secreted from the pituitary gland, so it is questionable how much of a role the vasopressin system has in maintenance of blood pressure under normal circumstances. Atrial natriuretic peptide (ANP) is released from atrial myocytes when they are stretched by increased central

*Table 1. Major hormones affecting function of the cardiovascular system*

| Hormone | Specific effect | Effect on blood pressure |
|---|---|---|
| Epinephrine | Peripheral vasoconstriction<br>Increased heart rate and force | Increased MABP, through increased TPR and increased CO |
| Angiotensin II | Peripheral vasoconstriction<br>Aldosterone secretion<br>Vasopressin secretion<br>Thirst | Increased MABP, through increased TPR and increased CO |
| Aldosterone | Increased salt and water reabsorption | Increased MABP, through increased CO |
| Vasopressin | Increased salt and water reabsorption<br>Vasoconstriction | Increased MABP, through increased CO and increased TPR |
| Atrial natriuretic peptide | Reduced salt and water reabsorption<br>Reduced renin, aldosterone and vasopressin secretion | Reduced MABP, through reduced CO and reduced TPR |

CO: cardiac output; MABP: mean arterial blood pressure; TPR: total peripheral resistance.

blood volume. As its name suggests, ANP promotes sodium (and thus water) excretion in the kidneys, thereby lowering blood volume. However, ANP also inhibits renin release and aldosterone and vasopressin secretion. ANP secretion is more of a mechanism for modulating central blood volume rather than blood pressure.

# D7 INVESTIGATING CARDIOVASCULAR FUNCTION

## Key Notes

**Cyanosis and peripheral pulses**

One of the simplest ways of investigating cardiovascular function is to observe the subject. A blue tinge to the nail bed, lips or tongue is a sign either that inadequate blood is reaching these areas, or that although enough blood is reaching these areas, it is not well-oxygenated. If the area is blue and warm, this is a sign that there is some problem with oxygenation of the blood, rather than inadequate perfusion. Another non-invasive way of investigating cardiovascular function is to palpate the peripheral pulses. Such investigations can elicit a lot of information.

**Blood pressure**

Blood pressure can be measured relatively easily using a blood pressure cuff, a stethoscope and a sphygmomanometer. Blood pressure is normally recorded at the brachial artery at the elbow. Hypertension (high blood pressure) or hypotension (low blood pressure) can lead to health problems, and both are usually treated medically, although hypertension is often considered to be more dangerous. Hypertension is treated with a range of drugs that are designed to slow heart rate, dilate peripheral blood vessels, and promote fluid loss in the urine.

**ECG and heart sounds**

Electrocardiography is an important method by which the pattern of electrical activation of the heart can be examined. The heart may have an abnormal pattern of activity (arrhythmia), or changes in the ECG can give clues as to what may have happened to heart muscle (e.g. a myocardial infarction). One can also listen to heart sounds using a stethoscope. Two heart sounds, $S_1$ and $S_2$, are normally heard, but other heart sounds can be a sign of pathology with the valves in the heart, or with the vasculature.

**Doppler and echocardiography**

More complicated investigations of cardiovascular function are based upon using high-frequency sound waves to visualize the heart, and to detect movement of blood in the heart and in the blood vessels. Such tests can help diagnose problems with the function of heart valves, the heart muscle itself, or vascular disease.

**Angiography and catheterization**

Invasive techniques for investigation of cardiovascular function include angiography and catheterization. In angiography, radio-opaque dyes are injected into blood vessels to allow visualization of narrowing of the arteries; this is most commonly done in subjects in whom one suspects narrowing of the coronary arteries. In cardiac catheterization (which is rarely done), doctors can directly measure the pressure in the various chambers of the heart.

| **Related topics** | Overview of the heart and great vessels (D1) | The cardiac cycle, blood pressure, and its maintenance (D6) |
|---|---|---|
| | Electrical activity of the heart – the ECG (D4) | |

## Cyanosis and peripheral pulses

One of the functions of the cardiovascular system is to supply the tissues of the body with oxygenated blood. **Hemoglobin** appears red when oxygen is bound, and blue when it is not. Hence, one way of testing a person's cardiovascular function is simply to look at them. If someone's nail beds appear blue (**cyanosis**), this might either be a sign that there is inadequate blood supply to the fingers or toes, or that although blood is reaching this area, it is carrying less oxygen. Touching the affected part of the body and looking under the person's tongue are useful discriminatory tests. If the fingers are blue and cold, but the tongue is not blue, it is likely that the person has some degree of **peripheral vascular disease**. If, on the other hand, the fingers are warm and the tongue appears blue, the person has central cyanosis and should receive prompt medical treatment. **Capillary refill tests** are also useful; pressing the nail bed for 5–10 s and then relieving the pressure should result in prompt (1–2 s) return of the pink tinge to the nail bed.

As blood is pumped around the body by the heart, the actions of systolic and diastolic pressures (see Topic D6) can be felt in some arteries that are near the surface of the body. This is known as palpation of **peripheral pulses**, and is a key element of any investigation of cardiovascular function. At some point in the past, we have all taken our pulse to work out our heart rate. This is most commonly done by placing the fingertips (not the thumb) over the **radial artery** as it passes close to the surface of the skin above the radius at the wrist. However, under normal circumstances, one can also palpate peripheral pulses in the **carotid artery** in the neck, the **brachial artery** in the crook of the elbow (the **antecubital fossa**), the **femoral artery** in the groin, the **popliteal artery** behind the knee, and the **dorsalis pedis artery** as it runs on the front of the foot.

In addition to describing rate, one can also describe the character of these pulses. For example, in someone with an increased blood volume, the radial pulse might well be described as 'full' and 'bounding', whereas in someone who has compromised cardiac function, it may be 'weak' or 'thready'. The relative strength of, or absence of, specific peripheral pulses can also be a sign of peripheral vascular disease. For example, in a subject with a normal left femoral pulse, a weak left popliteal pulse, and an absent left dorsalis pedis pulse, one should suspect peripheral vascular disease such as **atherosclerosis** or **thrombosis** between the femoral and popliteal arteries on the left. This is especially true if peripheral pulses on the right are normal. Relatively cold and/or blue extremities (**cyanosis**) can also be a sign of peripheral vascular disease, as this is a sign of under-perfusion (see above). One can also describe whether there is a 'delay' in palpation of either the same artery on either side, or of different arteries on the same side. For example, if one detects a significant delay between the radial pulse on the left side as compared with the right (**radioradial delay**), or between the radial and femoral arteries on same side (**radiofemoral delay**), then this might suggest some anatomical abnormality such as **coarctation of the aorta**.

## Blood pressure

**Arterial blood pressure** is normally measured in the brachial artery. Although it is often done nowadays with an automatic blood pressure monitor, it is of value to attempt the following auscultatory (listening) procedure and relate it to physiology (*Fig. 1*). Place a blood pressure cuff around the upper arm of the subject, so that when it is inflated, it can occlude blood flow in the brachial artery, and then locate the brachial artery in the antecubital fossa by palpation. While palpating the brachial artery, estimate **systolic blood pressure** by inflating the blood pressure cuff to the level where pulsation can no longer be felt. Note the value shown on the **sphygmomanometer** (mercury column or aneroid dial) and deflate the cuff. Next, place the **stethoscope** over the brachial artery and inflate the cuff to 30 mmHg above the estimate of systolic blood pressure by palpation. Deflate the cuff slowly (2–3 mmHg $s^{-1}$) and listen for sounds in the brachial artery using the stethoscope. The point at which **Korotkoff sounds** appear for two or more consecutive beats is the systolic blood pressure. Continue listening with the stethoscope while deflating the cuff. The point at which Korotkoff sounds can no longer be heard is the **diastolic blood pressure**. Deflate the cuff fully.

When the pressure of the cuff is above systolic blood pressure, there is no flow in the artery, and hence no sound. Below systolic blood pressure, but above diastolic blood pressure, flow is turbulent and sounds can be heard. Below diastolic blood pressure, flow is smooth and no sounds are heard. The reason that it is important to estimate systolic blood pressure by palpation before performing this procedure is that, in some subjects, there is an '**auscultatory gap**' between systolic and diastolic blood pressure (Korotkoff sounds appear, die away and then return as one lowers the pressure on the artery). If one does not estimate

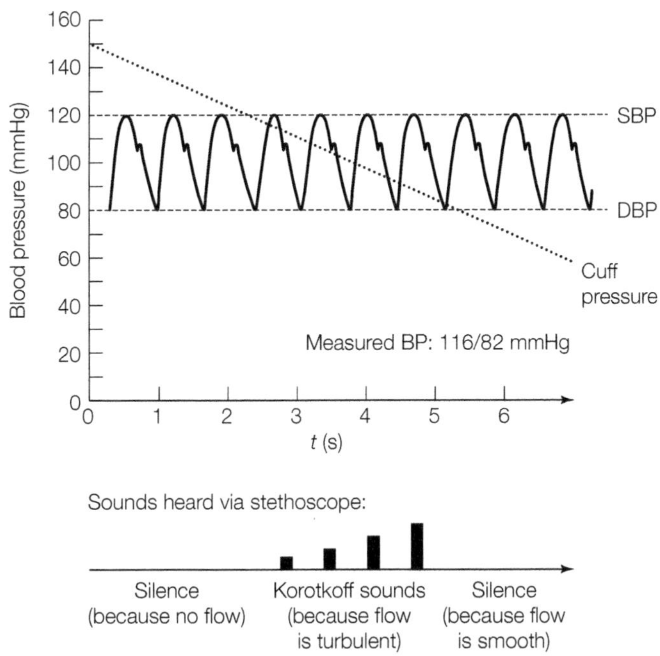


*Fig. 1. The auscultatory method for measuring blood pressure (BP) in the brachial artery. DBP, diastolic blood pressure; SBP, systolic blood pressure.*

systolic blood pressure by palpation, one may significantly underestimate systolic blood pressure by the auscultatory method.

Bear in mind that aortic blood pressure may be slightly different from this usual blood pressure measurement in the brachial artery, but it is important to have an international standard measurement of arterial blood pressure that is non-invasive and therefore well-tolerated by subjects. **Normal blood pressure** is often quoted as 120/80 mmHg, but it is unlikely that any given subject will have this blood pressure. Blood pressure is determined by many factors including age, sex and body size, so 'normal' blood pressure is actually best defined as a range of blood pressure measurements within which 95% of the general population can be categorized. Therefore, the cut-off point above which someone is diagnosed as having high blood pressure (**hypertension**) is constantly changing. Current values are around 140/85 mmHg. This means that someone with blood pressure above 140 mmHg systolic and/or 85 mmHg diastolic on three consecutive occasions is likely to receive a diagnosis of hypertension and be offered treatment, usually drugs like **angiotensin-converting enzyme (ACE) inhibitors** and **β-blockers**. However, blood pressure measurement is a difficult skill to master, and if one is likely to be using this method regularly, it is important to understand the various pitfalls and common errors that can arise. For example, if one underestimates a subject's blood pressure, then that person may be denied treatment for hypertension and be exposed to the risks associated with that disease. Equally, if one overestimates blood pressure, then the person may be treated for hypertension and possibly be unnecessarily exposed to the side-effects associated with blood pressure-lowering treatments. In many cases, subjects may be asked to wear an **ambulatory blood pressure** measuring device for a period of 24 h to determine the mean awake blood pressure and mean asleep blood pressure. So-called '**white coat hypertension**', where the blood pressure is elevated simply as a result of being in a doctor's surgery, can also be detected in this way. It is important to obtain a good estimate of baseline blood pressure before prescribing antihypertensive medication for patients.

## ECG and heart sounds

The theoretical basis for the **electrocardiogram (ECG)** is covered in Topic D4. The ECG is an important diagnostic tool in investigating cardiovascular function. By examining the ECG, it is possible to diagnose a range of conditions affecting the electrical activity of the heart. This includes **heart block** (where the conduction system of the heart is in some way defective), **atrial fibrillation** (where the atria are more active than they ought to be), **ventricular fibrillation** (where the ventricles do not contract and relax in a coordinated fashion and cannot pump blood effectively), and varying degrees of cardiac **ischemia**, or **heart attack**. However, it should be emphasized that one should use a systematic approach to ECG interpretation, since again there is a range of normal values for each of the features of the ECG. If one's attention is immediately drawn to the most obvious abnormality, one may inadvertently overlook more subtle, yet dangerous, problems.

As indicated in Topic D6, the various events of the **cardiac cycle** are associated with opening and closing of valves between the chambers of the heart, and between heart chambers and blood vessels. As these valves close, the vibration of the valves themselves and of the walls of the heart result in the two major heart sounds, $\mathbf{S_1}$ and $\mathbf{S_2}$. When using a stethoscope to listen to the chest, one can hear these heart sounds as the 'lub–dup' of the heartbeat, and this is a very useful clinical tool. Sounds other than these clear first and second heart sounds are known as murmurs, and these are often associated with pathology of the **heart valves**. A **systolic murmur** (between $S_1$ and $S_2$) may indicate **mitral regurgitation** (blood flow from ventricle to atrium) or **aortic stenosis** (inadequate opening of the aortic valve). A **diastolic murmur** (heard between $S_2$ and $S_1$) is usually caused by **mitral stenosis**.

The various heart valves can be heard best at different points on the chest wall. For example, the **mitral valve** (closure of which results in $S_1$) can be heard best at the apex of the heart (usually in the mid-clavicular line, in the space between the fifth and sixth ribs), and the **aortic valve** (closure of which results in the 'dup' sound) can be heard best just to the right of the sternal edge, in the space between the second and third ribs. This means that one can try to distinguish between a systolic murmur caused by mitral regurgitation and one caused by aortic stenosis by listening at these two sites. A mitral regurgitation murmur will be louder at the apex than at the right sternal edge, whereas the converse should be true for an aortic stenosis murmur. Nonetheless, definitive diagnosis of these conditions requires echocardiography (see below).

## Doppler and echocardiography

There are a number of techniques based on the use of ultrasound that are useful in the investigation of cardiovascular function. The two most important methods are **Doppler ultrasound**, and **echocardiography**. Doppler ultrasound is a method for using ultrasound waves to detect blood flow in peripheral arteries. This is also the basis of **fetal heartbeat** monitors that are commonly used when expectant mothers visit their doctor; the pulsatile flow of blood through the fetal heart can be detected as early as 10 weeks into the pregnancy. Doppler ultrasound is used as a method of detecting peripheral vascular disease by analyzing the **ankle:brachial pressure index (ABPI)**. In this technique, the amount of pressure that has to be applied via a blood pressure cuff to the brachial artery in order to stop flow is noted. The same is done with the tibial arteries, or the dorsalis pedis artery. These figures give an estimate of systolic blood pressure in the arteries of the ankle and in the brachial artery. Under normal circumstances, systolic blood pressure is usually slightly higher in the ankle than in the arm, so the ABPI is normally $>1.0$. If ABPI $< 0.95$, this indicates some degree of peripheral vascular disease. ABPI can be as low as 0.5 in people with pain in their legs at rest, and an ABPI of $<0.3$ suggests that gangrene and ulceration may be likely.

Echocardiography is a technique that uses ultrasound waves to visualize the heart as it moves during each cardiac cycle, and is similar to the technique used in expectant mothers in order to check for fetal abnormalities and for accurate dating of the pregnancy. With echocardiography, all of the chambers of the heart and the valves in the heart can be visualized. This technique is the definitive method for diagnosis of **valvular heart disease**, **septal defects**, problems associated with abnormally thickened myocardial walls, and **pericardial effusion**. It can also be twinned with Doppler ultrasound so that the speed and direction of blood flow through valves and through septal defects can be analyzed, and can also be useful in visualizing **pulmonary edema** and raised central **venous pressure**.

## Angiography and catheterization

In contrast to the methods for investigating cardiovascular function mentioned above, **angiography** and **cardiac catheterization** are invasive methods. They involve passing catheters (surgical tubes) into the heart and blood vessels under X-ray guidance. These catheters can then be used to introduce radio-opaque dyes into the blood vessels for angiography, or can be connected to pressure transducers to measure blood pressure in the various cardiac chambers in cardiac catheterization.

Coronary angiography is an essential preliminary investigation before **coronary artery bypass grafting**, in order to determine the degree and location of the narrowing in the coronary arteries. With angiography, it is possible to visualize where a given **coronary artery** may be narrowed and/or blocked and, in some cases, to inflate a balloon in an effort to open the narrowed blood vessel and allow more blood to reach that part of the heart. This is a fairly common procedure in patients who have **ischemic heart disease**.

Cardiac catheterization is much less commonly done now than in the past, mainly because it has been superseded by non-invasive echocardiographic techniques. However, if one needs to know what is happening to pressure in cardiac chambers and the aorta (as opposed to the brachial artery) during the cardiac cycle, then catheterization is still the most accurate method for doing this.

# D8 Cardiovascular response to exercise

## Key Notes

**Cardiac response to increased demand**

Cardiac output must increase in response to increased demand of the tissues for oxygen supply and carbon dioxide removal. Since cardiac output is the product of heart rate and stroke volume, increases in either of these two variables can produce increases in cardiac output. Cardiac output can increase by as much as fourfold during exercise, facilitated by an increase in heart rate at low-to-moderate exercise levels and by an increase in both heart rate and stroke volume at heavy exercise levels. During severe exercise, however, stroke volume falls somewhat because of the reduction in diastolic filling time at extremely high heart rates. This, together with a maximal heart rate of 180–200 beats $min^{-1}$, means that cardiac output is maximal at around 20 l $min^{-1}$. Denervation does not markedly affect the ability of the heart to increase output to match demand; hormonal influences and intrinsic heart mechanisms are important.

**Frank–Starling law of the heart**

One of the reasons why the heart is able to match output to demand is encapsulated in the Frank–Starling law of the heart; as ventricular end-diastolic volume (or ventricular fiber length at the end of diastole) increases, this is matched by increased contractility of the ventricular myocardium. During exercise, increased venous return, brought about by limb movements and increased breathing rate, is matched to increased output by this mechanism.

**Vascular response to increased demand**

Coronary blood flow is greatest during ventricular relaxation (diastole), and diastole becomes shortened as heart rate increases. Cardiac myocytes are only capable of aerobic respiration, so during periods of increased heart rate it is important that coronary arterioles are dilated in order that coronary blood flow is able to meet the increased oxygen demands of the cardiac muscle. Such changes in arteriolar diameter are brought about by metabolic factors released from respiring cardiac myocytes. The same is true of skeletal muscle blood flow during exercise. Dilatation of blood vessels supplying the skin is more of an active process involving nervous and hormonal (including local hormonal) influences.

**Isometric versus dynamic exercise**

Increased blood flow to skeletal muscle during exercise is essential, otherwise the muscle must switch over to anaerobic respiration and build up an oxygen debt. Increasing muscle lactate is one factor that leads to fatigue of muscle. If muscles are undertaking isometric exercise in which they contract and remain contracted, blood flow will be suboptimal even though the arterioles supplying the muscle fibers may be dilated. In dynamic exercise, blood flow to exercising muscles is less of a limiting factor than in isometric exercise.

| | | |
|---|---|---|
| **Related topics** | The cardiac cycle, blood pressure, and its maintenance (D6)<br>Utilization of $O_2$ and production of $CO_2$ in tissues (E5) | Respiration in exercise (E7)<br>Properties of skeletal muscle fiber types (I4)<br>Exercising muscle (I5) |

## Cardiac response to increased demand

When we exercise, it is essential that those muscle groups that are active receive an adequate blood supply to meet demand. **Oxygen** and nutrients must be delivered in sufficient quantities, and waste products (**carbon dioxide** and metabolites) must be removed at a fast enough rate. One of the major ways of meeting increased demand is to increase **cardiac output** (CO). CO in a healthy lean male at rest is around 5 l $min^{-1}$. The heart beats around 70 times per minute and ejects around 70 ml of blood per beat. During **exercise**, CO can be increased by increasing either of these variables; during the most extreme forms of exercise, **heart rate** may rise to as much as 200 beats $min^{-1}$ (depending upon age), and CO may be increased to as much as 20 l $min^{-1}$. This means that during the most extreme forms of exercise, **stroke volume** (the volume of blood ejected from the left ventricle per beat) must rise from 70 ml to 100 ml. Thus, in order to increase CO to four times its resting value, heart rate (HR) increases to 2.9 times and stroke volume (SV) to 1.4 times its resting value. It is clear, therefore, that the major contributor to increased CO during exercise is HR. However, the relationship between these various factors changes depending upon the amount of work being done (*Fig. 1*).

During light exercise, almost all of the increased CO is determined by a concomitant increase in HR. During moderate and heavy exercise, both HR and

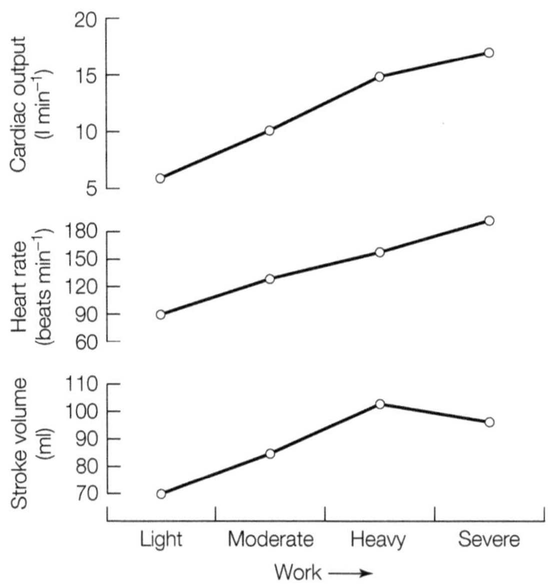


*Fig. 1. Graphs of the effect of exercise of varying severity on cardiac output, heart rate and stroke volume.*

SV are increased, although SV reaches its maximum level with heavy exercise. In the most extreme forms of exercise, HR is maximal, but SV falls off somewhat. This is because as HR increases to nearly three times its resting level, the time for ventricular filling (**diastole**) becomes critically shortened. This causes a reduction in ventricular **end-diastolic volume** and a concomitant reduction in SV. Increases in HR during exercise are accomplished by increased **sympathetic** drive and reduced **parasympathetic** outflow to the SA node, as outlined in Topic D4. Increases in SV are brought about by two mechanisms: increased sympathetic outflow to ventricular muscle (see Topic D4) and increased **venous return**.

When we exercise, blood flow back to the heart from the venous reservoir that exists at rest (about 70% of blood volume) increases. The rhythmic action of the large muscles of the legs (the '**muscle pump**') and of breathing ('the **respiratory pump**', see Topic D5) help to maintain venous return under normal circumstances. During exercise, the increased venous return brought about by the increased **respiratory rate** and by the movement of the limbs must be matched by increased CO, or the heart would become congested. Under normal circumstances, distention of the right atrium by increased venous return results in an increase in heart rate through neural input; this is the basis of **sinus arrhythmia**, the normal variation of heart rate with respiration. However, the heart is able to match venous return to CO very closely, even when the heart has had its nerve supply cut. Thus, the ability of the heart to vary output depending on return must be an integral property of the heart muscle itself.

## Frank–Starling law of the heart

If one were to plot a graph of **stroke volume** against ventricular **end-diastolic fiber length**, the graph would look something like the one in *Fig. 2*. As end-diastolic fiber length (comparable to ventricular end-diastolic volume) increases, stroke volume in the subsequent systole increases also. This is true over a wide range of initial fiber lengths, until fibers are stretched too much, and muscle filaments are stretched so far that they cannot interact to produce cell shortening (see Topic I3). This ability of the ventricles to match stroke volume to end-diastolic volume closely is often referred to as Starling's law of the heart, but there is less confusion in the minds of students between this law and **Starling's forces** (see Topic D5) when the law is given its proper name, the **Frank–Starling law of the heart**. Otto Frank (1865–1944) and Ernest Starling (1866–1927) independently elucidated the physiological mechanisms on which it is based. *Fig. 2* is a Frank–Starling curve or ventricular function curve.

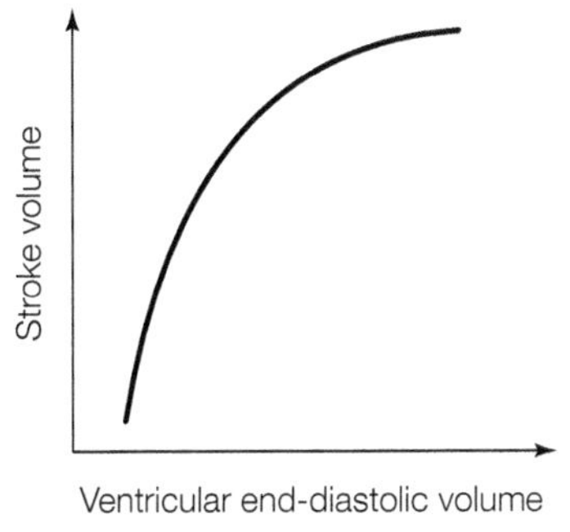


*Fig. 2. A Frank–Starling (ventricular function) curve.*

The Frank–Starling mechanism describes the process whereby right and left ventricular output are matched over a given period of time. If the right ventricle ejects more blood than the left ventricle in any given beat, left atrial pressure (and thereby the filling pressure of the left ventricle) will be greater in the subsequent diastole. This increased left ventricular filling pressure will result in an increase in end-diastolic fiber length in the left ventricle (if ventricular compliance is consistent from one beat to the next). The increased left ventricular end-diastolic fiber length will result in a greater stroke volume in the subsequent beat. If this results in too much blood being ejected, left atrial pressure will drop, and a compensatory reduction in stroke volume will be effected in the next beat. Over the course of several beats, this iterative process ensures that the output of the right and left ventricles is matched.

During moderate-to-heavy exercise, the limbs and respiratory system begin to work harder, venous return increases, and stroke volume increases in an effort to match output to return. Heart rate at this point is probably around 140 beats $min^{-1}$, and diastolic filling time is shortened, but not critically so. However, as heart rate increases to its maximal level with extreme exercise, the heart cannot spend as much time in diastole as would be necessary to accommodate venous return, and SV inevitably falls slightly (*Fig. 1*). Hence, the inability of the heart to increase SV over the entire range of HR is the limiting factor in determining exercise performance, in a cardiac sense. Muscle physiology and respiratory physiology are obviously also important (see Topics E5, E7, I4 and I5).

More details of changes that occur in the cardiovascular system in response to exercise are given in *Instant Notes in Sport and Exercise Physiology*.

## Vascular response to increased demand

During exercise, **cardiac muscle** and **skeletal muscle** are working much harder, consuming up to 6 times and 70 times as much oxygen, respectively, as they do at rest. However, if the blood vessels supplying muscle tissue run through the muscle itself, how is this oxygen delivery achieved, given that the muscles must be squeezing the vessels shut for a significant proportion of the time? The answer to this question is that there exist specific mechanisms that dilate blood vessels in response to requirements in both cardiac muscle and skeletal muscle, although the mechanisms are slightly different in each case.

Let us first consider cardiac muscle and the **coronary vasculature**. As outlined in topic D1, the majority of coronary arterial blood flow takes place during diastole, because the cusps of the **aortic valve** cover the inlets to the **coronary arteries** during **systole**, and the vessels are being squeezed during this phase of the **cardiac cycle**. In view of the fact that diastole shortens markedly during exercise, it should become apparent that the only way in which demand for oxygen in the **myocardium** can be met during exercise is if the vessels dilate. However, cardiac function is not markedly impaired by denervation of the heart, so it is clear that neural mechanisms for **vasodilatation** are not all-important. In fact, the coronary blood supply **autoregulates** its own blood flow. That is to say, in the absence of neural and hormonal input to the heart, the vessels can still dilate or constrict to match supply to demand.

Local metabolic factors predominate; if cardiac muscle is underperfused for just a short period of time, **vasoactive metabolites** such as potassium ions, $H^+$ ions, carbon dioxide and adenosine build up locally and lead to vasodilatation of the coronary arterioles, allowing more of the metabolites to be flushed away. In the absence of significant levels of these metabolites, the coronary arterioles will be slightly constricted, yet capable of vasodilatation. In cases of **myocardial**

**ischemia** (e.g. in **angina**), exercise leads to dilatation of the arterioles, but the larger coronary arteries are blocked and so supply fails to meet demand; these toxic metabolites build up in the heart tissue and cause pain. **Endothelial cells** lining the luminal surface of the blood vessels also release factors that act locally, the most notable of which is the gas **nitric oxide** (**NO**, see *Fig. 3*). NO is synthesized *de novo* from **L-arginine** in response to a number of circulating factors in the blood (e.g. **thrombin**, **bradykinin**, **acetylcholine**), and by **shear stress** on the endothelial cells. As a small lipid-soluble gas molecule, it is able to diffuse readily towards smooth muscle cells, enter them, and activate the enzyme **guanylyl cyclase**. This enzyme catalyzes the formation of **cyclic guanosine monophosphate (cGMP)** from **guanosine triphosphate (GTP)**, and cGMP produces relaxation of the smooth muscle cell. NO has a very short half-life and so its effect is short-lived. Thus it is only produced when required, and only acts for a short period of time. Drugs such as **sodium nitroprusside** and **glyceryl trinitrate (GTN)** are administered in angina because they mimic the effects of NO on smooth muscle cells and produce coronary vasodilatation, relieving pain.

Skeletal muscle receives around 20% of CO at rest and, during exercise, can receive up to 17 times more blood per minute (80% of CO). Thus, some degree of vasodilatation in skeletal muscle must also occur in response to exercise. Although metabolic factors and circulating epinephrine (see Topic D6) play an important role in mediating the increase in skeletal muscle blood flow that is observed during exercise, there is debate about whether neurogenic vasodilatation takes place. Whatever the mechanisms involved, it is clear that if skeletal muscle is receiving 80% of CO and the vascular bed is dilated, then the contribution that skeletal muscle makes to **total peripheral resistance (TPR)** must be high, and that TPR should fall during exercise. As explained in Topic D1, if TPR falls, then **mean arterial blood pressure (MABP)** should fall, since MABP = CO × TPR. However, the changes in TPR during exercise are compensated for by

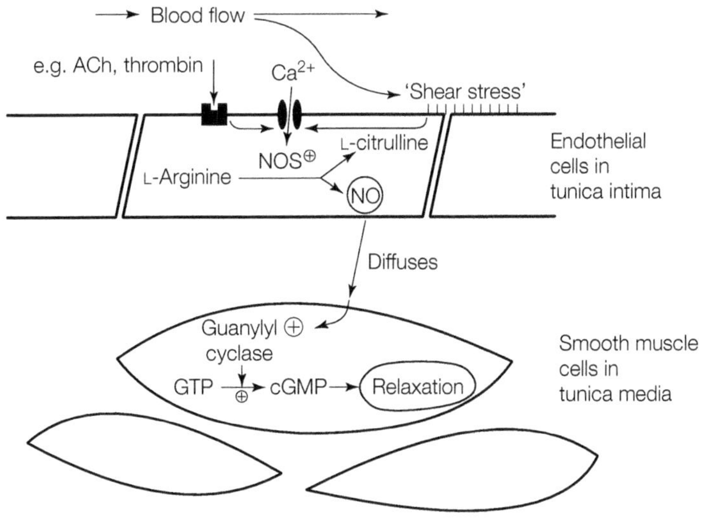


*Fig. 3. Mechanism of nitric oxide-induced vasodilation. ACh, acetylcholine; cGMP, cyclic guanosine monophosphate; GTP, guanosine triphosphate; NO, nitric oxide; NOS, nitric oxide synthase.*

increases in CO, and the resultant effect on MABP is minimal. During moderate exercise, MABP is usually increased by only around 20 mmHg, and during heavy exercise, by only 10 mmHg, as activity-induced vasodilatation begins to predominate. **Systolic blood pressure (SBP)** tends to rise during exercise, and **diastolic blood pressure (DBP)** tends to fall slightly, resulting in a much smaller change in MABP than one might expect.

Finally, in describing the cardiovascular response to exercise, it is important that the blood flow to the skin is mentioned. Although the arterioles feeding the splanchnic and mesenteric vascular beds (gut blood supply) can be constricted in an effort to direct more blood towards exercising muscle, those feeding the skin should be able to be dilated in an effort to lose more heat via that route. Intense exercise may increase the amount of heat being produced by the body by up to 10-fold. An increase in body temperature causes increased outflow of sympathetic nerve impulses to skin blood vessels, where these nerves release acetylcholine (ACh). This ACh has two effects: it promotes sweat production by **sweat glands**, and it causes breakdown of the plasma protein **kininogen** to form bradykinin, a potent NO releaser, and therefore a potent vasodilator (*Fig. 3*). Increased temperature near skin blood vessels can also produce direct vasodilatation of those vessels.

## Isometric versus dynamic exercise

Most fit people can maintain a moderate level of **dynamic exercise** (e.g. cycling, swimming) for at least 10 min. However, it is much more difficult to maintain isometric contractions (e.g. standing with the back against a wall, with the knees flexed at 90°) for anything like that period of time. Try to do this for just 2 min, and see how difficult it is. This is principally because in **isometric exercise**, the lack of rhythmic contraction and (more importantly) relaxation means that muscle blood vessels are constantly being squeezed, limiting blood flow. So although HR, CO and MABP increase, the muscles rapidly build up an **oxygen debt**, and their contraction fails. *Figure 4* shows the results of an experiment comparing the cardiovascular response to a few minutes of isometric exercise versus 10 min of moderate dynamic exercise.

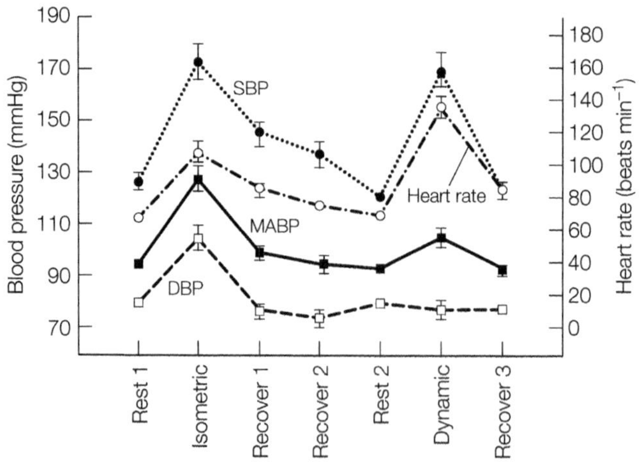


*Fig. 4. Graph of the response of DBP (diastolic blood pressure), MABP (mean arterial blood pressure), SBP (systolic blood pressure) and heart rate before, during, and after periods of isometric and dynamic exercise. The error bars are standard errors.*

There is an anticipatory (cognitive) response before exercise starts, where HR is elevated above normal resting levels. Both isometric and dynamic exercise produce increases in HR, SBP and MABP. However, isometric exercise also has a significant effect on DBP, and consequently a slightly larger effect on MABP. The differences can be explained by the nature of the two types of exercise:

- in dynamic, rhythmic exercise, the blood vessels contributing to TPR in active muscle groups are able to dilate in order to meet demand. Thus the effect of the increased HR and stroke volume are somewhat cancelled out by an overall reduction in TPR.
- in isometric exercise, the resistance vessels in the exercising muscle may well be dilated, but the larger vessels feeding them are most likely squeezed closed by the contracting muscle, which stays in position. Thus the effect on DBP and MABP is more marked (until the subject stops exercising), and blood flow to the exercising muscle is restored.

# E1 LAYOUT AND STRUCTURE OF THE RESPIRATORY TRACT

## Key Notes

**Introduction**

The respiratory tract is designed to ensure that oxygen is delivered to the blood and that carbon dioxide is removed from the blood. The process of breathing is distinct from that of (cellular) respiration. Breathing facilitates gas delivery, whereas respiration uses oxygen and produces carbon dioxide via metabolic processes. There are two lobes in the left lung and three lobes in the right, each of which is ventilated via a system of air passages known as the tracheobronchial tree. Inhalation of air is caused by contraction of the muscles of respiration (normally the diaphragm and intercostal muscles), but although breathing is an active process, it is controlled by regions of our brain that respond to the concentration of various gases in our bloodstream.

**The tracheobronchial tree and airway function**

The tracheobronchial tree is composed of the trachea, bronchi and bronchioles. These parts of the airways are known as the conducting portion, whereas exchange of gases between the inhaled air and the bloodstream takes place in the exchange portion of the airways, the respiratory bronchioles, and alveoli. The airways are held open by connective tissue in their walls, and inhaled air is moistened and warmed by secretions produced by epithelial cells.

**Epithelium, smooth muscle, and cartilage**

Epithelial tissue in the airways takes two main forms: ciliated pseudostratified columnar and simple cuboidal. Goblet cells and other epithelial cells arranged into simple glands produce mucus and watery secretions, respectively. Cilia help to beat mucus containing trapped bacteria and dirt from the lungs back into the pharynx. In the exchange portion of the airways, the epithelial cells are squamous (flattened); this means that the distance for diffusion of gases between alveolar air and blood and vice versa is small. Some parts of the airways have smooth muscle in the wall. This smooth muscle can contract or relax in response to neural and other input, and can have a marked effect on air movement in and out of the lungs. Bronchospasm, for example, is common in asthma. The larger airways such as the trachea are well endowed with cartilage; this helps keep the upper airways open to facilitate airflow.

**Goblet cells and glandular epithelium**

Goblet cells in the airway epithelium produce mucus that acts to trap bacteria and dirt that have been inhaled. Other epithelial cells are arranged into glandular tissue that produces more watery (serous) secretions. Proper function of airway epithelial cells is important in preventing lung infections.

**The role of exocrine secretion**

Mucous and serous secretions in the airway form a moist, sticky barrier to pathogens, and also moisten the air that enters the lungs. Moistening the air is very important in order to maximize gas exchange at the alveolus. It is important that enough serous secretion is produced; when this doesn't occur, for example in cystic fibrosis, the mucus becomes excessively thick and sticky. This means that it cannot be cleared from the lungs by the ciliary escalator, and lung infections can be common.

**Alveolar structure**

Exchange of gases between the alveolar air and the bloodstream (and vice versa) takes place across a thin, moist, large surface area in the terminal portions of the tracheobronchial tree. Here, squamous type I pneumocytes, interstitial space, and the endothelial cells of pulmonary capillaries form a barrier that is only a few micrometers thick. Cuboidal type II pneumocytes secrete surfactant that is important in preventing collapse of the alveoli. Elastic protein fibers are abundant, and these cause lung tissue to 'spring back' at the end of inhalation. For this reason, exhalation is normally a much more passive process than inhalation.

**Related topics**

Epithelia and connective tissue (C2)
The mechanics of breathing (E2)
Alveolar exchange and gas transport (E4)
Chemoreceptors and respiratory acid–base balance (E6)
Innervation of smooth muscle (I7)

## Introduction

The purpose of the respiratory tract is to ensure that adequate amounts of oxygenated air come into close enough contact with adequate amounts of blood, so that **oxygen** can be taken up by the blood and **carbon dioxide** can be exhaled. It is important to realize the distinction between **breathing** and **respiration**, although often these two terms are used interchangeably. In this text, the term 'breathing' refers to the work done by the respiratory system to ensure adequate ventilation of the lungs with air and adequate removal of stale air and carbon dioxide. 'Respiration' is used to refer to the cellular processes that utilize oxygen, and produce carbon dioxide and other waste products. By this method, all of the cells of the body convert foodstuffs like carbohydrate to carbon dioxide and water, and produce the cell's main storage chemical for energy, **adenosine triphosphate (ATP)**. This is covered in more detail in Topics B3 and E5.

The lungs are composed of several lobes, three in the right lung (upper, middle and lower), and two in the left lung (upper and lower). Each of these lobes is supplied with air by its own tube, the **lobar bronchus**. The chest cavity is separated from the abdominal cavity by a sheet of muscle known as the **diaphragm**. Normal quiet breathing is often called '**diaphragmatic breathing**'. When the diaphragm contracts and moves downwards, the reduced pressure in the chest cavity (**intrathoracic pressure**) causes air to rush into the lungs from the atmosphere. When the diaphragm relaxes, intrathoracic pressure rises and air leaves the lungs. Other muscles, such as those between the ribs (**intercostal muscles**), in the back (e.g. **quadratus lumborum**), and in the neck (e.g. **sterno-cleidomastoid**) can be used to aid breathing. As such, these muscles are often collectively termed the **'accessory' muscles of breathing**.

All breathing is dependent upon muscle contraction and is therefore an active process. Although we are not generally aware of 'deciding' to breathe, breathing is voluntary in that we can choose to hold our breath for a period of time. However, if we hold our breath for a long period of time, there will come a point (the '**break point**') at which our brain overrides this voluntary control. In general, the brain controls breathing in such a way as to maintain the oxygen content of arterial blood at a relatively high level, and the carbon dioxide content of arterial blood at a relatively low level. These mechanisms are explained in more detail in Topics E2 and E6.

## The tracheobronchial tree and airway function

When we inhale, air passes from the mouth or nose into the **larynx** and then into the **trachea** (which can be palpated in the front of the throat). The trachea bifurcates into the **bronchi** that supply air to both lungs, and again subdivides into the bronchi that supply the various lobes of the lungs. The bronchi further subdivide into **bronchioles**, tubes of even smaller diameter. Similar divisions of the **tracheobronchial tree** supply all of the lung tissue with air. The lobes of the lungs are each composed of 2–5 **bronchopulmonary segments**, each supplied by its own bronchus and branch of the **pulmonary artery** and drained by its own tributary of the **pulmonary vein**. In this way, all of the lung tissue is supplied with air and pulmonary arterial blood (which is deoxygenated), and drained of pulmonary venous blood (which is oxygenated). In circumstances where only a few bronchopulmonary segments are affected by disease, surgical removal of only those segments may be possible.

In order for normal breathing to take place, it is obvious that the airway must not be obstructed. This can occur when parts of the upper respiratory tract (e.g. the **pharynx**) become congested, closing off the supply of air. Equally, an inhaled foreign body can block one or more bronchi and limit air supply. In order to prevent the airways collapsing, their walls have features that help keep them open. As air passes into the lungs, it is warmed and moistened by secretions produced by the secretory **epithelial tissue** in the airways. The distinct structural features of the walls of the airways are covered in the next few paragraphs.

## Epithelium, smooth muscle, and cartilage

If one takes a section of the airway and slices it in cross-section so that the specimen is circular, one would be able to make an educated guess as to its source just by examining the extent and type of **epithelium**, **smooth muscle**, and **cartilage** in the wall. The general arrangement is always the same (*Fig. 1*): epithelium lining the lumen, submucous tissue containing glandular tissue below, smooth muscle, and then cartilage. However, the amount of each type of tissue and its specialization changes as one goes up the tracheobronchial tree (*Table 1*).

There are two main types of epithelium in the airways: pseudostratified ciliated columnar, and simple cuboidal (see Topic C2). The pseudostratified columnar epithelium has **cilia** so that these small hair-like structures can beat

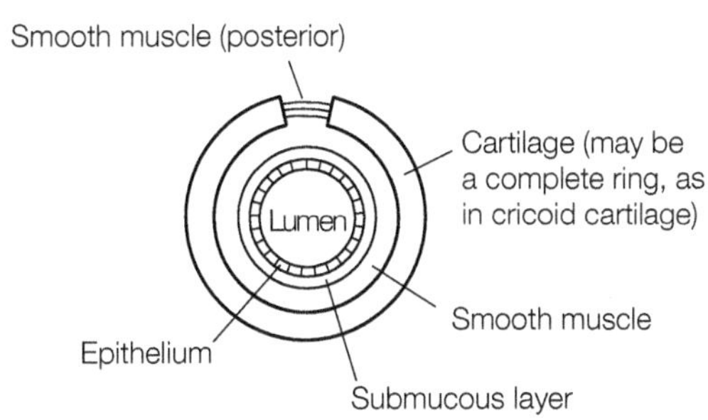


*Fig. 1. Schematic diagram of the arrangement of tissue layers in the airways.*

*Table 1. Composition of the wall of the airway in various parts of the tracheobronchial tree*

| | Trachea | Bronchi | Bronchioles | Alveoli |
|---|---|---|---|---|
| Epithelium | Pseudostratified ciliated columnar | Pseudostratified ciliated columnar | Pseudostratified ciliated columnar | Simple squamous |
| Goblet cells | Numerous | Numerous | Scattered | Absent |
| Glands in submucosa | Numerous | Numerous | Absent | Absent |
| Smooth muscle | Extensive | Extensive | Present | Only a few fibers |
| Cartilage | Present – incomplete rings | Present – plates | Absent | Absent |

the mucus being produced by epithelial cells upwards towards the trachea. Dust and bacteria are trapped in the mucus and it is cleared from the lungs by this **'ciliary escalator'**. There is a great deal of glandular tissue in the epithelium and submucous layer. **Goblet cells** in the epithelium produce mucus, and serous (watery) secretions are produced by more extensive epithelial tissue. Neither of these types of epithelium are especially well suited to the exchange role of the respiratory system; that role is taken on by the flattened (squamous) simple epithelium in the **alveoli** (the blind-ended sacs at the top of the tracheobronchial tree).

Smooth muscle in the walls of the airways is innervated by both the **parasympathetic** and **sympathetic** branches of the **autonomic nervous system**. When more ventilation is required (e.g. during exercise), outflow of the sympathetic nervous system and circulating **epinephrine** cause smooth muscle relaxation (**bronchodilation**). **Norepinephrine** and epinephrine act primarily on **$\beta_2$-adrenoceptors** in airway smooth muscle to increase the levels of the second messenger **cyclic adenosine monophosphate (cAMP)**. This cAMP causes relaxation of the smooth muscle (see Topic I6). **Acetylcholine** released from parasympathetic nerve endings acts on **muscarinic receptors** to reduce intracellular cAMP levels, thus causing **bronchoconstriction**. Smooth muscle in the walls of the airways can cause problems; spasm of bronchial smooth muscle can limit airflow, causing the symptoms of asthma. The symptoms of asthma can be relieved by inhalation of drugs that stimulate $\beta_2$-adrenoceptors or that block muscarinic receptors.

In the trachea, we can feel the cartilage, which exists in almost a complete ring. It is only absent at the back of the trachea, where the **esophagus** passes posteriorly. This allows food to pass down the esophagus unhindered, yet gives the trachea great structural integrity; the cartilage is strong and does not allow the trachea to be kinked and blocked. As one goes up the tracheobronchial tree, the cartilage becomes more sparse (arranged as plates rather than rings), until it runs out entirely by the time one has reached the level of the bronchioles. In fact at this level, it is important that the bronchioles are able to be collapsed, since this allows more air to be expelled during forceful exhalation.

## Goblet cells and glandular epithelium

These cells are so-called because, on tissue sections, they often resemble wine goblets (*Fig. 2*). This is because during processing of the tissue for histology, the mucus has been dissolved away, leaving behind a space that does not take up stain. The structure of glandular tissue is more complex; although the cells from

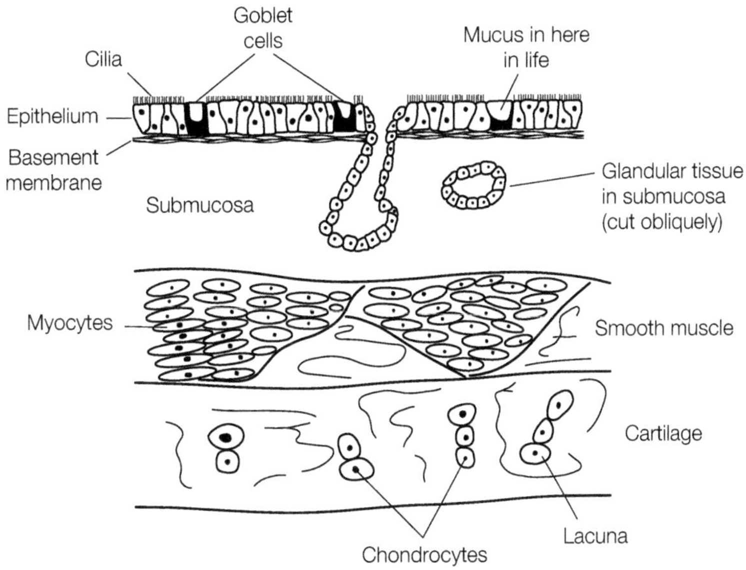


*Fig. 2. Diagram of the wall of the airway (trachea), showing goblet cells, glandular tissue in the submucosa, smooth muscle, and cartilage.*

which it is composed are in contact with the outside world and are epithelial, this is often not clear on tissue sections (*Fig. 2*). This is because histological slides are a two-dimensional slice through what is a three-dimensional structure; some elements will have been sampled at an oblique angle. The function of goblet cells and the other cells that make up airway epithelium is critical to respiratory function.

There are numerous other cell types (e.g. **mast cells** that produce **histamine**) in the airways that influence the production of mucous secretions by epithelial cells. Often, these cells are activated by some sort of allergen and stimulate production of secretions as well as constriction of bronchial smooth muscle. It is this exaggerated allergic response to certain triggers that characterizes **asthma**.

**The role of exocrine secretion**

Anyone who has ever suffered from the common cold understands the quantity of secretions that can be produced by the respiratory tract. All of these secretions are produced by epithelial cells lining the airways and have the purpose of allowing us to interact with the outside world in a healthy fashion. For example, mucus is secreted by specialized epithelial cells known as goblet cells with the express intention of trapping foreign particles that could cause lung infections if they are not prevented from entering the lungs, or if they are not cleared by the ciliary escalator. More watery, serous secretions are produced by other glandular epithelial tissue in the airways, and these serous secretions and the mucus mix to form a relatively thin epithelial coating that can be moved easily by the ciliary escalator.

Atmospheric air has relatively low water vapour content; the purpose of the serous secretions is to moisten the air as it passes down into the alveolus from the atmosphere, so that gas exchange is facilitated across the moist alveolar membrane. The amount and composition of these airway secretions can be affected by a number of different influences. For example, the production of serous secretions is increased by sympathetic nervous system activation and activation of β-adrenoceptors on glandular epithelial cells. This is a sensible response because under conditions where the sympathetic nervous system is activated, ventilation of the lungs will also be increased; if the production of secretions did not also increase, the lungs would start to dry out. This serous secretion is formed by the movement of ions across the cell membrane and the osmotic movement of water to match ion content. The serous secretion is generally thin and watery. One of the important proteins controlling the movement of ions in respiratory epithelium is called **CFTR (cystic fibrosis transmembrane conductance regulator)**. In **cystic fibrosis (CF)** the function of this protein is impaired, and this means that the secretions in the airways become thicker and sticky. This makes it difficult for the cilia to move the secretions up towards the pharynx, causing frequent lung infections in people with CF. Physiotherapy and inhaled drugs can be used in an effort to clear this thick mucus from the lungs of CF patients, but even so, health problems are common.

## Alveolar structure

Exchange of gases between the blood and the air in the lungs takes place in the alveoli. These are blind-ended sacs at the top of the tracheobronchial tree. The last few bronchioles may have a few alveoli which project from their walls. Thus these are often called '**respiratory bronchioles**' and are part of the respiratory component of the airways. The airways up to this point are known as the conducting portion of the airways. Alveolar structure is perfectly designed for exchange; the total surface area for exchange is around 70 $m^2$, roughly the size of a tennis court, and alveolar air and blood are separated by a barrier only a few thousandths of a millimeter thick (*Fig. 3*). Two distinct types of epithelial cells make up the alveolar walls; squamous **type I pneumocytes**, and cuboidal **type II pneumocytes** (which secrete the **surfactant**, see below). The squamous type I pneumocytes are separated from the **endothelial cells** that make up the walls of the pulmonary capillaries by a very thin layer of interstitial fluid, and the endothelial cells themselves are flattened. The **pulmonary capillaries** are of small diameter; this means that erythrocytes must 'file through' and come into very close contact with the walls of the capillary (rather than sitting in the center). All of these features are important in gas exchange.

In the alveoli, there are approximately equal numbers of both of type I and type II pneumocytes, but type I pneumocytes make up around 70% of the surface area of the alveoli. The type I pneumocytes are flattened (squamous) epithelial cells that are involved in the exchange of gases between the alveolar air and the bloodstream. As mentioned above, the distance for diffusion of gases between the blood and the alveolar air (and vice versa) is normally no more than a few micrometers. Note (*Fig. 3*) how thin the layer of interstitial fluid is between the type I pneumocytes and the capillary endothelial cells. If this interstitial space becomes thickened (e.g. in **pulmonary fibrosis**), this can adversely affect gaseous exchange and lead to inadequate oxygen delivery to the body tissues.

The type II pneumocytes, on the other hand, are cuboidal in shape and are therefore not specialized for exchange. Their role is to secrete a substance known as surfactant that helps to lower the **surface tension** of the alveolar walls. If aqueous solvents are placed on a clean, dry surface, the solvent will tend to aggregate together, much like a few milliliters of water spilled on a glass surface. The same is true in the alveoli, where the secretions tend to pull

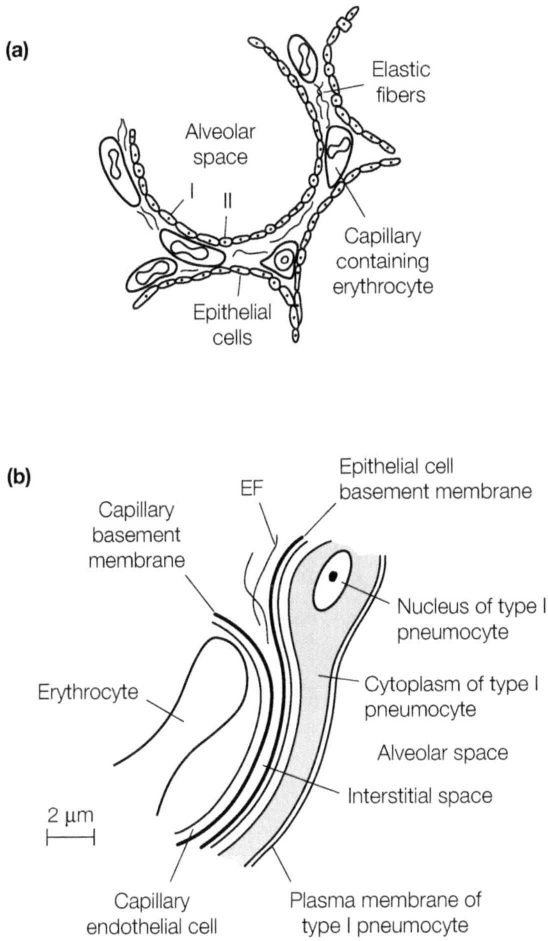


*Fig. 3. Structure of the alveolar wall at low power (a) and high power (b). I, type I pneumocytes; II, type II pneumocytes; EF, elastic fibers.*

together because of surface tension; this means that the liquid tends to collapse the alveoli. Surfactant is principally made up of **dipalmitoylphosphatidylcholine (DPPC)** and other phospholipids, and acts like a detergent. By lowering surface tension, pulmonary surfactant helps to keep the alveoli open on exhalation, increasing the compliance of the alveolar walls, thereby reducing the amount of work that needs to be done by the muscles of inspiration in order to inflate the lungs. In some premature babies (below about 35 weeks' gestation), the type II pneumocytes have not yet fully developed and their lungs do not have sufficient surfactant. This can mean that premature babies may have **respiratory distress syndrome**, in which the child has to struggle to inflate his/her lungs with each breath, and may appear blue (**cyanotic**). In premature babies, this is a dangerous condition and is usually treated by some form of mechanical ventilation.

The fine elastic protein fibers are another important structural feature of the walls of the alveoli. When the diaphragm relaxes, air rushes out of the lungs and is helped on its way by the elastic recoil of the alveolar walls. In diseases where the **elastic recoil** of the alveolar walls is reduced, exhalation is difficult and a large amount of air can be left in the lungs after each breath. This then means that a large amount of stale air is present in the lungs at any one time, which can cause problems with oxygen delivery to the alveoli, and consequently to the tissues.

# E2 THE MECHANICS OF BREATHING

## Key Notes

**Pleural membranes**

Two pleural membranes sit between lung tissue and the chest wall. The visceral pleura is tightly adherent to lung tissue and the parietal pleura is adherent to the chest wall. Between the two membranes there exists a potential space, the pleural space. This space is normally very small and is filled with a serous secretion that serves to lubricate the two membranes so that the lungs can move within the chest cavity without being damaged. The pressure in the pleural space is slightly lower than atmospheric pressure, because of the pulling action of the highly elastic lung tissue on the visceral pleura.

**Gas pressures**

Each gas in a mixture of gases contributes to the pressure exerted by that gas mixture. In atmospheric air at sea level, for example, 21% of the air is oxygen ($O_2$), meaning that although atmospheric pressure is 760 mmHg, the $O_2$ partial pressure ($P\text{O}_2$) is 160 mmHg ($0.21 \times 760$ mmHg). Alveolar air has different $P\text{O}_2$ from atmospheric air.

**Relative pressures**

Boyle's law states that the pressure of a gas is inversely proportional to volume. At a constant temperature, if volume increases, pressure will decrease. As lung volume increases, pressure in the thoracic cavity decreases and air enters the lungs, driven by the slightly higher atmospheric pressure. The pressure in the thoracic cavity is said to be negative, relative to atmospheric pressure. It is these pressure differences that cause air to enter and leave the lungs during the breathing cycle. Pleural pressure (see above) is lower than atmospheric pressure and alveolar pressure by around 5 $cmH_2O$. This keeps the lungs inflated slightly at rest. If the pleural membranes are ruptured (pneumothorax), the underlying lung tissue can collapse.

**Alveolar size and compliance**

Pleural pressure is slightly lower in the top of the lungs that at the bottom, because of gravity. Alveoli at the top of the lungs are more open than those at the bottom, but this actually means that they receive a lower percentage volume of fresh air than the alveoli at the bottom with each normal breath. However, the situation is different for different breathing patterns.

**Related topics**

Basic structure of the body (C1)
Investigating lung function (E3)
Alveolar exchange and gas transport (E4)

## Pleural membranes

As mentioned in Topic C1, the lungs are surrounded in the chest by two membranous layers, derived from embryonic mesothelium. These membranes, the pleural membranes, allow the lungs to move inside the chest, but prevent damage through friction because the space between the two membranes (the pleural space) is filled with fluid that acts as a lubricant (*Fig. 1*). Try to imagine two sheets of glass with a thin layer of fluid between; the two sheets of glass will move smoothly over one another and will also tend to stick together, defying gravity. The inner layer of pleural membrane is known as the visceral pleura and is tightly adherent to lung tissue. The outer layer of pleural membrane is known as the parietal pleura and is tightly adherent to the inside of the chest wall. Given that the small space between the two layers of pleural membrane is occupied by the lubricating fluid, it is easy to see why expanding the chest wall would naturally lead to a degree of expansion of the lung tissue. When we breathe in, the ribs move upwards and outwards, expanding the lung tissue.

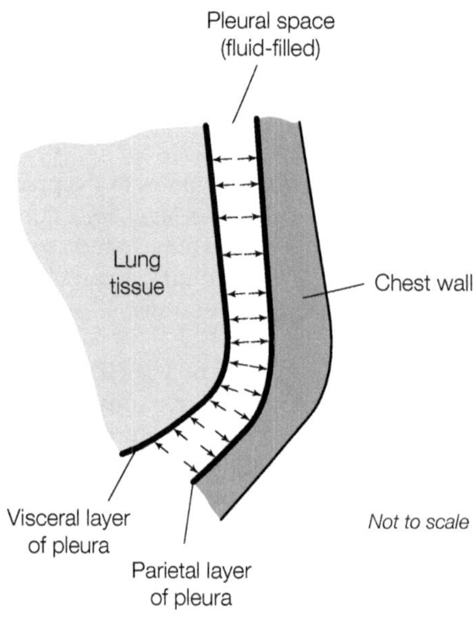


*Fig. 1. Schematic diagram of the arrangement of the pleural membranes, lung tissue, and the chest wall.*

One other important point to note is that because the lung tissue contains elastic protein fibers, it 'recoils' away from the chest wall. This, combined with the 'pulling' action of the chest wall on the parietal pleura, results in a slightly lower pressure in the pleural space than in the alveoli at rest.

## Gas pressures

The atmospheric air that we breathe exerts a given level of pressure upon us, known as atmospheric or barometric pressure ($P_B$). When we hear about high and low pressure in weather reports, this is referring to $P_B$. At sea level, $P_B$ is approximately 760 mmHg (SI units: 101 kPa; 7.5 mmHg = 1.0 kPa), and each of the gases in atmospheric air exert some of that pressure (their individual gas pressure). Dalton's law states that $P_B$ is equal to the sum of the partial pressures of each of the gases in atmospheric air:

$$P_B = PN_2 + PO_2 + PCO_2 + PH_2O + P_{other}$$

Atmospheric air at sea level contains 21% $O_2$ and 79% $N_2$ (levels of $CO_2$ and $H_2O$ vapour are negligible, as are those of the various noble gases), which means that $P\text{O}_2$ is 160 mmHg (0.21 × 760 mmHg) and $P\text{N}_2$ is 600 mmHg (0.79 × 760 mmHg). These values represent the partial pressures (also known as gas tension) that each gas would exert if all of the others were removed from atmospheric air. As air enters the lungs, it is warmed and moistened, and some mixing with stale air takes place. This results in a change in the various partial pressures of the gases, so that in alveolar air $P\text{O}_2$ and $P\text{N}_2$ are lower (102 and 571 mmHg, respectively), and $P\text{CO}_2$ and $P\text{H}_2\text{O}$ are higher (40 and 47 mmHg, respectively).

**Relative pressures**

Boyle's law states that, at constant temperature, the pressure of a gas ($P$) is inversely proportional to the volume ($V$) that it fills:

$$P \propto 1/V$$

Thus, if pressure or volume change and temperature remains the same, the arithmetic product of pressure and volume will remain constant:

$$P_1V_1 = P_2V_2$$

In general, the temperature in the lungs does not change significantly from one moment to the next, so if lung volume increases, pressure in the chest cavity will decrease. These changes in pressure result in air entering the lungs from the atmosphere because atmospheric pressure is then higher than alveolar pressure by around 1.5 mmHg. In view of the fact that the changes in pressure brought about by inhalation and exhalation are so small, they are generally expressed in centimetres of water ($\text{cmH}_2\text{O}$); 1 $\text{cmH}_2\text{O}$ is equal to 0.74 mmHg. When referring to pressure changes in the lungs during breathing, we usually refer to relative pressures. In this system, $P_B$ is set at 0 $\text{cmH}_2\text{O}$; a pressure of –5 $\text{cmH}_2\text{O}$ means that the pressure is 5 $\text{cmH}_2\text{O}$ lower than $P_B$, and a pressure of +5 $\text{cmH}_2\text{O}$ means that the pressure is 5 $\text{cmH}_2\text{O}$ higher than $P_B$. *Figure 2* shows the important relative pressures that one must consider when thinking about pressure changes during inhalation and exhalation, and how these change over the course of one breathing cycle.

As described above, when no air movement is taking place, alveolar pressure ($P_A$) is 0 $\text{cmH}_2\text{O}$. However, because the pressure in the pleural space ($P_{pl}$) is lower than that in the alveolus (see above), transpulmonary pressure ($P_L$) is +5 $\text{cmH}_2\text{O}$. $P_L$ is therefore a 'distending pressure', serving the function of keeping the lungs inflated slightly, even at rest. $P_{pl}$ can fall to –8 $\text{cmH}_2\text{O}$ during normal breathing, resulting in airflow into the lungs. This results in a slightly negative $P_A$ (less than 1 $\text{cmH}_2\text{O}$ lower than $P_B$). As the diaphragm relaxes and the chest wall moves in and down, $P_{pl}$ returns to –5 $\text{cmH}_2\text{O}$, $P_A$ rises to around +1 $\text{cmH}_2\text{O}$, and air is forced out of the lungs. When $P_A$ returns to 0 $\text{cmH}_2\text{O}$ (i.e. $P_A = P_B$), airflow stops.

**Alveolar size and compliance**

One factor that has not been discussed thus far is the influence of gravity on lung pressures. When standing, gravitational force has a greater influence on those parts of the lungs that are closer to the ground. This actually means that $P_{pl}$ is variable, rather than being –5 $\text{cmH}_2\text{O}$ uniformly throughout the lungs. $P_{pl}$ is actually higher than at the bottom (base) of the lungs (–3 $\text{cmH}_2\text{O}$) than at the top (apex) of the lungs (–10 $\text{cmH}_2\text{O}$). This means that it is easier for alveolar pressure to keep the alveoli open at the apex than at the base. During normal breathing, the alveoli at the apex are more compliant, but the magnitude of

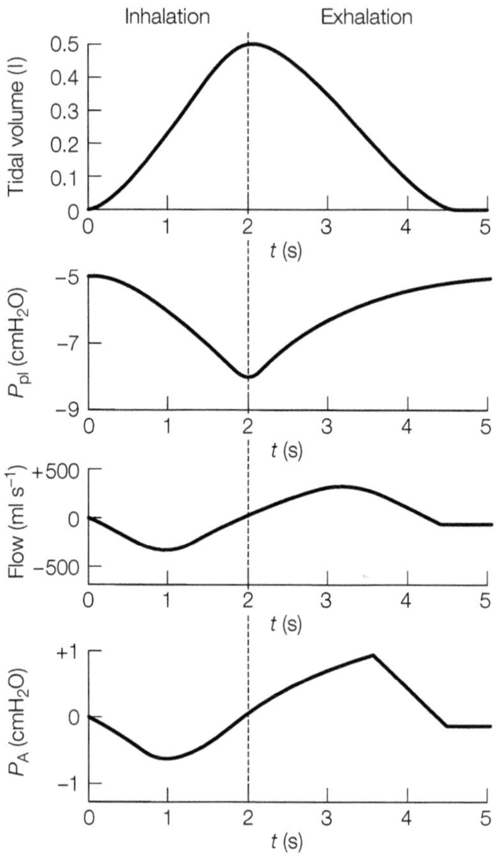


*Fig. 2. Changes in pressures ($P_{pl}$, pleural pressure; $P_A$, alveolar pressure), airflow, and lung volume during one cycle of normal breathing.*

volume change possible is lower than for those at the base. Thus, during quiet breathing, the apex receives less ventilation than the base. However, at low lung volumes (after a full exhalation), $P_{pl}$ at the base may actually be positive (around +3 cm$H_2O$), and the alveoli may be collapsed; as the chest expands, air does not enter the alveoli at the base until $P_{pl} < 0$ cm$H_2O$. This means that, in contrast to quiet breathing, the apex receives more ventilation than the base. The secretion of surfactant by type II pneumocytes (see Topic E1) helps to equalize these regional variations in ventilation. As surface tension of the alveoli is reduced by surfactant, this helps to prevent alveolar collapse (see Topic E1).

# E3 INVESTIGATING LUNG FUNCTION

## Key Notes

**The normal ventilatory pattern**

Total lung volume is influenced by sex, physical size and age. During quiet breathing, we actually only use a small proportion (10–20%) of our lung volume. Significant extra volumes of air can be inhaled (inspiratory reserve volume) or exhaled (expiratory reserve volume) if we fill our lungs to capacity. Residual volume is the amount of air left in the lungs after we have exhaled fully; the lungs can never be emptied completely.

**Lung function tests**

Lung function tests fall into three main categories: peak flow tests, vitalograph tests, and flow–volume tests. Each of these tests can be used to examine different components of airflow and/or lung volume, and are normally used in a clinical setting to investigate patients with respiratory diseases such as asthma, pulmonary fibrosis or chronic obstructive pulmonary disease.

**Obstructive versus restrictive lung disease**

Many lung diseases can be categorized as obstructive (in which airflow is limited but lung volume is relatively normal) or restrictive (in which lung volume is low but airflow is relatively unaffected). Peak flow tests can detect obstructive pulmonary disease but cannot distinguish between the various forms. Vitalograph tests can detect both restrictive and obstructive lung disease, but are not generally able to distinguish between the various forms of obstructive lung disease. Flow–volume tests can distinguish between almost all of the various forms of lung disease.

**Related topics**

Layout and structure of the respiratory tract (E1)

The mechanics of breathing (E2)

## The normal ventilatory pattern

The **total lung capacity** of a healthy 70 kg man is around 6 l (*Fig. 1*). However, 6 l of air are not moved in and out of the lungs in any breath that we normally take, whether at rest or during exercise. The average amount of air that is moved in and out of the lungs in any breath during normal quiet breathing (called the **tidal volume**) is around 0.5 l, and because much of the airways are conducting airways (not involved in exchange), it is rare for more than 65% of this tidal volume to be involved in gas exchange in any one breath. This 35% 'dead space' in the airways is known as **anatomical dead space**. If we inhale maximally, we may fill the lungs with a further 3 l (the **inspiratory reserve volume**), and if we exhale fully, we can also expel a further 1.2 l of air (the **expiratory reserve volume**). However, even with maximal exhalation, we are unable to empty our lungs totally. The volume that remains (1.2 l) is the **residual volume**. Hence, our total lung capacity is made up of residual volume + expiratory reserve volume + tidal volume + inspiratory reserve volume

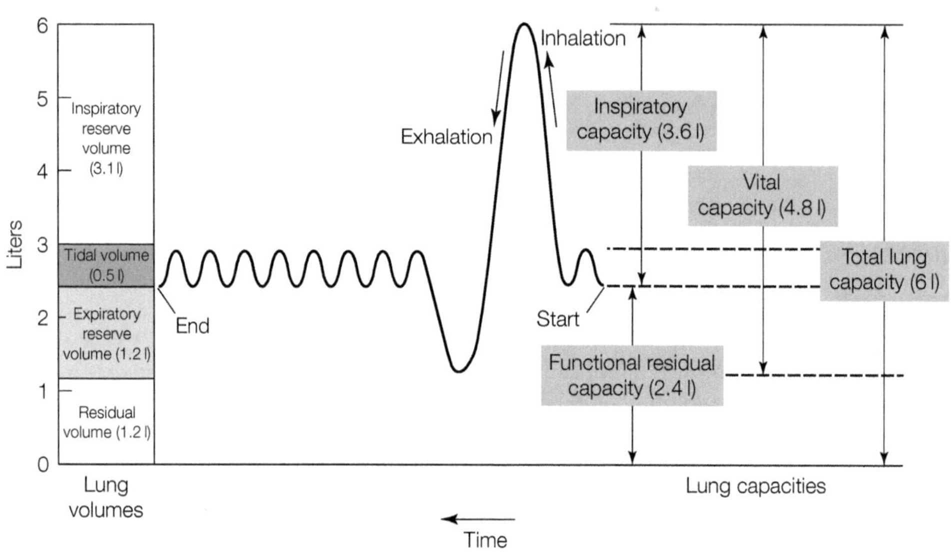


*Fig. 1.* *Spirogram showing the various lung volumes and the related lung capacities.*

(1.2 + 1.2 + 0.5 + 3.0 l = 5.9 l). The relationship of these various lung volumes to one another is shown in *Fig. 1*.

A **spirogram** is shown in *Fig. 1*. These tracings may be done in a hospital respiratory medicine department in the investigation of lung capacities in patients with respiratory disease. The only one of these volumes that cannot be measured by **spirometry** is residual volume, and this must be calculated by other methods. However, all of the other volumes and capacities can be measured using this simple test. The **vital capacity (VC)** of the lungs is equivalent to the sum of the inspiratory reserve volume, the tidal volume, and the expiratory reserve volume. In a 70 kg man, this figure is usually between 4 and 5 l. In diseases where the vital capacity is reduced (e.g. by scarring and stiffening of the alveoli in fibrotic lung disease), this can fall markedly, to around 2 l in some cases. If we want to know more about a person's ventilatory function, however, we must use other laboratory methods to calculate the relevant parameters.

## Lung function tests

Although it is not the purpose of this textbook to help teach clinical medicine, it is worthwhile exploring clinically relevant lung function tests as these are often the best way of explaining what is happening in the airways under normal circumstances. Three main lung function tests are used to investigate ventilatory function. These are:

- peak flow measurements
- vitalograph
- flow–volume loops.

**Peak flow** measurements are the simplest of the three investigations to perform. In this test, the subject is asked to inhale fully and then to exhale as quickly as they can into a peak flow meter (usually a simple plastic tube with a sliding meter that indicates the maximal speed of exhalation). For a healthy 70 kg man, peak flow measurements are usually between 400 and 500 l $min^{-1}$. In diseases where the airways are constricted in some way (e.g. asthma), peak flow measurements may be substantially lower than this (100–200 l $min^{-1}$). This test is so simple to perform that subjects are usually able to take the **peak flow meter** home with them and monitor their peak flow measurements several times a day. In this way, clinicians may be able to notice a pattern in changes in peak flow (e.g. worse in the morning than in the afternoon, or worse when exposed to pets) that might help in treatment of the condition.

A more complicated ventilatory function is **vitalograph** testing (*Fig. 2a*). In vitalograph testing, the subject is asked to inhale fully and to then breathe out as quickly as possible, for as long a time as possible. The vitalograph then plots a graph of total volume exhaled against time. Two main measures can be obtained from vitalograph traces: **forced vital capacity (FVC)**, and **forced expiratory volume in 1 second ($FEV_1$)**. FVC is a measure of vital capacity and $FEV_1$ indicates how much air can be exhaled with maximal effort in the first second of the

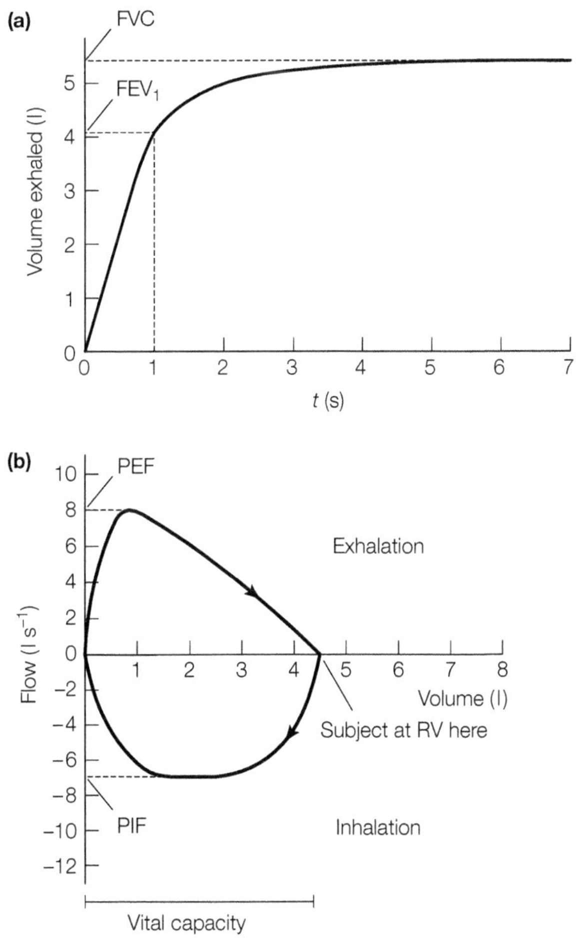


*Fig. 2. (a) Vitalograph testing; (b) flow–volume loop testing. $FEV_1$, forced expiratory volume in one second; FVC, forced vital capacity; PEF, peak expiratory flow; PIF, peak inspiratory flow; RV, residual volume.*

exhalation. In a healthy 70 kg man, FVC should be around 4.5 l and $FEV_1$ should be around 3.7 l. In addition, the ratio of **$FEV_1$/FVC** is an important measure of ventilatory function. An $FEV_1$/FVC ratio of less than 0.75 is indicative of some type of airway obstruction.

A third, still more complicated ventilatory function test is **flow–volume loop** testing (*Fig. 2b*). In these tests, the subject is asked to perform maximal inhalation and maximal exhalation as quickly as possible. Modern electronic equipment monitors flow at the mouthpiece and the volume of air entering and exiting the **spirometer** and flow (l $s^{-1}$) is then plotted against volume (l) for both the inspiratory and expiratory portion of the cycle of breathing. Such tests can measure many more ventilatory parameters, such as **peak expiratory flow (PEF)**, and **peak inspiratory flow (PIF)**, in addition to $FEV_1$ and FVC. The shape of the curve that is plotted by the spirometer also has distinct diagnostic advantages (see below).

## Obstructive versus restrictive lung disease

The three lung function tests outlined above are the most commonly used tests in clinical practice. Each one has its distinct advantages and drawbacks. In order to demonstrate this, let us consider four subjects:

- subject A – healthy 70 kg male
- subject B – 70 kg male with **asthma** (**obstructive lung disease**)
- subject C – 70 kg male with **interstitial lung disease** (a form of **restrictive lung disease**)
- subject D – 70 kg male with **chronic obstructive pulmonary disease** (COPD, a form of obstructive lung disease).

With simple peak flow measurements, subjects A and C would be indistinguishable from each other (both would have normal peak flow readings), as would subjects B and D (both would have reduced peak flow readings). Peak flow meters only detect reductions in expiratory flow, and do not distinguish between the various forms of obstructive lung disease.

With vitalograph testing, the lung disease in subjects B, C and D would be apparent. While FVC would be likely to be normal or close to normal in subjects A, B and D, and reduced in subject C, $FEV_1$ would most likely be reduced in subjects B, C and D. $FEV_1$/FVC would be likely to be reduced in subjects B and D, but might be elevated in subject C. Vitalograph testing detects changes in vital capacity (subject C), and also distinguishes between obstructive lung disease (subjects B and D) and restrictive lung disease (subject C), since the pattern of $FEV_1$/FVC is very different in these diseases. $FEV_1$/FVC is reduced in obstructive lung diseases because obstruction of the airways means that air cannot be forced out as quickly as normal. However, lung capacity is usually not affected to the same extent. In restrictive lung disease (subject C), FVC is reduced because the lungs are small and stiff due to scarring of the interstitial space, and because lung compliance may be reduced, they may have more recoil than normal, resulting in an increase in $FEV_1$/FVC. The different forms of obstructive lung diseases cannot, however, usually be distinguished by vitalograph testing.

In flow–volume loop testing (*Fig. 3, Table 1*), characteristic patterns are seen for all three diseases.

The major diagnostic value of these tests is in the character of the trace. In the example given above, it is easy to see the difference between the two forms of obstructive lung disease (in contrast to vitalograph testing). The early collapse in

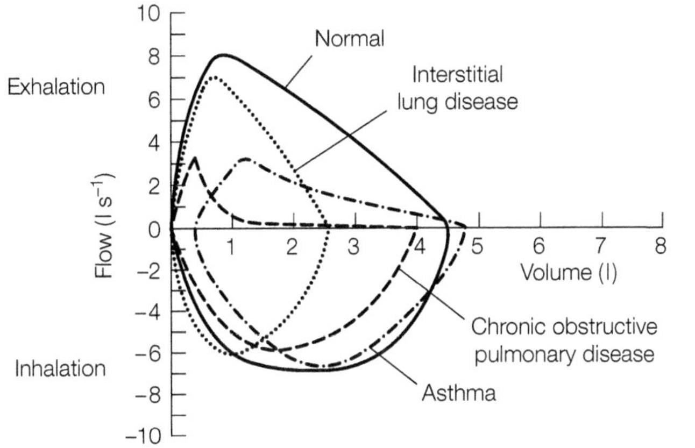


*Fig. 3. The results of flow–volume loop testing in various lung diseases.*

*Table 1. Flow–volume loop parameters in a normal subject (A) and subjects with asthma (B), interstitial lung disease (C) and chronic obstructive pulmonary disease (D)*

| Subject | PEF (l $s^{-1}$) | PIF (l $s^{-1}$) | VC (l) | Character of expiratory flow | Character of inspiratory flow |
|---|---|---|---|---|---|
| A | 8.0 | 6.9 | 4.5 | Normal | Normal |
| B | 3.3 | 6.8 | 4.4 | Curvilinear | Normal |
| C | 6.8 | 6.0 | 2.5 | Normal | Normal |
| D | 3.4 | 5.7 | 4.0 | 'Scooped out' | Normal |

flow in subject D compares to the smooth curvilinear character of the expiratory portion of the trace in subject B, even though all other parameters are similar. This is because COPD often causes collapse of the airways in expiration, whereas asthma does not. This reflects differences in the etiology of the two diseases. It is therefore easy to see why flow–volume loops are of most important diagnostic value in obstructive lung diseases. In restrictive lung disease (subject C), the total exhaled volume is small, and the flow–volume curve is horizontally compressed, indicative of the smaller, stiffer lungs in such subjects.

# E4 Alveolar exchange and gas transport

## Key Notes

**Gas pressures**

Any gas in a mixture of gases exerts a partial pressure or gas tension. In analysis of the amount of gas physically dissolved in a fluid (e.g. in plasma), the partial pressure of that gas is the partial pressure of that gas in the gas mixture above the solution. The normal partial pressures of oxygen ($P\text{O}_2$) or carbon dioxide ($P\text{CO}_2$) in arterial blood are 95–100 mmHg and 40 mmHg, respectively. In venous blood, the $P\text{O}_2$ is normally around 40 mmHg, and the $P\text{CO}_2$ is normally around 46 mmHg. Gas exchange occurs between the air in the alveolus and blood in the pulmonary capillaries, and between capillary blood and respiring tissues.

**Alveolar exchange**

Exchange of gases between alveolar air and pulmonary capillary blood (in both directions) is facilitated by three main factors: the short distance for diffusion of gases; the large surface area over which diffusion can take place; and the difference in gas partial pressure between the two compartments. These factors are incorporated into Fick's law for diffusion of gases. Exchange of gases is promoted by partial pressure differences (60 mmHg in the case of $O_2$), the large alveolar surface area (around 70 $m^2$), short distance for diffusion (2–3 μm), and solubility of the gas in aqueous solvents ($CO_2 > O_2$).

**Structure of hemoglobin and $O_2$ transport**

Hemoglobin acts as the oxygen-carrying protein in erythrocytes; around 200 ml oxygen can be carried by 1 l of normal human blood. Hemoglobin is composed of four globin chains, each with an iron-containing heme moiety at its center that binds a molecule of diatomic oxygen. The partial pressures of oxygen or carbon dioxide quoted for blood reflect dissolved oxygen, fed by the large store bound to hemoglobin. The characteristics of binding of oxygen to the four globin chains of hemoglobin result in a sigmoidal oxygen–hemoglobin dissociation curve. These characteristics also maximize the likelihood that oxygen will be picked up by hemoglobin in the lungs and delivered to respiring cells in the peripheral tissues. Some other factors such as blood $P\text{CO}_2$ also affect the oxygen–hemoglobin dissociation curve in favour of oxygen delivery in the tissues (the Bohr effect).

**$CO_2$ transport**

Carbon dioxide is transported in the blood in three forms: physically dissolved in plasma, as bicarbonate ions ($HCO_3^-$), and bound to proteins (including hemoglobin). The interactions between these various forms of carbon dioxide transport are complex and are also influenced by oxygen transport (the Haldane effect).

| **Related topics** | | |
|---|---|---|
| | Phenotypic diversity of cells (B5) | Utilization of $O_2$ and production of $CO_2$ in tissues (E5) |
| | Blood and immunity (C6) | Acid–base balance (M4) |
| | Layout and structure of the respiratory tract (E1) | |

**Gas pressures**

As outlined in Topic E3, the pressure that would be exerted by a particular gas in a mixture of gases is related to the proportion of the gas mixture that is contributed by that particular gas. This means that, in atmospheric air at sea level, where oxygen makes up 21% of the gas mixture and atmospheric pressure ($P_B$) is 760 mmHg, the oxygen **partial pressure** is 160 mmHg ($0.21 \times 760$ mmHg). When one is referring to the amount of a particular gas that is physically dissolved in a solution, the partial pressure of the gas in the solution is equal to the partial pressure of the gas in the gas mixture above the solution. So, for example, as the partial pressure of the oxygen in a gas mixture above a solution increases, the amount of oxygen dissolved in the solution will also increase. Thus, the concentration of a dissolved gas in a solution can also be expressed in terms of pressure.

**Gas tension** is a term that is sometimes used interchangeably with partial pressure. In trying to understand this term, it is sometimes useful to imagine that the gas in solution is straining to get out of solution all the time. One can visualize this by thinking about an unopened bottle of carbonated drink, like fizzy mineral water; when the bottle is unopened, shaking the bottle will release no bubbles of gas because the liquid is under pressure. However, if one were then to open the bottle, there would be an immediate release of bubbles as the pressure on top of the liquid was reduced. The higher the gas tension or partial pressure, the more likely it is that the gas will come out of solution (or move to a region of lower partial pressure). This principle of diffusion of dissolved gas can be applied to alveolar air, blood, interstitial fluid or even intracellular fluid.

The partial pressures of oxygen (**$PO_2$**) and carbon dioxide (**$PCO_2$**) are very different in blood from different sources. *Table 1* shows $PO_2$ and $PCO_2$ in mixed venous blood, alveolar air, and arterial blood. **Venous blood** that enters the lungs via the **pulmonary arteries** is relatively low in $O_2$ and relatively high in $CO_2$, because $O_2$ has been used and $CO_2$ has been produced by cellular **respiration** (see Topic E5). As it passes the $O_2$-rich air of the alveoli, the blood takes up $O_2$ and gives up $CO_2$, to become rich in $O_2$ and relatively less rich in $CO_2$. This blood then returns to the heart in the **pulmonary veins**, and is then pumped round the rest of the body to meet $O_2$ demand and to take up $CO_2$, at which point the whole process starts over again.

The SI unit of pressure is the Pascal (Pa) and 1 kPa = 7.5 mmHg. Blood gas partial pressures are sometimes quoted in kPa.

*Table 1. Normal mean $PO_2$ and $PCO_2$ of venous blood, alveolar air and arterial blood*

| | Venous blood (pulmonary arteries) | Alveolar air | Arterial blood (pulmonary veins) |
|---|---|---|---|
| $PO_2$ | 40 mmHg (5.3 kPa) | 102 mmHg (13.6 kPa) | 95 mmHg (12.7 kPa) |
| $PCO_2$ | 46 mmHg (6.1 kPa) | 40 mmHg (5.3 kPa) | 40 mmHg (5.3 kPa) |

**Alveolar exchange** As discussed in Topic E1, the barrier to diffusion of gases at the alveolus is extremely thin. The alveolar air and the blood in the capillary are separated by a distance of only a few micrometers. This means that, under normal circumstances, it is extremely easy for the gases in alveolar air and pulmonary capillary blood to equilibrate completely. *Table 1* shows that this is so. The $P_{O_2}$ of arterial blood is very close to that of the alveolar air, even though the $P_{O_2}$ of the blood in the pulmonary arteries is >60 mmHg lower. The $P_{CO_2}$ of arterial blood is the same as that of the alveolar air. The factors governing the volume of gas that can diffuse from one compartment to another are presented in **Fick's law**, which states that the volume of gas diffusing over a given period of time ($V$) is proportional to the partial pressure difference for that gas on either side of the barrier ($\Delta P$), and the area over which diffusion can take place ($A$), but is inversely proportional to the distance over which diffusion must take place ($D$):

$$V = (s \times A \times \Delta P)/D$$

where $s$ is the diffusion coefficient of the gas (solubility/$\sqrt{}$molecular weight).

In considering diffusion of $O_2$ in the lungs (*Fig. 1*), the relevant features are the extremely large surface area ($A$) over which gas can diffuse (approximately 70 $m^2$), the large partial pressure difference ($\Delta P$) between alveolar air and pulmonary arterial blood (>60 mmHg), and the short distance for diffusion (2–3 μm). The diffusion coefficient ($s$) for $O_2$ is 20 times lower than that for $CO_2$. Thus, although the driving force for diffusion ($\Delta P$) is greater for $O_2$ than for $CO_2$ (60 mmHg vs. 6 mmHg), $CO_2$ is still able to equilibrate fully with the gas in the alveolus.

However, notice that the $P_{O_2}$ of arterial blood is slightly lower than that of the alveolar air (95 vs. 102 mmHg). This is because not all of the blood returning to

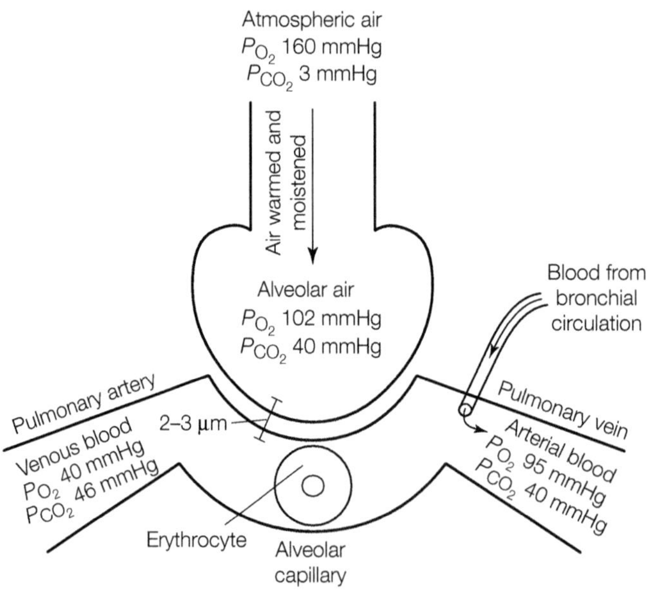


*Fig. 1. Schematic diagram showing the interrelationship between gas tensions in the venous blood, alveolar air and arterial blood.* $P_{CO_2}$*, partial pressure of carbon dioxide;* $P_{O_2}$*, partial pressure of oxygen.*

the heart in the pulmonary veins is oxygenated; some of it has been used to supply the lung tissue with oxygen (a **physiological shunt**), and so the $PO_2$ of the blood in the pulmonary veins is slightly lower than that which has been involved in gas exchange.

## Structure of hemoglobin and $O_2$ transport

**Hemoglobin** is the oxygen-carrying protein that is present in erythrocytes. The synthesis of **erythrocytes** is stimulated by the release of the hormone **erythropoietin** by the kidneys in response to hypoxia. Every liter of human blood contains around 150 g of hemoglobin (within some $5 \times 10^{12}$ erythrocytes), and this much hemoglobin is able to carry up to 200 ml of $O_2$. Hemoglobin exists in adults as four protein subunits; two α-globin chains of 141 amino acids, and two β-globin chains of 146 amino acids (*Fig. 2a*). The structure of hemoglobin was elucidated by Max Perutz and John Kendrew, work which received the ultimate recognition of the Nobel prize in chemistry for 1962. In the fetus, hemoglobin is usually made up of two α-globin chains and two γ-globin chains, and the affinity of this form of hemoglobin for $O_2$ is greater than that of adult hemoglobin. Near the center of each globin chain there exists a **heme** group (an iron-containing molecule), and it is this heme that actually binds the oxygen.

The $PO_2$ of blood reflects the amount of $O_2$ physically dissolved in the plasma. At around 100 mmHg, this correlates to around 3 ml $O_2$ per liter of blood. Thus,

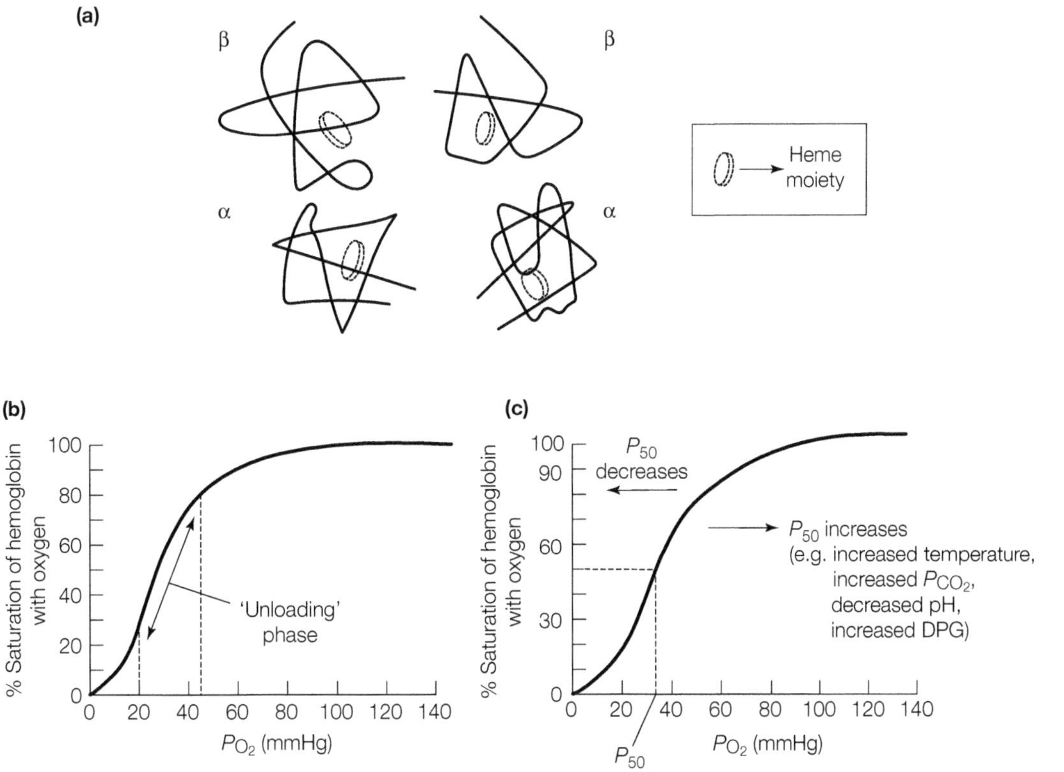


*Fig. 2. (a) The structure of hemoglobin; (b) the oxygen–hemoglobin dissociation curve; (c) effects of various factors on $P_{50}$ (the partial pressure of oxygen at which hemoglobin is 50% saturated with the gas). DPG, 2,3-diphosphoglycerate; $PCO_2$, partial pressure of carbon dioxide; $PO_2$, partial pressure of oxygen.*

in arterial blood, >98% of the oxygen content is bound to hemoglobin, and only a tiny fraction is dissolved in the blood. The hemoglobin acts as a store for $O_2$, feeding the dissolved $O_2$ that is available to the tissues. Therefore, in order for the tissues to be supplied with $O_2$, hemoglobin must give up its bound $O_2$.

The interaction between $O_2$ and hemoglobin is not linear. That is to say, saturation of the hemoglobin molecule does not follow a simple proportional relationship to $PO_2$. In fact, the relationship is sigmoidal (*Fig. 2b*). When none of the subunits of an individual hemoglobin molecule have $O_2$ attached, it is extremely difficult for the molecule to bind even one $O_2$; a lot of $O_2$ must be added before some will bind. However, once one of the hemoglobin subunits has bound $O_2$, it becomes much easier for another two $O_2$ molecules to bind to two other hemoglobin subunits. Binding of a total of three $O_2$ molecules results in the fourth hemoglobin subunit having low affinity for $O_2$, meaning that $PO_2$ must rise to extremely high levels (600–700 mmHg) before all four hemoglobin subunits have an $O_2$ molecule attached (although at arterial $PO_2$ of around 95 mmHg, >95% of the molecules of hemoglobin are fully saturated with $O_2$). This phenomenon is known as **cooperativity** and explains the sigmoidal shape of the **$O_2$–hemoglobin dissociation** curve.

The changes in the affinity of hemoglobin for oxygen can be thought of as changes in the shape of the molecule that are associated with $O_2$ binding. With no $O_2$ bound, all of the binding sites for $O_2$ are 'hidden', but as one molecule binds, two more become readily 'visible'. Once those are occupied by $O_2$ molecules, the fourth binding site becomes 'hidden' again.

Notice that the steep unloading phase of the $O_2$–hemoglobin dissociation curve occurs at $PO_2$ measurements between 20 and 40 mmHg, which are similar to the $PO_2$ of respiring tissues. Thus, the interaction between $O_2$ and hemoglobin is ideally designed to release $O_2$ to respiring tissues, and to receive $O_2$ in the lungs. The level of $PO_2$ at which hemoglobin is 50% saturated is known as the $\boldsymbol{P_{50}}$ and this is normally around 28 mmHg, but several other factors influence $P_{50}$ (*Fig. 2c*). For example, if temperature rises, $PCO_2$ rises, or pH falls, this raises the $P_{50}$, meaning that $O_2$ will dissociate from hemoglobin at higher $PO_2$ than normal. This is caused by $O_2$ being forced off the hemoglobin by $CO_2$ and $H^+$ ions, both of which bind to hemoglobin. This means that in active tissues, more $O_2$ can be delivered, since tissue activity is associated with an increase in temperature and $PCO_2$, and a reduction in pH. This effect of $PCO_2$ and pH on the $P_{50}$ is known as the **Bohr effect**. Another factor that increases $P_{50}$ is a rise in levels of **2,3-diphosphoglycerate (DPG)** in the erythrocytes. DPG is a by-product of **glycolysis** (the first set of reactions for metabolism of glucose in cells, see Topic B3), and cellular DPG increases in response to exercise and hypoxia.

## $CO_2$ transport

$CO_2$ is transported in the blood in three forms. Around 10% is physically dissolved in plasma. The majority (60–70%) is carried as **bicarbonate ions** ($HCO_3^-$) in the plasma and in erythrocytes. The remaining 20–30% is carried as carbamino proteins, one of which is **carbaminohemoglobin**. This interaction of $CO_2$ and hemoglobin is readily reversible, so an individual hemoglobin molecule may be carrying $CO_2$ as it goes towards the lungs and carrying $O_2$ as it leaves the lungs (*Fig. 3a*). Blood $PCO_2$ reflects the amount of $CO_2$ physically dissolved in plasma, and so $HCO_3^-$ and carbamino proteins act as extra storage media. As the blood passes active tissues, it will pick up $CO_2$. Some of this $CO_2$ remains dissolved in the plasma (10%), but some diffuses into the erythrocytes. The $CO_2$ that enters erythrocytes can then take one of two pathways. Firstly, it

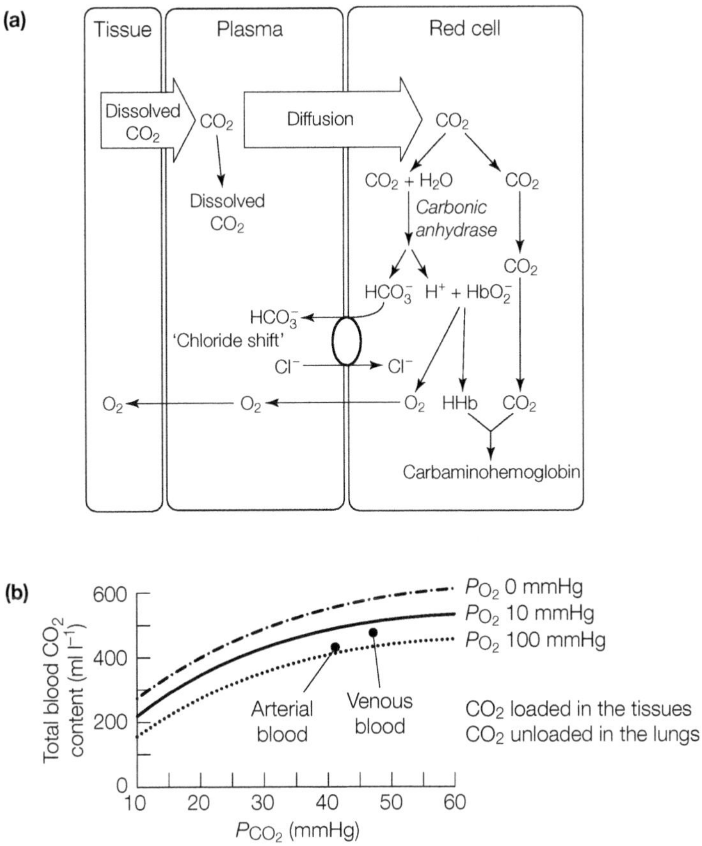


*Fig. 3. (a) Carbon dioxide ($CO_2$) transport in the blood; (b) the $CO_2$ equilibrium curve.*

may be combined with water using the enzyme **carbonic anhydrase**, to form $H^+$ and $HCO_3^-$ ions (some of these $H^+$ and $HCO_3^-$ ions can recombine to form **carbonic acid** ($H_2CO_3$), although at physiological pH, very little $H_2CO_3$ is present). The $HCO_3^-$ can then be exchanged for $Cl^-$ ions using a **$HCO_3^-/Cl^-$ exchanger**. Thus the effect of elevated $P_{CO_2}$ is to cause a **'chloride shift'**; $Cl^-$ ions enter red cells at higher $P_{CO_2}$. The $H^+$ ions that are formed by carbonic anhydrase can be used to help displace $O_2$ from hemoglobin, forming **reduced hemoglobin (HHb)**, and the $O_2$ can diffuse from the erythrocyte, into the plasma, and then into the respiring tissue. The $CO_2$ in the erythrocyte that is not involved in the carbonic anhydrase reaction can then bind to HHb to form carbaminohemoglobin. In the lungs, this whole process can be reversed, with the erythrocytes taking up $O_2$ and releasing $CO_2$. One can construct a graph of blood $CO_2$ content against $P_{CO_2}$ (*Fig. 3b*).

In this case, the factor which has a major effect on the capacity of the blood to carry $CO_2$ is $P_{O_2}$. At higher $P_{O_2}$, more $CO_2$ is given up than at low $P_{O_2}$. This means that in the lungs, where $P_{O_2}$ rises steeply and $P_{CO_2}$ falls, $CO_2$ is readily given up to be exhaled. In the tissues, where $P_{O_2}$ is low and $P_{CO_2}$ is rising, more $CO_2$ can be loaded into the blood. This effect of $P_{O_2}$ in altering the $CO_2$-carrying capacity of the blood is known as the **Haldane effect**.

# E5 Utilization of $O_2$ and production of $CO_2$ in tissues

## Key Notes

**Oxygen delivery**

Oxygen is carried to the tissues bound to hemoglobin. The gas diffuses from regions of high partial pressure to regions of low partial pressure (capillary blood to respiring cells). Oxygen is used primarily in the oxidation of substrates in the mitochondrion to form carbon dioxide and water; some of the energy released is used to make adenosine triphosphate (ATP). The terminal phosphate–phosphate bond of ATP is a high-energy phosphate bond; splitting ATP into adenosine diphosphate (ADP) and inorganic phosphate yields energy that can be utilized in many cellular reactions.

**Oxygen use**

Oxygen is used in the metabolism of fats and carbohydrates, which are the major metabolic substrates of body cells. When oxygen is used in the metabolism of fats, more energy is released than when carbohydrates are being metabolized. When oxygen is used as the terminal electron acceptor in the mitochondrial electron transport chain, some of the energy yielded is used to phosphorylate ADP to ATP, but the majority is released as heat energy to keep the body warm.

**Carbon dioxide production**

Carbon dioxide is produced in the tricarboxylic acid cycle in the mitochondrion. The gas diffuses from regions of high partial pressure to regions of low partial pressure (respiring cells to capillary blood). Carbon dioxide is combined with water to form $H^+$ ions and $HCO_3^-$ (bicarbonate) ions, in a reaction catalyzed by carbonic anhydrase (see Topic E4). Approximately 70% of the carbon dioxide content of the blood is transported as $HCO_3^-$. $H^+$ and $HCO_3^-$ can combine to form carbonic acid ($H_2CO_3$), although very little $H_2CO_3$ exists at physiological pH. Some carbon dioxide is physically dissolved in the plasma, and some is bound to proteins, including hemoglobin. Carbon dioxide is sometimes referred to as 'respiratory acid' because of the tendency for $H_2CO_3$ to be produced. The gas can be blown off at the lungs, where it diffuses down a partial pressure gradient of around 6 mmHg between pulmonary arterial (venous) blood and alveolar air.

**Pathology of hypoxia**

When blood supply to cells is interrupted, the resultant lack of oxygen (hypoxia) means that only small amounts of ATP can be generated by glycolysis. This is usually insufficient to maintain cellular function, and cells become damaged and may die as a result. In some cases, reperfusion of the tissue can exacerbate cellular damage by producing reactive oxygen species; for example in situations where adenosine is produced from ATP, ADP and adenosine monophosphate (AMP), in an effort to harness as much energy from phosphate bonds as possible. Reactive oxygen species damage cells by attacking double bonds in cellular macromolecules.

**Related topics**

Metabolic processes (B3)
Membrane potential (B4)
Alveolar exchange and gas transport (E4)
Chemoreceptors and respiratory acid–base balance (E6)

## Oxygen delivery

Oxygen ($O_2$) is delivered to tissues by **hemoglobin** and the amount of $O_2$ delivered is dependent on the cardiac output and the oxygen content in systemic arterial blood. Whole-body energy expenditure can be assessed by measuring $O_2$ uptake and carbon dioxide ($CO_2$) production by **indirect calorimetry**. The exchange of gases between the lung and tissue and tissue mitochondria is a two-way process. Atmospheric $O_2$ passes eventually into the **mitochondria** having moved through the lungs, blood, extracellular fluid, interstitial fluid, and the cell cytosol. In this journey, oxygen moves down a gradient of decreasing partial pressure of about 160 mmHg in atmospheric air, to 1.5–15 mmHg in the mitochondrion (*Fig. 1*).

Inspired air 160 mmHg → Alveolar gas 102 mmHg → Arterial blood 95 mmHg → Cell cytosol 15–37 mmHg → Mitochondrion 1.5–15 mmHg

*Fig. 1. Oxygen cascade: partial pressure of oxygen from air to mitochondrion.*

The major metabolic use of $O_2$ is in the oxidation of fuels in the mitochondrion for the generation of energy either in the form of chemical bond energy in **adenosine triphosphate (ATP)** or as heat. This results in the production of $CO_2$ and water. $CO_2$ is eliminated from the body in expired air from the lungs, and reaches the lungs from mitochondria by a reversal of the oxygen cascade route.

The oxygen *content* of arterial blood is related mainly to that carried by hemoglobin, although there is also a small contribution from dissolved oxygen (see Topic E4). This latter component can be increased dramatically in hyperbaric oxygen conditions. The high **partial pressure of $O_2$** in the pulmonary (lung) alveoli leads to almost complete saturation of hemoglobin (Hb) with $O_2$, and forces protons from the Hb molecules. In the capillary beds however, especially in metabolically active tissues, the pH is slightly lower and Hb:$O_2$ will accept protons and release $O_2$ for uptake by the tissue. Delivery of $O_2$ from capillary blood into the tissues depends on the difference in partial pressure of $O_2$ between the capillary and the cell. $O_2$ utilization through fuel oxidation by the tissue will increase this gradient, and thereby also increase the amount of $O_2$ taken up by the tissue. Under normal conditions, the delivery of $O_2$ is sufficient to satisfy the metabolic needs of the tissues. However, when $O_2$ delivery is limited through decreased blood flow, $O_2$ uptake by the tissue is proportional to the rate of $O_2$ delivery.

## Oxygen use

The major components of 24-hour energy expenditure are given in *Table 1*.

In an average, moderately active individual with a **daily energy expenditure** of 10 MJ (megajoules), approximately 60% of energy expenditure is due to basal metabolism, 'keeping the body ticking over', together with the essential energy

*Table 1. Oxidation of major metabolic substrates*

| Substrate | $O_2$ used (l (g substrate)$^{-1}$) | $CO_2$ produced (l (g substrate)$^{-1}$) | Respiratory quotient (RQ)[a] | Energy yield (kJ (l $O_2$)$^{-1}$) |
|---|---|---|---|---|
| Glucose | 0.83 | 0.83 | 1.0 | 21.1 |
| Fat | 2.02 | 1.43 | 0.71 | 19.6 |
| Protein | 0.97 | 0.78 | 0.80 | 19.3 |
| Ethanol | 1.46 | 0.98 | 0.67 | 19.9 |

[a]RQ = amount of $CO_2$ produced / amount of $O_2$ used.

expenditure of the ingestion, absorption, distribution and storage of food. The other contributions to energy expenditure come from physical activity and postural effects such as sitting or standing.

The **$O_2$ utilization** by various tissues and organs in a resting, non-obese individual in the normal state is shown in *Table 2*. The figures presented assume a whole-body resting $O_2$ consumption of about 300 ml min$^{-1}$, equivalent to an energy expenditure of about 6 kJ min$^{-1}$. The table illustrates the different substrates that can be used as sources of energy for metabolism in the individual tissues and organs. **Ketones** refer to **acetoacetate** and **3-hydroxybutyrate** derived from **excessive β-oxidation** of **fatty acids** such as is seen in starvation and untreated type I diabetes mellitus. In the fed state, the brain uses **glucose** almost exclusively as respiratory fuel, oxidizing almost 129 g glucose per day. During an overnight fast, much of the glucose (75%) oxidized by the brain is derived from **hepatic glycogen**, with the rest coming from **gluconeogenesis** (see Topic B3). The ability of the brain to adapt to oxidation of ketone bodies during prolonged fasting reduces the glucose requirement of the body, and hence reduces both the rate at which body protein is degraded and the energy expenditure on gluconeogenesis. Mature **erythrocytes**, which lack mitochondria and thus cannot carry out **oxidative phosphorylation**, have a continual need for glucose (about 16 g daily) as they derive all of their ATP from **glycolysis**. Other tissues will oxidize glucose when this is abundant, for example, after a meal, but will oxidize fatty acids at other times. Fuel use by the heart is discussed in Topic D3.

The energy content of the major food components is related to their structure and in particular the extent to which the molecule is already oxidized. Thus

*Table 2. Tissue fuel consumption in a 70 kg individual*

| Tissue | Wet weight of tissue (kg) | $O_2$ consumption (ml min$^{-1}$) | Fuel[a] |
|---|---|---|---|
| Skeletal muscle | 20–30 | 70 | G, F, K |
| Adipose tissue | 9–13 | small | G, F |
| Gastrointestinal tract | 2.6 | 58 | G, F, K |
| Blood | 5.5 | small | G |
| Liver | 1.7 | 75 | G, F |
| Brain | 1.5 | 46 | G, K |
| Lung | 1.0 | 12 | G, F, K |
| Heart | 0.3 | 27 | G, F, K |
| Kidney | 0.3 | 16 | G, F, K |

[a]F, fatty acids; G, glucose; K, ketone bodies. These are not in order of preferential use by the tissue.

**carbohydrate**, with a general structure of $C_6H_{12}O_6$, is already half oxidized, compared to saturated fatty acids such as **palmitic acid**, $CH_3(CH_2)_{14}COOH$, in which only 1 carbon out of 16 is oxidized. The energy yield on **oxidation of fat** thus greatly exceeds that of carbohydrate (*Table 3*).

*Table 3. Energy content of major food components*

| Energy-yielding nutrients | Energy content | |
|---|---|---|
| | kJ $g^{-1}$ | kCal $g^{-1}$ |
| Carbohydrate | 17 | 4 |
| Fat | 38 | 9 |
| Protein | 17 | 4 |
| Ethanol | 29 | 7 |

$O_2$ is the terminal electron acceptor in the electron transport chain on the inner membrane of the mitochondrion. Reduced cofactors (three molecules of **reduced nicotinamide adenine dinucleotide (NADH)** and one of **reduced flavine adenine dinucleotide ($FADH_2$)**) are produced at the four oxidative steps in the **tricarboxylic acid (TCA) cycle** per molecule of **acetyl-CoA** oxidized to $CO_2$. Other sources of reduced nucleotides in the mitochondrion are oxidation of **pyruvate** to $CO_2$ and β-oxidation of fatty acids. It is during the redox reactions that take place as electrons flow down the electron transport chain from the reduced cofactors to oxygen that sufficient energy is released to drive **ATP synthesis**. In terms of efficiency, approximately 40% of the energy released from the redox reactions is stored in the **high-energy phosphate bonds** of ATP, while the rest is released as heat and helps maintain body temperature at 37°C, well above ambient temperature. The high-energy phosphate bonds are used in other energy-requiring processes in the cell, including maintenance of transmembrane ionic gradients, movement and enzyme reactions.

## Carbon dioxide production

$CO_2$ is released from the two oxidative decarboxylation reactions (catalyzed by **isocitrate dehydrogenase** and **α-ketoglutarate dehydrogenase**) of the TCA cycle, and this loss of two carbons from the cycle is balanced by the two atoms of carbon of acetate entering the cycle at the **citrate synthase** step. $CO_2$ is also generated in the mitochondrion from oxidative decarboxylation of pyruvate to acetyl-CoA by **pyruvate dehydrogenase**:

$$\text{Pyruvate} \rightarrow \text{acetyl-CoA} + CO_2 + \text{NADH}$$

The passage of $CO_2$ from mitochondrion to the lungs is the reverse of that described for oxygen, and is down a $CO_2$ partial pressure gradient which is opposite to that for $O_2$. However, $CO_2$ is much more water soluble than $O_2$, and undergoes much more rapid tissue diffusion. Much of the $CO_2$ produced in the metabolic reactions described is converted to bicarbonate by the action of the enzyme carbonic anhydrase, an enzyme which is present in large amounts in the erythrocyte and catalyzes the following reaction:

$$CO_2 + H_2O \rightarrow H^+ + HCO_3^-$$

It is in this form, $HCO_3^-$, that most (70%) of the $CO_2$ is present in blood. Bicarbonate is in equilibrium with free $CO_2$ in solution (10%) and $CO_2$ bound to protein, especially hemoglobin (4%). Carbamino compounds are formed between $CO_2$ and the N-terminal amino groups of proteins, and the binding of $CO_2$ to deoxygenated Hb is three times greater than that of oxygenated Hb. Deoxygenated Hb is also less acidic than oxygenated Hb, and this allows it to act as a proton acceptor and increase the formation of bicarbonate ions, thereby facilitating the removal of $CO_2$ from peripheral tissues. These processes are reversed at the lung and promote the offloading of $CO_2$ into the pulmonary tissue for expiration. Besides this transport of $CO_2$ as $HCO_3^-$ dissolved in blood, it is also transported as $HCO_3^-$ in erythrocytes. Here the hydrogen ions are buffered by Hb and the bicarbonate moves out of the erythrocyte, down a concentration gradient in exchange for chloride (see Topic E4, *Fig. 3*).

## Pathology of hypoxia

The unloading of $O_2$ from the blood into the tissues is affected by three major pathological factors:

- cellular ischemia
- inflammatory mediators
- free radical injury.

**Cellular ischemia** is probably the main cause of cell death in situations with low cardiac output. Inadequate perfusion, such as is seen in hypovolemic **shock**, reduces the $O_2$ supply to the tissues such that ATP production now comes from anaerobic glycolysis in the cytosol rather than from the TCA cycle and oxidative phosphorylation in the mitochondria, which is an extremely inefficient use of glucose (2–4 molecules of ATP per glucose in glycolysis versus >30 molecules of ATP per glucose via the oxidative route). This will inevitably lead to a depletion of tissue energy stores and cellular ATP. **Lactate** will accumulate and produce an intracellular **acidosis**. A lack of ATP will result in an inability to maintain transmembrane ionic gradients (e.g. for $K^+$ and $Ca^{2+}$), and decreased activity of energy-dependent enzymes. Eventually this causes ultrastructural damage to the cell, particularly in the mitochondria, and cell death.

The inflammatory response is a complex system that acts in response to tissue injury or infection to try to restore homeostasis. A number of inflammatory mediators, including **cytokines**, stimulate host defense mechanisms and dilate blood vessels to enhance blood flow to the site of damage or infection. However, in more severe cases of trauma and infection, the release of cytokines such as **tumor necrosis factor** and **interleukin-1** may have an opposing effect and decrease perfusion of damaged tissues, and even cause direct damage to the cells themselves.

Somewhat paradoxically, restoring blood flow to a previously ischemic tissue can lead to cell death. As mentioned above, ischemia can lead to a failure to maintain ion gradients across the cell membrane. The intracellular calcium concentration is submicromolar compared with the extracellular concentration of approximately 2.5 mmol $l^{-1}$. An increase in intracellular calcium due to inefficient pumping can activate **calcium-dependent proteases**, including that required for the activation of **xanthine dehydrogenase** to **xanthine oxidase**. At the same time, ATP stores become depleted and adenosine is metabolized to **hypoxanthine** via **inosine**. If oxygen now enters the cell, xanthine oxidase is able to convert hypoxanthine to **uric acid**, with the concomitant production of **reactive oxygen species** including **superoxide radicals**, **hydroxyl radicals**, and **hydrogen peroxide**. The **free radicals** will attack double bonds in all macromolecules, inducing damage to proteins and DNA and **peroxidation** of membrane lipids. This will eventually lead to cell death and tissue injury.

# E6 Chemoreceptors and respiratory acid–base balance

## Key Notes

**Acids, bases and buffers**

$H^+$ ion concentrations in extracellular fluid are maintained within a tight normal range (35–45 nmol $l^{-1}$, pH 7.4). An acid is a molecule that, in solution, can dissociate to donate a $H^+$ ion. A base is a molecule that, in solution, can accept an $H^+$ ion to form an acid. Buffers are molecules that can accept or donate $H^+$ ions to maintain $H^+$ ion concentrations within a narrow range. The most common physiological buffers are bicarbonate ions, phosphate ions, and proteins.

**Chemoreceptors**

Chemoreceptors in the aortic body, carotid body and brainstem provide information to the central nervous system about the partial pressure of oxygen and carbon dioxide and the pH of blood. If carbon dioxide partial pressure ($P\text{CO}_2$) rises or pH falls, breathing is stimulated. If the partial pressure of oxygen ($P\text{O}_2$) falls, this also stimulates breathing, but $P\text{O}_2$ is less important in chemoreceptor mechanisms than $P\text{CO}_2$ or pH. The activation of chemoreceptors by these factors is interrelated; $P\text{O}_2$ can affect the response of chemoreceptors to $P\text{CO}_2$, for example. Hyperventilation, induced by rising arterial $P\text{CO}_2$, blows off more carbon dioxide at the lungs, reflexly reducing $P\text{CO}_2$. Peripheral and central chemoreceptors respond to different factors.

**Brainstem breathing centers**

Several groups of neurons in the brainstem are known to be involved in control of breathing. The majority of these neurons are in the pons and medulla (the most phylogenetically ancient part of the CNS). Some influence only inspiration, whereas others influence both inspiration and expiration. Chemoreceptor input from the glossopharyngeal and vagus nerves is integrated here, together with sensory input from pulmonary stretch receptors.

**$CO_2$ as 'respiratory acid'**

Carbon dioxide ($CO_2$) can be combined with water (using the enzyme carbonic anhydrase) to form bicarbonate ions and $H^+$ ions, thus affecting blood pH. $CO_2$ is often called 'respiratory acid'. If $P\text{CO}_2$ rises and breathing is stimulated, more $CO_2$ will be blown off at the lungs, resulting in a slight increase in pH as more $H^+$ ions combine with bicarbonate to form carbonic acid, which can then be broken down by carbonic anhydrase to $CO_2$ and water. Arterial pH, $P\text{CO}_2$ and bicarbonate ion are used to determine the type and extent of acid–base imbalance in the body.

| **Related topics** | Homeostasis, and integration of body systems (A2)<br>The mechanics of breathing (E2)<br>Alveolar exchange and gas transport (E4) | Parts of the brain (G1)<br>Cranial nerves (G4)<br>Acid–base balance (M4) |
|---|---|---|

**Acids, bases, and buffers**

Intracellular and extracellular hydrogen ion ($H^+$) concentrations are normally kept within a narrow range. However, disturbances in the mechanisms controlling this homeostatic range may occur in a number of situations encountered in critical care, for example.

An **acid** is a molecule that, in solution, dissociates to a **proton** and a **base**:

$$HA \rightleftharpoons H^+ + A^-$$

A base is a molecule which can accept a proton:

$$B^- + H^+ \rightleftharpoons BH$$

A useful means of expressing the $H^+$ concentration $[H^+]$ is **pH**, the negative logarithm to base 10 of the $H^+$ concentration in molar concentration. Expressed this way, as acidity increases (i.e. $H^+$ increases) the pH decreases. Neutral pH is defined as pH 7.0; when water has a pH of 7.0, $H^+$ and $OH^-$ ion concentrations are equal at $10^{-7}$ M. However, since more water is ionized at body temperature, neutral pH at body temperature is 6.8, the average intracellular pH. Normal **extracellular pH** is 7.4.

Buffers are molecules which can accept or donate $H^+$ and thereby minimize changes in $H^+$ ion concentration and pH. They consist of a combination of a **weak acid** and its conjugate base, and are most effective at the **p$K_a$** of the acid; the pH at which the acid is 50% ionized. About 80% of the buffering capacity of any buffer occurs in the range of ±1 pH unit around its p$K_a$, and thus the ideal physiological buffer would have its p$K_a$ value close to 7.4, the physiological pH. However, only inorganic phosphate has its p$K_a$ close to 7.4 and, of amino acid residues on proteins, the side chains of histidine and cysteine have their p$K_a$ closest to 7.4. However, the effectiveness of a physiological buffer also depends on its availability, so neither of these are viable options for the majority of buffering that has to be carried out in the body. The most important extracellular buffers are **bicarbonate ions** ($HCO_3^-$), inorganic **phosphate ions** ($HPO_4^{2-}$), and **plasma proteins**, in that order.

**Chemoreceptors**

The partial pressure of $CO_2$ (**arterial $P_{CO_2}$**) is continually sensed by central and peripheral **chemoreceptors**. The peripheral chemoreceptors are in the **carotid bodies** and **aortic bodies** and respond to changes in $P_{CO_2}$, pH and partial pressure of oxygen (**$P_{O_2}$**). These chemoreceptors are innervated by the **glossopharyngeal** and **vagus** nerves, respectively. A decrease in blood pH or an increase in blood $P_{CO_2}$ results in breathing being stimulated. However, these chemoreceptors are less sensitive to changes in $P_{O_2}$; the relationship between ventilation rate and alveolar $P_{CO_2}$ is linear (*Fig. 1a*), whereas alveolar $P_{O_2}$ has to fall below about 80 mmHg before this has any marked effect on ventilation (*Fig. 1b*). However, there is significant interaction between these two variables; for example the response to rising $P$CO$_2$ is magnified if $P$O$_2$ also falls (slope of line 1 in *Fig. 1a* is 1.9 l min$^{-1}$ (mmHg)$^{-1}$; slope of line 2 is 3.4 l min$^{-1}$ (mmHg)$^{-1}$), and other substances such as **cyanide ion** and other respiratory poisons also influence the chemoreceptors. Also, in order to determine the relationship between **arterial $P$O$_2$** and minute ventilation, it is necessary to maintain alveolar **$P$CO$_2$** at around 40 mmHg. Otherwise, as decreasing alveolar $P$O$_2$ stimulates ventilation, more $P$CO$_2$ is 'blown off', thereby reducing ventilation.

By increasing ventilation rate (**hyperventilation**) in response to reduced blood pH and/or elevated blood $P$CO$_2$, the body is able to 'blow off' more $CO_2$ and thus reduce arterial $P$CO$_2$ and increase pH to within normal limits. The converse is true for increased blood pH and reduced $P$CO$_2$.

The central chemoreceptor neurons are in a region of the brainstem known as the **medulla**, although their exact location has yet to be elucidated. These receptors are only affected by $P$CO$_2$ of the blood, because $H^+$ ions cannot cross the **blood–brain barrier**, and so can only influence peripheral chemoreceptors. For this reason, brain $P$CO$_2$ and pH are more susceptible to **acid–base disturbances** of respiratory origin than those of metabolic origin.

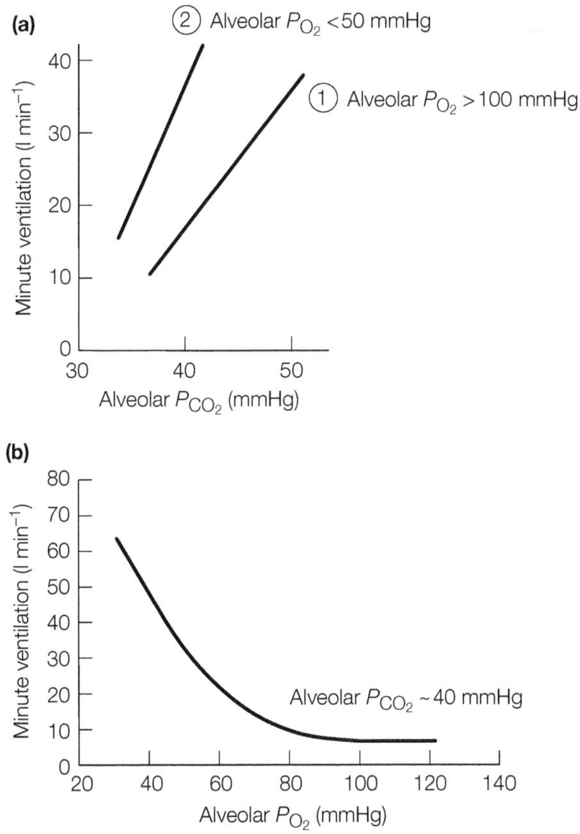


*Fig. 1. Effects of alveolar gas composition on ventilation.* $P$CO$_2$*, partial pressure of carbon dioxide;* $P$O$_2$*, partial pressure of oxygen.*

## Brainstem breathing centers

As mentioned in Topic E1, breathing is essentially a **reflex** response over which levels of blood gases and other factors have influence. Two sets of neurons in the medulla control the basic rhythm of breathing, namely the **dorsal respiratory group (DRG)**, and the **ventral respiratory group (VRG)**. Neurons in the DRG mainly influence inspiratory muscles, whereas those of the VRG influence muscles of both inspiration and expiration. These neurons, together with other groups of cells in the medulla and the **pons**, act in a coordinated fashion to produce a general on–off switch to inspiratory muscles.

A group of neurons in the medulla known as the **central inspiratory activity (CIA)** integrator is normally under inhibition from 'inspiratory off-switch' neurons, which receive excitatory input from the **pontine respiratory group**. Thus, cells in the pons 'dampen down' respiratory drive. Three external factors influence breathing: **chemoreceptor** input, **pulmonary irritant receptor** input, and **pulmonary stretch receptor** input. Of these, the first two, when activated, produce an increase in the activity of the CIA integrator, leading to ventilatory drive. Chemoreceptors also inhibit the activity of the 'inspiratory off-switch' neurons, leading to an increase in ventilatory drive. Activation of pulmonary stretch receptors, on the other hand, increase the activity of the 'inspiratory off-switch' neurons, reducing ventilatory drive; the **Hering–Breuer reflex**. A further important point to note is that CIA integrator activity also leads to stimulation of 'inspiratory off-switch' neurons, thus limiting the level of CIA activity.

## $CO_2$ as 'respiratory acid'

Disturbances in acid–base balance have many causes, but these fall into the two main categories of respiratory and metabolic. This section will consider only respiratory acid–base balance, but Section M has more details about metabolic acid–base balance. Carbon dioxide ($CO_2$) is called '**respiratory acid**' since it is formed from cellular respiration (burning carbon-rich molecules in $O_2$), and combines *in vivo* with $H_2O$ to produce $H^+$ and $HCO_3^-$, a reaction catalyzed by the enzyme **carbonic anhydrase**. Carbonic anhydrase is present in many cells of the body, but most abundant in erythrocytes (see Topic E4). As mentioned above, the $P\text{CO}_2$ and pH of the blood are maintained within relatively tight limits by altering ventilation rate. This is an 'open buffer' system, where breathing has an important role to play in modulating one of the important factors; increased ventilation can help remove $CO_2$ from the body and counteract **acidosis**.

The major buffer system for $CO_2$ in the blood is **bicarbonate** ($HCO_3^-$). When $H^+$ and $HCO_3^-$ ions combine, they form carbonic acid ($H_2CO_3$) although, given that the $pK_a$ of $H_2CO_3$ is 6.1, relatively little $H_2CO_3$ is present at physiological pH. The dissociation constant for $H_2CO_3$ ($K_a$) is defined thus:

$$K_a = [H^+] \times [HCO_3^-]/[H_2CO_3] \qquad (1)$$

Rearranging this equation yields the following:

$$[H^+] \times [HCO_3^-] = K_a \times [H_2CO_3] \qquad (2)$$

If we take into account the solubility of $CO_2$ in plasma and correct for the necessary alterations in units, Equation (2) can be changed to a simple equation that describes the normal relationship between $H^+$, $HCO_3^-$ and $P\text{CO}_2$, when body temperature and blood **hemoglobin** levels are within normal physiological limits:

$$H^+ \text{ (nM)} \times HCO_3^- \text{ (mM)} = K \text{ (constant)} \times P\text{CO}_2 \text{ (mmHg)} \qquad (3)$$

Thus at normal blood pH (7.4, $H^+$ of 40 nM), normal blood $HCO_3^-$ (24 mM) and normal **arterial** $\boldsymbol{P}\mathbf{CO_2}$ (40 mmHg):

$$40 \times 24 = 24 \times 40$$

The relationship between blood $HCO_3^-$, $PCO_2$ and pH can be shown graphically (*Fig. 2*).

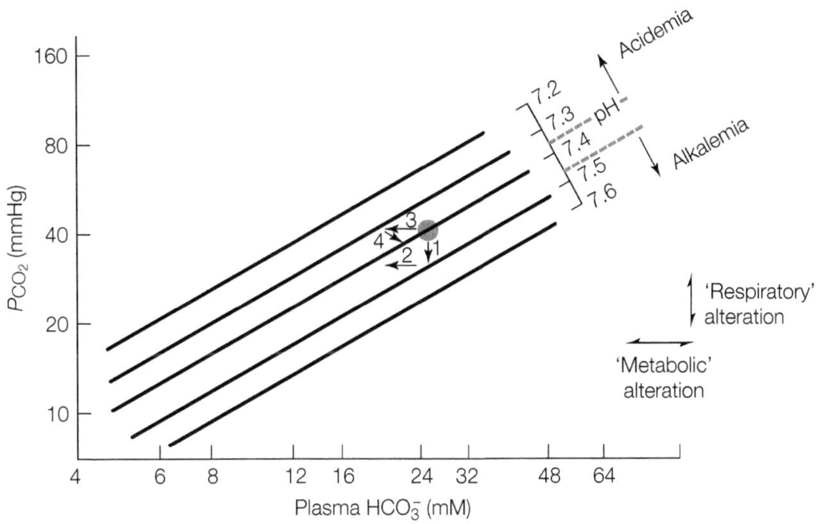


*Fig. 2. Interrelationship between plasma $HCO_3^-$, blood $P$CO$_2$, and blood pH.*

In order to maintain blood pH at 7.4, it is necessary that as $P$CO$_2$ increases, blood $HCO_3^-$ also increases. However, if $P$CO$_2$ falls as a result of hyperventilation (arrow 1 in *Fig. 2*), this might result in **alkalemia** (blood pH > 7.45) until the kidneys are able to increase net $HCO_3^-$ excretion (see Topic M4) and bring blood pH within normal limits (arrow 2 in *Fig. 2*). This is known as **respiratory alkalosis** with **metabolic compensation** because the initial change in blood pH was caused by a change in ventilation rate. If, on the other hand, large amounts of alkaline fluids are being lost (e.g. intestinal secretions in diarrhea, arrow 3 in *Fig. 2*), this will result in a fall in blood pH (**acidemia** if pH < 7.35), and a drive to increase ventilation rate. This will, in turn, lead to a reduction in arterial $P$CO$_2$, as $CO_2$ is blown off in the lungs. A further consequence of acidemia is that the kidneys will excrete more acid than normal. Both of these compensatory responses attempt to raise blood pH (arrow 4 in *Fig. 2*). This is known as **metabolic acidosis** with **mixed compensation**, because the initial change in blood pH was not caused by an alteration in ventilation rate.

# E7 Respiration in exercise

## Key Notes

**Ventilation–perfusion matching**

Oxygen delivery and carbon dioxide removal at the lungs is dependent on two factors: lung ventilation and lung perfusion. If lung ventilation increased and pulmonary blood flow remained the same, oxygen delivery to the tissues would be no greater than at rest. If pulmonary blood flow increased and ventilation remained the same, oxygen delivery would be adversely affected. Although the ventilation–perfusion ratio ($\dot{V}/\dot{Q}$) is around 1.0 at rest, it rises to around 3.0 at moderate exercise levels. The purpose of such increases in $\dot{V}/\dot{Q}$ with activity is to maintain partial pressures of oxygen in the alveoli and in arterial blood at high levels and prevent build-up of carbon dioxide in the arterial circulation.

**Physiological dead space**

Around 35% of the air we breathe at rest is not involved in alveolar exchange; this is the anatomical dead space in the conducting portion of the airways. Alveolar dead space is contributed by alveoli that are collapsed, closed off, or which have inadequate blood flow. Physiological dead space is the sum of anatomical dead space and this alveolar dead space. Blood flow to alveolar capillaries is controlled by the partial pressure of oxygen in the pulmonary blood of the alveoli; when this falls, blood flow to that region becomes is reduced by vasoconstriction to shunt blood to other, better ventilated, areas.

**Adaptive responses to exercise**

Unlike the cardiovascular system, the respiratory system itself is little affected by exercise or training, although the effects on associated body systems can be substantial. Secretion of erythropoietin from the kidney can increase the hematocrit (the proportion of the blood taken up by blood cells), and increased levels of 2,3-diphosphoglycerate in erythrocytes increase the likelihood that oxygen will be unloaded in respiring tissues. Some forms of training can increase a person's lactate threshold, allowing them to continue exercising by reducing the likelihood of significant acidosis.

**Related topics**

Metabolic processes (B3)
Cardiovascular response to exercise (D8)
Layout and structure of the respiratory tract (E1)
Alveolar exchange and gas transport (E4)
Utilization of $O_2$ and production of $CO_2$ in tissues (E5)
Chemoreceptors and respiratory acid–base balance (E6)
Exercising muscle (I5)

## Ventilation–perfusion matching

Oxygen ($O_2$) consumption rises dramatically during exercise. At rest a healthy 70 kg man utilizes around 300 ml $O_2$ $min^{-1}$, but this can rise to more than 4000 ml $O_2$ $min^{-1}$ during exercise in untrained individuals and more than 5500 ml $O_2$ $min^{-1}$ in trained aerobic athletes such as distance runners. As outlined in Topic

D8, **cardiac output ($\dot{Q}$)** can rise to around 20 l min$^{-1}$ from a baseline of 5 l min$^{-1}$. **Minute ventilation ($\dot{V}$)** can rise to around 100 l min$^{-1}$ from a baseline of around 6 l min$^{-1}$. This increase in $\dot{V}$ is brought about by increasing **tidal volume** and/or **ventilation rate**. An important variable in oxygen delivery is the **ventilation–perfusion ratio ($\dot{V}/\dot{Q}$)**. There would be no point in either increasing $\dot{V}$ if $\dot{Q}$ remained the same, or vice versa. At rest in healthy individuals, $\dot{V}/\dot{Q}$ is normally around 1, and may rise to around 5 during heavy exercise to keep **arterial $O_2$ tension ($P_{aO_2}$)** high under circumstances where active muscles are using a lot of $O_2$ and producing a lot of $CO_2$. *Table 1* shows what happens to some cardiorespiratory variables during exercise.

*Table 1. Effects of exercise on various ventilatory parameters*

| Intensity of exercise | $O_2$ consumption (ml (kg body weight)$^{-1}$ min$^{-1}$) | $\dot{V}$ (l min$^{-1}$) | $\dot{Q}$ (l min$^{-1}$) | $\dot{V}/\dot{Q}$ | $P_{AO_2}$ (mmHg) | $P_{aO_2}$ (mmHg) | $P_{aCO_2}$ (mmHg) | pH of arterial blood |
|---|---|---|---|---|---|---|---|---|
| Rest | 3.5 | 6 | 5 | 1.2 | 102 | 98 | 40 | 7.40 |
| Mild | 20 | 30 | 10 | 3 | 102 | 98 | 40 | 7.40 |
| Moderate | 50 | 60 | 15 | 4 | 105 | 98 | 35 | 7.40 |
| Heavy | 70 | 100 | 20 | 5 | 108 | 98 | 28 | 7.33 |

$\dot{V}$, ventilation; $\dot{Q}$, cardiac output; $\dot{V}/\dot{Q}$, ventilation–perfusion ratio; A: alveolar; a: arterial.

Note that, as the intensity of exercise increases, $\dot{V}/\dot{Q}$ and $P_{AO_2}$ (**alveolar $O_2$ tension**) increase, and **arterial carbon dioxide tension ($P_{aCO_2}$)** falls. However, the degree of increment in $\dot{V}/\dot{Q}$ is actually greatest on going from rest to mild exercise, even though $P_{aO_2}$, $P_{aCO_2}$ and arterial **blood pH** do not change between these two states. These parameters are the normal 'drivers' for increased ventilation, which begs the question why, if these parameters are unchanged between rest and mild exercise, there is such a large increase in $\dot{V}$ and $\dot{V}/\dot{Q}$. The perceived wisdom is that such changes in $\dot{V}$ and $\dot{V}/\dot{Q}$ are 'anticipatory' to some extent, and that muscle and joint receptors (see Topic H3), when stimulated, provoke an increase in $\dot{V}$ that is independent of changes in parameters sensed by arterial **chemoreceptors**. It is only at the heaviest levels of exercise that accumulation of **lactic acid** has any effect on blood pH, provoking the large increase in $\dot{V}$ that is observed, in order to blow off more $CO_2$ and try to maintain a normal arterial blood pH. This point is called the **lactate threshold**, and is dependent upon the person's level of training.

## Physiological dead space

As shown in Topic E3, a significant amount of the air that enters the lungs during quiet breathing is not used in gas exchange. In fact, of 500 ml tidal volume at rest, perhaps only 325 ml enters the respiratory parts of the airways. The remaining 175 ml is known as the **anatomical dead space** (*Fig. 1*), and stays in the conducting portion of the airways (nose, pharynx, larynx, trachea, bronchi, and bronchioles). Similarly, not all parts of the lungs are adequately perfused. Regional differences in both **alveolar ventilation** (*Fig. 1b*) and blood flow (*Fig. 1c*) mean that the apices of the lungs have a high $\dot{V}/\dot{Q}$ (>3.0), whereas the $\dot{V}/\dot{Q}$ at the bases of the lungs is much lower (~0.6). For whatever reason, some alveoli do not receive a good blood supply. What this means is that even if those alveoli are extremely well ventilated, their $\dot{V}/\dot{Q}$ will be too high for

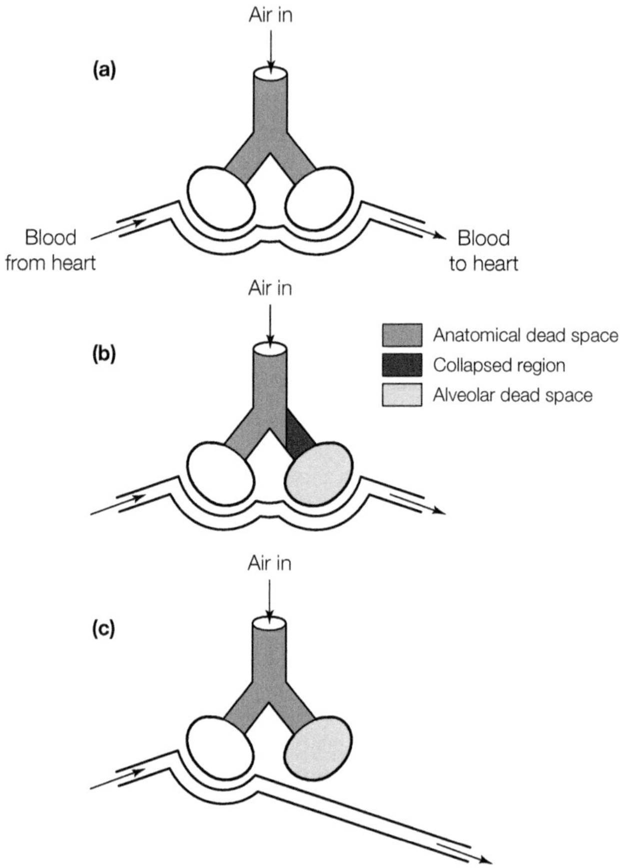


*Fig. 1. Contributions to physiological dead space in the lungs: (a) normal alveolar ventilation and alveolar perfusion; (b) normal alveolar perfusion, but inadequate ventilation; (c) normal alveolar ventilation, but inadequate perfusion.*

adequate oxygenation of or removal of $CO_2$ from the blood. These alveoli, together with those that are collapsed, are part of the **alveolar dead space**. However, provided that there are not too many such alveoli, this has little effect on total gas exchange. When total alveolar ventilation falls as a result of obstruction or disease, this has a much greater effect on gas exchange. The **physiological dead space** is the sum of the anatomical dead space and alveolar dead space.

The factors controlling the flow of blood in pulmonary capillaries are subtly different from those in the peripheral circulation. As cardiac output increases, pulmonary capillaries distend to accommodate this increased blood flow. This starts as a recruitment of collapsed vessels, then leads on to distention of all of the capillaries. This can help to prevent **pulmonary edema**, since an increase in **hydrostatic pressure** in the pulmonary capillaries would tend to force more fluid into the **interstitial space** (see Topic D5).

Another striking difference between the pulmonary and peripheral circulations is that low $P_{O_2}$ in the alveoli or in the pulmonary circulation causes constriction of small pulmonary arteries (hypoxia-induced pulmonary vasoconstriction, HPV), in contrast to the vasodilatatory response to **hypoxia** in the peripheral circulation. The mechanism of HPV has not been elucidated, but one can envisage that under conditions of mild regional **hypoxemia**, this would help increase $\dot{V}/\dot{Q}$ and allow more time for $O_2$ and $CO_2$ diffusion across the alveolar exchange epithelium, while at the same time diverting more blood to alveoli that are better ventilated.

## Adaptive responses to exercise

In contrast to its effects on the cardiovascular system, training has little effect on the respiratory system, except at exercise intensities approaching maximal aerobic performance. Almost all of the effects of training on our respiratory capabilities in exercise can be ascribed to changes in other systems, either the ability of the blood to carry more oxygen, or the reduced lactate production brought about by training. In terms of the ability of the blood to carry $O_2$ and deliver it to the tissues, the most important elements in adaptation to exercise are the circulating level of **erythropoietin (EPO)** and the amount of **2,3-diphosphoglycerate (DPG)** in erythrocytes.

EPO is released by the kidneys in response to hypoxia or (by virtue of the lower atmospheric pressure and $Po_2$ of inhaled air) living and training at altitude. EPO causes **polycythemia** by stimulating the production of **erythrocytes** in the bone marrow and increasing **hematocrit**. As erythrocyte numbers increase, so does the $O_2$-carrying potential of the blood. Some athletes have been known to take synthetic forms of EPO in an effort to improve their aerobic performance. However, in view of the fact that the hormone increases hematocrit (and blood viscosity), it can have catastrophic effects on **hemodynamics** (blood flow characteristics), especially in the pulmonary system. It is not unheard of for seemingly fit athletes with low resting heart rates to die mysteriously in their sleep.

DPG levels in erythrocytes increase in response to exercise and hypoxia. As outlined in Topic E4, higher levels of DPG lead to a rightward shift in the $O_2$–hemoglobin dissociation curve, meaning that the hemoglobin gives up $O_2$ at higher $Po_2$ than normal. Exercise, especially at altitude, leads to changes in both these parameters and in the $O_2$-delivery capability of the blood. Thus, exercise can increase the functionality of the cardiovascular system to cope with increased demand for oxygen, without affecting the lungs themselves.

As mentioned above, the lactate threshold is another factor that influences the likelihood that someone will be able to exercise at a high level for a long period of time. As shown in *Table 1*, $O_2$ delivery is not a limiting factor. In dynamic exercise like swimming, cycling, and running, the muscles start off exercise very easily on **oxidative metabolism** of **carbohydrates**. With more training, muscle cells can adapt to use fat as well. This adaptation decreases the likelihood of lactic acid production in muscle and the consequent **acidosis** that can result. It also increases the likelihood that the ADP/ATP ratio will remain low, as oxidative metabolism produces much more ATP per molecule of substrate than **anaerobic metabolism** does. It is not certain to what extent pain from tired muscles (with lactic acid accumulation) or intracellular events in the muscle (reductions in **ATP synthesis**) lead to fatigue and exhaustion. Pain may be a decided demotivating factor on top of failure of ATP synthesis. Nevertheless, adaptation of muscle as a result of training will increase a person's ability to continue with heavy exercise for longer periods. See Topics I4 and I5 for more details of muscle fiber types and the response of skeletal muscle to exercise.

# F1 Neurons and glia II

## Key Notes

**Morphology and location of neurons and glia**

The nervous system is composed of two types of cells, neurons and glia, which take many forms and have many different roles. Neurons have extensive processes, but glial cells do not; neurons are used in intercellular communication within the nervous system and between the nervous system and peripheral tissues. The processes of some neurons are insulated with myelin, which helps to increase the speed of conduction of nerve impulses. Glial cells have a number of different trophic and support roles in the nervous system, and some classes of glial cells (Schwann cells and oligodendrocytes) myelinate neuronal processes. Neurons are organized into groups with a common function, known as nuclei, and specific neural pathways exist in the nervous system.

**Schwann cells and myelination**

Myelination of neuronal processes speeds up conduction of nerve impulses. The glial cells in the peripheral nervous system that myelinate neurons are known as Schwann cells. The cell membrane of Schwann cells is wrapped around the neuronal process (the axon), and the phospholipid is modified to become a fatty covering known as myelin. This myelin insulates the axon, except where small gaps (nodes of Ranvier) occur between Schwann cells, and it is at these small gaps that the electrical activity associated with nerve impulses takes place.

**Related topics**

Neurons and glia I (C4)
Action potentials (F2)
Structure and function of peripheral nerve (H3)

## Morphology and location of neurons and glia

Nerve cells (**neurons**) and their supporting cells (glial cells or **glia**) are present throughout the nervous system. As described in Topic C4, neurons and glia take a multitude of forms and have many different roles. Section A of *Instant Notes in Neuroscience* gives a more detailed description of these brain cells than is necessary here, but it is necessary for the understanding of later chapters that a brief description is given here. There are between 300 and 500 billion neurons in the nervous system, and these take on many forms. However, the features of neurons that enable them to be distinguished from glial cells are their long cellular processes known as **neurites**, which extend from the cell body (*Fig. 1*). Glial cells do not have these processes. An individual neuron will have many **dendrites**, which receive input from other neurons, but normally only one **axon**, which sends output to other neurons. Axons can be extraordinarily long. Those from motor neurons in the lower parts of the spinal cord may have axons that are more than 1 m long. Dendrites and axons are also clearly distinguishable from the point of view of structure. For example, axons may be **myelinated** (insulated with a fatty covering), whereas dendrites are not. Axons contain only mitochondria, but dendrites, being shorter and therefore much closer to the cell body, contain all organelles, even Golgi apparatus.

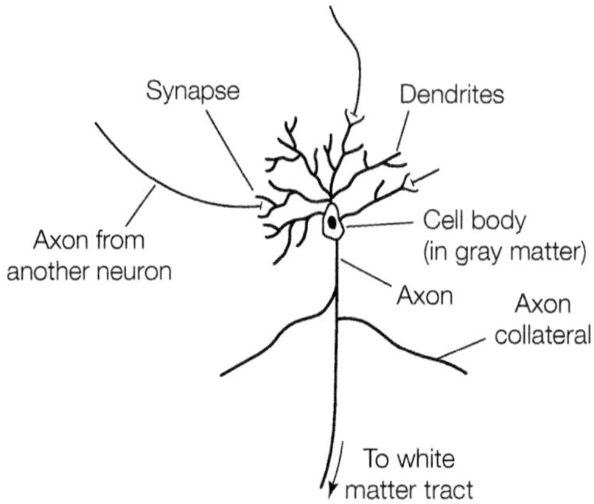


*Fig. 1. Basic structure of a neuron.*

Glial cells are, in the main, much smaller than neurons. They lack the long processes that neurons possess and are able, therefore, to influence only the local environment in which they reside. Glial cells are important in supporting the neurons of the nervous system. The function of important classes of glial cells (**oligodendrocytes**, **Schwann cells**, **astrocytes** and **microglia**) are outlined in Topic C4.

Although neurons and glia are present throughout the nervous system, the neurons are arranged into nuclei (collections of neuronal cell bodies with a similar function and which project to similar areas of the nervous system or the periphery). For example, the neurons that are responsible for initiating movement are collected in a discrete band of **cerebral cortex** known as the primary motor cortex in the frontal lobes of the brain (*Fig. 2a*). The cell bodies of neurons that form the final excitatory output pathway to voluntary muscle are present throughout the spinal cord, but they are always collected in the ventral (front) horn of gray matter (*Fig. 2b*). The cell bodies of those neurons that have receptors in the periphery for touch, pain and other **sensory modalities** are always present in collections of cell bodies known as ganglia, outside of the central nervous system (*Fig. 2c*). In this sense, the nervous system is highly organized.

## Schwann cells and myelination

Neurons have to communicate with each other and with cells in blood vessels, muscle (smooth and skeletal), and glands throughout the body. Given that the distances involved are vast on a cellular scale, it is important that the information is encoded and conveyed in such a way that it is not lost or garbled when it reaches its destination. Neurons use electrical conduction to send their messages to their effector cells, and it is important that this information is conveyed speedily. There is no point, for example, in the motor cortex sending a message to run away from a dangerous situation if it takes several seconds for the information to be conducted to the muscles of the legs. Neurons like these ensure rapid conduction of information (60–90 m $s^{-1}$) by insulating their axons with myelin.

The cells in the peripheral nervous system that myelinate axons are known as Schwann cells. These cells exist throughout peripheral nerve, and one Schwann cell myelinates a single axon. However, several Schwann cells are required to myelinate the entire length of an individual axon, as shown in *Fig. 3a*.

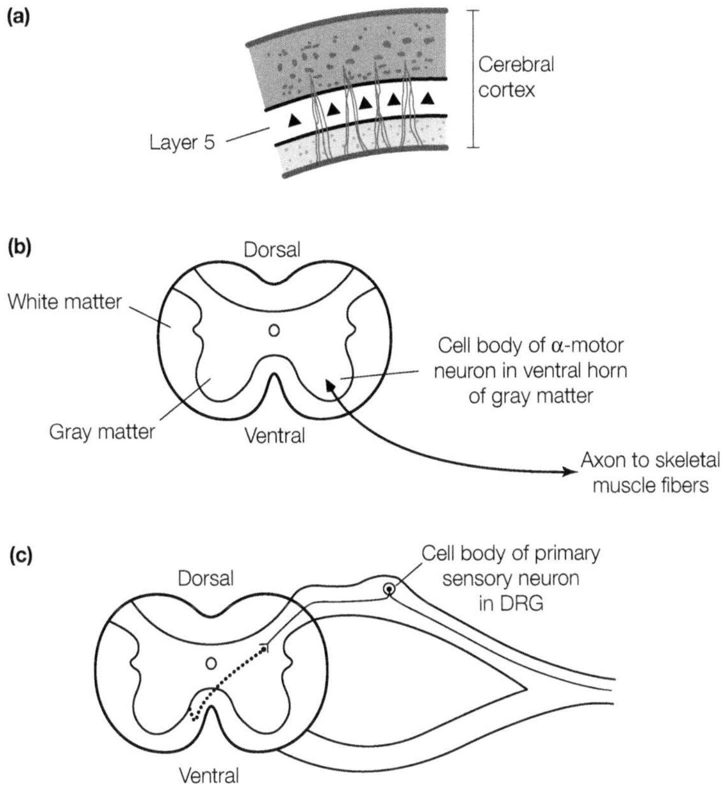


*Fig. 2. (a) Organization of the cerebral cortex into layers. Layer 5 is the location of the pyramidal upper motor neurons that control voluntary movement (Betz cells); (b) cross-section of spinal cord showing the location of α-motor neuron cell bodies; (c) cross-section of spinal cord and spinal nerve roots showing the location of the cell bodies of primary sensory neurons in the dorsal root ganglion (DRG).*

An individual Schwann cell myelinates between 0.15 and 1.5 mm of axon (*Fig. 3a*). Where the myelin sheath is wrapped around the axon, no current flow is possible between the extracellular fluid and the intracellular fluid of the axon (the **axoplasm**). This is akin to an insulating cover on an electrical wire. However, small regions of myelinated axons have no covering. This occurs in the gaps between each Schwann cell that myelinates a particular region of the axon. At these points, known as the **nodes of Ranvier**, the axonal membrane is exposed to the extracellular fluid, meaning that current can flow across the membrane in myelinated axons only at the nodes of Ranvier. Myelination has important implications for function, as outlined in topic F2.

**Myelin** is not simply a fatty substance that is secreted by Schwann cells. It is highly organized and is in fact made up of several layers of Schwann cell membrane, wrapped tightly around the axon (*Fig. 3b*). If a myelinated axon is viewed in cross-section at high magnification, it is actually possible to make out the many layers of Schwann cell membrane.

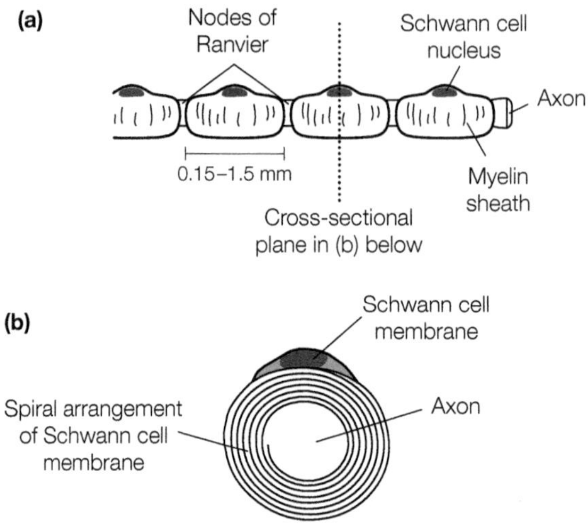


*Fig. 3. Myelination of peripheral nerve fibers by Schwann cells. (a) Longitudinal view; (b) cross-sectional view.*

# F2 ACTION POTENTIALS

## Key Notes

**Membrane potential**

The concentration of various ions is different on either side of the cell membrane. Such differences are set up by ion pumps in the cell membrane, and one of these pumps, the $Na^+/K^+$ ATPase, is electrogenic. This, together with the small potential differences set up close to the cell membrane by the potential diffusion of ions through ion channels, causes the inside of the cell to be slightly negatively charged compared to the outside of the cell. This potential difference is known as membrane potential, and can be calculated using the Goldman equation. The resting membrane potential of the cell is determined primarily by the higher permeability of the cell membrane to $K^+$ ions at rest. As membrane permeability to various ions changes, so does membrane potential.

**Neuronal ion channels**

The permeability of cell membranes to ions is determined by ion-specific protein channels. There are several types of channel for each of the major cations, and these are influenced to open ('gated') by different factors; some are gated by alterations in membrane potential and are known as voltage-sensitive ion channels. Excitable tissues such as neurons and muscle cells can rapidly switch their membrane potential by opening and closing voltage-sensitive ion channels such a $Na^+$ channels and $K^+$ channels.

**The neuronal action potential**

The rapid, organized fluctuations in membrane potential that take place in neurons and muscle cells in response to excitation are known as action potentials. These potentials can be measured using intracellular electrodes. Action potentials are brought about by predictable alterations in the permeability of the cell membrane to ions, via voltage-sensitive ion channels. In neurons, depolarization of the membrane is caused by opening of voltage-sensitive $Na^+$ channels, and repolarization is caused by opening of voltage-sensitive $K^+$ channels. At certain points in the neuronal action potential, no new action potential can be elicited (the refractory period); this has implications for neuronal function.

**Propagation of the action potential**

Nerve impulses are action potentials being propagated down a nerve fiber. Action potentials are propagated down axons by spread of the charge associated with movement of charged ions across the cell membrane. Action potentials could be conducted in both directions down an axon, but the refractoriness of the area of membrane that has just undergone an action potential means that conduction proceeds in one direction only.

**Effects of diameter and myelination**

Larger-diameter axons have faster conduction velocities, as do those that are myelinated. As axon diameter increases, the resistance to current flow down the cell is reduced, speeding up conduction velocity. Myelination increases conduction velocity of nerve impulses by preventing current leak, and by forcing current flow across the cell membrane to take place at the gaps in the myelin covering known as the nodes of Ranvier. Therefore, action potentials are conducted in myelinated nerve fibers by saltatory conduction.

**Related topics**

Membrane potential (B4)
Neurons and glia I (C4)
Neurons and glia II (F1)
Structure and function of peripheral nerve (H3)

## Membrane potential

As described in topic B4, the **plasmalemma** is only selectively permeable to ions through ion channels. Different concentrations of $K^+$, $Na^+$ and $Cl^-$ ions are maintained across the cell membrane by the action of **ion pumps**, which require **adenosine triphosphate (ATP)** to pump these ions against their concentration gradients. The most important example of these ion pumps is the **$Na^+/K^+$ ATPase**, which pumps out three $Na^+$ ions and pumps in two $K^+$ ions for every cycle of the pump. It is the action of this pump that sets up the different concentrations of $K^+$ and $Na^+$ ions on either side of the cell membrane – $K^+$ high inside and $Na^+$ high outside the cell. The pump is also electrogenic, in that more positive charge is moved out of the cell in a given period of time than is moved into the cell. This, together with the relatively high permeability of all cells to $K^+$ ions (there are many open $K^+$ 'leak' channels in most cells compared to $Na^+$ channels), means that the inside of all cells in the body is negative with respect to the outside of the cell. As described in topic B4, the theoretical **membrane potential ($E_m$)** of cells can be calculated using the **Goldman equation**:

$$E_m(\text{mV}) = -61\log_{10}\left(\frac{P_K[K^+]_o + P_{Na}[Na^+]_o + P_{Cl}[Cl^-]_i}{P_K[K^+]_i + P_{Na}[Na^+]_i + P_{Cl}[Cl^-]_o}\right) \quad (1)$$

Using known values for ion concentrations on either side of the cell membrane in neurons (*Table 1*), one can calculate $E_m$ for neurons at rest, when the permeability of the membrane to $K^+$ ions is 100-fold higher than to $Na^+$ or $Cl^-$ ions (i.e. $P_{Na}$ and $P_{Cl}$ in Equation (1) above are 0.01, compared to $P_K$ of 1.0). The actual resting $E_m$ in neurons is around –70 mV.

What is apparent from Equation (1) is that, as the relative permeability of the neuronal cell membrane to the various ions changes, $E_m$ should also change. If the membrane becomes more permeable to $Na^+$ ions, then $E_m$ will tend to move towards a more positive value (**depolarization**), and if it becomes more permeable to $K^+$ ions, it will tend to move towards a more negative value (**hyperpolarization**). During nerve activity, both of these things happen at certain times, through the gating of voltage-sensitive ion channels (see below).

*Table 1. Intracellular and extracellular concentrations of various ions in a nerve cell*

| Ion | Extracellular fluid concentration (mM) | Axoplasm concentration (mM) |
|---|---|---|
| $K^+$ | 2.5 | 115 |
| $Na^+$ | 145 | 14 |
| $Cl^-$ | 90 | 6 |
| $Proteins^-$ | Negligible | 100 |

## Neuronal ion channels

The permeability of neuronal cell membranes to ions is controlled by ion channels. These are proteins embedded in the neuronal cell membrane that form aqueous pores through the **phospholipid bilayer**, but which have integral 'gates' that block off the aqueous pore and prevent ions entering or leaving the neuron. Neurons have a multitude of different ion channels, most of which are selective for one particular ion over another. However, not all of these ion channels are identical, even if they are selective for the same ion. For example, $Na^+$ channels in neurons share common structural features, but are genetically diverse, being encoded by many different genes. The same is true of $K^+$ channels. This molecular diversity means that different (genetic) channels that regulate conductance of the membrane to the same ion may be affected by different extracellular and intracellular factors, one of which is $E_m$. Channels that are triggered to open or close by changes in $E_m$ are known as **voltage-sensitive (VS) ion channels**. Evidence suggests that the grouping of charged amino acid residues at certain points in the protein is critical for voltage sensitivity, and one can envisage that a change in voltage across the membrane could move these charged residues from one position to another, resulting in opening or closure of the channel. More information on the molecular character of VS ion channels is given in *Instant Notes in Neuroscience*, Section B.

In terms of understanding what is happening during nerve cell activity, it is most important to understand the differences between three types of ion channel on neuronal cell membranes. These are **'leak' $K^+$ channels**, **VS $Na^+$ channels** and **VS $K^+$ channels**. 'Leak' $K^+$ channels are open almost continually and are the reason why the cell membrane of neurons and other cells are more permeable to $K^+$ at rest. VS $Na^+$ channels are triggered to open (and thus allow $Na^+$ ions to flow into the cell, down their concentration gradient) when $E_m$ reaches a 'threshold' level (around –55 mV). VS $Na^+$ channels open very quickly, but close and become inactivated when $E_m$ rises above about 0 mV. The VS $K^+$ channels, when open, allow $K^+$ ions to leave the cell. VS $K^+$ channels are also triggered to open at the same 'threshold' $E_m$ as VS $Na^+$ channels, but they open and close much more slowly than VS $Na^+$ channels.

## The neuronal action potential

When a section of a nerve axon is active, an electrode introduced into the cell to record $E_m$ shows that $E_m$ changes in a characteristic manner (*Fig. 1*). This is known as the **action potential (AP)**.

APs are propagated along axons from the neuronal cell body to the end of the axon, where the neuron communicates with either another neuron or an effector cell (e.g. a skeletal muscle fiber) at a nerve terminal. What this means is that, when an AP is recorded experimentally from one region of an axon as in *Fig. 1*, the portion of axon immediately before that region (i.e. closer to the cell body) has already undergone an AP, and the portion immediately after (closer to the nerve terminal) is at rest. Within a few milliseconds, the AP will have moved a little closer to the nerve terminal. The changes in the state of VS ion channels that take place during the AP are shown in *Table 2*.

Note the changes in the state of the VS $Na^+$ channels during the AP. During phases 2 and 3 of the AP, these ion channels are open, which means that the permeability of the membrane to $Na^+$ ions is much higher than it was at rest (phase 0) and $E_m$ increases (depolarization). However, once $E_m$ rises above 0 mV, these ion channels become closed and inactivated, which means that they cannot be opened again, no matter the stimulus. Thus as VS $K^+$ channels open and $E_m$ falls again (phase 4, **repolarization**), $E_m$ passes –55 mV (the threshold for

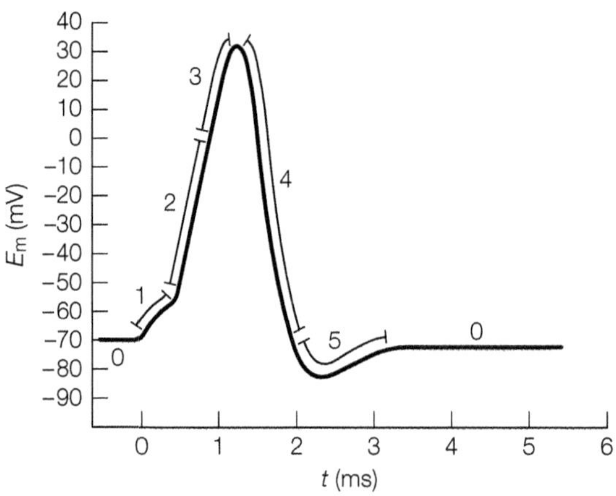


*Fig. 1. The neuronal action potential, as recorded using an intracellular electrode. $E_m$, membrane potential. For details of numbered phases, see Table 2.*

*Table 2. Phases of the action potential (AP) in neurons*

| Phase of AP | $E_m$ range (mV) | VS $Na^+$ channels | VS $K^+$ channels |
|---|---|---|---|
| 0 | –70 | Closed | Closed |
| 1 | –69 to –55 | Small subset of channels open | Closed |
| 2 | –55 to 0 | All channels open | Triggered, but closed (slow to open) |
| 3 | 0 to +30 | Some channels closing | Open |
| 4 | +30 to –70 | Closed and inactivated | Open |
| 5 | –70 to –85 | Closed | Some channels open, most closing |

$E_m$, membrane potential; VS, voltage-sensitive.

opening of VS $Na^+$ channels), but the VS $Na^+$ channels remain closed, allowing $E_m$ to keep falling. In phase 5, the VS $Na^+$ channels have been reset, but because some of the VS $K^+$ channels are still open (due to their slow kinetics), $E_m$ dips below –70 mV (hyperpolarization). This means that although VS $Na^+$ channels could theoretically be activated during this phase of the AP, it takes a greater stimulus to reach threshold (20–30 mV as opposed to 15 mV) and produce another AP.

These subtle changes in the opening states of the various channels mean that during phases 2–4, another AP cannot be initiated no matter how strong the depolarizing stimulus. The timespan of phases 2–4 (usually 2–3 ms) is known as the **absolute refractory period** of the neuron. During phase 5 of the AP, another AP can be initiated only if the depolarizing stimulus is significantly greater than normal. Phase 5 (usually 1–2 ms) is known as the **relative refractory period** of the neuron. These refractory periods have implications both for the direction of propagation of the AP (from cell body to nerve terminal) and for the maximal frequency of activation of an individual neuron.

### Propagation of the action potential

APs are, under normal circumstances, propagated from the cell body of a neuron to the axon terminal, as befits the role of neurons in passing on 'information' to other neurons and to effector cells. However, if a neuron is artificially excited (depolarized) somewhere along the axon, an AP can be propagated in both directions – towards the central nervous system (CNS) and out towards the periphery. This can be demonstrated by tapping the **ulnar nerve** at the elbow (sometimes if we knock this nerve, we say that we've knocked our 'funny bone'); involuntary movement of the muscles of the hand and a sensation of tingling can be experienced. The movement of the muscles of the hand is caused by APs being propagated from the elbow to the muscles of the hand and causing contraction, and the tingling sensation is caused by propagation of APs from the elbow to the CNS, where the brain assumes that the parts of the hand normally innervated by those neurons have in some way been stimulated. Thus, **AP propagation** in axons has the potential to be bidirectional, but is normally unidirectional. **Sensory neurons** are excited by stimuli in the periphery, and this information is conveyed to the CNS. **Motor neurons** are activated by descending input from other parts of the CNS, and this information is conveyed to the periphery. Conduction of an AP in the 'proper' direction is known as **orthodromic conduction**, and conduction of an AP in the 'incorrect' direction is known as **antidromic conduction**.

A diagram of the current flow involved in phase 3 of the AP in an axon is shown in *Fig. 2*. In the active portion of the axon, the inside of the cell has become positive with respect to the extracellular environment. The trigger for this depolarization was current flow from the refractory portion (closer to the cell body in the case of a motor neuron), which was then in phase 3, but is now in phase 4 or 5. The local circuit current associated with the active zone is not strong enough to reach areas of nerve cell axon that (although upstream of the AP) are now in phase 0.

When current flow from the active zone is large enough, this will depolarize the resting portion (closer to the effector cell in the case of a motor neuron),

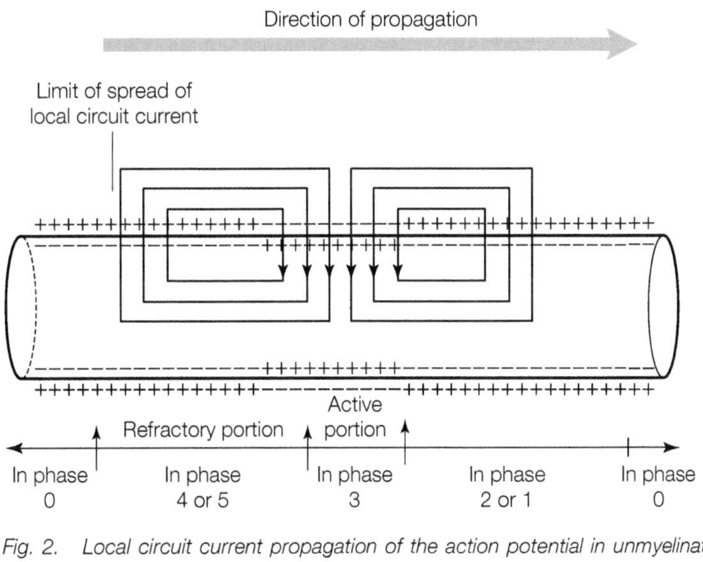


*Fig. 2. Local circuit current propagation of the action potential in unmyelinated axons. For details of numbered phases, see Table 2.*

which will eventually move from phase 0 to phase 1 or phase 2, but will not depolarize the refractory portion because either the VS $Na^+$ channels in the refractory portion are closed and inactivated, or the cell membrane is hyperpolarized. In this way, propagation of the AP occurs in the correct direction (from cell body to axon terminal in the case of a motor neuron). The currents involved in propagation of the AP are known as local circuit currents.

## Effects of diameter and myelination

Axon diameter is an important factor in determining the speed of AP propagation. As diameter increases, the resistance to current flow down the axoplasm is reduced and local circuit currents flow more quickly. The relationship between **conduction velocity (θ)** and axon diameter ($d$) in **unmyelinated axons** can be expressed thus:

$$\theta = kd^{1/2} \qquad (2)$$

where $k$ is a constant.

The unidirectional conduction of APs is true in both slow unmyelinated and fast myelinated axons. Even in the slowest unmyelinated axons, APs are conducted along the length of the axon at around 0.5 m $s^{-1}$ (0.5 mm $ms^{-1}$). Thus by the time the refractory zone has moved into phase 0 (2–3 ms), the active zone has moved some 1–1.5 mm down the axon. The current flow across the membrane associated with phase 2 of the AP is actually very small and decreases with distance, so it is most unusual for APs to 'jump' backwards in unmyelinated axons.

In myelinated axons, the neuronal cell membrane is extremely well insulated between the nodes of Ranvier, but membrane resistance is several thousand times lower at the nodes of Ranvier. This means that there is a saltatory flow of current through the membrane, with APs 'jumping' from one node of Ranvier to the next. This process is known as **saltatory conduction** (*Fig. 3*). When current flow reaches an upstream region where VS $Na^+$ channels are inactivated, current cannot flow across the membrane, preventing retrograde propagation of the AP. In comparison to the flow of current in unmyelinated axons, which depends upon local circuits to depolarize the next region of membrane, this method of

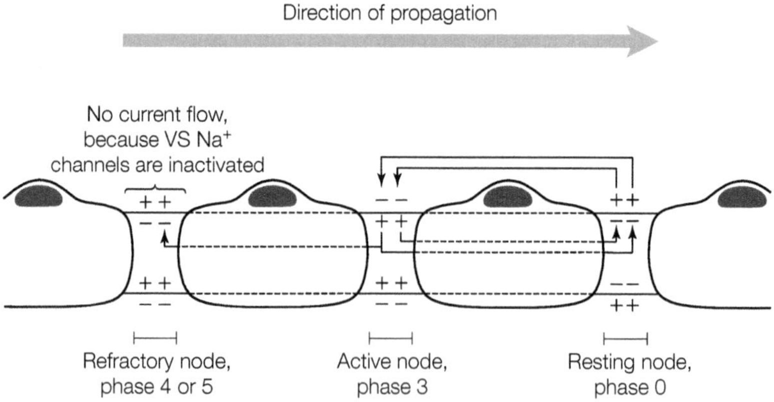


*Fig. 3. Saltatory propagation of the action potential in myelinated axons. For details of numbered phases, see Table 2.*

propagation is much quicker. The relationship between θ and $d$ in myelinated axons can be expressed thus:

$$\theta = kd \quad (3)$$

This means that in an axon with a diameter (for example) of 9 μm, myelination increases the speed of conduction of APs by about threefold (substitute 9 for $d$ in Equation (2) and Equation (3)).

# F3 Synaptic transmission

## Key Notes

**What is synaptic transmission?**

Synaptic transmission is the process by which information is communicated onward to other neurons or effector cells. Neurons communicate at electrical or chemical synapses. Electrical synapses (gap junctions) are intercellular ion channels that couple the electrical activity of the two cells. In the chemical synapse, neurotransmitters diffuse and activate or inhibit the postsynaptic cell by binding to receptors on its surface.

**The four stages of synaptic transmission**

Synaptic transmission involves (1) synthesis and storage of neurotransmitter; (2) release of neurotransmitter; (3) activation of postsynaptic receptors; and (4) termination of the action of the transmitter. Most neurotransmitters are stored in vesicles until needed for release. When an action potential invades a presynaptic terminal, vesicles discharge their contents into the synapse, where it interacts with receptors on the postsynaptic membrane. After the neurotransmitter has activated its target receptors, its action is terminated by reuptake or enzymatic degradation.

**Types of synapse**

Synapses may be categorized according to the site of interaction with the target neuron. Most often, synapses are classified on the basis of their morphology. Type I synapses typically have a cleft width of 30 nm and are as much as 2 μm in diameter. These synapses are asymmetric, and commonly excitatory. Type II synapses are narrower, smaller, symmetrical and generally inhibitory.

**Volume transmission**

In certain parts of the nervous system, presynaptic terminals (as defined by the presence of vesicles) exist without evidence of postsynaptic specialization. Because the channel of communication is outside the synapse, this process is known as extrasynaptic or volume transmission.

**Related topics**

Action potentials (F2)
The panoply of transmitters (F4)
Action of transmitters on effector cells – receptors (F5)
Degradation and reuptake of transmitters (F6)

## What is synaptic transmission?

In Topic F2, we described how a neuron propagates electrical impulses (**action potentials**) from their point of generation at the cell body (the **axon hillock**), along the axon, to the nerve terminals. But what happens when the action potentials reach the terminals? How is this electrical impulse conveyed onward to its target or to the next cell in a sequence? **Synaptic transmission** is the process by which the information carried by the action potential is communicated onward to other neurons, or to effector cells such as muscles or glands.

Neurons communicate with other cells, most commonly at specialized points of contact or apposition called **synapses**, a term coined by the English physiologist Charles Sherrington in 1897. The word synapse is derived from the Greek *synapsis*, and means junction, a good description of what is essentially a handshake between cells. Synapses may be either electrical or chemical. Electrical synapses (also called **gap junctions**) are those in which information from the input (**presynaptic**) neuron is conveyed to the output (**postsynaptic**) cell directly through intercellular ion channels that bridge the gap and couple the electrical activity of the two cells involved. These are efficient means of communication, and are much more widespread in invertebrates than vertebrates, although electrical synapses have also been found in the mammalian brainstem, among other locations.

However, by far the most common means of intercellular communication is the **chemical synapse** in which **neurotransmitters** (see Topic F4) are released from a presynaptic terminal, diffuse across the **synaptic cleft**, and activate the postsynaptic cell by interacting with specific proteins (**receptors**) on its surface. Although chemical synapses are larger and slower than their electrical counterparts, they allow a greater range of effect upon the postsynaptic cell. For instance, they may either excite or inhibit the cell, and these effects may be rapid or prolonged.

## The four stages of synaptic transmission

Conceptually, synaptic transmission can best be considered as a four-step process involving (1) synthesis and storage of the neurotransmitter; (2) release of neurotransmitter into the synaptic cleft; (3) activation of receptors on the postsynaptic membrane; and (4) termination of the action of the transmitter. A schematic nerve terminal is shown in *Fig. 1*. These steps are, of course, generalizations and not rules to which all neurotransmitters adhere. Some neurotransmitters are not stored but are synthesized and released on demand. Others do not act on postsynaptic receptors, or may have no active process by which their action is terminated. Some of these are mentioned in Topic F4.

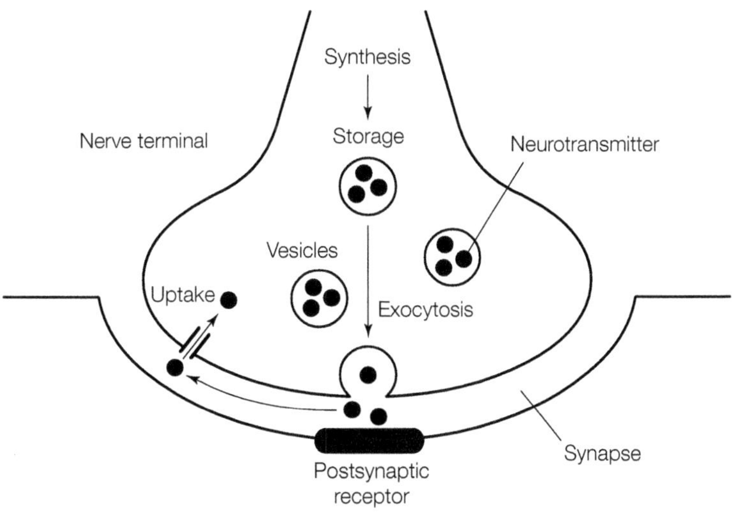


*Fig. 1. Schematic nerve terminal showing the key stages of neurotransmission from synthesis and storage through to termination of action (by reuptake in this case).*

The means by which neurotransmitters are synthesized differ as much as the molecules themselves. Representative examples are given in Topic F4, and will not be covered here. Nevertheless, once synthesized, most neurotransmitters are packaged into small bags or '**vesicles**' between 30 and 150 nm in diameter, where they are stored until needed for release. Synaptic vesicles (sometimes also called storage granules) use ATP-dependent uptake processes to both take up and concentrate neurotransmitters within. Most vesicles contain other molecules too that may assist in the efficient packaging of the neurotransmitter. For instance the synaptic vesicles of noradrenergic neurons contain, in addition to **norepinephrine**, **adenosine triphosphate (ATP)** and **chromogranins**. These are released along with the neurotransmitter.

For many years it was held to be true that synaptic vesicles could only store a single neurotransmitter. This 'one neuron, one transmitter' doctrine was proposed by the British pharmacologist Sir Henry Dale in 1935. We now know that many, if not most, neurons contain multiple neurotransmitters. For instance **dopaminergic neurons** in the **nigrostriatal pathway** also contain **cholecystokinin**, a peptide neurotransmitter first discovered in the gastrointestinal tract. Storage of multiple neurotransmitters in the same neuron is known as **cotransmission**.

When an action potential invades a presynaptic terminal, a sequence of events is set in motion that triggers the release of neurotransmitter into the synaptic cleft. This process, generally involving fusion of the synaptic vesicles with the presynaptic membrane is called **exocytosis**. Central to exocytosis is the elevation of intracellular $Ca^{2+}$ level within the nerve terminal. $Ca^{2+}$ enters the terminal through **voltage-sensitive calcium channels (VSCCs)**. At normal resting membrane potential, VSCCs are closed but, once opened by the wave of depolarization, $Ca^{2+}$ travels down its concentration gradient into the nerve terminal cytosol. Vesicles that are ready for immediate release are clustered in active zones of the presynaptic terminal in close proximity to VSCCs.

The precise mechanism by which intracellular $Ca^{2+}$ evokes exocytosis remains at least partly conjectural, but is thought to involve binding of $Ca^{2+}$ to **synaptotagmin** which, in turn, promotes an interaction of **synaptobrevin** (on the vesicle) and **syntaxin** (on the presynaptic membrane) that facilitates vesicular docking to the cell membrane. In any case, the mechanics are outside the scope of this topic (see *Instant Notes in Neuroscience* Topic C5). Vesicles that have discharged their contents may be recycled by **endocytosis**.

When a synaptic vesicle fuses with the presynaptic membrane, the entire contents are discharged into the synapse where they diffuse across the synaptic cleft to the postsynaptic membrane. Each vesicular packet is considered to represent a single 'quantum' of neurotransmitter, and this means of release is sometimes called **quantal transmission**. The process of chemical neurotransmission introduces a **synaptic delay** between presynaptic and postsynaptic events, in contrast to electrical transmission. The duration of this delay is determined mainly by the width of the synapse. Many synaptic clefts in the brain are narrow (about 30 nm across), and neurotransmitter traverses this gap quickly. In the **autonomic nervous system**, synapses can be an order of magnitude wider, and synaptic transmission here is commensurately slower.

Once the neurotransmitter reaches the far side of the synapse, it interacts with receptors on the postsynaptic membrane where its effect on the postsynaptic cell is determined. This effect may be excitatory or inhibitory, large or small, fast or slow, discrete or prolonged. This is mostly determined by the number and types of postsynaptic receptor. Detailed description of postsynaptic receptors and their function can be found in Topics F4 and F5. Once the neurotransmitter has reached and activated its target receptors, its action must be terminated. This prevents continuous stimulation of the receptors, and returns the synapse to a state where further neurotransmitter release events can occur and exert their effects. The action of neurotransmitter may be terminated by various means. Many neurotransmitters have active **reuptake** systems. A neurotransmitter may be taken up by the terminal from which it was released, or by adjacent terminals. **Glial cells** may also mop up neurotransmitters from the extracellular space. For other neurotransmitters, their action may be terminated by **enzymatic degradation**, or even by diffusion away from the synapse. These processes are discussed in detail in Topic F6.

## Types of synapse

In simple terms, the synapse may be considered as a tripartite structure consisting of presynaptic terminal and a postsynaptic membrane separated by the synaptic cleft. The presynaptic terminal is essentially a swelling of the axon that encloses the synaptic vesicles. The postsynaptic membrane contains the receptors upon which a neurotransmitter acts. The synaptic cleft is an extracellular space between the pre- and postsynaptic structures.

Synapses may be categorized in many different ways. For instance, they may be categorized according the site of interaction with the target neuron. A junction onto another neuron's dendrite is an **axodendritic** synapse. If the synapse is formed with the cell body, it is described as **axosomatic**. Sometimes **dendrodendritic** synapses occur between the dendrites of adjacent cells.

Most often, synapses are classified on the basis of their morphology. Perhaps the most enduring classification of this kind was made by Gray in 1959, who found two morphologically distinct types of synapse in the cerebral cortex. These were classified as types I and II. **Type I synapses** typically have a cleft width of 30 nm and are as much as 2 μm in diameter. These synapses are asymmetric, having a greater postsynaptic than presynaptic thickening. Type I synapses are commonly excitatory.

**Type II synapses** are a little narrower (~20 nm) and smaller (less than 1 μm in diameter) and with similar presynaptic and postsynaptic membrane thickness and flattened oval vesicles. Type II synapses are generally inhibitory. *Fig. 2* shows both types.

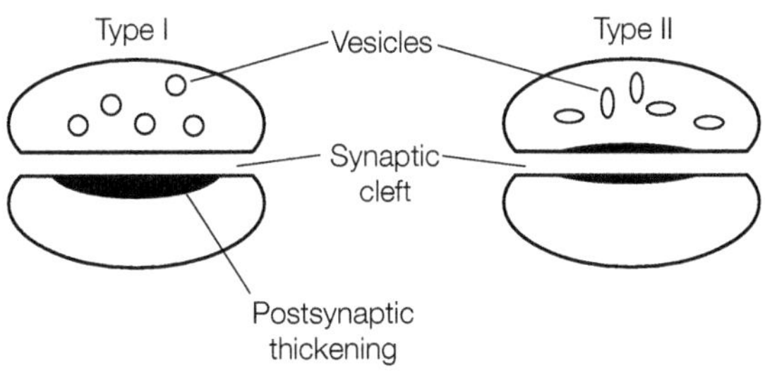


*Fig. 2. Gray's Type I and II synapses.*

## Volume transmission

Not all communication between neurons is effected through synaptic transmission in the strict sense. In certain parts of the nervous system, presynaptic terminals (as defined by the presence of vesicles) exist without evidence of postsynaptic specialization – in other words, without synapses. This is particularly common for the **monoamine** neurotransmitters where the receptors are located at some distance from the sites of neurotransmitter release. This means that the monoamines must diffuse through the extracellular space, sometimes substantial distances, to reach their target receptors. This feature, distinct from classical synaptic transmission, enables the monoamines to interact with many other neurons. Because the channel of communication is outside the synapse, this process is known as extrasynaptic or **volume transmission**.

# F4 THE PANOPLY OF TRANSMITTERS

## Key Notes

**Acetylcholine**

Acetylcholine (ACh) is synthesized from acetyl coenzyme A, and choline by the enzyme choline acetyl transferase, and stored in vesicles within cholinergic nerve terminals. On release, ACh can act on nicotinic or muscarinic receptor types to affect the activity of postsynaptic neurons or effector cells. Nicotinic ACh receptors are ligand-gated ion channels (ionotropic receptors) and muscarinic ACh receptors are G-protein-linked metabotropic receptors. ACh is degraded in the synapse by acetylcholinesterase to acetate and choline, and choline is taken up into the cholinergic nerve terminal by a choline transporter protein.

**Amines**

Several different (mono)amines are used as neurotransmitters in the nervous system. These include the catecholamines dopamine, norepinephrine and epinephrine, synthesized from tyrosine, and the indoleamine serotonin (5-hydroxytryptamine), synthesized from tryptophan. The monoamines are released from synaptic vesicles, and have effects on receptors that are almost exclusively metabotropic. Their action on postsynaptic cells is limited by reuptake, degradation and diffusion.

**Amino acids**

The amino acid neurotransmitters are glutamate, γ-aminobutyric acid (GABA), aspartate, and glycine. Glutamate and GABA are subject to vesicular release and have excitatory and inhibitory effects on postsynaptic cells, respectively. Both amino acids act on a mixture of ionotropic and metabotropic receptors, although ionotropic effects predominate. They are subject to high-affinity reuptake mechanisms that limit their action on postsynaptic cells.

**Peptides**

Some neurons utilize peptides such as the enkephalins as their neurotransmitters. Peptide neurotransmitters are synthesized near the cell body, and are transported to the nerve terminals via axonal transport. Their actions are commonly neuromodulatory and are terminated by degradative peptidases and by diffusion away from the synapse.

**Nitric oxide**

Nitric oxide (NO) is a lipophilic gas neurotransmitter that is synthesized on demand in presynaptic neurons from L-arginine using the enzyme NO synthase. Unlike other neurotransmitters, it is not stored in vesicles, but diffuses out of the neuron, across the synaptic cleft and into a postsynaptic cell, where it increases the production of cyclic guanosine monophosphate (cGMP) by guanylyl cyclase. cGMP then has metabolic effects in the postsynaptic cell. NO has a short half-life, and so no specific mechanisms for its degradation or reuptake are necessary.

| Related topics | | |
|---|---|---|
| | Gene transcription and protein translation (B2) | Degradation and reuptake of transmitters (F6) |
| | Action of transmitters on effector cells – receptors (F5) | The locomotor system (G2) |
| | | The limbic system (G3) |

## Acetylcholine

**Acetylcholine (ACh)** is a small-molecule transmitter (*Fig. 1*) synthesized by **cholinergic neurons** in both the central and peripheral nervous systems. ACh is synthesized near the nerve terminal from acetyl coenzyme A and choline by the enzyme **choline acetyl transferase (ChAT)**. It is stored in vesicles in nerve terminals, and released on depolarization of the nerve terminal. Cholinergic transmission in the CNS is closely linked to memory and cognition.

When released, ACh acts on two distinct receptor types, **nicotinic ACh receptors (nAChRs)** and **muscarinic ACh receptors (mAChRs)**. nAChRs are composed of five protein subunits arranged around a central cation channel which conducts both $Na^+$ and $K^+$ ions. When two molecules of ACh bind to the nAChR, the ion channel opens, and some $Na^+$ flows into and some $K^+$ flows out of the target cell. The net effect of this ion flow is depolarization (excitation) of the target cell. Such receptors are known as **ionotropic receptors**. This is a form of what is known as **'fast' neurotransmission**, because the ion flow that results has an almost instantaneous effect on **membrane potential ($E_m$)** in the target cell.

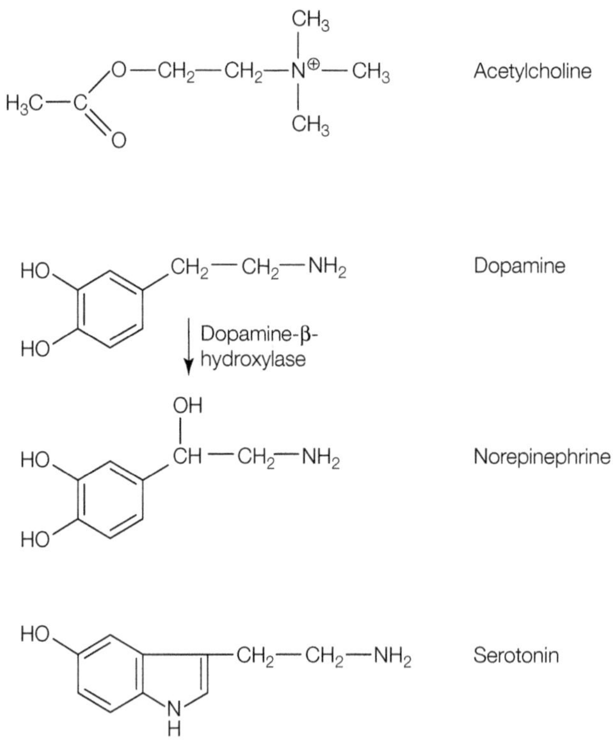


*Fig. 1. Molecular structures of acetylcholine, dopamine, norepinephrine and serotonin.*

mAChRs are **G-protein-linked receptors** with seven putative transmembrane domains. When ACh binds to mAChRs, this 'switches on' or 'switches off' a cellular enzyme. Most commonly, mAChRs are linked through **G-proteins** to the enzyme **adenylyl cyclase**, the enzyme that catalyzes the conversion of **adenosine triphosphate (ATP)** to **cyclic adenosine monophosphate (cAMP)**, which alters activity of other enzymes in the cell, some of which change the likelihood of certain ions channels being open. Such receptors are known as **metabotropic receptors**. Altering the activity of adenylyl cyclase therefore has the capacity to change intracellular concentrations of cAMP and alter the permeability of the membrane of the target cell to ions and alter $E_m$. The effects of ACh can be either excitatory or inhibitory, depending on whether the G-protein is stimulatory or inhibitory to the linked enzyme, although in most cases ACh acts to excite (depolarize) the target cell. This is a form of what is known as **'slow' neurotransmission**, because the effect of binding of the transmitter to the receptor is not immediately linked to changes in $E_m$. Nonetheless, the effects of ACh acting through mAChRs (and other transmitters that act through G-protein-linked receptors) still take only a few milliseconds to be transmitted into a change in $E_m$ in the target cell.

ACh is degraded by the enzyme **acetylcholinesterase** to form acetate and choline. There are high levels of acetylcholinesterase in the synaptic cleft between cholinergic neurons and their target cells. This is the major method for limiting the action of ACh action on target cells. Choline is then transported back into the cholinergic neuron by specific **choline uptake transporter** proteins embedded in the presynaptic membrane, and is used in the subsequent synthesis of more ACh.

## Amines

Another class of molecule commonly used by the nervous system is the (mono)amines, the most abundant of which are **dopamine**, **norepinephrine** and **serotonin** (*Fig. 1*). Dopamine and norepinephrine are catecholamines synthesized near the nerve terminal from the amino acid tyrosine in a synthetic pathway that is interlinked (e.g. dopamine is converted to norepinephrine by the action of the enzyme **dopamine-β-hydroxylase**). Serotonin (**5-hydroxytryptamine**) is an indoleamine synthesized near the nerve terminal from the amino acid tryptophan in a two-step pathway catalyzed by the enzymes **tryptophan hydroxylase** and **5-hydroxytryptophan decarboxylase**. Monoaminergic transmission in the CNS is closely linked to mood and reward pathways.

All of the amines are stored in vesicles and released upon depolarization of the nerve terminal. With one exception, all of the receptors for the monoamines that have so far been discovered are G-protein linked (*Table 1*), so their potential effects on target cells are diverse and can be excitatory or inhibitory. In all cases, the major method for limiting the action of the monoamines is by uptake of the

*Table 1. Types of receptors for monoamine neurotransmitters*

| Monoamine | Receptor type | Ionotropic or metabotropic |
|---|---|---|
| Dopamine | $D_1$–$D_5$ | Metabotropic |
| Norepinephrine and epinephrine | α- and β-adrenoceptors | Metabotropic |
| Serotonin | 5-$HT_3$ | Ionotropic |
| All others | (5-$HT_1$, 5-$HT_2$, 5-$HT_4$, 5-$HT_5$, 5-$HT_6$, 5-$HT_7$) | Metabotropic |

molecules into presynaptic terminals via specific **transporter proteins**. These transporter proteins are the targets for many drugs, some of which are legal (e.g. **Prozac** on serotonin transporters) and some of which are illegal (e.g. **cocaine** on dopamine transporters). A minor route for limiting the action of amines is the action of catalytic enzymes such as **catechol-*O*-methyltransferase** in the synaptic cleft.

## Amino acids

Several different amino acids are utilized by the nervous system as transmitters. The most important of these are **glutamate** and **γ-aminobutyric acid (GABA)**. The majority of fast excitatory transmission in the CNS is carried out by glutamate, and the major inhibitory transmitter is GABA (*Fig. 2*).

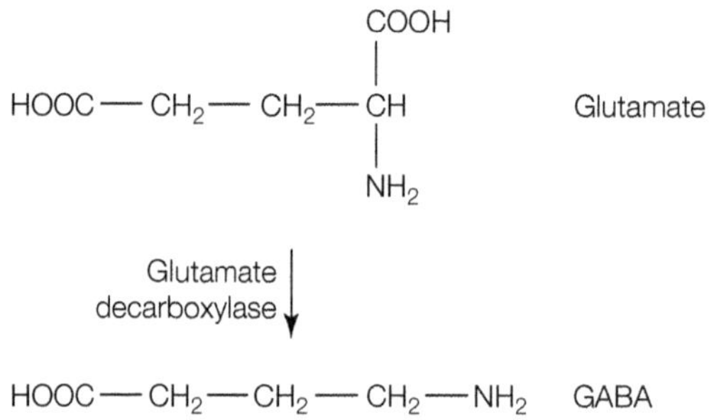


*Fig. 2. Molecular structures of glutamate and GABA (γ-aminobutyric acid).*

Glutamate is synthesized near the nerve terminal from **glutamine** and **α-ketoglutarate**, and is stored in vesicles that are released on depolarization of the nerve terminal. Some glutamate comes from dietary sources, and is transported into neurons and glial cells by specific uptake proteins. Glutamate receptors on target cells are most often ionotropic receptors of the **AMPA** (α-amino-3-hydroxyl-5-methyl-4-isoxazole-propionic acid) and **NMDA** (*N*-methyl-D-aspartate) types (meaning that they are activated by certain synthetic compounds as well as by glutamate). In both cases, activation of these receptors by glutamate results in excitation of the target cell. A very few glutamate receptors are metabotropic, and are not activated by either AMPA or NMDA. The major method for limiting the action of glutamate in the synapse is removal of glutamate by high-affinity transporters on the presynaptic membrane and on astrocytes. If taken up into the presynaptic terminal, the glutamate can be repackaged into vesicles, but in the glial cells, it is normally converted to **glutamine**, before being used by the neurons to make more glutamate. In view of its major excitatory role and the flux of $Ca^{2+}$ ions into the cell caused by NMDA receptor activation, glutamate has been implicated in mediating neuronal cell death in a number of **neurodegenerative diseases**, and in anoxic conditions such as stroke. This is known as the **excitotoxicity** hypothesis in which high intracellular $Ca^{2+}$ levels switch on a number of degradative enzymes inside the cell.

GABA is synthesized from glutamate near the nerve terminal of GABAergic neurons by action of the enzyme **glutamate decarboxylase** (*Fig. 2*). It is stored in vesicles that are released on depolarization of the presynaptic membrane terminal, and acts on postsynaptic GABA receptors. The majority of these receptors are of the **$GABA_A$** type, and are ionotropic receptors that conduct $Cl^-$ ions. On activation, the postsynaptic $E_m$ tends towards the equilibrium potential for $Cl^-$. This usually results in hyperpolarization of the target cell, making it less easily activated by excitatory input. Thus the action of GABA on $GABA_A$ receptors is inhibitory neurotransmission. **$GABA_B$** receptors are also inhibitory, but are G-protein linked (metabotropic) and open $K^+$ channels in target cells. The major route for limiting action of GABA is reuptake into GABAergic nerve terminals via a specific transporter protein. As GABA is the major inhibitory transmitter in the CNS, some drugs that elevate the levels of GABA in the brain are used to prevent **epilepsy**, a condition in which some regions of the brain are hyperexcitable.

## Peptides

In contrast to all other transmitter classes, peptides are not synthesized near the nerve terminal. This is because peptide transmitters are synthesized in the **endoplasmic reticulum** and **Golgi apparatus** (see Topic B1), and this does not extend into nerve cell **axons** (see Topic F1). Thus, peptide transmitters use **axonal transport** processes, meaning that they are ferried down the **cytoskeleton** to be used as transmitters at nerve terminals. Peptide transmitters synthesized by neurons in the CNS are not always subject to synaptic transmission. Some are released directly into the bloodstream to affect other cells in the CNS, or to affect cells in the periphery. Such peptides are part of what is known as the **neuroendocrine system** (see Sections K and L). An example of one class of peptide that is used in synaptic transmission is the **enkephalins** and these are described in more detail below.

Enkephalins are used as synaptic transmitters in some parts of both the central and peripheral nervous systems. Like all other peptide neurotransmitters, they are stored in dense-cored **secretory granules** made by the Golgi apparatus, and transported to the nerve terminal, where they are released by exocytosis when high-frequency action potentials reach the nerve terminal. Enkephalins act on **μ- and δ-opioid receptors** on target cells, which are seven transmembrane domain, G-protein linked receptors. Most often, the target cells for enkephalinergic neurons in the CNS are involved in pain pathways and activation of μ- and δ-opioid receptors result in a 'dampening' of **pain sensation**. Drugs that act on opioid receptors, such as **morphine** and **heroin**, have the same analgesic effect. Unlike other transmitters, no specific presynaptic uptake processes for the enkephalins or other peptide neurotransmitters exist. They are degraded by the action of **synaptic peptidases** into shorter, inactive peptides, or their constituent amino acids.

## Nitric oxide

One molecule that is used as a neurotransmitter that is entirely different from all of those described above is the gas **nitric oxide** (NO, *Fig. 3*). NO was originally discovered to be a relaxing factor released from endothelial cells in blood vessels (see Topic D8), but in recent years it has become apparent that it is used as a neurotransmitter in the nervous system. NO is synthesized *de novo* from **L-arginine** by the enzyme **nitric oxide synthase**, in response to rises in intracellular $Ca^{2+}$ ion concentrations. As a small lipophilic gas molecule, it is able to diffuse across the presynaptic cell membrane, enter target cells, and activate the enzyme **guanylyl cyclase**. Thus it is not stored in vesicles and is not released by exocytosis; neither are there any postsynaptic membrane-bound receptors for NO.

Guanylyl cyclase catalyses the formation of **cyclic guanosine monophosphate (cGMP)** from **guanosine triphosphate (GTP)**, and cGMP produces its effects on

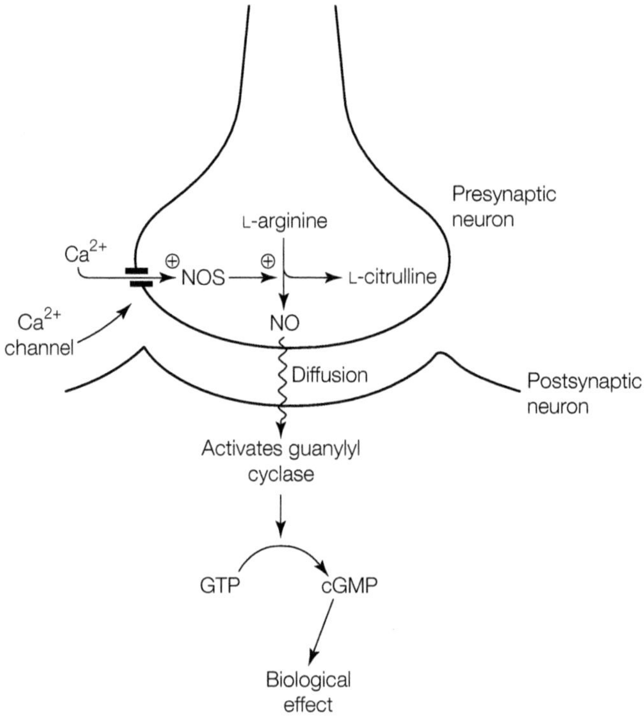


*Fig. 3. Synthesis and action of nitric oxide (NO). cGMP, cyclic guanosine monophosphate; GTP, guanosine triphosphate; NOS, calcium-dependent NO synthase.*

the target cells. Thus it acts through a second messenger system, but as NO has a very short half-life its effect is short-lived. It is only produced when required, only acts for a short period of time, and does not require degradative enzymes or reuptake to limit its effects. Conversion of NO to a highly reactive anion **peroxynitrite ($ONOO^-$)** has also been implicated in neurodegeneration, especially in response to anoxia.

# F5 Action of transmitters on effector cells – receptors

## Key Notes

**Receptor plurality**

Some neurotransmitters such as serotonin can interact with as many as 15 distinct receptors. This combination of neurotransmitter and receptor diversity allows an almost infinite range of communication between neurons.

**The development of receptor classification**

The concept of receptors was outlined by Langley in 1905. Much of the basic knowledge of receptors comes from the work of Gaddum, Clark, Schild, and others. Receptor classification is now based upon molecular biology. Most receptors have been cloned. Receptors may mediate very fast responses or act over seconds. Receptors may be ionotropic or metabotropic.

**Ionotropic receptors**

Ionotropic receptors mediate neurotransmission over milliseconds. These receptors consist of various subunits clustered around a central ion channel. When activated, ions enter to depolarize or hyperpolarize the cell. Neurotransmitters that act via ionotropic receptors include glutamate, acetylcholine, and glycine.

**Metabotropic receptors**

Most neurotransmitters act upon metabotropic receptors in which the neurotransmitter recognition site is coupled, via G-proteins to a transducer – a second messenger system or an ion channel. These receptors are monomeric with seven transmembrane domains. G-protein coupling is achieved via a long intracellular loop. Transduction systems associated with metabotropic receptors include adenylyl cyclase/cAMP and phospholipase C/inositol triphosphate systems.

**Receptor location**

Many receptors are found on the postsynaptic membrane where they are exposed to the highest concentration of neurotransmitter. Others are located outside the synaptic cleft and mediate slower forms of communication. Receptors found on the nerve terminal are said to be presynaptic, and respond to neurotransmitter released either from other nearby terminals (heteroceptors), or from the terminal itself (autoreceptors). Autoreceptors regulate the release and synthesis of neurotransmitter.

**Related topics**

Synaptic transmission (F3)
The panoply of transmitters (F4)
Degradation and reuptake of transmitters (F6)

## Receptor plurality

In Topic F3 we examined how **synaptic transmission** can be subdivided into four stages comprising synthesis and storage of the neurotransmitter, its release into the synapse, the activation of **receptors**, and finally, its removal by uptake or enzymatic degradation. The nature of communication between origin and target cell is most profoundly influenced by the interaction of two factors – the neurotransmitter released and the receptors upon which it acts. The known range of neurotransmitter molecules is enormous, extending from simple gases (e.g. **nitric oxide, NO**) through to large, complex peptides. Some examples are given in Topic F4. Supplementing this panoply of neurotransmitters is an equally magnificent array of receptors. While some **neurotransmitters** may act upon a single receptor, others such as **serotonin** can interact with as many as 15 distinct receptors. This combination of neurotransmitter and receptor diversity allows an almost infinite range of communication between neurons.

## The development of receptor classification

The concept of specific receptors on the cell surface was outlined by Langley in 1905, but was not given substance until the 1970s, when biochemical techniques were able to extract and purify receptor proteins and allow their characterization *ex vivo*. Previously, receptors were classified *in vivo* or in isolated tissues on the basis of their relative responses to series of **agonists** and **antagonists**. Much of what we know about the basic properties of receptors and about agonists and antagonists comes from the work of Gaddum, Clark, Schild, and others during that period. Classification of receptors during this period was based solely upon the **pharmacology** of tissue responses.

The principal drawback of such 'bioassays' of receptor activity is that it is not possible to separate the interaction of neurotransmitter and receptor (recognition) from its transduction into the tissue response. However, in the 1970s, receptor labeling methods allowed the biochemistry of the receptors to be examined in isolation from their function, and provided much more information about the recognition component.

**Receptor classification** nowadays is based upon **molecular biology**. Most receptors have been cloned, and their amino acid sequence elucidated. The discovery of their genes has often unmasked even greater diversity than was previously anticipated (see below). Sometimes, it has revealed that what were previously considered different receptors have the same genetic source, and are merely species homologs. Occasionally, receptors have been found for which there is currently no known endogenous ligand. These are termed **orphan receptors**.

There are many ways of subdividing and categorizing receptors. For instance, receptors may either excite or inhibit their target cell. Thus the action of a given neurotransmitter is dependent upon the receptor on the postsynaptic cell. Receptors may mediate very fast responses (within milliseconds), or act over a much longer time frame (seconds). Receptors may be subdivided according to the processes with which they are associated: they may open ion channels or be coupled to intracellular biochemical cascades. These are **ionotropic** and **metabotropic** receptors respectively.

## Ionotropic receptors

Ionotropic receptors, or **ligand-gated ion channels**, typically mediate **fast synaptic neurotransmission**, occurring over milliseconds. These receptors consist of various subunits clustered around a central pore or ion channel. Each subunit consists of a long sequence of amino acids that crosses the cell membrane several times. These crossing points are called **transmembrane**

**domains**, and their number differs from subunit to subunit. For instance, the subunits of the **GABA$_A$ receptor** each have four transmembrane domains, while those of the **NMDA receptor** contain only three true transmembrane domains. The general features of ionotropic receptors are shown in *Fig. 1*.

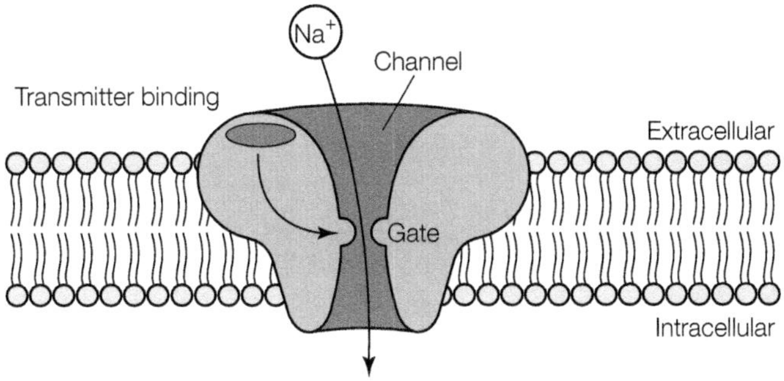


*Fig. 1. Schematic cross-section of an ionotropic receptor showing the transmitter binding site which, when activated, opens the gate to permit ion (in this case $Na^+$) flow via the channel into the cell.*

The building blocks of the GABA$_A$ receptor are the α-, β-, γ, δ-, ε-, and ρ-subunits which are assembled, in various permutations, into a pentameric structure surrounding a $Cl^-$ ion pore. When activated by GABA, the pore opens and $Cl^-$ ions enter to hyperpolarize the cell. The α- and β-subunits are essential for functional activity. Notably, there are many subtypes of each basic subunit. For instance, there are six known α-subunits, and therefore enormous receptor diversity is possible. In many respects, the GABA$_A$ receptor should be considered less as a single receptor than as an entire class of related receptors.

The subunits of ionotropic **glutamate receptors** (such as NMDA) have only three true transmembrane domains. These subunits are organized into a tetrameric structure that surrounds a non-specific cation channel. When opened by glutamate or aspartate, $Na^+$, $K^+$, and, to a lesser extent, $Ca^{2+}$ flow down their concentration gradients. The net effect is one of depolarization, with the inside of the neuron becoming more positive than at rest. Again, there are many subunit variants and thousands of possible combinations, theoretically at least.

Other neurotransmitters that act via ionotropic receptors include acetylcholine (**nicotinic ACh receptors**), glycine (**Gly-R**) and serotonin (**5-HT$_3$ receptors**). Acetylcholine and serotonin (but not glycine) also act on metabotropic receptors elsewhere.

## Metabotropic receptors

Most neurotransmitters act upon metabotropic receptors. These include norepinephrine, dopamine, serotonin and histamine, as well as most of those transmitters that also have ionotropic receptors such as glutamate, acetylcholine and GABA.

Metabotropic receptors are those in which the neurotransmitter recognition site is coupled, via **guanosine nucleotide binding proteins (G-proteins)** to a transducer – a **second messenger system** or an **ion channel** (*Fig. 2*). Unlike the ionotropic receptors, these receptors are monomeric rather than composite entities. Each metabotropic receptor is serpentine with seven transmembrane

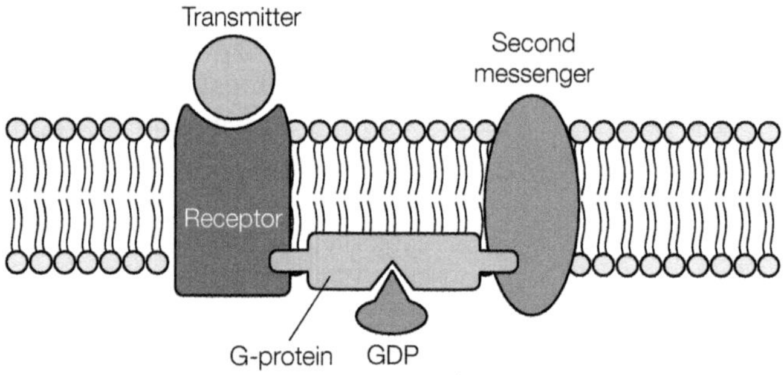


*Fig. 2. Schematic of a metabotropic receptor showing the G-protein (guanosine nucleotide-binding protein) linkage between the receptor and a second messenger system. In its associated state, the G-protein is bound to GDP (guanosine diphosphate).*

domains weaving in and out of the cell. Each of these is composed of a run of hydrophobic amino acid residues and is helical in structure. The receptors are therefore often referred to as **heptahelical**. Depending on the ligand, its binding site is either associated with the extracellular N-terminal loop or with the cleft between the transmembrane helices.

G-protein coupling between receptor and transducer is achieved via a long intracellular loop. In its resting state, the G-protein consists of closely associated α-, β-, and γ-subunits, bound to **guanosine diphosphate (GDP)** and anchored to the cell membrane. Activation of a metabotropic receptor causes replacement of the GDP by intracellular **guanosine triphosphate (GTP)**, and dissociation of the α-subunit from the βγ components of the G-protein. The α-GTP complex dissociates from the receptor and binds to the target transducer system. Binding to the target transducer increases the **GTPase** activity of the α-subunit, causing hydrolysis of GTP to GDP and re-association of the α- and βγ-subunits.

G-proteins exist in three main classes **$G_s$, $G_i$ and $G_q$**. Despite being a common biochemical transduction mechanism for so many different receptors, specificity for individual transmitter receptors is mainly achieved by as many as 20 α-subunit variants.

Various transduction systems are associated with metabotropic receptors, and it is beyond the scope of this (physiology) book to discuss each in detail. Such **second messenger systems** include the **adenylyl cyclase**/cAMP and the **phospholipase C**/inositol triphosphate systems. Second messenger molecules couple the binding of a hormone or neurotransmitter to an intracellular response, and also act to amplify the signal because although only one transmitter or hormone molecule may be required to activate a receptor, far more second messenger molecules are generally produced. Thus, the release or circulation of relatively low concentrations of transmitter or hormone can be transduced into powerful effects on target cells. More details of second messenger systems can be found in *Instant Notes in Biochemistry*, Topic E5.

## Receptor location

The most efficient way of translating a neurotransmitter release event into a response occurs when the receptor responsible is located close to the sites where the neurotransmitter is released. Not surprisingly, many receptors are found within synapses, on the postsynaptic membrane (see Topic F3). In this location, the receptors are exposed to the highest concentration of neurotransmitter and can rapidly translate this into a response.

Other receptors are located outside the synaptic cleft. Here they are exposed only to neurotransmitter that has escaped the synapse. These receptors are typically more sensitive than synaptic receptors, and can respond to the much lower levels of neurotransmitter that circulate in the extracellular space. Neurotransmitter takes longer to diffuse to these receptors, and consequently these receptors most often mediate slower forms of communication. This type of receptor is involved in volume transmission (see Topic F3).

Receptors can occur in both postsynaptic and presynaptic locations. Receptors found on the nerve terminal are said to be presynaptic. These presynaptic receptors may respond to neurotransmitter released either from other nearby terminals or from the terminal itself. When the presynaptic receptor responds to a different transmitter from that released by the terminal, it is called a **heteroceptor**. Conversely, presynaptic receptors that respond to the same transmitter as released by the terminal are termed **autoreceptors**. These terms are derived from the Greek words *heteros* and *auto*, meaning 'other' and 'self' respectively. Both may be either excitatory or inhibitory. Autoreceptors regulate the release and synthesis of neurotransmitter.

# F6 DEGRADATION AND REUPTAKE OF TRANSMITTERS

## Key Notes

| | |
|---|---|
| **Termination of neurotransmitter action – general principles** | The action of neurotransmitters is terminated by reuptake, enzymatic degradation or diffusion away from the synapse. Some neurotransmitters may be subject to all three processes. |
| **Reuptake of neurotransmitters** | Reuptake of neurotransmitter by transporters is the most common means of clearance of transmitters (such as the monoamines). Monoamine uptake involves energy-dependent cotransport of $Na^+$ and $Cl^-$ into the terminal. Once taken up, monoamine may be metabolized or repackaged into vesicles. Further neurotransmitter reuptake systems are found on glial cells in the brain, or upon muscle in the periphery. The physiological function of these transporters is not clear. |
| **Enzymatic breakdown of neurotransmitters** | The action of acetylcholine is terminated by enzymatic degradation. Acetylcholinesterase is found in synapses and at the neuromuscular junction, where it cleaves acetylcholine into acetate and choline. Most choline is taken back into the cholinergic nerve terminal. |
| **Diffusion from the synapse** | Diffusion of neurotransmitter from the site of action is probably the principal means by which the postsynaptic effects of the peptides are curtailed. |
| **Targets of drug action** | The above processes are important sites of drug action in conditions where it is desirable to increase the concentration of neurotransmitter in the synapse. These include myasthenia gravis, Alzheimer's disease, depression and Parkinson's disease. |

**Related topics** Synaptic transmission (F3) The panoply of transmitters (F4) Action of transmitters on effector cells – receptors (F5)

### Termination of neurotransmitter action – general principles

As previously outlined (Topic F3), synaptic transmission is essentially a sequential process consisting of neurotransmitter synthesis and storage, release into the synapse, and receptor activation, followed by some means of terminating the action of the neurotransmitter. This mechanism, whether for removal or degradation, is essential to synaptic transmission since it allows the process to begin again. Without a means of removing neurotransmitter from the synapse, postsynaptic receptors would potentially remain activated, and any further transmitter released by the nerve terminal would be unable to act upon the receptors. Thus neurotransmitter clearance does not limit so much as permit postsynaptic receptor activation.

There are three principal means by which the action of neurotransmitters is terminated. These are **reuptake**, **enzymatic degradation**, and **diffusion** away from the synapse. The actions of the **monoamines** are largely concluded by active reuptake. By contrast, the effects of **acetylcholine** are terminated by enzymatic degradation. **Acetylcholinesterase**, present in the synaptic cleft, splits the neurotransmitter into choline and acetate. Simple diffusion of neurotransmitter from the synapse appears to be the main mechanism by which the action of larger molecules such as the peptides is curtailed.

No process acts in isolation. For instance, neurotransmitters that have been removed from the synapse by reuptake will still be metabolized to biologically inactive products. Similarly, transmitters that escape the synapse by diffusion will ultimately be removed from the extracellular milieu by reuptake or metabolism, even if this process is not engaged in terminating their immediate postsynaptic effects. Some neurotransmitters may be subject to all three processes in varying degrees. The various means of transmitter inactivation are shown in *Fig. 1*.

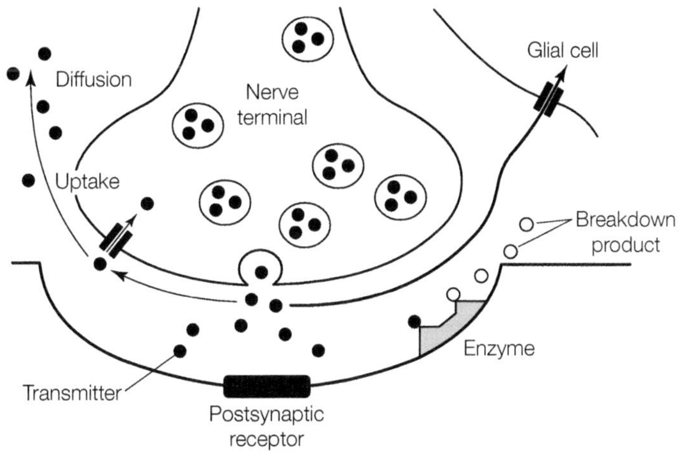


*Fig. 1. Transmitter inactivation processes. Transmitter may diffuse away from the synapse, be broken down by enzymes or taken up by transporters into neurons or glia.*

## Reuptake of neurotransmitters

Reuptake of neurotransmitter by saturable transporters is probably the most common means by which the actions of neurotransmitter are terminated. **Neurotransmitter transporters** are broadly location specific: for instance, **serotonin transporters** are only found on serotonergic neurons. However, the transporters are not necessarily substrate specific. The **norepinephrine transporter**, for instance, has a higher affinity for **dopamine** than for norepinephrine, and there is some degree of promiscuity among all of the monoamine transporters.

The monoamine transporters have all been cloned, and their structure is established. Each has 12 hydrophobic transmembrane domains with intracellular N- and C-terminals. The dopamine and norepinephrine transporters have similar amino acid sequences but differences in substrate specificity.

Although the precise mechanism by which the monoamine transporters sequester amines remains uncertain, it is known to involve **cotransport** of $Na^+$ and $Cl^-$ into the terminal. Typically $K^+$ is transported in the opposite direction.

This is an energy-dependent process, and the ionic gradients needed to maintain transport are generated by membrane **$Na^+/K^+$ ATPase**.

Once taken up by the transporter, monoamine may be subject to metabolism or to vesicular repackaging. The proportions that undergo each process may vary in different neuronal locations. The principal intracellular enzyme is **monoamine oxidase (MAO)**, which deaminates the monoamines. Dopamine is converted to **3,4-dihydroxyphenylacetic acid (DOPAC)** while **5-hydroxyindoleacetic acid (5-HIAA)** is the deaminated metabolite of serotonin. Other neurotransmitters whose actions are terminated by uptake include glutamate, aspartate, GABA, and epinephrine.

Further neurotransmitter reuptake systems are found on **glial cells** in the brain or upon muscle in the periphery. It is possible that there are others that are, as yet, undiscovered. Being located some way from the synapse, the function of these transporters is less clear, and it is uncertain whether they have a physiological role in terminating the effect of neurotransmitters. For instance, these transporter systems often have a low affinity (but often a high capacity) for neurotransmitters, and it is possible that they are only called into play during circumstances of pathological neuronal activity that result in significant overflow of transmitter from the synapse. Glial cells possess transport systems for glutamate, and it is thought that these have a role in limiting **excitotoxicity** following ischemic insults.

### Enzymatic breakdown of neurotransmitters

The action of acetylcholine is terminated by enzymatic degradation. Acetylcholinesterase is found in synapses and at the neuromuscular junction, mainly attached to the cell membrane where it cleaves acetylcholine into acetate and choline by hydrolysis with 'flashlike suddenness' as the pharmacologist Sir Henry Dale described it. The efficiency of acetylcholinesterase is such that the average lifespan of an acetylcholine molecule in the synapse is less than a millisecond, consistent with its role in mediating 'fast' neurotransmission at nicotinic synapses. Much of the choline generated by hydrolysis is taken back into the cholinergic nerve terminal, by a high-affinity transporter. The choline is then reused in synthesis of acetylcholine.

Enzymatic degradation plays a part, albeit smaller, in the clearance of **catecholamines**. Although the synaptic actions of the catecholamines are most often thought to be terminated by reuptake, at least some catecholamine is subject to direct enzymatic degradation by **catechol-*O*-methyltransferase (COMT)** within the synapse.

### Diffusion from the synapse

Diffusion of neurotransmitter from the site of action, the synapse, is the simplest way in which the action of a neurotransmitter is terminated. It is probably the principal means by which the postsynaptic effects of the peptides are curtailed. Various **peptidases** may also contribute to the degradation of peptides, but are probably not important in terminating postsynaptic effects.

Diffusion may also be relevant for other transmitters. For instance, recent studies have suggested that dopamine transporters are located at some distance from dopamine synapses, and thus that they may only take up dopamine that has already escaped from the synaptic cleft by diffusion.

### Targets of drug action

The above processes are also important sites of drug action in circumstances where one wishes to increase the concentration of neurotransmitter in the synapse. For instance, the symptoms of **myasthenia gravis**, an autoimmune disease, are caused by damage to the postsynaptic **nicotinic acetylcholine receptors** at the **neuromuscular junction**. Since there are fewer receptors to mediate a response, a successful compensatory treatment approach has been to prolong the presence of acetylcholine in the synaptic cleft by preventing its breakdown. Anticholinesterase drugs (such as **pyridostigmine**) allow acetylcholine to persist for longer, and help to counteract the effects of nicotinic receptor loss. A similar approach has been used in **Alzheimer's disease**, a neurodegenerative disease characterized by loss of cholinergic neurons in the brain. Again, anticholinesterases help to counterbalance the loss of acetylcholine release.

By the same token, neurotransmitter transporters are also useful sites of action for drugs. For instance, depression is considered to be due to a deficiency in brain noradrenergic and/or serotonergic neuronal function. Drugs that block the norepinephrine or serotonin transporter (such as **amitryptiline** or **fluoxetine** respectively) increase the levels of these transmitters in the synapse, and help to alleviate the symptoms of depression. In **Parkinson's disease**, where there is a loss of dopaminergic cells in the **nigrostriatal pathway**, dopamine uptake inhibitors such as **benztropine** and **nomifensine** have been used as adjunctive treatments.

# F7 Integration of nerve cell function

## Key Notes

**What is synaptic integration?**

In the CNS, nerve cells are subject to a constant bombardment of information from a huge range of input sources. The output of a neuron is dependent on the relative strength, timing, and sign of these inputs. This process is synaptic integration.

**EPSPs and IPSPs**

A neuron receives thousands of synaptic contacts on its dendrites and soma. At each of these, neurotransmitter release activates postsynaptic receptors and causes excitatory or inhibitory postsynaptic potentials (EPSPs and IPSPs respectively). Certain transmitters such as glutamate always evoke EPSPs. Others such as GABA only induce IPSPs. Many transmitters can elicit either.

**Basics of neuronal integration**

On its own, an EPSP is unable to elicit an action potential in a neuron. Neurons generate action potentials in the axon hillock, where inputs are integrated. Four aspects of postsynaptic potentials determine whether an action potential will occur. These are their frequency and sign, as well as their spatial and temporal localization.

**Temporal summation**

Temporal summation is a process by which individual postsynaptic potentials may add together (summate) when they arrive within a short interval of each other. A volley of EPSPs arriving in close succession can increase the membrane potential to the point where the axon hillock depolarizes, and an action potential ensues.

**Spatial summation**

A neuron must also simultaneously integrate information arriving at different points and from different sources. When spatially separated inputs are activated simultaneously, the EPSPs are additive. This is spatial summation.

**Related topics**

Neurons and glia II (F1)
Action potentials (F2)
Synaptic transmission (F3)
The panoply of transmitters (F4)
Action of transmitters on effector cells – receptors (F5)

## What is synaptic integration?

The basic currency of neuronal information transfer is the **action potential**. In Topic F2, we describe how the action potential is propagated along the axon, and in Topics F3–F6 we examine the processes by which this information is communicated to the next cell in a sequence.

In simple **synapses**, such as the **neuromuscular junction** of skeletal muscle, there is reliable one-to-one transfer of presynaptic information to the postsynaptic cell. Each action potential that invades the presynaptic terminal evokes an action potential in the postsynaptic cell (see Topic I2). However, in the CNS, the situation is more complex, and such direct information transfer is rare. More often, a nerve cell is subject to a constant bombardment of information from a huge range of input sources. The output of a neuron is dependent on the relative strength, timing and sign of these inputs.

If we think of neuronal communication as a form of speech, a simple analogy helps. Each neuron hears a range of sometimes conflicting views – some softly, others forcibly expressed. Some are fleeting whispers, while others amount to persistent barracking. The neuron must listen to these viewpoints and weigh argument and counter-argument before deciding which way it is swayed. This process is **synaptic integration**.

## EPSPs and IPSPs

The action potential is essentially an all-or-nothing event, but the probability of an action potential being generated is determined by local fluxes of membrane potential in the vicinity of the **axon hillock**. A typical neuron receives thousands of synaptic contacts from other neurons, mostly on its **dendrites** and cell body (**soma**). At each of these synapses, neurotransmitter release activates postsynaptic receptors and causes a local change in **membrane potential**. These miniature potentials are known as excitatory or inhibitory postsynaptic potentials (**EPSPs** and **IPSPs** respectively), and are shown in *Fig. 1*.

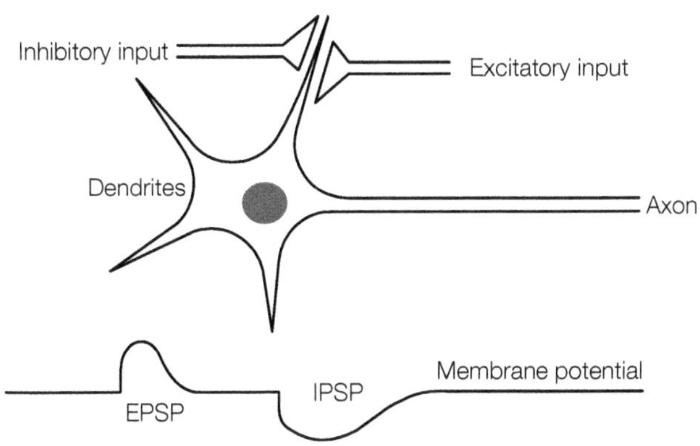


*Fig. 1. Excitatory and inhibitory inputs to a neuron cause EPSPs and IPSPs respectively. Two inputs to the neuron are shown – a fast excitatory EPSP caused by stimulation of an ionotropic receptor, and a slow IPSP induced by activation of an inhibitory metabotropic receptor. EPSP, excitatory postsynaptic potential; IPSP, inhibitory postsynaptic potential.*

The nature of the postsynaptic potential is determined by the neurotransmitter–receptor combination involved. Certain transmitters such as **glutamate** are always excitatory, and thus evoke EPSPs by facilitating $Na^+$ entry. Others such as **GABA** open $Cl^-$ channels, and only induce IPSPs. Many transmitters such as **acetylcholine** or **dopamine** can elicit either type of postsynaptic potential, depending on the receptor upon which they act. The type of receptor determines not only the sign (positive or negative) but also the persistence of the postsynaptic potential. **Ionotropic** receptors mediate faster responses than **metabotropic** receptors, for instance (see Topic F5).

## Basics of neuronal integration

On its own, an EPSP is unable to elicit an action potential in a neuron. Nor is a single IPSP able to prevent an action potential in an already active neuron. EPSPs and IPSPs only determine action potential generation (neuronal firing) when they act in concert. Neurons generate action potentials in a specialized region known as the axon hillock or initial segment of the axon. This part of the neuron has a particularly high density of **voltage-sensitive $Na^+$ channels**, and is therefore highly sensitive to changes in membrane potential. The axon hillock is the point at which the various inputs are integrated, and an action potential is either generated or prevented.

Four aspects of postsynaptic potentials determine whether an action potential will be generated. These are their frequency and sign (EPSP or IPSP), as well as their spatial and temporal localization. For instance, inputs must arrive at sufficiently high frequency and be of the right sign (excitatory) in order to increase the likelihood of an action potential. The localization of the synapses within the **dendritic tree** is also important. Postsynaptic potentials decay over distance, and those synapses closer (proximal) to the axon hillock exert a greater influence than those found in more distant (distal) dendritic branches. Similarly, since postsynaptic potentials decay with time, the interval between synaptic potentials is important. If a second potential is generated before the first has decayed, the effects are additive. This is known as **temporal summation**.

## Temporal summation

Temporal summation is a process by which individual postsynaptic potentials may add together (summate) when they arrive within a short interval of each other. This is illustrated in *Fig. 2*. When an excitatory input is activated, an EPSP is generated in the dendrites of the postsynaptic neuron. This EPSP decays with time and distance from its origin. A second incoming action potential arriving some time later generates another EPSP similar to the first. However, if the second EPSP occurs before the first has completely dissipated, the result is additive and a compound response occurs. A volley of EPSPs arriving in close succession can increase the membrane potential to the point where the axon hillock depolarizes and an action potential ensues. The same holds true for IPSPs, which separately induce brief periods of **hyperpolarization**. Sufficient

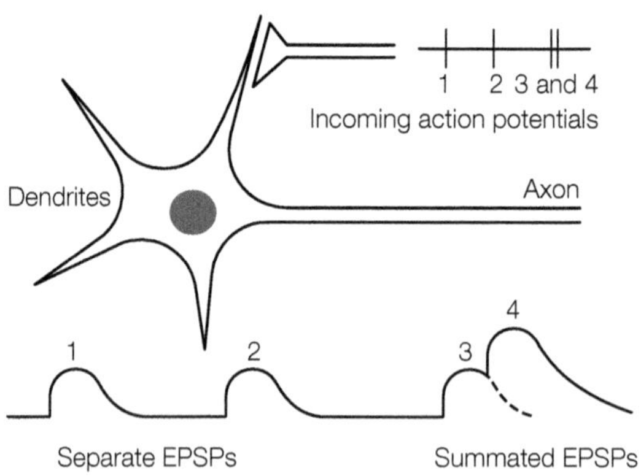


*Fig. 2. Temporal summation of EPSPs. The upper panel shows activation of an excitatory input by four incoming action potentials. The first two arrive separately, while the last two arrive in close succession. The lower panel shows summation of EPSPs when their arrival is close together. EPSP, excitatory postsynaptic potential.*

closely spaced IPSPs can prevent the generation of an action potential by the axon hillock.

**Spatial summation** In addition to processing information with respect to time, a neuron must also simultaneously integrate information arriving at different points and from different sources. This is **spatial summation**, and is illustrated in *Fig. 3*. Two separate excitatory inputs each separately induce an EPSP. When these inputs are activated at different times, the resulting EPSPs are distinct events. When they are activated simultaneously, the EPSPs summate in the same way as in temporal summation.

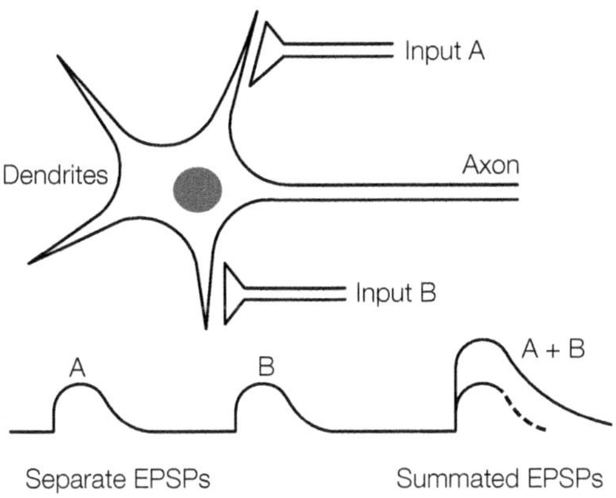


*Fig. 3. Spatial summation of EPSPs. The upper panel shows activation of two excitatory inputs (A and B) at different points in the dendritic tree. The lower panel shows summation of EPSPs from the two sources when they are simultaneously activated. EPSP, excitatory post-synaptic potential.*

Neuronal integration, in the form of temporal and spatial summation, is a dynamic process. A neuron is constantly balancing and assessing a range of changing inputs. Its output reflects the balance of excitatory and inhibitory inputs at any one time. In the same way that a neuron may have a thousand or more synapses upon its dendrites, its own axon terminals may form another thousand contacts with other neurons.

# F8 METHODS FOR ANALYZING NEUROCHEMISTRY

## Key Notes

**Brain tissue preparations**

Neurotransmission may be studied at various levels from the intact whole brain down to individual nerve terminals. Brain slices retain complete neuronal circuits, and allow the influence of pathways upon each other to be studied. Individual neonatal cells of the nervous system can be kept alive in a specialized culture medium for several weeks, allowing pre- and postsynaptic processes to be studied. Organotypic cultures have features of both brain slices and cell cultures. A further degree of reductionism is offered by the synaptosome. Synaptosomes have little neuronal structure, generate no action potentials, and form no synapses. Much of what we know about exocytosis has nonetheless been gleaned from synaptosomes.

**Techniques for studying neurochemistry**

Voltammetry is an electrochemical technique suitable for detection of the monoamine neurotransmitters at microelectrodes. A voltage gradient oxidizes any transmitter present on the electrode surface, releasing electrons. Voltammetric measurements last only milliseconds. A complementary method is microdialysis, in which a probe surrounded by a semipermeable membrane is perfused with an artificial cerebrospinal fluid. Neurotransmitters (and other small molecules) pass through the membrane and are collected in the perfusate for analysis, usually by chromatography.

**Studying neurotransmission in man**

In positron emission tomography (PET) scanning, a subject ingests a positron-emitting isotope prior to scanning. Clusters of gamma cameras surrounding the head of the person then reveal the spatial distribution of the isotope within the brain. Use of a dopamine receptor ligand enables visualization of dopamine receptors and determination of their activation by dopamine. If glucose analogs such as 2-deoxyglucose are used, one can localize areas of high brain activity involved in the performance of particular cognitive tasks. In functional magnetic resonance imaging (fMRI), subjects receive no radioisotopes. The technique is based on the capacity of nuclei to flip polarity when exposed to high-frequency changes in magnetic field direction. As with PET, fMRI can detect areas of neuronal activity. By not using radioisotopes, it is more suited to repeated usage.

**Related topics**

Neurons and glia II (F1)
Action potentials (F2)
Synaptic transmission (F3)
The panoply of transmitters (F4)
Action of transmitters on effector cells – receptors (F5)
Degradation and reuptake of transmitters (F6)
Integration of nerve cell function (F7)

**Brain tissue preparations**

To be able to put in context what we know about the processes of **neurotransmission**, it is helpful to have some understanding of the techniques by which the information was gleaned. For instance, much of what is known about the anatomy of nerves and nerve tissue has been gathered from post mortem techniques, in sections of animal and human brain tissues, using microscopy in some guise. However, neurotransmission is a dynamic process, and therefore much of our current grasp of the subject has been accrued, in some manner, from the study of living nervous tissue. Inevitably this has meant the use of animals, although many of the methodological approaches used here have since spawned clinical applications and led to new diagnostic tools and improvements in patient care.

**Neurochemistry** in general, and neurotransmission in particular may be studied at various levels, from the intact whole brain down to individual nerve terminals. The brain may be studied *in situ* (*in vivo*) or after removal from the body (*in vitro*). Living brain tissue may be kept alive *in vitro* in sections (**brain slices**), or as individual cells grown in a dish (**neuronal culture**). **Synaptosomes** (pinched off nerve endings) offer a still more reductionist slant on neurotransmission.

Each approach has complementary value. Brain slices, sectioned using a vibratome, are about a third of a millimeter thick, and retain at least some synaptic integrity. These can be kept alive for many hours *in vitro* by superfusion with oxygenated artificial cerebrospinal fluid. The rat hippocampal brain slice (*Fig. 1*) retains complete neuronal circuits and allows the influence of pathways upon each other to be studied. **Microelectrodes** can be used to investigate the activity of individual synapses. Such data could theoretically be obtained *in vivo*, but the animal would need to be anesthetized, and anesthesia has disruptive effects on neurotransmission. Brain slices provide a simple alternative.

Individual cells of the nervous system can be studied *in vitro*. Brain tissue is homogenized or digested with enzymes, and individual neurons separated from others. Often this separation process shears off axons and dendrites. However, the cells remain viable and can be kept alive in a specialized culture medium for several weeks, during which the cells develop new processes and can form synapses, allowing pre- and postsynaptic processes to be studied. The principal

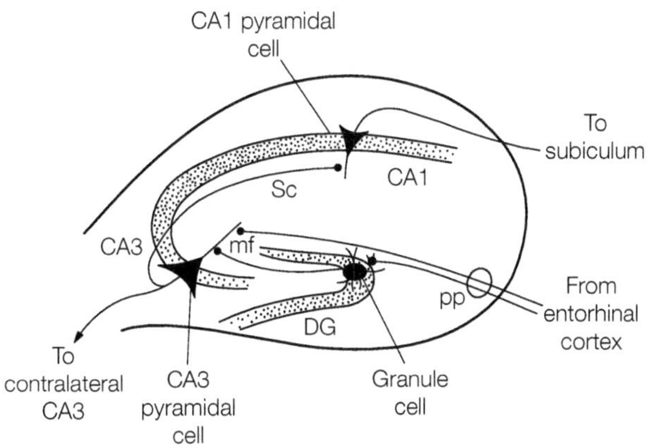


*Fig. 1. Rat hippocampal slice showing the interaction of afferent and efferent circuitry. CA, cornu ammonis fields; DG, dentate gyrus; mf, mossy fiber; pp, perforant pathway; Sc, Schaffer collateral pathway.*

caveat to neuronal culture is that it is necessary to use embryonic or neonatal brain tissue as a starting point.

**Organotypic cultures** have features of both brain slices and cell cultures. In this case, small pieces of tissue similar to slices are cultured, often with pieces from other parts of the brain. Over several days, the tissue thins to a monolayer and, depending on the brain regions and their distance, connections may form between the two.

A further degree of reductionism is offered by the synaptosome. Synaptosomes are formed by homogenization of neural tissue, often from a specific brain area, followed by differential centrifugation and fractionation of the homogenate. Under these conditions, most neuronal architecture is destroyed, and nerve terminals are separated from axons and cell bodies. Synaptosomes are these pinched off nerve terminals in which the membrane reseals. They are resuspended in an oxygenated buffer and remain viable *in vitro* for hours. Synaptosomes have little neuronal structure, generate no action potentials, and form no synapses. Nevertheless, they have proven valuable tools for studying the basic processes of neurotransmission. Much of what we know about **exocytotic release** of neurotransmitters (see Topic F3) has been gleaned from synaptosomes.

## Techniques for studying neurochemistry

In combination with the above brain tissue preparations, techniques have been developed to monitor the neurochemistry of neurotransmission. For many years it was impossible to measure neurotransmitter release directly, and it was necessary to infer its release from changes in levels of transmitter metabolites or by '**radiolabeling**' the transmitter pool with tritiated neurotransmitters, on the assumption that release of these mirrored release of the endogenous transmitter. More recently, techniques such as **microdialysis** or **voltammetry** have been developed that circumvent many of these drawbacks and allow the endogenous transmitter to be measured. These techniques provide complementary information about neurotransmission.

Voltammetry is an electrochemical technique suitable for detection of the **monoamine neurotransmitters** at microelectrodes. Essentially a **carbon microelectrode** is implanted into the brain tissue close to the synapse. This technique may be used *in vivo* or *in vitro*. It is particularly well suited to brain slices, where the microelectrode may be easily located in a brain region of interest. A voltage gradient is applied to the microelectrode and, at a given level, characteristic of the particular neurotransmitter being studied, any transmitter present on the electrode surface oxidizes, releasing electrons. The electrons are measured in the form of a current that is directly proportional to the amount of transmitter present. Voltammetric measurements last only milliseconds and can follow transmitter release and reuptake events faithfully.

A complementary method is microdialysis, in which a probe is implanted into a brain region of interest and the neurotransmitter is collected for analysis. The microdialysis probe is essentially a narrow tube surrounded by a semipermeable membrane. The tube is perfused with an **artificial cerebrospinal fluid**. Neurotransmitters (and other small molecules) pass through the membrane, and are collected in the perfusate for analysis, usually by **chromatography**. Large molecules such as proteins are excluded. Thus acetylcholine but not acetylcholinesterase would be collected. Microdialysis has the capacity to measure many neurotransmitters at the same location, but has limited time resolution. Measurements take several minutes.

## Studying neurotransmission in man

Most of our basic understanding of neurotransmission has been derived using the above techniques, applied either in animals or to animal brain tissue *in vitro*. More recently, microdialysis has been used in humans, albeit predominantly in intensive care, to study metabolic status in damaged tissue, and to measure the release of glutamate, a key mediator of excitotoxicity following stroke or head injury.

However, neurotransmission in a different sense has been investigated in man. Techniques such as **functional magnetic resonance imaging (fMRI)** and **positron emission tomography (PET)** can be used to study neurotransmission, or the consequences thereof, in a spatial domain. The advantages of these methods over, say, voltammetry and microdialysis are twofold. Firstly, the techniques are non-invasive and, secondly, they examine neurotransmission at a whole-brain level, localizing it to active areas. The main disadvantage is the inability to look at the time course of neurotransmission, except in a limited way.

In PET scanning, a subject ingests a positron emitting isotope prior to scanning. When positrons are emitted by the isotope, the positron is rapidly annihilated by spare electrons and gamma rays of identical energy are emitted at 180° to each other. Clusters of gamma cameras surrounding the head of the person (*Fig. 2*) then reveal the spatial distribution of the isotope within the brain. The source of radioactivity can be localized along a straight line between facing detectors

The nature of the isotopes used in such experiments determines the data obtainable. For instance, $^{11}C$-raclopride, a dopamine receptor ligand, can not only allow visualization of dopamine receptors within the brain, but also determine the extent to which they are activated by dopamine; in circumstances of high dopamine neurotransmission, dopamine will prevent $^{11}C$-raclopride from binding to the receptors and weaken the signal. If glucose analogs such as **2-deoxyglucose (2-DG)** are used, one can localize areas of high brain activity. 2-DG is taken up in the same way as glucose, and its uptake therefore reflects areas of neuronal activity. This approach has been used to illuminate brain areas involved in the performance of particular cognitive tasks.

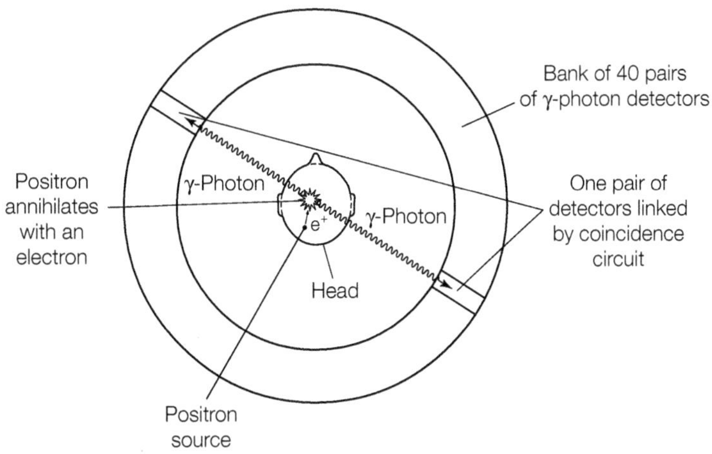


*Fig. 2. Principle of positron emission tomography (PET) scanning.*

fMRI differs from PET in that subjects receive no radioisotopes. The technique is based on the capacity of atomic nuclei to flip polarity when exposed to high-frequency changes in magnetic field direction. Oxyhemoglobin and deoxyhemoglobin have different magnetic resonance 'signatures', and can be distinguished by fMRI; when a given area of the brain is active, the ratio of oxyhemoglobin to deoxyhemoglobin changes, and this is detected by the scan. As with PET, fMRI can be used to detect areas of neuronal activity. By not using radioisotopes, it is more suited to repeated usage and can be used to examine neuronal responses engendered by learnt tasks.

# G1 Parts of the brain

## Key Notes

**Divisions of the brain**

The brain develops from the anterior portion of the neural tube. By embryological day 28, the anterior neural tube has divided into the hindbrain, midbrain and forebrain. By birth, the forebrain has further subdivided into the diencephalon and telencephalon.

**The brainstem (hindbrain and midbrain)**

The brainstem contains the midbrain, medulla and pons. The brainstem is responsible for the physiological housekeeping of the body, including feeding and drinking. The medulla controls breathing, regulates the circulation, and directs proprioceptive input to the cerebellum. The pons lies between the medulla and the midbrain. The ventral pontine nuclei relay information from the cerebral cortex to the cerebellum. Nuclei in the dorsal pons are involved in control of sleep. The cerebellum coordinates movement planning and the learning of motor skills. The brainstem tegmentum is involved in motor control and is the main point of entry and exit of the cranial nerves. The tectum consists of the superior and inferior colliculi. The superior colliculus receives mainly visual inputs. The inferior colliculus receives and processes auditory signals. Both project to the thalamus. Also located within the brainstem is the reticular formation, a basketwork of nuclei responsible for behavioral arousal.

**The diencephalon**

The diencephalon is the caudal part of the forebrain and contains two main groups of nuclei – the thalamus and hypothalamus. The thalamus is a staging point in exchange of ascending information between the periphery and the cortex, and acts as a sensory filter. The hypothalamus is engaged in homeostasis of heart rate, blood pressure, osmolality, and glucose levels. The hypothalamus also controls body temperature. Its effects are largely exerted through the autonomic nerves and the endocrine system.

**The telencephalon**

The telencephalon comprises the basal ganglia, amygdala and hippocampus, as well as the highly specialized cerebral cortex. The hippocampus has links with adjacent parts of the cerebral cortex and has a key role in long-term memory. The posterior part of the hippocampus is associated with spatial navigation. The amygdala reinforces memory by associating emotional importance to memories. The amygdala is most commonly associated with fear and the response to fear. The basal ganglia comprise the caudate nucleus and putamen (collectively called the striatum), and the globus pallidus. The basal ganglia and cerebral cortex form a motor circuit to initiate and control movement.

The cortex receives sensory information and initiates a motor response. Afferent inputs to the cortex project to the relevant primary sensory cortices. Adjacent cortical association areas receive inputs from the primary sensory cortices and modulate the sensory information. The primary motor cortex also has an association area that prepares the motor output of the primary motor cortex. Cortical cell bodies send axons horizontally between primary and association cortices. Other fibers cross the midline to communicate with the contralateral cortex, while the projection fibers form connections with other nuclei or the spinal cord.

**Related topics**

The locomotor system (G2)
The limbic system (G3)
Cranial nerves (G4)
Vision (G5)
Audition (G6)
The vestibular system and balance (G7)

## Divisions of the brain

Since the earliest anatomists first dissected the human brain, there has been a desire to describe, categorize and classify the structures observed. It is possible to subdivide the brain according to numerous anatomical or functional criteria. Functional subdivisions often encompass structures from different anatomical brain divisions. For the purpose of this topic, we will focus solely on brain anatomy, indicating function where appropriate.

The brain develops from the anterior portion of the neural tube. By embryological day 28, the anterior neural tube has divided into three discernible segments. These are the hindbrain (rhombencephalon), midbrain (mesencephalon) and forebrain (prosencephalon). By birth, the forebrain has been further subdivided into the caudal diencephalon and rostral telencephalon, and has become much bigger. The main subdivisions of the postnatal brain are shown in *Fig. 1*.

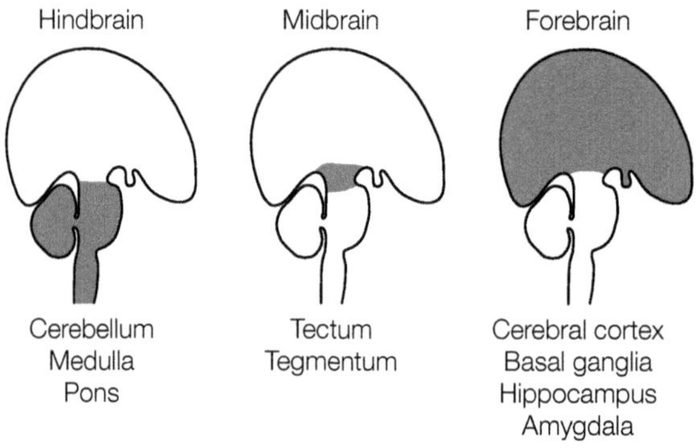


*Fig. 1. The principal anatomical subdivisions of the human brain.*

## The brainstem (hindbrain and midbrain)

Many texts complicate definitions by discussing the '**brainstem**'. The brainstem is a cross-structural definition and includes the midbrain and parts of the hindbrain (the **medulla** and the **pons**). The brainstem is found, in some form, in all vertebrates. It is responsible for much of the physiological housekeeping of the body, including feeding and drinking. The brainstem has many organizational similarities to the spinal cord with which it is connected. For instance, motor pathways pass through the ventral part of the brainstem, while sensory inputs ascend in the dorsal brainstem.

The medulla (or **medulla oblongata**) is the most caudal subdivision of the hindbrain, located between the spinal cord and the pons. Nuclei within the medulla control breathing and regulate the circulation. The medulla also helps to control balance by directing sensory and proprioceptive input to the **cerebellum**. Not surprisingly, damage to the medulla (such as occurs following a stroke in an artery that is a branch of the **basilar artery**) is often fatal.

Ascending rostrally, the pons (from the Latin meaning *bridge* and reflecting its role in linking the two cerebellar hemispheres) lies between the medulla and the midbrain. The ventral pontine nuclei relay information from the cerebral cortex to the cerebellum. Nuclei in the dorsal pons are involved in control of sleep.

The third major subdivision of the hindbrain, situated dorsal to the pons, is the cerebellum, a symmetrical bilateral structure shaped like a cauliflower. Like the **cerebral cortex** (see below), the cerebellum is highly folded. The cerebellum is mainly (but not wholly) a motor structure. In this context, the role of the cerebellum is to coordinate but not initiate body movements. The cerebellum plays a key part in the planning of movement and in the learning of motor skills. There is evidence that the cerebellum also has a role in language and cognition. Neuronal fibers enter and leave the cerebellum via three peduncles (from the Latin meaning *little feet*); the inferior, middle and superior **cerebellar peduncles**.

The midbrain comprises the **tegmentum** and **tectum**. The tegmentum is situated ventrally and is mainly involved in motor control, and nuclei within the tegmentum form relay points that link the cerebral hemispheres and the cerebellum. The midbrain tegmentum is also the point of exit of cranial nerve III, the **oculomotor nerve** (see Topic G4). The tectum (from the Latin, meaning *roof*) is situated dorsally and consists of the superior and inferior colliculi. The **superior colliculus** receives mainly visual inputs, but also some nociceptive and tactile information. The **inferior colliculus** receives and processes auditory signals. Both send projections to the **thalamus** that are then routed to the cerebral cortex. Cranial nerve IV, the **trochlear nerve** (see Topic G4), exits the brainstem from the tectum just inferior to the inferior colliculus. It is the only brainstem cranial nerve that does this; all others enter or exit the brainstem via its ventral aspect.

Also located within the brainstem is the **reticular formation**, a basketwork of nuclei that send extensive diffuse ascending and descending projections. The reticular formation is responsible for behavioral arousal as well as being influential in motor control, breathing and nociception.

## The diencephalon

The diencephalon is the caudal part of the forebrain and contains two main groups of nuclei – the thalamus and **hypothalamus**. The thalamus is the more dorsal of the two and consists of a cluster of several functionally distinct nuclei separated by narrow myelinated nerve fiber bands. The thalamus is most often described as a relay station between the periphery and the cerebral cortex, but this understates its pivotal role in sensory processing. Although the thalamus is indeed a staging point in ascending information exchange (both sensory and motor) between the periphery and cortex, it also acts as a vital sensory filter. When you cease to be aware of a dripping tap, this is due to thalamic **sensory filtration**. The thalamus comprises many nuclei, each specialized by sensory modality, and fibers leaving the thalamus project to the cortex via the **internal capsule**.

Located ventral to the thalamus is the hypothalamus, which is engaged in **homeostasis** (see Topic A2). As with the thalamus, the hypothalamus is a collection of interconnected nuclei. The hypothalamus represents only 1% of the volume of the brain, yet exerts disproportionately large effects. The **blood–brain**

**barrier**, which acts to separate vascular and neuronal compartments, is patchier in the hypothalamus, allowing this brain region to receive direct information about plasma osmolality and glucose levels. The hypothalamus is therefore uniquely positioned to control, for example, feeding, drinking and the ionic composition of the blood. The hypothalamus also controls body temperature. Its effects are exerted through the **autonomic** nerves (see Topic H5), the **endocrine system** (see Topic K5) and, indirectly, by altering motivation and drive. The hypothalamus plays a part in sexual behavior and is often considered to be part of the **limbic system**.

## The telencephalon

The telencephalon is by far the largest brain subdivision in man, and comprises the bilateral structures of the cerebral hemispheres. These include the **basal ganglia**, **amygdala**, and **hippocampus** as well as the highly specialized cerebral cortex.

Arcing bilaterally from the inferior horn of the **lateral ventricle** into the underside of the temporal lobe is the hippocampus, its name derived from its supposed resemblance to a seahorse. The hippocampus has links with adjacent parts of the cerebral cortex and has a key role in memory formation, particularly long-term memory. The posterior part of the hippocampus is associated with spatial navigation. A study published in 2000 by a research group at University College London supports this theory; they showed that the posterior hippocampus of London taxi drivers was larger than that of control subjects, and that the degree of increase in size of this region of the hippocampus was positively correlated with the length of time spent working as a taxi driver (see Further Reading p. 440 for a reference). Damage to the hippocampus prevents new memories being formed, although older memories are often intact, indicating that the hippocampus is part of the process by which new memories are laid down. **Temporal lobectomy** (which removes much of the hippocampus) was, for many years, an accepted practice for the treatment of intractable **epilepsy**.

Also located within the temporal lobe, and with connections to the hippocampus, is the amygdala, a key component of the limbic system. The amygdala also has a role in memory, where it is thought to reinforce the process by associating emotional importance to memories. The amygdala is most commonly associated with fear and the response to fear. Activity of the amygdala can trigger a 'fight or flight' response. However, it is now known that the amygdala also mediates aggression and anger. Many of these responses occur at an unconscious level.

More centrally located within the forebrain are the basal ganglia. Together these form the **extrapyramidal motor system**. The nuclei of the basal ganglia are the **caudate nucleus** and **putamen** (collectively called the **striatum**), and the **globus pallidus**. Also included in the basal ganglia, although not located in the telencephalon are the **subthalamic nucleus** (diencephalon) and **substantia nigra** (midbrain). The basal ganglia and cerebral cortex form a motor circuit whose main role is the initiation and control of movement.

Key components of this circuit are a dopaminergic projection from the substantia nigra to the striatum, and a reciprocal GABAergic (neurons utilizing $\gamma$-aminobutyric acid (GABA) as a neurotransmitter) pathway from the striatum to the substantia nigra. Damage to these pathways disrupts motor control in different ways. In **Huntington's disease**, **striatonigral** GABAergic neurons are lost, thereby removing much of the inhibitory control within the motor circuit.

Patients experience hyperkinetic uncontrolled movements and tics. In **Parkinson's disease, nigrostriatal** dopaminergic neurons are lost, and patients experience akinesia (an inability to initiate movement) and muscular rigidity.

Encircling these nuclei are the cerebral cortices. These are large folded structures grossly divided into four lobes each named after the bone plates of the skull, under which they sit. These are shown in *Fig. 2*. The folding of the cerebrocortical surface into **gyri** and **sulci** increases the surface area of the cortex many times. Covering this surface, to a variable depth of 2–4 mm, is the **gray matter** itself, the cell layer.

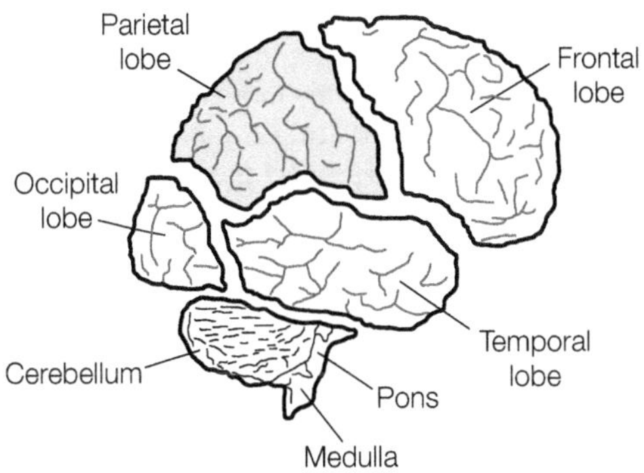


*Fig. 2. Exploded view of the cerebral hemispheres, showing the frontal, parietal, temporal, and occipital lobes of the cortex. The cortex is highly folded into troughs (sulci) and ridges (gyri).*

The cortex is responsible for receiving sensory information from a range of different modalities and for initiating motor response. The cortex determines how we feel and what we do. It allows us to plan actions, lets us see and imagine, read, and write. The cortex holds our memories and makes us recognize the familiar and learn about the unfamiliar.

The cerebral cortex has many different levels of organization. It can be divided by sensory input (auditory, somatosensory, visual) for instance. The **primary auditory cortex** is located in the **temporal lobe**, while **primary visual cortex** is at the base of the **occipital lobe**. Afferent inputs to the cortex project to the relevant primary sensory cortices. Adjacent to each primary cortex are the **cortical association areas**. These receive inputs from the primary sensory cortices and modulate the sensory information.

**Primary motor cortex** is located in the **frontal lobe**, along the boundary with the **parietal lobe**, and separated from the **somatosensory cortex** by the **central sulcus**. The primary motor cortex also has an association area that prepares the motor output of the primary motor cortex.

Cortical cell bodies send axons in three main directions. Many axons run horizontally between primary and association cortices. Other longer fibers cross the midline in the **corpus callosum** and **anterior commissure** and communicate with their opposite numbers, while the projection fibers leave the cortex to form connections with other nuclei or neurons in the **spinal cord**.

Although visually indistinguishable, and linked to each other, we know that the left and right halves of the cortex perform very different roles. For instance, language is localized primarily to the left hemisphere, while the right hemisphere is engaged in abstract reasoning.

# G2 THE LOCOMOTOR SYSTEM

## Key Notes

**Overview of motor control**

Movement is either reflexive or voluntary. Reflexive movement is an involuntary response to an external stimulus and is not subject to significant conscious modulation. Voluntary movement requires no external stimulus. Voluntary motor control is exerted chiefly through the synergistic effects of the basal ganglia, the cerebellum, and the motor cortex. The motor cortex initiates movement and sends descending projections to the spinal cord and brainstem via the corticobulbospinal tract. The basal ganglia act to select movement options and convey these to the motor cortex. The cerebellum modulates the output of the brainstem by acting as a comparator between intended and actual motor response.

**The motor cortex**

Not only does activation of the primary motor cortex induce contralateral movements, but this is somatotopically organized. Areas that induce wrist movement are close to those that move the hand and so on. Motor somatotopy is not however proportional to the size of the muscle groups innervated, but to the degree of control exerted by the motor cortex. The main collective output of the motor cortex is the corticobulbospinal tract. Upper motor neurons from the body region of the cortex form the corticospinal (pyramidal) pathway projecting to the spinal cord, while cells from the face region (corticobulbar neurons) project to the brainstem, on to cranial nerve motor nuclei.

**The basal ganglia**

The basal ganglia act to select appropriate movements, constantly filtering and comparing inputs. The basal ganglia consist of four main nuclei – the striatum (the caudate nucleus and putamen), the globus pallidus (comprising the external and internal pallidum), the substantia nigra, and the subthalamic nucleus. The function of the basal ganglia is a balance between two opposing output pathways from the striatum. The first 'direct' pathway from the striatum to the internal pallidum and substantia nigra increases thalamic activation of the motor cortex. Balanced against this is an 'indirect' pathway from the striatum to the external pallidum which, via the subthalamic nucleus, reduces thalamic activation of the cortex. The selection and execution of movement thus reflects a dynamic balance between the direct excitatory and indirect inhibitory pathways.

**The cerebellum**

The cerebellum acts as a motor comparator, receiving information on motor intention from the cerebral cortex, which it relates to afferent proprioceptive information, making corrections as needed. The cerebellum also has a role in learning complex motor behavior. The cerebellum is divided into three functional subdivisions (vestibulocerebellum, spinocerebellum, and neocerebellum). The vestibulocerebellum is the oldest part of cerebellum and receives inputs from the vestibular nerve, vestibular nuclei, and neck proprioceptors, via the inferior cerebellar peduncle. The principal role of the vestibulocerebellum is to control head balance and gaze during movement. The spinocerebellum receives sensory and proprioceptive inputs and has two main output pathways. A projection from vermal Purkinje cells to the vestibular nuclei and reticular formation integrates proprioceptive information to maintain balance and posture, and the paravermal Purkinje cells project to the red nucleus and thalamus to coordinate and correct movements. The neocerebellum initiates limb movements and helps the planning of complex motor behaviors.

**Other motor pathways**

The red nucleus in the midbrain receives inputs from the contralateral deep cerebellar nuclei and the ipsilateral motor cortex. The rubrospinal tract is involved in the control of both the flexor and extensor muscles. The reticulospinal pathway is really two counteracting pathways modulating muscle tone. Medial medullary neurons project to the spinal cord in the lateral column, where they facilitate activation of the motor neurons of the flexor muscles, while simultaneously inhibiting those of the extensors. Conversely, lateral pontine neurons facilitate the motor neurons to the extensor muscles, while inhibiting the flexor motor neurons.

**Related topics**

Parts of the brain (G1)

The vestibular system and balance (G7)

## Overview of motor control

Movement is either reflexive or voluntary. **Reflexive movement** is an involuntary response to an external stimulus and is not subject to significant conscious modulation. Much reflexive movement is mediated at the level of the spinal cord. **Voluntary movement** requires no external stimulus but is entirely self-generated. Voluntary movement is nonetheless subject to sensory input to direct it.

Voluntary motor control is exerted chiefly through the synergistic effects of the **basal ganglia**, the **cerebellum** and the **motor cortex**. A schematic representation of some of the major levels and directions of influence is shown in *Fig. 1*. Note that these do not represent monosynaptic connections.

The motor cortex initiates movement and sends descending projections to the **spinal cord** and **brainstem** via the **corticobulbospinal tract**. The basal ganglia act to select movement options and convey these to the motor cortex. The cerebellum modulates the output of the brainstem by acting as a comparator between intended and actual motor response. The cerebellum is also responsible for the learning and control of skilled complex movements.

## The motor cortex

First postulated by Paul Broca for speech, the notion of **cortical specialization** was supported by Hughlings Jackson, who observed that in patients with focal epilepsy, the seizures spread throughout the limbs in an orderly fashion, leading him to conclude that there must be a sequence of adjacent motor foci within the brain. The location and role of the motor cortex was first inferred in the mid-19th century during wartime, when physicians had unprecedented access to open head injuries. Touching the brain immediately anterior to the **central sulcus** (in the **precentral gyrus**) evoked movements on the contralateral side of

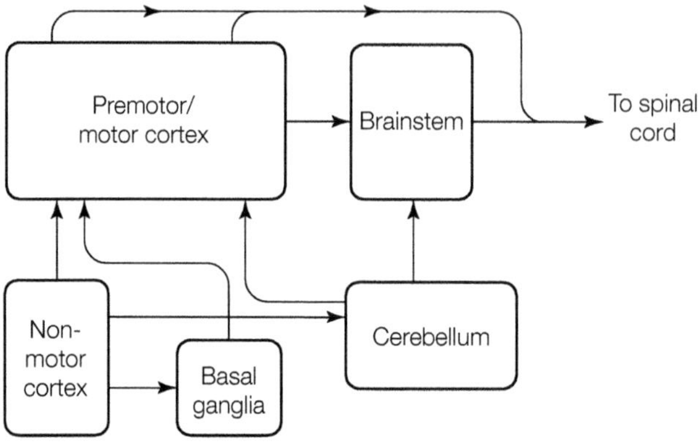


*Fig. 1. Major functional influences of the principal components of the locomotor system. Afferent sensory input is not shown.*

the body. Subsequent experiments in animals, and later in humans, established that stimulation of the **primary motor cortex** not only induced contralateral movements, but that this was **somatotopically organized**. Areas that induced wrist movement were close to those that moved the hand and so on. Motor somatotopy is not, however, proportional to the size of the muscle groups innervated, but to the degree of control exerted by the motor cortex: one cortical cell may determine the activation of several muscles related to the same movement. Thus the figure mapped onto the precentral gyrus, the **homunculus**, is distorted. The motor homunculus is shown in *Fig. 2*.

The primary motor cortex does not act in isolation. The motor cortex consists also of the **supplementary motor area** and **premotor cortex**. The supplementary and premotor cortices are connected with each other and with the primary motor cortex.

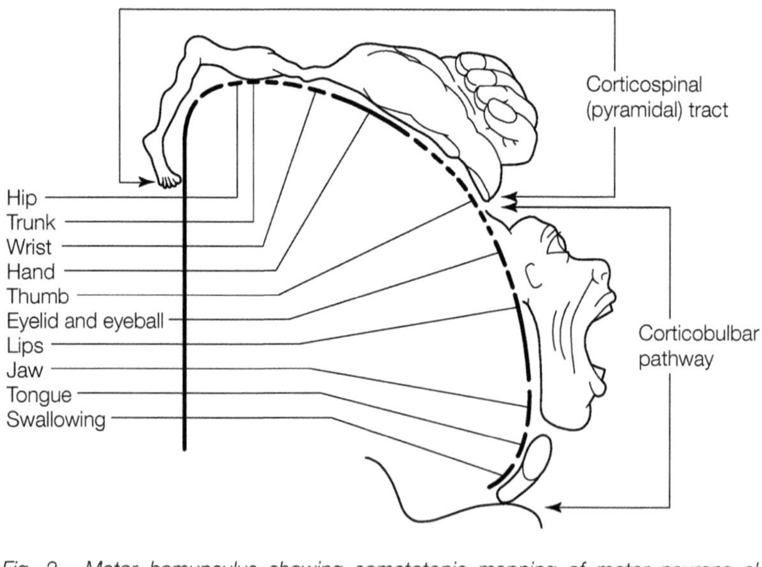


*Fig. 2. Motor homunculus showing somatotopic mapping of motor neurons along the precentral gyrus.*

The main collective output of the motor cortex is the corticobulbospinal tract. This has two components. **Upper motor neurons** from the body region of the cortex form the corticospinal (pyramidal) pathway projecting to the spinal cord, while cells from the face region (**corticobulbar** neurons) project to the **brainstem**. Both pathways are channelled through the **internal capsule**. Neurons of the corticobulbar component cross the midline (decussate) at the brainstem and make connections with **cranial nerve motor nuclei**. The remainder pass via the ventral **pyramids** of the medulla, at which point the majority decussate to become the **lateral corticospinal tract**. The remaining ipsilateral fibers form the ventral (**anterior**) **corticospinal tract**. Corticospinal neurons synapse either directly, or via **interneurons**, with **α-motor neurons** in the **ventral horn** of spinal cord gray matter. Corticospinal efferents generally excite flexor and inhibit extensor muscles.

Much of what we know about the function of the motor cortex comes from studies in which specific areas are damaged. Lesions of the supplementary motor area result in a contralateral **akinesia**. Bilateral damage to the premotor cortex impairs motor responses to visual or other sensory cues. Unilateral damage produces minimal symptoms, as the contralateral cortex is able to take over the functions of the damaged tissue. Damage to large areas of premotor or supplementary motor cortex lead to the clinical syndrome of **apraxia**, a difficulty in performing or sequencing complex motor tasks.

## The basal ganglia

The basal ganglia act to select appropriate movements, constantly filtering and comparing inputs. Most of the time, the basal ganglia act to suppress unwanted movement, only allowing the expression of appropriately selected motor subsets.

The basal ganglia consist of four main nuclei – the **striatum** (the **caudate nucleus** and **putamen**), the **globus pallidus** (comprising the external and internal pallidum), the **substantia nigra**, and the **subthalamic nucleus**. These act as a neural network to process the information received from the cortex which sends a massive, topographically mapped, projection to the striatum. Inputs are from all parts (sensory, association and motor) apart from primary visual or auditory cortices.

Direct inhibitory outputs from the substantia nigra and internal pallidum project to the ventrolateral and ventral anterior nuclei of the **thalamus** respectively, which also receive input from the cerebellum (see below). The thalamus, in turn, sends excitatory projections to the motor cortex, and thence to the efferent motor neurons. A further inhibitory pathway projects from the substantia nigra to the **superior colliculus** in the **midbrain**. This is shown schematically in *Fig. 3*.

The function of the basal ganglia, as a unit, can be thought of as a balance between two opposing output pathways from the striatum. The first is a **direct pathway** consisting of the inhibitory projections from the striatum to the internal pallidum and substantia nigra. Activation of these pathways inhibits further inhibitory efferent neurons that project to the thalamus. The net effect of this is to increase thalamic activation of the motor cortex. Balanced against this is an **indirect pathway** consisting of an inhibitory projection from the striatum to the external pallidum, which inhibits a second inhibitory path to the subthalamic nucleus. The output of the subthalamic nucleus is excitatory and projects to the internal pallidum and substantia nigra. The net effect of activation of the indirect circuit is to reduce thalamic activation of the cortex. The selection and

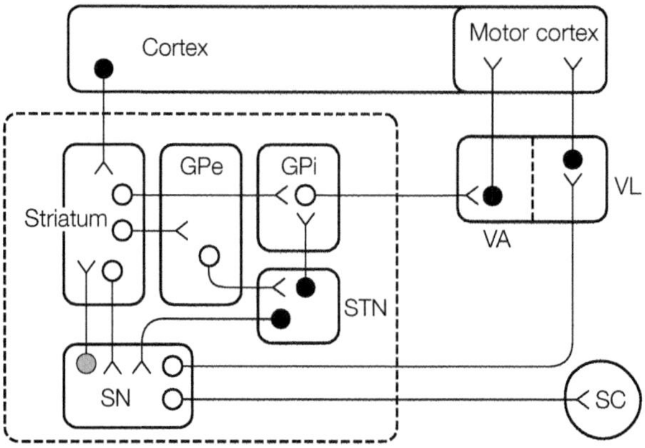


*Fig. 3. Intrinsic and extrinsic connections of the basal ganglia (within the dotted lines). Excitatory projections are shown with black cell bodies, while inhibitory pathways have white cell bodies. Gray cell bodies show projections with both excitatory and inhibitory components (e.g. dopamine-producing cells in the substantia nigra that either excite the 'direct' pathway through activation of $D_1$ dopamine receptors, or inhibit the 'indirect' pathway through activation of $D_2$ dopamine receptors). GPe, external globus pallidus; GPi, internal globus pallidus; SC, superior colliculus; SN, substantia nigra; STN, subthalamic nucleus; VA, ventral anterior thalamus; VL, ventrolateral thalamus.*

execution of movement thus reflects a dynamic balance between the direct excitatory and indirect inhibitory pathways.

Our understanding of the roles of the individual nuclei stems from patients with localized damage to individual components. For instance, damage to the globus pallidus gives rise to **athetosis** (a slow writhing), while lesions of the subthalamic nucleus are associated with **ballism** (violent, erratic movements of the limbs). Loss of the dopamine-containing cells of the substantia nigra causes **rigidity**, **akinesia** (lack of movements), and **resting tremor** while degeneration of the GABAergic neurons of the striatum (utilizing GABA as a neurotransmitter) gives rise to **chorea** (sudden, jerky movements). These last two examples occur in **Parkinson's disease** and **Huntington's disease**, respectively.

## The cerebellum

The cerebellum acts as a motor comparator, receiving information on motor intention from the cerebral cortex, which it relates to afferent proprioceptive information, making corrections as needed. The cerebellum also has a role in learning complex motor behavior such as playing the piano or riding a bicycle.

The cerebellum is divided into folia (the equivalent of cortical gyri but finer), and contains more neurons than the entire remainder of the brain. The outermost layer of the **cerebellar cortex** is the molecular layer, consisting mainly of axons and dendrites. Beneath is a narrow layer of large **Purkinje cells**, and below these are tightly packed granule cells. The three deep **cerebellar nuclei** – the fastigial, interpositus and dentate nuclei – are located beneath the cerebellar cortex. These, along with the **vestibular nuclei**, constitute the main output relays of the cerebellum.

Axonal traffic enters and exits the cerebellum via the three (inferior, middle, and superior) **cerebellar peduncles**. The cerebellum has three main inputs – from the spinocerebellar mossy fibers, olivocerebellar climbing fibers, and pontocerebellar mossy fibers relaying information from cerebral cortex.

Cerebellocortical input is channeled into the granule cell layer which sends excitatory projections to the Purkinje cells and to local inhibitory interneurons (stellate or basket cells). These make inhibitory connections with Purkinje cells and other interneurons. The output of the cerebellar cortex is an inhibitory GABAergic projection from the Purkinje cells to the deep cerebellar nuclei. Unlike the cerebrum, whose fibers cross the midline to innervate the contralateral side of the body, the cerebellum mostly innervates ipsilateral structures, and so lesions of the cerebellum cause ipsilateral signs and symptoms.

The cerebellum has three anatomical lobes (anterior, posterior and flocculonodular) and is divided into three functional subdivisions (**vestibulocerebellum**, **spinocerebellum**, and **neocerebellum**).

The vestibulocerebellum is the oldest part of cerebellum and receives inputs from the vestibular nerve, vestibular nuclei and neck proprioceptors, via the **inferior cerebellar peduncle**. The principal role of the vestibulocerebellum is to control head balance and gaze during movement. Vestibular information about head movement is sent initially to the vestibular nuclei (see Topic G7). This is integrated with information coming from the muscles in the neck about head movements, and forwarded to the vestibulocerebellum.

The vestibulocerebellum determines the muscle actions needed to keep the head balanced. Its principal outputs go to the fastigial and vestibular nuclei, and thence to the muscles (via the medial vestibulospinal tract (MVST) and to the limb extensors (via the lateral vestibulospinal tract (LVST)). The vestibulocerebellum forms a 'gain' system to modulate the vestibular ocular reflex (the reflex that ensures that gaze is fixed even when the head is moved, see Topic G7).

The spinocerebellum (comprising the vermis and anterior lobe of the cerebellar cortex, and the fastigial and interpositus nuclei) receives sensory (tactile, auditory, visual) and proprioceptive inputs. The spinocerebellum sends two main output pathways: a projection from vermal Purkinje cells goes via the fastigial nucleus to the vestibular nuclei and reticular formation. This controls the vestibulospinal and reticulospinal projections. The paravermal Purkinje cells project via the interpositus nuclei and **superior cerebellar peduncle** to the contralateral **red nucleus** and thalamus. The vermal pathway integrates proprioceptive information to maintain balance and posture. It also generates ambulatory motor patterns, while the paravermal outputs coordinate and correct movements.

The neocerebellum also projects to the contralateral red nucleus and thalamus. Its role is to initiate limb movements and to help determine the planning of complex motor behaviors.

Vestibulocerebellar damage is characterized by vertigo, an ataxic gait, loss of control of axial muscles and **nystagmus**. Spinocerebellar lesions (of the anterior lobe) manifest themselves in the form of ataxia, hypotonia and gait instability. Purkinje cell degeneration is particularly common in alcoholics. Neocerebellar damage leads to a loss of coordinated movement of the hands and face and an intention tremor.

## Other motor pathways

Three other pathways – two efferent, one afferent – should also be mentioned. The red nucleus (nucleus ruber) within the rostral midbrain receives inputs from the contralateral deep cerebellar nuclei and the ipsilateral motor cortex. The red nucleus is linked to the contralateral spinal cord via the **rubrospinal tract** which passes adjacent to the lateral corticospinal tract. The rubrospinal tract is involved in the control of both the flexor and extensor muscles, although its significance in man is unclear. Since the motor cortex projects to the red nucleus, the rubrospinal tract is, in many ways, an indirect corticospinal pathway.

The **reticulospinal pathway** is really two counteracting pathways and plays a pivotal role in modulation of muscle tone. Medial medullary neurons project to the spinal cord in the lateral column where they activate the motor neurons of the flexor muscles, while simultaneously inhibiting those of the extensors. Conversely, lateral pontine neurons facilitate the motor neurons to the extensor muscles, while inhibiting the flexor motor neurons. The excitatory drive from the cortex projects more to the medulla than the pons, which is mainly excited by ascending sensory fiber collaterals.

The spinocerebellum receives proprioceptive afferents through the posterior and anterior **spinocerebellar tracts**. Of these, the posterior spinocerebellar tract, which carries information from the trunk and arms, provides the larger contribution. The anterior spinocerebellar tract is mostly concerned with the control of posture, and sends proprioceptive information from the legs.

# G3 THE LIMBIC SYSTEM

## Key Notes

**What is the limbic system?**

One of the most controversial classifications of brain areas is the limbic system. The name was first coined in the 19th century by Pierre Paul Broca to define the group of structures that surrounded the brainstem. The first functional description of a limbic system was by James Papez in 1937, as an anatomical substrate for emotion. The hippocampus, amygdala, hypothalamus, and cingulate cortex are key areas, although some definitions include even the afferent midbrain projections to the forebrain.

**The hippocampus**

The principal role of the hippocampus (see Topic G1) is in the registration of memory, specifically the transfer of short- to long-term memory. However, because of its role in declarative memory, the hippocampus also has an explicit effect on emotion.

**The amygdala**

The amygdaloid complex is particularly associated with fear and aggression. When the amygdala is stimulated, animals typically respond with aggression and fear. Although fear is the most common emotion observed in man when the amygdala is stimulated, some patients find the experience causes pleasure. If the amygdala is destroyed, aggression and fear are both reduced.

**The hypothalamus**

The posterior hypothalamus is responsible for anger. Stimulation in animals evokes a rage. The hypothalamus is also responsible for modulating responses to pain, sexual satisfaction and satiety. The hypothalamus receives inputs from a number of sources. These include the hippocampus, brainstem and amygdala. Much of the output of the hypothalamus is expressed via the autonomic nervous system and neuroendocrine axis.

**The cingulate gyrus**

The cingulate gyrus has complex roles in emotion. The frontal portion appears to associate sensory input (olfactory and visual) with memories of previous emotions. Sadness activates the cingulate gyrus. This region is also engaged in the emotional response to pain, and in the regulation of aggressive behavior.

**Other limbic areas**

The ventral tegmental area is the source of the ascending dopamine pathways to the amygdala, prefrontal cortex, and nucleus accumbens. These pathways determine responses to novelty and one's search for novel experiences. They play a part in determining addiction.

**Related topics**

Parts of the brain (G1)

## What is the limbic system?

As described in Topic G1, classification of brain structure can be made at various levels. Classifications based upon definable brain structures are supported by anatomy. Functional subdivisions which may encompass structures from different anatomical brain divisions require evidence of both anatomical and functional connectivity.

One of the most controversial classifications of brain areas is the **limbic system**. The word limbic was first coined in the latter part of the 19th century by the French anatomist Pierre Paul Broca who used the terms *grand lobe limbique* (from the Latin *limbus* meaning border) to define the group of structures that surrounded the brainstem. He included several cortical regions – the **cingulate gyrus**, **subcallosal gyrus**, and **parahippocampal gyrus**, as well as the **hippocampus** itself. He focused on the inner edge or border of the cerebral hemispheres. As an anatomist, he described the structures and the connections between them, but did not address their function.

The first functional description of anything akin to a limbic system was by the neurologist James Papez who, in 1937, proposed a circuit that linked some of the regions described by Broca, as well as the hypothalamus, into what he considered to be an anatomical substrate for the expression of emotion. Papez had observed that many of his patients with damage to the hippocampus or cingulate cortex were subject to sudden emotional episodes. He reasoned, since emotions are consciously perceived, and higher cognitive functions can modify emotions, that each function must be reciprocally connected. Therefore, cortical structures might influence the functioning of the **hypothalamus**, an area felt by others to be involved in the expression of emotion. The **Papez circuit** is shown in *Fig. 1*.

The pivotal role of the **temporal lobe** in emotion was also shown in 1937, by Paul Bucy and Heinrich Klüver. Bilateral temporal lobe ablation in animals (which removes both the hippocampus and amygdala) causes a behavioral syndrome characterized by hypersexuality, an absence of fear, and visual agnosia. A similar pattern of behavior occurs rarely in man (**Klüver–Bucy syndrome**). This and other findings led Paul MacLean to extend the Papez circuit to encompass the **amygdala**, the **nucleus accumbens**, and more of the cortex and hypothalamus. MacLean also first coined the term 'limbic system'.

Despite more than half a century of research, there is still controversy over exactly what nuclei constitute the limbic system. Although the hippocampus,

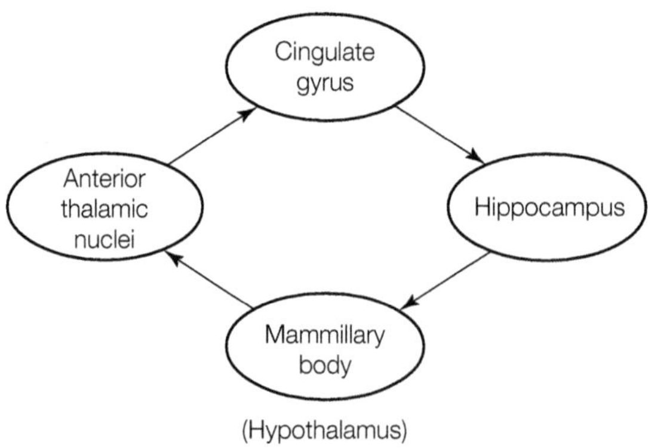


*Fig. 1. Papez circuit, showing interlinking of cortical and subcortical regions.*

amygdala, hypothalamus, and cingulate cortex are considered key areas, some definitions include even the afferent midbrain projections from the **ventral tegmental area** to the **prefrontal cortex**, and nucleus accumbens. There is even controversy over the existence of a limbic system *per se*. Nevertheless, irrespective of whether these nuclei act in an integrated fashion as a circuit, each undoubtedly plays a role in the expression of emotion, and there are defined connections between many of them.

### The hippocampus

The hippocampus is a bilateral horn-shaped structure in the temporal lobe that tapers from the amygdala, at its most rostral extent, in an arc back to the splenium of the **corpus callosum**. The principal role of the hippocampus (see Topic G1) is in the registration of memory, specifically the transfer of short- to long-term memory. If the hippocampus is damaged, a person cannot create new memories (anterograde amnesia), although previously acquired memories are unaffected. However, because of its role in declarative memory, the hippocampus also has an explicit effect on emotion. In conditioning, learning of an association between cue and stimulus is dependent on an intact hippocampus. Damage to the hippocampus obliterates explicit knowledge of that association, although autonomic correlates of emotion are still observed. This is the mirror image of the response seen with the amygdala (see below).

The hippocampus receives inputs mainly from the cingulate gyrus, **entorhinal cortex** (via the perforant pathway), septum, and hypothalamus (via the **fornix**), as well as some input from the contralateral hippocampus. Hippocampal outputs go to the entorhinal cortex, septum, cingulate cortex, anterior thalamic nuclei, and **mammillary bodies** of the hypothalamus.

### The amygdala

The amygdala (from the Greek *amygdale*, meaning almond) is a small bilateral structure located in the temporal lobe between the thalamus and the hippocampus. Although often described as a single nucleus, the amygdala is in fact a basketwork of connected nuclei (including the central, paracentral, medial, and mediobasal nuclei *inter alia*) within the **amygdaloid complex**.

Overall, the amygdaloid complex has reciprocal connections with the **prefrontal cortex**, **olfactory bulb**, cingulate gyrus, anterior and ventromedial hypothalamus, dorsomedial thalamus, and septum, as well as several brainstem nuclei, including the **locus ceruleus**, **substantia nigra**, and **raphe nuclei**. The amygdala also sends outputs to the autonomic nervous system. Many of the afferent and efferent connections are highly targeted to specific nuclei within the amygdaloid complex. For instance, the olfactory bulb afferents project to the dorsomedial amygdala which, in turn, has connections with the ventrolateral amygdaloid complex .

The amygdala is implicated in the expression of a range of emotions, but is particularly associated with fear and aggression. Experiments in animals have concentrated on the effects of stimulation or destruction of the nucleus. When the amygdala is stimulated, animals typically respond with aggression and fear. This is associated with autonomic symptoms such as an increase in heart rate, breathing and sweating. When the same is done in humans, a more complex range of emotion is observed, reflecting the many different emotional roles of the amygdala, and the heterogeneity of the structure. Although fear is again the most common emotion observed, some patients find the experience causes pleasure. In either case, there are emotionally appropriate associated autonomic effects.

If the amygdala is destroyed, aggression and fear are both attenuated. Previously aggressive animals are tamed. Generally this is seen as an indifference to previously emotional stimuli. In humans, a rare lipoid proteinosis (**Urbach–Wiethe disease**) occurs in which there is bilateral deterioration of the amygdala. Patients with the condition have a reduced emotional response to normally emotive cues, and do not appear to experience anger or fear. This absence of response is mirrored at the autonomic level – there is no galvanic skin response, for instance.

## The hypothalamus

The hypothalamus is a linked group of nuclei located just below the thalamus on both sides of the third ventricle (see Topic G1). As mentioned previously, its chief role is **homeostasis**, mediated via the **autonomic nervous system** and **neuroendocrine axis**. These aspects are discussed extensively elsewhere (Topics H5, K4 and K5).

The posterior part of the hypothalamus is responsible for anger responses. Stimulation of this region in animals evokes a form of rage. The hypothalamus is also responsible for modulating responses to pain, sexual satisfaction, and satiety.

What we often see as the physiological accompaniments of emotion (sweating, increased heart rate, etc.) are responsible, at the very least, for enhancing emotional perception. According to the James–Lange hypothesis, they may even be the emotion itself rather than its passive accompaniment. Although this cannot explain the fact that similar physiological changes occur with opposing emotions, it introduces the idea of **interoceptive cues**. For instance, some anti-anxiety drugs such as β-blockers probably do not need to enter the brain to ameliorate anxiety. They remove the symptoms such as tremor (interoceptive cues), and thereby alleviate the condition. The hypothalamus is a site where interoceptive information may be processed. For instance, as previously mentioned, the hypothalamus responds to heart rate, and it is possible that a role of the hypothalamus in emotion may be as an interface between interoceptive cues and physiological response.

The hypothalamus receives inputs from a number of sources. These include the hippocampus, brainstem and amygdala. Hypothalamic efferent fibers innervate the anterior thalamus. As previously mentioned, much of the output of the hypothalamus is expressed via the autonomic nervous system and neuroendocrine axis.

## The cingulate gyrus

The cingulate gyrus has complex roles in emotion. The frontal portion appears to associate sensory input (olfactory and visual) with memories of previous emotions. Interestingly, the subgenual anterior cingulate gyrus is reported to be smaller in subjects with familial mood disorder. **Positron emission tomography (PET)** scans have shown that sadness activates the cingulate gyrus. **Functional magnetic resonance imaging (fMRI)** studies support this association; a high level of activation is observed in the cingulate gyrus when subjects are shown negative (sad) images.

This region is also engaged in the emotional response to pain and in the regulation of aggressive behavior. As with the amygdala, lesions to the cingulate gyrus reduce aggression in animals. Interestingly, unilateral lesion of the cingulate gyrus (cingulectomy) appears to reduce depression and anxiety in humans.

The cingulate gyrus receives inputs from the parietal and temporal lobes, hippocampus, prefrontal cortex, septum and anteroventral thalamus. Cingulate efferents go to the prefrontal cortex, hippocampus, parietal lobe, anteroventral and dorsomedial thalamus, and subiculum, as well as many brainstem nuclei (including the **superior colliculus**, **periaqueductal gray**, **midbrain tegmentum**, and locus ceruleus).

**Other limbic areas**

The ventral tegmental area is the source of the ascending dopamine pathways that project to the amygdala, prefrontal cortex and nucleus accumbens. There is good evidence that these pathways mediate hedonic responses to stimuli. These pathways determine responses to novelty and a person's search for novel experiences. They play a part in determining addiction.

The prefrontal cortex is also closely linked to the limbic system. The prefrontal cortex is involved in planning and taking action. It too receives an input from the ventral tegmental area and plays a role in pleasure and addiction.

# G4 CRANIAL NERVES

## Key Notes

| | |
|---|---|
| **Twelve cranial nerves** | Cranial nerves emerge from the skull via various apertures (foramina), and primarily innervate structures in the head and neck. However, some cranial nerves innervate other structures such as the heart, airways and gut via long projections. The 12 cranial nerves are distinct from the spinal nerves that emerge from the spinal cord between the vertebrae. Some cranial nerves have purely motor function, some have purely sensory function, and some have a mixed motor and sensory function, like spinal nerves. |
| **Cranial nerve motor nuclei** | Cranial nerves III, IV, V, VI, VII, IX, X, XI, and XII have some motor function and five of these (III, IV, VI, XI, and XII) are solely motor. The cell bodies of the neurons that make up these motor cranial nerves are in the brainstem. Some of the motor outflow from the parasympathetic branch of the autonomic nervous system is conveyed by cranial nerves III, VII, IX and X. |
| **Cranial nerve sensory ganglia and nuclei** | Cranial nerves I, II, V, VII, VIII, IX, and X have sensory function. Like the sensory neurons making up spinal nerves, the cell bodies of these neurons reside in ganglia outside the CNS, and they project to synapse with other neurons in the brain, which reside in sensory nuclei, such as the chief sensory nucleus of the trigeminal nerve in the pons. |
| **Functional correlates** | Cranial nerves subserve important functions, especially in the head and neck. The function of cranial nerves can be tested reasonably easily by examining smell, taste, visual reflexes, sensation on the face, balance, speech, and swallowing. |

**Related topics**

Parts of the brain (G1)
Vision (G5)
Audition (G6)
The vestibular system and balance (G7)
Taste and olfaction (G8)
Structure and function of the autonomic nervous system (H5)
Innervation of smooth muscle (I7)

## Twelve cranial nerves

The **central nervous system (CNS)** innervates the entire body. This includes structures that are immediately apparent from the outside such as skin and muscle, but also glands inside the body and visceral organs. The CNS innervates the body by way of two systems: **cranial nerves** and **spinal nerves**. The cranial nerves, as their collective name suggests, emerge from the cranium (mostly from the brainstem (see Topic G1)), and the spinal nerves emerge from the **spinal cord** in the spaces between the vertebral bones that make up the spine. All spinal nerves have a mixed motor and sensory function; they conduct information from the CNS to the periphery and vice versa, with separate nerve fibers subserving each of these functions. However, of the cranial nerves, some have a mixed function, some are sensory only, and some are motor only. There are 12 cranial nerves, numbered roughly from the front of the brain to the back (rostral to caudal) and named according to their function or their pattern of innervation. *Table 1* gives names and ascribes broad functions to the cranial nerves.

**Cranial nerve motor nuclei**

Of the 12 cranial nerves, nine (III, IV, V, VI, VII, IX, X, XI, and XII) have some motor function and of these, five (III, IV, VI, XI, and XII) are purely motor. This motor output may be under voluntary control, such as movement of the eyes

*Table 1.* *Cranial nerves*

| Cranial nerve number and name | Emerges | Motor, sensory or both | Broad function(s) |
|---|---|---|---|
| I Olfactory | From the forebrain via the cribriform plate | Sensory | Olfaction |
| II Optic | From the forebrain via optic canal | Sensory | Vision |
| III Oculomotor | From the midbrain, then via the superior orbital fissure | Motor | Movements of the eyeball and upper eyelid, constriction of the pupil, focusing on close objects |
| IV Trochlear | From midbrain tectum, then via superior orbital fissure | Motor | Turning the eye down and out (superior oblique muscle only) |
| V Trigeminal | From the pons, then via the superior orbital fissure (Va), foramen rotundum (Vb) and foramen ovale (Vc) | Both | Sensory to the eyeball, anterior scalp, face, nasal cavities, chin, teeth and ear. Motor to the muscles of chewing |
| VI Abducens | From the pontomedullary angle, then via the superior orbital fissure | Motor | Turning the eye out (lateral rectus muscle) |
| VII Facial | From the cerebellopontine angle, then via the internal acoustic meatus, facial canal and stylomastoid foramen | Both | Muscles of facial expression, some salivary secretions, taste |
| VIII Vestibulocochlear | From the cerebellopontine angle, then via the internal acoustic meatus | Sensory | Balance and hearing |
| IX Glossopharyngeal | From the medulla, then via the jugular foramen | Both | Taste at the back of the tongue, sensation in the oropharynx, some salivary secretion |
| X Vagus | From the medulla, then via the jugular foramen | Both | Swallowing, motor and sensory to the heart, lungs and gut, sensation from the ear, pharynx and larynx |
| XI (spinal) Accessory | From the medulla and upper end of the spinal cord, then via the jugular foramen | Motor | A few neck and shoulder movements |
| XII Hypoglossal | From the medulla, then via the hypoglossal canal | Motor | Movement of the tongue muscles (i.e. chewing and speech) |

(III, IV, and VI) or may be involuntary, such as **pupillary constriction** in response to light (III), **salivation** (VII, and IX) or decreasing heart rate in response to elevated blood pressure (X). Note that the involuntary motor output of the cranial nerves forms part of the output of the **parasympathetic** branch of the **autonomic nervous system** (see Topic H5). **Cranial nerve motor nuclei** are collections of neuronal cell bodies in the brainstem, the axons of which form the motor fibers of those cranial nerves. *Table 2* gives the location of the cranial nerve motor nuclei in the **brainstem** (midbrain, pons, and medulla, see Topic G1).

*Table 2. Functions of cranial nerve motor nuclei*

| Cranial nerve motor nucleus | Controls | Location in brainstem |
|---|---|---|
| III | Pupillary constriction, lens, eye and eyelid movements | Periaqueductal gray matter of the midbrain |
| IV | Eye movement | Near the periaqueductal gray matter of the midbrain |
| V | Some elements of chewing | Pons |
| VI | Eye movement | Pons |
| VII | Facial expression, some salivary secretion | Pons |
| IX | Some salivary secretion | Medulla |
| X | Output to the heart, lungs and gut (*inter alia*) | Medulla |
| XI | Some head and neck movements | Upper parts of the cervical spinal cord |
| XII | Movement of the tongue, chewing, speech | Medulla |

Of these nuclei, some are very clearly delineated (such as those for III, IV, V, VI, and VII), but others are more scattered down the length of the pons (IX, X, and XII) or the upper portions of cervical spinal cord (XI). In general, motor nuclei (e.g. the motor nucleus of cranial nerve XII) are closer to the midline, whereas sensory nuclei (e.g. the vestibular and cochlear nuclei for balance and hearing) are more lateral.

## Cranial nerve sensory ganglia and nuclei

A ganglion (plural: ganglia) is a collection of nerve cell bodies. In the case of sensory ganglia, there is no synaptic connection between two neurons in these ganglia; the cell bodies have two processes, one that goes to the periphery (and which ends in a receptor for a particular stimulus like touch or temperature), and another that projects to the CNS. In common with the spinal nerves (see Topic H1), the cell bodies of these primary sensory neurons are located outside the CNS. In the case of the spinal nerves, the cell bodies of such neurons reside in dorsal root ganglia near the spinal cord, but outside the CNS. In the case of the cranial nerves with sensory function (I, II, V, VII, VIII, IX, and X), the cell bodies of the sensory neurons are present in:

- the olfactory epithelium (in the case of I – see Topic G8)
- the **retina** (in the case of II – see Topic G5)
- the **trigeminal ganglion** (in the case of V)

- the **geniculate ganglion** (in the case of VII)
- the **cochlear (spiral) ganglion** and **vestibular ganglion** (in the case of VIII)
- the superior and petrosal (inferior) glossopharyngeal ganglia (in the case of IX)
- the jugular (superior) and nodose (inferior) ganglia of vagus (in the case of X).

Do not confuse these sensory ganglia with sensory *nuclei* of cranial nerves; these exist in the brainstem and are the collections of secondary sensory neuron cell bodies that the primary sensory neurons synapse onto. These secondary sensory neurons have projections up to higher centers in the CNS for processing of sensory information. Like some of the motor nuclei of cranial nerves, these sensory nuclei are more scattered in the brainstem (e.g. the **spinal nucleus of V** runs almost the entire length of the medulla and down into the upper parts of the cervical spinal cord), although it is not fully understood why this should be so.

## Functional correlates

In view of the high level of anatomical organization of the cranial nerves, their motor nuclei, sensory ganglia, and sensory nuclei, it is possible to test the function of cranial nerves and predict where problems might have occurred. For example, since the parasympathetic components of the output of III control constriction of the pupil in response to shining a light in the eye, problems with this reflex response can arise by several routes (see Topic G5). Firstly, the problem may exist in the retina or in II, so that light is not being sensed. However, a conscious subject would be likely to tell the person testing the reflex that they couldn't see. The most likely cause is a problem with one or more parts of the **midbrain**, either the **pretectal area** which projects to those neurons that make up the parasympathetic output of III (the **Edinger–Westphal nucleus**), or with III itself. **Raised intracranial pressure** caused by a head injury can lead to compression of III and loss of the **pupillary light reflex** in a head-injured patient. Fixed, dilated pupils are a possible indicator of brainstem death.

Another possibility is loss of the blood supply to a region of the brainstem containing certain cranial nerve nuclei (motor or sensory), caused for example by a **stroke**. The blood supply to the various parts of the brainstem is highly organized, coming from branches of the **basilar arteries**. For example, occlusion of the right posterior inferior cerebellar artery that supplies the lateral medulla and the lower cerebellum leads to problems with chewing, swallowing, and speaking on the same side because the **nucleus ambiguus** (IX, X, and XI) and hypoglossal nucleus (XII) are affected.

As some cranial nerves have sensory function, problems with cranial nerves, their ganglia or nuclei can manifest as sensory problems. One example is where the chickenpox virus (**Varicella zoster**) has lain dormant in the trigeminal ganglion since the first infection, only to emerge as a painful lesion in the sensory distribution of the nerve, called **shingles** (herpes zoster).

# G5 Vision

## Key Notes

**The eyeball**

The eyeball is lined with light-sensitive neurons. At the front of the eye, the cornea allows light into the eye and refracts that light. The eye has two chambers separated by a lens. Contraction or relaxation of the ciliary muscles determines the shape of the lens, and therefore the focusing of the eye. In front of the lens is the iris, that controls the amount of light reaching the retina. Contraction of the annular muscle reduces pupil size, while contraction of the radial muscle dilates the pupil. Pupil diameter is controlled largely by the autonomic nervous system.

**The retina**

The retina contains two types of photoreceptors – rods and cones. Rods are responsible for achromatic vision, while cones produce color vision. Three types of cone are known, responsive to blue, green, and red light. The photoreceptor cells form synapses with small bipolar neurons, which form connections with the ganglion cells whose axons form the optic nerve. Each bipolar cell has inputs from several photoreceptors, but projects to a single ganglion cell which may, in turn, receive inputs from several adjacent bipolar cells. Each ganglion cell has a small receptive field. In the outer plexiform layer, horizontal cells modulate information exchange between the photoreceptors and the bipolar cells. In the inner plexiform layer, amacrine cells serve the same role between bipolar and ganglion cells.

**The optic pathways**

The two optic nerves converge in the optic chiasm where nasal fibers cross to the contralateral side of the brain. The optic tract sends retinal fibers to three different locations, each controlling a different aspect of visual function. The principal retinal projection is to the lateral geniculate nucleus (LGN) of the thalamus. The LGN projects retinotopically, via the optic radiation, to the primary visual cortex. This projection is responsible for visual perception, particularly shape perception. The visual cortex is organized in a series of interlocking columns, cells within each column responding to a quality (movement, contrast, orientation) of the visual stimulus. Two further visual pathways project from the retina to the midbrain. One projects from the retina to the superior colliculus and is responsible for visual tracking and the generation of saccadic eye movements. The second links the retina and the pretectum, and is responsible for the generation of pupillary movements and reflexes.

**Related topics**

Parts of the brain (G1) Cranial nerves (G4)

## The eyeball

The human eyeball is a near-spherical structure, approximately 3½ cm in diameter, lined with light-sensitive (photoreceptive) neurons (*Fig. 1*). Over much of its surface, the eyeball has three main layers – the **sclera**, **choroid**, and **retina**.

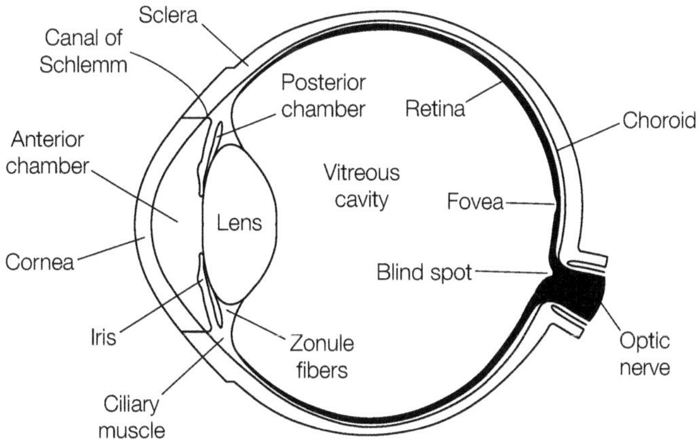


*Fig. 1. The human eyeball.*

The robust outermost coat is the sclera, composed of connective tissue and forming the white of the eye. The muscles of the eye attach to this layer. Beneath the sclera is the choroid, a thin dark layer of melanized cells that serve to reduce internal reflection within the eye. Innermost is the retina, comprising the **photoreceptor cells**.

At the front of the eye, the sclera gives way to a circular, slightly more convex, transparent aperture, approximately 1 cm in diameter – **the cornea**. The cornea allows light into the eye and refracts that light.

The eye has three chambers. Between the cornea and the **iris** is the **anterior chamber**, and between the iris and the **lens** and **zonule fibers** is the **posterior chamber**. **Aqueous humor** is secreted by the **ciliary body** into the posterior chamber and finds its way into the anterior chamber by the narrow gap between the iris and the lens. The aqueous humor is drained via the **canals of Schlemm**, at the junction of the cornea and iris. Between the lens and the retina is the **vitreous cavity** or body, filled with more gelatinous **vitreous humor**.

Situated between the anterior and posterior chambers and the vitreous cavity is the lens, held in place around its perimeter by the zonule fibers, attached to the **ciliary muscles**. The tension of the zonule fibers determines the shape of the lens; if the zonule fibers are tense, the lens is pulled into a flatter shape, whereas if they are slack the lens springs back to its naturally more curved shape. The shape of the lens determines its refractive power and therefore the ability of the eye to focus. Relaxation of the ciliary muscle increases tension in the zonule fibers, flattening the lens and focusing the eye on more distant objects. Contraction of the ciliary muscle reduces tension in the zonule fibers, allowing the lens to become more curved and focusing the eye on nearby objects. Some problems with vision are caused by abnormalities in lens function, and others are caused by the eyeball being too large or too small. As a person ages, the lens becomes less elastic and flatter, making it more difficult to focus on nearby objects (**presbyopia**). This is the reason why most people need reading glasses when they get older. If the eyeball is too small in the anteroposterior diameter (i.e. the distance between the lens and the retina is shorter than normal), this causes problems in focusing on nearby objects (longsightedness or **hypermetropia**). If the eyeball is too large in the anteroposterior diameter, this causes shortsightedness (**myopia**), in which the person cannot focus adequately on objects that are far away. Thankfully, most such problems can be corrected by different types of lenses in glasses or by the use of contact lenses. Other problems with the lens are less easily solved; if the lens becomes cloudy or, ultimately, opaque, this is known as a **cataract**. Cataracts can only be treated surgically; the cloudy lens is removed and the person either wears glasses or has a prosthetic lens implanted.

In the anterior chamber, immediately in front of the lens is the iris, a pigmented muscular extension of the choroid layer that controls the amount of light reaching the retina, acting like the diaphragm on a camera. The iris has both radial and annular muscle layers. Contraction of the annular muscle reduces the aperture (pupil) size, while contraction of the radial muscle dilates the pupil. **Pupil diameter** is controlled largely by the **autonomic nervous system**.

## The retina

The apparatus of the eyeball serves to direct light appropriately to the retina, the innermost layer of the eye containing the photoreceptive cells. There are two types of photoreceptors – **rods** and **cones** – each with different properties. Rods are responsible for achromatic vision, while cones produce color vision. Rods are much more common than cones, and are several orders of magnitude more sensitive to light. Rods respond only slowly and contribute particularly to **night vision**. In daylight, rods are comparatively unresponsive.

Cones, on the other hand, are relatively insensitive to light and contribute mainly to **daylight vision**. More importantly, cones are responsible for color vision. Cones respond rapidly to changes. Three types of cone cell are known, responsive to short, medium, and long wavelengths of light. In essence, these are selectively responsive to blue, green, and red light. **Color blindness** is due to the absence of one or more cone type.

Cones and rods have different but complementary distributions across the retina. Cones are concentrated toward the center of the visual field, and are especially dense at the **fovea**, an area of high visual acuity. Rods are absent from the fovea, and are found more evenly across the retinal disk, with a higher concentration around the periphery of vision. This property accounts for the greater ability to see objects at night using averted gaze.

The photoreceptor cells form synapses with small **retinal bipolar neurons**, orientated perpendicular to the plane of the retina, in the **plexiform layer**, which respond either to an increase or a decrease in light. These bipolar neurons form connections with the **retinal ganglion cells** whose axons form the **optic nerve** (cranial nerve II). Each bipolar cell may receive inputs from several adjacent photoreceptors, but will project to a single ganglion cell. Again, each ganglion cell receives inputs from several adjacent bipolar cells. In this way, each ganglion cell has a small but defined receptive visual field.

Two further networks of inhibitory neurons, running parallel to the retinal surface, modify transmission of visual information. In the outer plexiform layer are **horizontal cells** that modulate information exchange between the photoreceptors and the bipolar cells. In the inner plexiform layer, **amacrine cells** serve the same role between bipolar and ganglion cells. Some of the connections between cells within the retina are shown in *Fig. 2*.

The axons of the ganglion cells converge topographically at a single point on the retina to form the optic nerve. This location, because of the high density of axons, has no photoreceptors and is thus the **blind spot**.

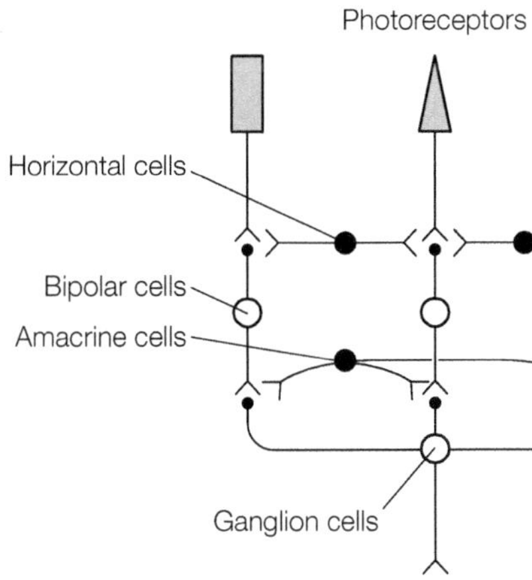


*Fig. 2. Cells and connections of the retina.*

## The optic pathways

The optic nerve is ensheathed by the **dura**, an extension of the sclera. The two optic nerves converge in the **optic chiasm**, at the base of the brain. At this point, nasal (medial) fibers undergo **decussation** (crossing over the midline) to the contralateral side of the brain, while temporal (lateral) fibers remain ipsilateral. Thus information from the right visual field projects to the left half of the brain and vice versa. Fibers leaving the optic chiasm form the **optic tract**. The optic tract sends retinal fibers to three different locations, each controlling a different aspect of visual function (*Fig. 3*).

The principal retinal projection is to the **lateral geniculate nucleus (LGN)** of the thalamus. The LGN has six layers, each receiving an input from only one eye. The two ventral tiers form the **magnocellular layers of the LGN**, while the remainder are known as the **parvocellular layers of the LGN**. These layers project, via the **optic radiation**, to layer 4 of the **primary visual cortex**, at the most caudal aspect of the **occipital lobe**. This projection is retinotopic, retaining the spatial mapping of the visual field, and is responsible for visual perception, particularly shape perception. The visual cortex is organized in a series of interlocking columns, each with a visual receptive field. Cells within each column have specific properties, responding not only to the presence of a visual stimulus, but to a quality (movement, contrast, orientation) thereof.

The fovea, at the center of the visual field, sends a particularly strong cortical projection. Although the fovea occupies less than 5% of the retina, its information is expanded to cover more than two-thirds of the primary visual cortex. This sensory over-representation is similar to the representation of the fingertips in the somatosensory cortex (see Topic G2).

Individuals with damage to the primary visual cortex have normal eye tracking, and can still respond to objects within the visual field but have no perception of doing so. This phenomenon is known as **blindsight**.

Two further visual pathways project from the retina to the **midbrain**. One, encompassing perhaps one-tenth of the retinal axons, projects from the retina to the **superior colliculus**, and is responsible for visual tracking and the generation of **saccadic eye movements**. The superior colliculus has direct connections via the tectobulbar pathway to the motor systems that coordinate eye movements

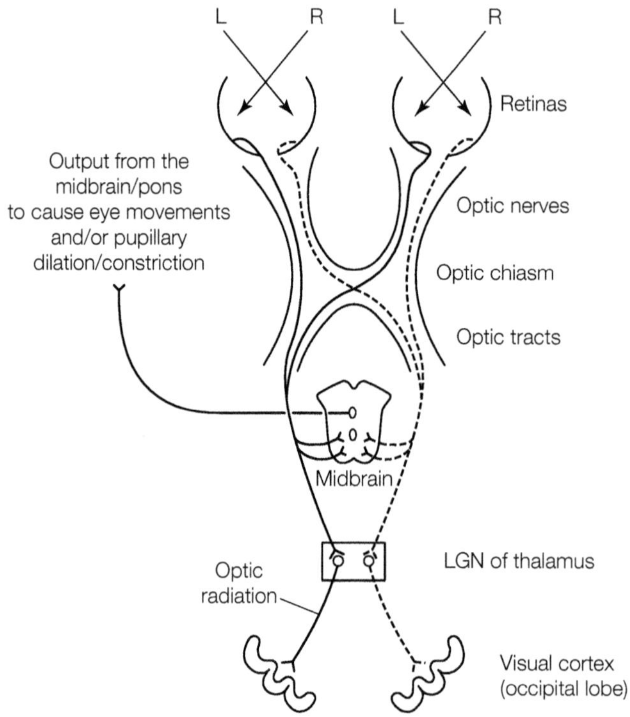


*Fig. 3. Visual pathways. See text for details. Pathways from the parts of the retina that receive input from the right visual field are shown with solid lines, whereas those from the left visual field are shown with broken lines. LGN, lateral geniculate nucleus; L, light from left visual field; R, light from right visual field.*

and reaching. The pathway is rapid, and is pivotal to accurate movements in ball games, for instance. This pathway is intact in blindsight.

A third pathway links the retina and the **pretectum**. This pathway is responsible for the generation of pupillary movements and reflexes. The **pupillary reflex** controls the amount of light entering the eye via a bilateral projection from the pretectum to the **Edinger–Westphal nucleus**. These neurons then send their **parasympathetic** preganglionic fibers out of the CNS in the **oculomotor nerve** (cranial nerve III). The fibers synapse with parasympathetic postganglionic neurons in the **ciliary ganglion**, which send cholinergic fibers to the pupillary muscles.

# G6 AUDITION

## Key Notes

**Sound and hearing** Audition, or hearing, is the means by which we translate sounds into neuronal impulses. Humans can typically hear frequencies between about 30 Hz and 15 kHz. Young children can sometimes hear frequencies as high as 20 kHz.

**The ear** The outer ear extends from the pinna to the tympanic membrane. The middle ear transmits tympanic membrane vibrations to the cochlea via three ossicles – the malleus, incus, and stapes. A small muscle, the tensor tympani, attaches the malleus to the inner surface of the middle ear, and dampens excessive vibration of the tympanic membrane.

**The inner ear** The inner ear comprises the semicircular canals and the cochlea. The curled cochlea is divided into three main chambers – the scala vestibuli, the scala media, and the scala tympani. Sound waves pass from the oval window, along the scala vestibuli, through the helicotrema and back along the scala tympani. On the upper surface of the basilar membrane, within the scala media, is the organ of Corti, the sensory apparatus. Sounds oscillate the basilar membrane. The organ of Corti is essentially a ribbon of columnar epithelium with sensory hair cells akin to those found in the vestibular apparatus. These hair cells have stereocilia that are sensitive to movement. Sound vibrations cause movements of the basilar membrane, making the stereocilia flex, and causing depolarization or hyperpolarization in the hair cell. Hair cells form synapses with auditory neurons that project tonotopically to the cortex.

**The auditory nerve projection** Auditory nerve fibers project initially to the cochlear nucleus then, via the brainstem, to the medial geniculate nucleus of the thalamus which sends fibers, via the auditory radiation, to the primary auditory cortex in the temporal lobe. These areas correspond very closely to the cortical areas responsive to speech.

**Related topics** Parts of the brain (G1) Cranial nerves (G4)

## Sound and hearing

Sound, in the form humans recognize it, consists of pressure changes in the air around the ear. These pressure changes are, in essence, waves, and therefore can be described in those terms. Sound waves thus have a wavelength and amplitude. Peals of thunder are long-wavelength, low-frequency sounds, while birdsong is a short-wavelength (high-frequency) sound. The amplitude of these waves corresponds to the loudness of the sound.

Audition, or hearing, is the means by which we translate sounds into neuronal impulses and, ultimately, patterns of cortical activation. Humans can typically hear frequencies between about 30 Hz and 15 kHz. Women generally hear higher frequencies than men, and the capacity to detect the very highest frequencies deteriorates with age. Young children can sometimes hear frequencies as high as 20 kHz.

Hearing, like many of the senses, serves to facilitate communication, and to alert the hearer to danger.

## The ear

The ear consists of three anatomical subdivisions – the outer, middle, and inner ear (*Fig. 1*). The **outer ear** extends from the **pinna** (the outwardly visible ear), via the external auditory canal to the **tympanic membrane** (or eardrum), and acts to channel sound waves to the eardrum, causing it to vibrate.

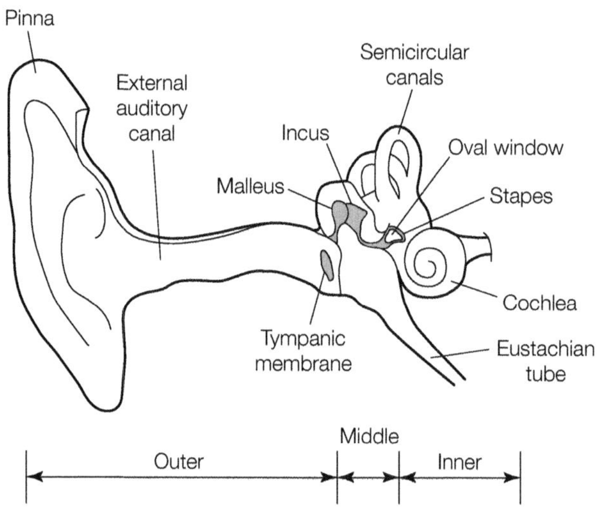


*Fig. 1. The structure of the ear.*

The **middle ear** transmits vibrations of the tympanic membrane to a second much smaller membrane on the **cochlea**. Three tiny articulated bones or **ossicles** – the **malleus, incus**, and **stapes** – act as a series of pivots to translate relatively large vibrations of the eardrum into smaller vibrations at the **oval window** of the cochlea.

The first bone, the malleus (from the Latin, *hammer*), is attached to the tympanic membrane at its 'handle' end, and to the incus (from the Latin, *anvil*) at the head. The incus connects, via a long pivoted process, to the stapes (from the Latin, *stirrup*), which is attached by a ligament to the oval window. The joints are articulated such that inward pressure on the tympanic membrane pulls the head of the malleus outward and, with it, the outer part of the incus. This pushes the long process of the incus inward and thus, via the stapes, applies pressure to the oval window of the cochlea. The middle ear is connected, via the **Eustachian tube**, to the **pharynx**, which ensures that the pressure on both sides of the tympanic membrane is the same. A small muscle, the **tensor tympani**, attaches the malleus to the inner surface of the middle ear, and acts to dampen excessive vibration of the tympanic membrane.

## The inner ear

The **inner ear** comprises the **semicircular canals** that determine balance (see Topic G7), and the cochlea. The cochlea, a fluid-filled tube approximately 35 mm long, is curled, much like a French horn, and divided, in cross-section, into three main chambers – the **scala vestibuli**, the **scala media**, and the **scala tympani** (*Fig. 2a*). The scala vestibuli is connected via a small gap at the apex of the cochlea (the **helicotrema**) to the scala tympani. Thus, sound waves pass along a path from the oval window, along the length of the scala vestibuli, through the helicotrema, and back along the length of the scala tympani to the **round window**.

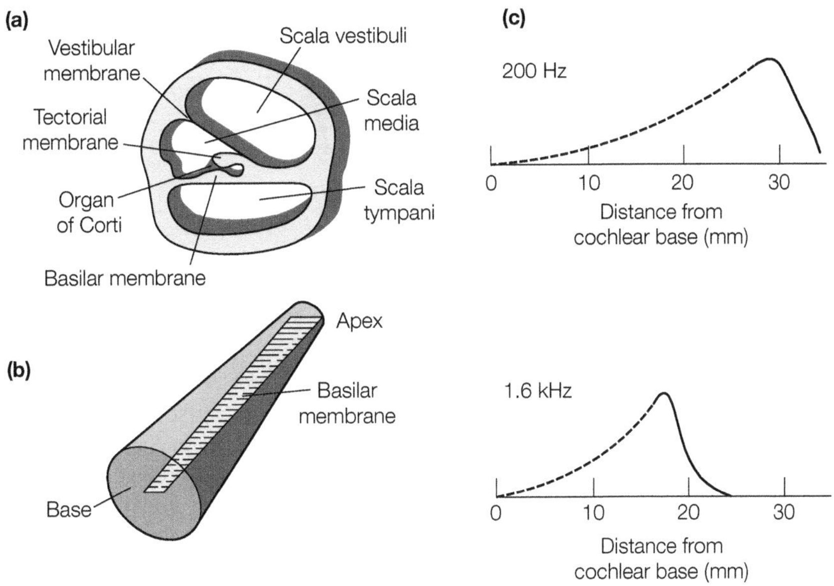


*Fig. 2. The structure of the cochlea in cross-section (a), illustrating the propagation of waves along the basilar membrane (b), and the distances of activation at different frequencies (c).*

Separating the scala media and scala tympani, and running along the entire length of the cochlea from base to apex, is the **basilar membrane**. On the upper surface of the basilar membrane, within the scala media, is the **organ of Corti**, the sensory apparatus. The basilar membrane can be thought of, in essence, as a vibration-sensitive sounding board extending from the narrow, stiff base of the cochlea, to the wider, floppier apex (*Fig. 2b*). The organ of Corti, running along it, could be conceptualized as a band of microphones. Together, these allow incoming sound to be dissected into its constituent frequency elements.

High-frequency sounds generate waves that oscillate the stiff base end of the basilar membrane, while lower frequencies can propagate much further along the basilar membrane and cause membrane displacements toward the apex (*Fig. 2c*). Thus the distance along the basilar membrane that a given frequency travels provides a place map with respect to frequency, akin to the distribution of keys on a piano. In view of this organizational map, mechanical damage to the cochlea by exposure to excessive levels of noise most often causes hearing loss preferentially affecting higher sound frequencies, since the base of the basilar membrane is closest to the ossicles and the tympanic membrane, and therefore more prone to damage.

The sensory apparatus of the organ of Corti thus provides a spatial map of the frequency distribution of the sound detected, and transduces this into the neural signals of the **auditory nerve** (part of cranial nerve VIII). The organ of Corti is shown in cross-section in *Fig. 3*.

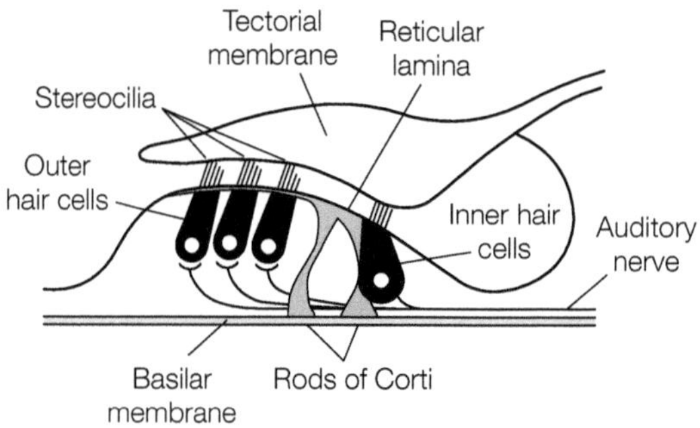


*Fig. 3. The organ of Corti.*

The organ of Corti is essentially a ribbon of columnar epithelium with sensory hair cells akin to those found in the vestibular apparatus. These hair cells have **stereocilia** that are sensitive to movement, often embedded in the **tectorial membrane**. The process by which we hear sounds is thus a form of **mechanoreception**. Sound vibrations cause movements of the basilar membrane relative to the tectorial membrane, making the stereocilia flex. This flexion and relaxation causes depolarization and hyperpolarization cycles in the hair cell, altering its output of glutamate, and thereby determining activation of the auditory nerve afferents.

Inner hair cells form synapses with many myelinated auditory neurons, and carry both high- and low-frequency information. Conversely, the many outer hair cells form synapses with fewer unmyelinated fibers and mainly transmit low-frequency data. At low frequencies (<400 Hz), auditory nerves are phase-locked to the depolarization cycles of the hair cells. At higher frequencies, the auditory nerve cannot fire at an equivalent rate, and the response ceases to be phase-locked to the stimulus.

## The auditory nerve projection

The auditory nerve projection is tonotopically organized throughout its path to the cortex. Auditory nerve fibers, part of cranial nerve VIII, project initially to the **cochlear nucleus**, which is divided into isofrequency strips, each containing a set of neurons responsive to a narrow frequency band. The auditory fibers then project to the **superior olivary nucleus** and, via the **lateral lemniscus**, to the **inferior colliculus**. This region of the **brainstem** is important in coordination of **auditory reflexes** such as turning the head towards the direction of sound. There is some **decussation** and mixing of auditory fibers in the brainstem nuclei, to allow localization of sound by contralateral delay. In the extremely rare case of a person with complete bilateral hearing loss caused by damage to both sides of the hearing pathway above the level of the brainstem, such auditory reflexes are preserved; the person may be startled by a loud noise, or turn their head towards the source of the sound, yet be unsure as to why they did this, since they did not actually hear the sound.

Ascending auditory fibers from the brainstem synapse in the **medial geniculate nucleus (MGN)** of the thalamus which then projects, via the **auditory radiation**, to the **primary auditory cortex** in the **temporal lobe**. The **tonotopic organization** of the auditory nerve is preserved in the auditory cortex with isofrequency columns selectively responsive to high- or low-frequency information. On the whole, columns responsive to higher frequencies are located more rostrally. These areas correspond very closely to the cortical areas responsive to speech, such as **Wernicke's area**.

# G7 THE VESTIBULAR SYSTEM AND BALANCE

## Key Notes

**Balance**

Balance integrates information from proprioceptive inputs, cutaneous afferents, visual information, and specialized input from the vestibular apparatus.

**The vestibular apparatus**

The vestibular apparatus comprises the otolith (utricle and saccule) and the semicircular canals. The otolith structures detect linear movement and position of the head in relation to gravity. The semicircular canals respond to rotational movements of the head in space.

Within both are specialized structures comprised of hair cells which form synapses with vestibular primary afferent neurons. In the ampullae, the hair cells are lined along the ampullary crest, encased in a gelatinous cupula. Movement of the head displaces the cilia, increasing the firing rate of the vestibular primary afferent neurons. The macula transduces movement into activity of the vestibular primary afferents by a similar mechanism.

**Central vestibular projections**

Vestibular primary afferent neurons join the cranial nerve VIII, and project initially to the vestibular nuclei. Descending neurons in the lateral and medial vestibular nuclei form the vestibulospinal system, sending projections to the motor neurons. The lateral vestibular nucleus receives input primarily from the utricle, while the medial vestibular nucleus receives input mostly from the semicircular canals. Vestibulospinal and vestibulocollic reflexes are activated in response to sudden changes in posture. The superior vestibular nucleus, along with cells from the medial and lateral vestibular nuclei, forms the vestibulo-ocular pathway. The vestibulo-ocular reflex uses visual and positional information to ensure a stable retinal image. Projections from the inferior vestibular nucleus form the vestibulocerebellar pathway and have a pivotal role in sensorimotor coordination. Relatively few fibers, mainly from the inferior vestibular nucleus, project to the face area of the somatosensory cortex.

**Related topics**

Parts of the brain (G1)
The locomotor system (G2)
Cranial nerves (G4)
Vision (G5)
Audition (G6)

## Balance

Balance is the dynamic coordination of sensory input and motor output to achieve continuous postural stability or equilibrium, whether stationary or moving. Balance integrates information from a range of sensory inputs – proprioceptive inputs from the joint and muscle **mechanoreceptors**, cutaneous afferents, visual information, and specialized input from the **vestibular apparatus**.

Visual information is particularly important. It is much harder to maintain a stable posture with one's eyes closed. Many roller coaster and flight simulators make good use of visual and vestibular information to create a realistic sensation of movement.

## The vestibular apparatus

The vestibular apparatus (**labyrinth**) is a specialized organ in the **inner ear**, adjacent to and contiguous, via the **ductus reuniens**, with the **cochlea** (see Topic G6). The labyrinth provides information about the position and movement of the head.

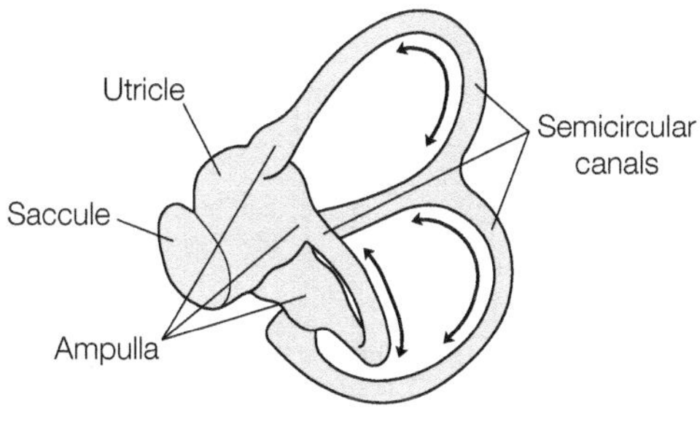


*Fig. 1. The vestibular apparatus.*

The inner ear comprises the **bony labyrinth**, which surrounds and encloses the **membranous labyrinth** within a bath of **perilymph**. The membranous labyrinth includes the **auditory labyrinth** (the **cochlea**, see Topic G6) and vestibular apparatus. The vestibular labyrinth comprises two main structures with similar but complementary roles – the **otolith** and the **semicircular canals**. The otolith is further subdivided into the **utricle** and **saccule**, a swelling contiguous with the cochlea. The otolith structures detect linear movement and position of the head in relation to gravity.

Extending from the utricle at three swellings (**ampullae**) are the three semicircular canals (anterior, posterior, and horizontal/lateral). The semicircular canals, at right angles to each other, act in concert to respond to rotational movements of the head in space, the signals from each describing a movement vector in the same way that a point may be described on a three-dimensional graph in terms of its $x$, $y$, and $z$ coordinates.

The auditory and vestibular labyrinths are filled with **endolymph**. Unusually, the endolymph, despite being an extracellular fluid, has a high potassium concentration (150 mM). Endolymph is formed by the **stria vascularis**, and drains via the **endolymphatic sac**. Since the cochlea and the vestibular organ are connected via the endolymph, overproduction of endolymph or reduced drainage typically affects both hearing and balance. **Ménière's disease**, in which too much endolymph is produced, is characterized by vertigo and tinnitus.

Within the otolith and semicircular canals are specialized structures for detection of movement, similar to those of the cochlea. As with the cochlea, these are comprised of hair cells which form synapses with vestibular primary afferent neurons (*Fig.* 2). In the utricle and saccule, the structure is the macula. In the semicircular canals, movement is detected by crest structures, located within the ampullae.

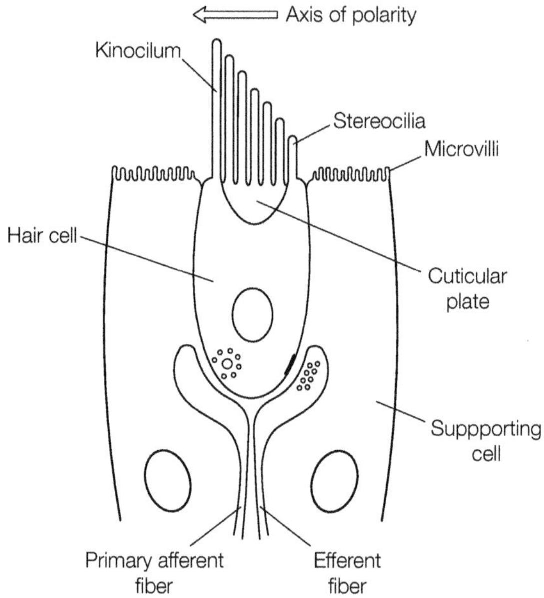


*Fig. 2. The hair cell of an otolith organ surrounded by supporting cells of the sensory epithelium.*

In the ampullae, the hair cells are lined, at intervals, along the **ampullary crest**, which runs at right angles to the direction of the canal. The hair cells along this ridge have **stereocilia** and a longer **kinocilium**, encased in a gelatinous cupula that extends to the roof of the ampulla and acts like a sail. Movement of the head creates a fluid inertia that displaces the cilia. Movement in one direction depolarizes the hair cell, causing transmitter release and increasing the firing rate of the vestibular primary afferent neurons. Movement in the opposite direction causes hyperpolarization. It is worth noting that movement that causes depolarization of primary afferents in one labyrinth will have the opposite effect in the labyrinth of the contralateral ear.

The **macula**, the mechanosensitive organ of the otolith, also uses hair cells, and transduces movement into activity of the vestibular primary afferents by the same mechanism. As in the ampulla, the stereocilia and kinocilium are encased in a gelatinous mass containing crystals of calcium salts. This has a greater density than the surrounding endolymph, and means that gravity or linear acceleration are easily detected. The **utricular macula** is located on the floor of the utricle, while the **saccular macula** is attached to the side of the saccule. Thus the utricular macula produces no stimulus to the vestibular afferents when the head is erect, but a strong stimulus when the head is lying on its side. The opposite is true of the saccular macula.

## Central vestibular projections

Approximately 20 000 vestibular primary afferent neurons join the cochlear auditory neurons to form the **vestibulocochlear nerve** (cranial nerve VIII). These bipolar cells, with cell bodies in the **vestibular ganglion**, project initially to the **vestibular complex** in the **brainstem**, located immediately below the **fourth ventricle**. The vestibular complex comprises the superior, lateral, medial, and inferior vestibular nuclei, and the vestibular afferents form connections with each. Each of the vestibular nuclei has a distinct pattern of onward connectivity and function. The afferent and efferent connections of the vestibular nuclei are shown in *Fig. 3*.

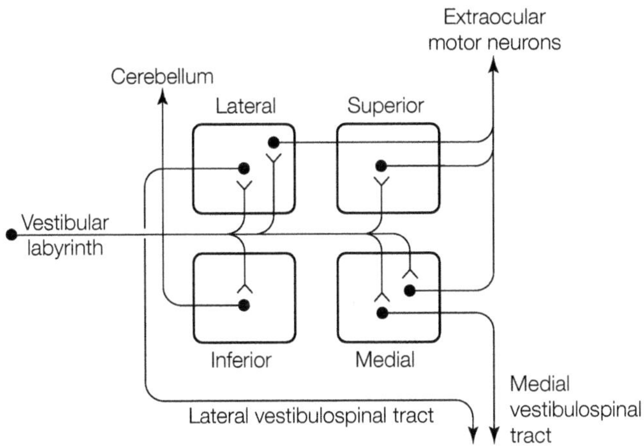


*Fig. 3. Principal vestibular nuclei connections.*

Descending neurons in the lateral and medial vestibular nuclei form the **vestibulospinal system**, sending projections to the motor neurons. The lateral vestibular nucleus receives input primarily from the utricle, and responds to tilting of the head. Parts of the nucleus have inhibitory afferents from the cerebellum (see Topic G2). Excitatory neurons leaving the lateral vestibular nucleus descend ipsilaterally via the lateral **vestibulospinal tract** to activate the motor neurons innervating the antigravity muscles of the limbs. The medial vestibular nucleus receives its vestibular input mostly from the semicircular canals. Output cells, both excitatory and inhibitory, converge initially into the **medial longitudinal fasciculus (MLF)**, before projecting down the medial vestibulospinal tract to motor neurons innervating the neck and trunk regions. Vestibulospinal and vestibulocollic reflexes are activated in response to sudden changes in posture. For instance, if one trips and falls forward, **vestibulocollic reflexes** keep the head upright, while **vestibulospinal reflexes** cause the arms to reach forward to break one's fall.

The superior vestibular nucleus also receives a mainly macular vestibular input. Along with a subset of cells from the medial and lateral vestibular nuclei, it forms the vestibulo-ocular pathway, a disynaptic projection that innervates the extraocular muscles via the **abducens nerve** (cranial nerve VI). This is subject to modulation by the **cerebellum** (see Topic G2).

The **vestibulo-ocular reflex** is controlled by vestibular and cerebellar activity. When the gaze is fixed on a given object and the head is then rotated, activity in the semicircular canals initiates compensatory eye movements in the opposite direction, in order to maintain the object at the focus of gaze. The reflex represents the integration of visual and positional information to ensure a stable retinal image. It is of particular value to ballet dancers who maintain gaze on a fixing point while spinning. Interconnections running in the brainstem fiber bundle, the MLF, are important in integration of vestibular, ocular and cerebellar reflexes; damage to the MLF can cause **nystagmus** (involuntary, jerky movements of the eyes when looking straight ahead). Clinical tests of the patency of vestibular function include the **caloric test**, in which warm or cold water is instilled into the outer ear; the subsequent warming or cooling of the fluid in the (predominantly) lateral semicircular canal sets up convection currents, mimicking rotational movement, and compensatory eye movements can be observed.

The inferior vestibular nucleus receives inputs from most of the labyrinth structure. Its projections form the **vestibulocerebellar pathway** along with some fibers from the other nuclei of the vestibular complex. The vestibular nuclei receive reciprocal connections from the cerebellum, and have a pivotal role in sensorimotor coordination.

Despite the wealth of information received by the vestibular nuclei, relatively little is transmitted to the cortex. A handful of fibers, mainly from the inferior vestibular nucleus, project, via synaptic relays in the ventral posterior nucleus of the **thalamus**, to the face area of the **somatosensory cortex** (parietal cortex).

# G8 TASTE AND OLFACTION

## Key Notes

**The chemical senses**

Taste (gustation) and smell (olfaction) are chemical senses, responding to a range of stimulants. For a chemical to have a taste, it must be soluble in water (saliva). To be smelt, the compound must be volatile.

**Taste buds**

The 'receptors' for taste are the 2000 to 5000 taste buds situated on the upper surface of the tongue. Taste buds are particularly prevalent on the papillae, small peg-like elevated areas of the tongue surface. Each taste bud consists of 50 or more taste receptor cells, which make synaptic contacts with gustatory primary afferent neurons.

**Taste modalities**

There are four main distinguishable tastes – sweet, sour, salt, and bitter. Different parts of the tongue are sensitive to the four tastes. The means of transduction of taste into neuronal responses differs for each taste. Some involve ion channels, while others involve more complex second messenger cascades. Taste receptor cells respond to all taste modalities, but with differing sensitivity. Cranial nerves VII and IX innervate the taste buds and project initially to the nucleus of the solitary tract, and thence to the ventral posterior medial nucleus of the thalamus. Thalamocortical fibers then project to the primary gustatory cortex and to the insula.

**Olfaction (smell)**

Air inspired via the nose contacts the olfactory epithelium, lined with bipolar olfactory receptor neurons. On each olfactory neuron, several cilia containing the odorant receptors extend. Odor molecules bind initially to special odorant-binding proteins, which convey odorants to the receptor sites. Odorant receptors are G-protein-coupled receptors. These receptors respond to many odorants. A given odor activates a particular set of olfactory receptors, each to a different degree.

**Olfactory circuitry**

The olfactory nerve projects to the olfactory bulb, making excitatory connections with mitral cells in synaptic 'glomeruli'. This connection is modulated by periglomerular interneurons. The mitral cells send axons directly to the pyriform and entorhinal cortices. The pyriform cortex is responsible for the perception of smell. The entorhinal cortex projects to the hippocampus (see Topic G3), and probably mediates episodic memory associations of smells and events.

**Related topics**

Parts of the brain (G1)
The limbic system (G3)
Cranial nerves (G4)

## The chemical senses

Taste (**gustation**) and smell (**olfaction**) are chemical senses, responding to a range of stimulants. For a chemical to have a taste, it must be soluble in water (saliva). To be smelt, the compound must be volatile.

Taste and smell serve to help a person choose nutritionally valuable foods and avoid those that might be dangerous. Taste also has strong links to the **autonomic nervous system**, and to the responses involved in digestion. The smell and taste of palatable foodstuffs cause **salivation** as well as an increase in gastrointestinal activity (see Section J).

## Taste buds

The 'receptors' for taste are the 2000 to 5000 **taste buds** situated on the upper surface of the tongue. Taste buds are found across the entire upper surface of the tongue on pitted and grooved indentations, but are particularly prevalent on the **papillae**, small peg-like elevated areas of the tongue surface.

There are three types of papillae – circumvallate, foliate and fungiform. **Fungiform papillae**, shaped like mushrooms, are mainly found at the front of the tongue, while foliate, leaf-like, papillae are located at the side (*Fig. 1a*). **Circumvallate papillae** are shaped like small pimples, and are found at the back of the tongue. Some papillae may contain hundreds of taste buds. **Foliate papillae**, by contrast, contain very few taste buds.

Each near-spherical taste bud, seated on the basement membrane, consists of 50 or more taste receptor cells surrounded by supporting cells. Beneath these are basal cells, which have the capacity to differentiate into new taste receptors, as needed. Taste receptor cells are short-lived, with a lifespan of 10–14 days. At the apical end of the taste receptors are **microvilli** which make contact, via taste pores, with saliva. Taste receptor cells make synaptic contacts with gustatory primary afferent neurons. Each primary afferent may receive contacts from several taste buds.

## Taste modalities

There are considered to be four main distinguishable taste modalities – sweet, sour, salt and bitter. More recently, it has been suggested that a fifth – '**umami**', from the Japanese word for savory – may exist.

Different parts of the tongue are sensitive to the four tastes (*Fig. 1b*). The tip of the tongue is preferentially sensitive to sweet, while the sides of the tongue respond better to sour and salty flavors. The back of the tongue is more sensitive to bitter tastes.

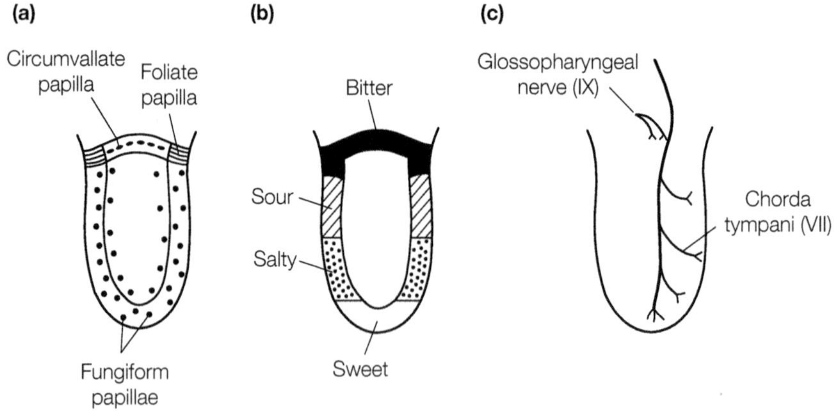


*Fig. 1. Distribution of (a) papillae; (b) taste modalities; and (c) nerve paths for taste sensation in the tongue.*

The means of transduction of taste into neuronal responses differs for each of the taste modalities. Saltiness is detected directly by a channel that lets cations, particularly sodium or potassium, enter the cell down their concentration gradient. This causes depolarization of the taste receptor cell, and transmitter release at the synapse with the gustatory primary afferent neuron. Sourness or acidity (e.g. vinegar) is detected directly by a pH-sensitive potassium channel. Hydrogen ions block the channel and reduce the normal outward flux of potassium ions, thus causing depolarization. Sweetness (such as sucrose, fructose or aspartame) is detected when sweet-tasting compounds bind to metabotropic receptors on the villi of the taste cell. These act, via **G-proteins**, to increase intracellular **cyclic adenosine monophosphate (cAMP)**, then closing potassium channels. Bitterness (such as quinine or caffeine), perhaps because of the chemical diversity of such agents, is mediated by several mechanisms including direct opening of ion channels and the activation of second messengers. Umami appears to involve activation of **metabotropic glutamate receptors** by monosodium glutamate.

The vast majority of taste receptor cells respond to all taste modalities. Nevertheless, the magnitude of their response and its polarity differs for each cell. Taste cells may be depolarized or hyperpolarized by a given taste. A taste cell may be preferentially responsive to sweet tastes, for instance. Gustatory afferent neurons are spontaneously active, but their activity is differentially modulated by the taste receptor cells from which they receive inputs. Afferents are inhibited by some tastes, or excited by others.

**Cranial nerve VII** and **cranial nerve IX** innervate the taste buds (*Fig. 1c*). Neurons from the majority of the tongue, encompassing the fungiform papillae, travel in the facial nerve (**chorda tympani**), while the glossopharyngeal nerve (cranial nerve IX) carries sensory information from the back of the tongue, the location of the foliate and circumvallate papillae. A handful of taste buds in the esophagus send neurons that join the vagus nerve (cranial nerve X). Taste afferents from all three cranial nerves project initially to the **nucleus of the solitary tract**, in the dorsal part of the medulla. From there, axons ascend to the ventral posterior medial nucleus of the **thalamus**. Thalamocortical fibers then project to the **primary gustatory cortex** (below the sensory face cortex), and to the taste areas of the **insula**.

## Olfaction (smell)

Air inspired via the nose is spun into vortices and eddies by the nasal **turbinate bones**. This forces the air into contact with the **olfactory epithelium**, located in the mucus membrane of the nasal cavity at the base of the skull. The human olfactory epithelium has a surface area of approximately 10 $cm^2$. The olfactory epithelium is lined with bipolar olfactory receptor neurons, separated by mucus-secreting support cells and glands (**Bowman's glands**), above a layer of basal cells (*Fig. 2*). These olfactory neurons, like the rods and cones of the **retina** (see Topic G5) and taste receptors of the tongue, are highly specialized. On each, a single dendrite extends between the support cells to form a small bulb or vesicle from which several cilia extend. These cilia contain the odorant receptors.

Odor molecules, usually organic, always lipid-soluble but often hydrophobic, are trapped in the mucus overlying the olfactory epithelium, and bind initially to special odorant-binding proteins in the mucus. These odorant-binding proteins help to convey relatively hydrophobic odorants to the receptor sites.

Once bound, the odorant–protein complex binds to odorant receptors on the cilia of the olfactory receptor neurons. Odorant receptors are G-protein-coupled

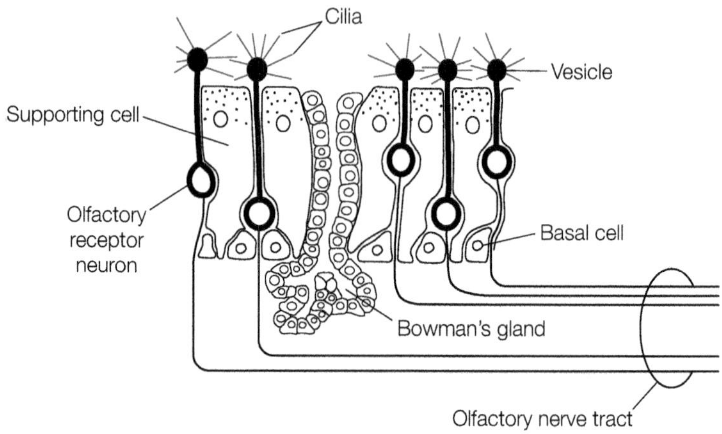


*Fig. 2. Olfactory epithelium, showing olfactory receptor neurons.*

receptors which, in turn, increase the levels of cAMP inside the cells. The rise in cAMP opens cation channels in the cilia, and action potentials are generated in the olfactory nerve axon. Many hundreds of olfactory receptors exist. These receptors are relatively promiscuous and respond to many odorants. Thus a given odor activates a particular set of olfactory receptors, each to a different degree. A second odor may activate a separate set, some similar, others different, depending on its similarity to the first.

**Olfactory circuitry**

The **olfactory nerve** (cranial nerve I) is very short, consisting of many small nerve bundles. These project, through the osseous fenestrations of the **cribriform plate**, into the **olfactory bulb**. Olfactory neurons make excitatory connections with mitral cells in synaptic 'glomeruli' within the olfactory bulb. As in the visual system (see Topic G5), the connection between the primary afferent (the olfactory receptor cell) and the second-order cells is modulated by interneurons. In the olfactory bulb these are the periglomerular cells, analogous to the horizontal cells in the retina.

Each olfactory glomerulus receives the integrated input of those olfactory receptor neurons that respond specifically to a given odor. Thus each glomerulus is odor specific. There are some 2000 olfactory glomeruli. **Norepinephrine** and **serotonin** axons innervate the olfactory bulb, and are thought to regulate sensitivity to odors. For instance, states of high arousal (e.g. fear, sexual desire) enhances olfactory sensitivity.

The mitral cells of the olfactory bulb glomeruli send axons directly to the various subdivisions of the olfactory cortex, especially the pyriform and entorhinal cortices. The **pyriform cortex** is responsible for the perception of smell. This pathway is unusual for a sensory input in that there is no thalamic relay for this smell pathway. Thus smell is not a synthesized sense like seeing or hearing. The entorhinal cortex projects to the hippocampus (see Topic G3), and probably mediates episodic memory associations of smells and events. For instance, the scent of a particular flower might invoke the memory of a childhood holiday.

# H1 SPINAL CORD STRUCTURE

## Key Notes

**Layout**

The spinal cord is the part of the central nervous system that connects the brain to peripheral nerve output. It sits in the vertebral column, although the lowest parts of the vertebral column do not contain spinal cord, but a collection of spinal nerves known as the cauda equina. Some spinal nerves emerge between vertebrae that are lower down than the region of the spinal cord from which they originate.

**White versus gray matter**

The spinal cord is composed of white matter (myelinated, with no neuronal cell bodies) and gray matter (unmyelinated, containing neuronal cell bodies). The gray matter lies towards the center of the spinal cord, and the white matter is confined more to the outside. The white matter is made up of the long processes of sensory and motor neurons, conducting nerve impulses from the brain to the periphery and vice versa.

**Sensory versus motor**

The dorsal (posterior) side of the spinal cord is more concerned with sensory functions; it receives input from the peripheral nervous system, and sensory fibers ascend to the brain in this region. The ventral (anterior) side of the spinal cord is more concerned with motor functions; motor output (somatic and visceral) emerges from the spinal cord via the ventral roots of spinal nerves. The cell bodies of primary sensory neurons are located in the dorsal root ganglia; these cells have processes that synapse with cells in the spinal cord or, in the case of certain sensory modalities, cells in the brainstem. Thus second-order sensory neuronal cell bodies are located in spinal cord gray matter or in the brainstem. Second-order sensory neurons project to the thalamus (after having crossed over to the other side of the central nervous system), where the cell bodies of third-order sensory neurons are located. These third-order sensory neurons project to the somatosensory cortex in the parietal lobe. Upper motor neurons for somatic motor function are located in the primary motor cortex in the frontal lobe. They project to lower motor neuron cell bodies in the ventral horn of spinal cord gray matter, and cross over to the other side of the central nervous system on the way. Both sensory input and motor output are therefore contralateral; sensation from the right side of the body is sensed in the left side of the brain (and vice versa), and the left side of the brain controls the right side of the body (and vice versa). Lower motor neurons of sympathetic and parasympathetic branches of the autonomic nervous system are present in lateral horns of spinal cord gray matter in some parts of the spinal cord.

**Ascending and descending tracts**

Some white matter tracts in the spinal cord are concerned with sensory function (these are ascending tracts), and others are concerned with motor function (these are descending tracts). Sensory input to the brain, and motor output from the brain are organized according to function. For example, the dorsal columns of spinal cord white matter conduct information about touch and proprioception, whereas another ascending tract, the spinothalamic tract, conducts information about pain and temperature. The anterior and lateral corticospinal tracts contain the processes of upper motor neurons that synapse with lower motor neurons in the ventral horn of spinal cord gray matter, but the anterior corticospinal tract conveys signals to lower motor neurons that act on muscles of the trunk, whereas the lateral corticospinal tract conveys signals to lower motor neurons that act on limb muscles. Different tracts conduct ipsilateral or contralateral information to or from the brain.

**Related topics**

Parts of the brain (G1)
The locomotor system (G2)
Spinal nerves and plexuses (H2)
Structure and function of peripheral nerve (H3)
Spinal reflexes (H4)
Structure and function of the autonomic nervous system (H5)
Skeletal muscle function and properties (I1)
Innervation of smooth muscle (I7)

## Layout

The **spinal cord** runs in the vertebral column, from the medullary part of the brainstem down to the level of the first or second lumbar vertebrae in adults. The **vertebral column** is composed of seven cervical, twelve thoracic, five lumbar, and five sacral vertebrae. In the spaces or foramina (singular: foramen) between the vertebrae, nerves emerge from the spinal cord to innervate the body. High up in the vertebral canal, these **spinal nerves** emerge from foramina that are roughly in line with the spinal cord segment from which they are derived (*Fig. 1*). Lower down (roughly from the fifth cervical vertebra), the spinal nerves emerge in foramina that are significantly lower than the spinal cord segment from which they are derived. Spinal nerves derived from the lowest spinal cord segments are collected inside the vertebral canal in a bundle of nerves known as the ***cauda equina*** (Latin: horse's tail). These spinal nerves emerge from foramina further down the vertebral column. It is for this reason that, in adults, sampling of the cerebrospinal fluid (CSF) that surrounds the spinal cord (**lumbar puncture**) is most often done by inserting a needle 1–2 vertebral segments below the second lumbar vertebra. This reduces the likelihood of injuring the spinal cord, as the nerves that make up the cauda equina are more free to move out of the way of the needle than the more solid spinal cord would be.

## White versus gray matter

If the spinal cord is removed and cut transversely, the interior structure can be revealed. A cross-section of part of the spinal cord in the thoracic region is shown in *Fig. 2*. In the very center of the spinal cord, there is a canal that, *in vivo*, contains CSF. This is known as **the central canal**, and is surrounded by **gray matter** (cell bodies and processes of neurons, the majority of which are unmyelinated). The gray matter has two **dorsal horns** (posterior *in vivo*) and two **ventral horns** (anterior *in vivo*). In some segments of the spinal cord (e.g. thoracic), a further **lateral horn** of gray matter exists. Outside the gray matter, there exists **white matter**. In common with other parts of the CNS, the white matter of spinal cord is composed of myelinated nerve fiber processes. Unlike gray matter, the

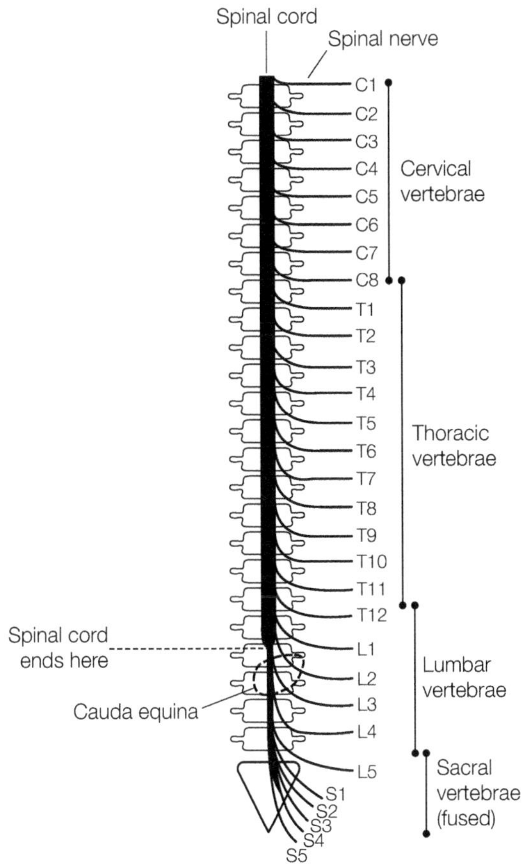


*Fig. 1. Layout of the spinal cord in the vertebral canal and exit foramina for spinal nerves.*

white matter of spinal cord does not contain any nerve cell bodies. The nerve fiber processes in spinal cord white matter are ascending to the brain, conducting sensory information, or descending to other segments of spinal cord, conducting motor information.

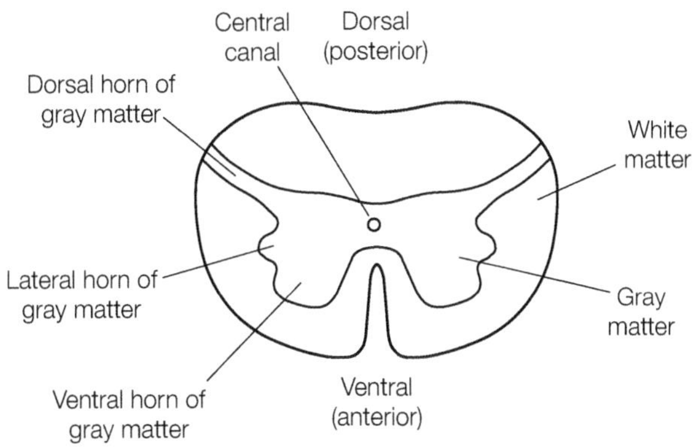


*Fig. 2. Cross-sectional view of a section of thoracic spinal cord.*

## Sensory versus motor

A cross-sectional diagram of the spinal cord can be marked with an approximate line dividing sensory and motor functions (*Fig. 3*). Structures dorsal to the central canal are more likely to be involved in the conduction and processing of sensory information, whereas structures ventral to the central canal are more likely to be involved in motor functions. Where spinal nerves enter and emerge from the spinal cord, one can see their dorsal and ventral roots. The **dorsal root** of a spinal nerve carries fibers that conduct sensory information, and the ventral root carries fibers that conduct motor information. Just before the dorsal root enters the spinal cord, there exists a small swelling or ganglion. This is the **dorsal root ganglion (DRG)**, containing the cell bodies of those neurons that have processes extending out to the periphery, and ending in a specialized sensory receptor. These neurons are known as **primary sensory neurons**. The cells in the DRG also have processes that extend towards the spinal cord. Some of these processes enter the dorsal horn of spinal cord gray matter and synapse with the cell bodies of second-order sensory neurons. Others ascend to the **medulla** before synapsing with the cell bodies of other second-order sensory neurons (*Fig. 3*). After synapses between primary and second-order sensory neurons have been made, **decussation** (crossing over) of fibers occurs (see below for the implications of this).

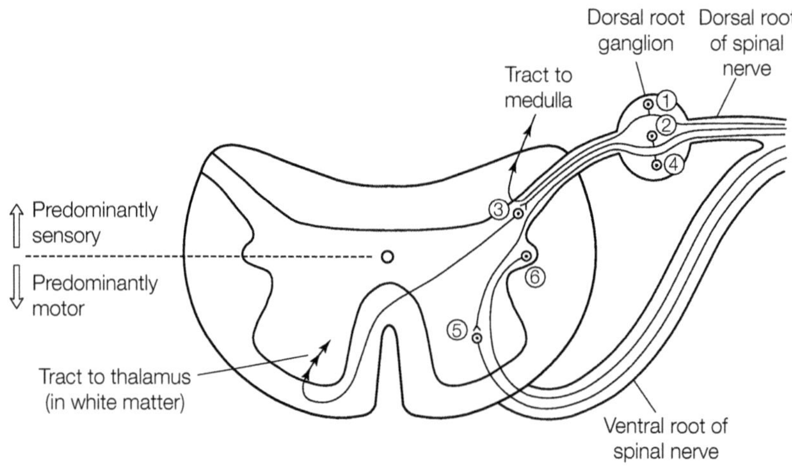


*Fig. 3. Cross-sectional view of the spinal cord, showing the location of the dorsal root ganglion and (1) cell bodies of primary sensory neurons that ascend in the spinal cord to the medulla; (2) cell bodies of primary sensory neurons that synapse with (3) second-order sensory neurons that decussate and ascend in the spinal cord to the thalamus; (4) cell bodies of primary sensory neurons that synapse with (5) lower motor neurons in the ventral horn of spinal cord gray matter; (6) cell bodies of lower (preganglionic) motor neurons of the autonomic nervous system.*

After these synapses between the processes of primary sensory neurons and the cell bodies of second-order sensory neurons, the processes of the second-order sensory neurons cross to the other side of the spinal cord or brainstem (depending upon where the synapse occurred), and project to the **thalamus**, where they make synapses with the cell bodies of third-order sensory neurons that project to the somatosensory cortex. In addition to those processes that ascend to the medulla, some primary sensory neurons have processes that synapse directly with voluntary motor neurons in other parts of spinal cord gray matter. These direct processes from sensory neurons to motor neurons are the basis for the **knee-jerk reflex**, where stretching of the quadriceps muscle group in the thigh leads to reflex contraction of the quadriceps group and inhibition of the hamstrings muscle group. This leads to extension of the knee joint (see Topic H4 for further details of spinal reflexes).

The cell bodies of **lower motor neurons (LMNs)** exist in two distinct places in the spinal cord (*Fig. 3*). LMNs that control the output to skeletal (voluntary) muscle have their cell bodies in the ventral horn of spinal cord gray matter. Those that control (involuntary) output to visceral smooth muscle, the heart and glands have their cell bodies in the lateral horn of spinal cord gray matter. These are the **preganglionic neurons** of the **autonomic nervous system** (see Topic H5), and are present in the thoracic and upper lumbar segments of spinal cord and in sacral spinal cord segments 2–4. All LMNs receive input from the brain via projections that descend in spinal cord white matter. Motor output from the spinal cord goes via the **ventral root** of spinal nerves.

When removed from the body, the spinal cord shows some very obvious enlargements, principally in the cervical and lumbosacral segments of the cord. This is because in both these regions, there are many LMN cell bodies innervating the large muscle groups of the arms and legs.

## Ascending and descending tracts

There is an absolute requirement for most functions of the CNS, whether they are sensory or motor, to have some sort of ascending input to higher centers, or descending input from the brain. This is done by placing all of this 'cabling' in spinal cord white matter. This 'cabling' is organized into specific tracts, each with a set of discrete functions (*Fig. 4*).

The **corticospinal tract**, for example, contains the processes of **upper motor neurons** that are descending the spinal cord to impinge upon LMNs. Their

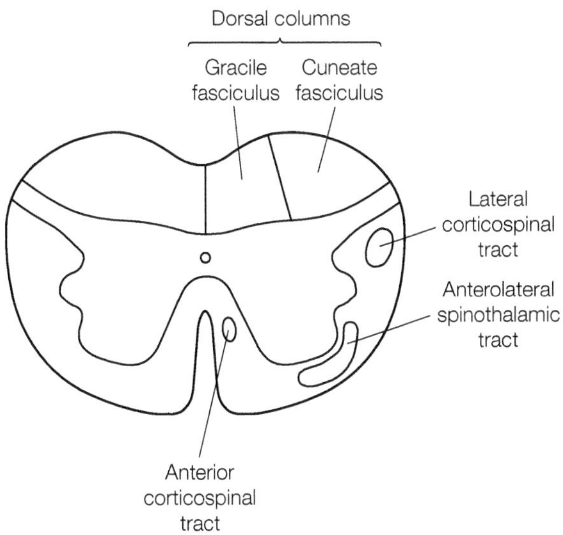


*Fig. 4. Cross-sectional view of the spinal cord, showing the location of some important white matter tracts.*

influence on LMNs may be either excitatory or inhibitory; this allows fine-tuning of movement. Damage to the corticospinal tract results in weakness of muscles below the level of the lesion. Ascending tracts include the **dorsal columns** and the anterolateral **spinothalamic tract**. The dorsal columns contain the central processes of those primary sensory neurons that subserve touch, vibration, and joint position sense. These ascend to the medulla before synapsing there with second-order sensory neurons that then decussate to the other side of the CNS. Thus, damage to the dorsal columns on one side results in loss of touch, vibration and joint position sense on that side (ipsilaterally), below the level of the lesion. The anterolateral parts of the spinothalamic tract contain the central processes of those second-order sensory neurons that have already decussated, having received input from primary sensory neurons in the spinal cord, and are ascending to the thalamus (*Fig. 3*). Thus when the anterolateral spinothalamic tract is damaged on one side of the spinal cord, pain and temperature sensation are lost on the other side (contralaterally), below the level of the lesion. The increasing number of fibers in white matter as one ascends the spinal cord is a further reason for the cervical enlargement of the spinal cord noted above. Further information on spinal cord tracts is given in *Instant Notes in Neuroscience*, Section G.

# H2 SPINAL NERVES AND PLEXUSES

## Key Notes

**Introduction**

Spinal nerves are the method by which the central nervous system communicates with the periphery. The spinal nerves emerge on either side of the vertebral column. They are all composed of the axons of motor and sensory neurons. Components of several spinal nerves can be combined to form a named nerve in the body (e.g. the radial nerve is composed of fibers contributed by spinal nerves C5–L1).

**The brachial plexus**

Spinal nerves from C5–L1 emerge from the spinal cord, and are organized into a plexus known as the brachial plexus. Five major named nerves emerge from the brachial plexus, and these nerves innervate the arm and hand.

**The lumbosacral plexus**

Spinal nerves from L1–S4 emerge from the spinal cord and are organized into two plexuses known as the lumbar plexus and the sacral plexus. The lumbosacral trunk connects the two plexuses and allows fibers from L4 and L5 spinal nerves to find their way into the sciatic nerve trunk. Eight major named nerves emerge from the lumbosacral plexus, and these nerves innervate the abdominal wall, pelvic structures, and the lower limbs and feet.

**Dermatomes and myotomes**

The territory innervated by fibers from a given spinal nerve is the dermatome or myotome of that spinal nerve. Dermatome refers to the area of skin innervated by sensory fibers from a particular spinal nerve, whereas myotome refers to the muscle or muscle group innervated by motor fibers from a particular spinal nerve. Dermatome patterns in humans are complicated in the limbs and hand.

**Related topics**

Neurons and glia II (F1)
Structure and function of peripheral nerve (H3)
Spinal reflexes (H4)

## Introduction

As described in topic H1, **spinal nerves** emerge from the vertebral column as paired nerves, one on either side. They leave the vertebral column by way of small spaces (foramina) between the individual bones (vertebrae) that make up the vertebral column. These foramina are extremely small (no more than a few millimeters in diameter), but can be made larger by flexion of the spine, either forwards or away from the side at which the foramen is situated. There are seven cervical, twelve thoracic, five lumbar, and five sacral vertebrae, conventionally numbered C1–C7, T1–T12, L1–L5 and S1–S5. There are eight cervical, twelve thoracic, five lumbar, and five sacral spinal nerves on each side. The first cervical spinal nerve emerges above vertebra C1, the second above vertebra C2 (below vertebra C1, *Fig. 1*) and so on, until the eighth cervical spinal nerve emerges below vertebra C7 (above vertebra T1). From then on down the vertebral column, the various spinal nerves emerge directly below their respective vertebrae (e.g. tenth thoracic nerve below vertebra T10).

It is important to realize that these are spinal nerves, and that their relationship to the various named nerves in the body (e.g. the **radial nerve**) is complex. This is because at several points in the body, most notably in the axilla (armpit) and posterior abdominal wall, the spinal nerves become mixed up into interchanges known as plexuses, and the nerves that emerge from these plexuses can contain fibers from several different spinal nerves. A useful analogy is traffic coming from different origins and heading to different destinations, but sharing a stretch of motorway for a period of time. What this means functionally is that the areas innervated by an individual spinal nerve from the point of view of sensory function (the **dermatome** of that spinal nerve) and motor function (the **myotome** of that spinal nerve) can be quite different.

## The brachial plexus

The **brachial plexus** is located in the axilla, in a region bounded by the head of the humerus (the bone between elbow and shoulder), the ribcage and the

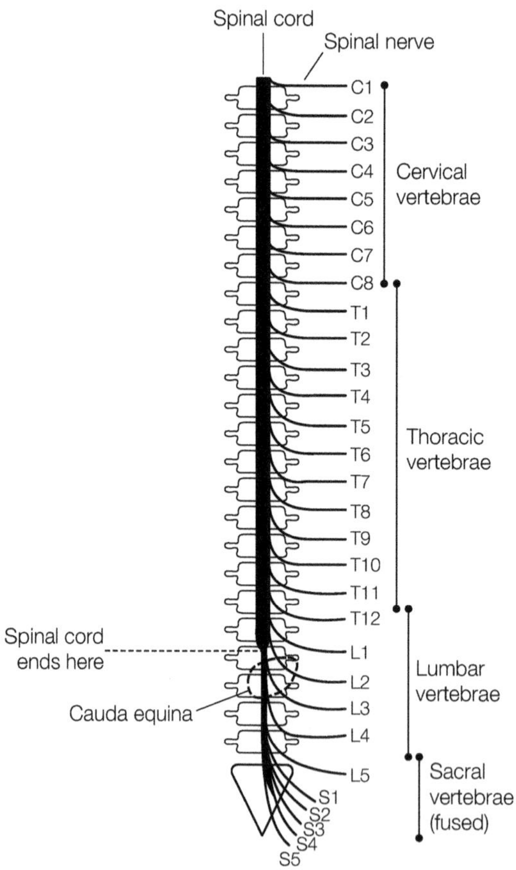


*Fig. 1. Layout of the spinal cord in the vertebral canal and exit foramina for spinal nerves.*

scapula (shoulderblade). This plexus is a distribution center for fibers from spinal nerves C5–T1. The major nerves that emerge from the plexus are the **axillary nerve**, the radial nerve, the **musculocutaneous nerve**, the **ulnar nerve**, and the **median nerve**. The approximate distribution of nerve fibers from the various spinal nerves (C5–T1) in these named nerves is as shown in *Fig. 2*.

*Table 1* lists the major structures that are innervated by the various major nerves emerging from the brachial plexus. Notice that in some cases (e.g. the musculocutaneous nerve), the muscle group innervated by a particular nerve is some way removed from the skin that the same nerve supplies.

**The lumbosacral plexus**

Another major interchange for spinal nerves in the body exists to distribute nerve fibers from various spinal nerves to the lower limbs and pelvic viscera.

*Table 1. Functions of the major nerves arising from the brachial plexus*

| 'Named' nerve | Spinal nerve(s) contributing | Sensory innervation | Motor innervation |
|---|---|---|---|
| Axillary | C5, C6 | Skin over deltoid muscle | Deltoid muscle |
| Radial | C5–T1 | Posterior skin of upper limb (including back of hand) | All arm joint extensor muscles |
| Musculocutaneous | C5, C6, C7 | Lateral skin of forearm | Elbow flexors |
| Median | C5–T1 | Skin on radial side of palm of hand (side nearer to thumb) | Most forearm (i.e. wrist) flexors, thenar muscles at base of thumb |
| Ulnar | C7, C8, T1 | Skin on ulnar side of hand (side nearer to little finger) | Some forearm (i.e. wrist) flexors, most hand muscles |

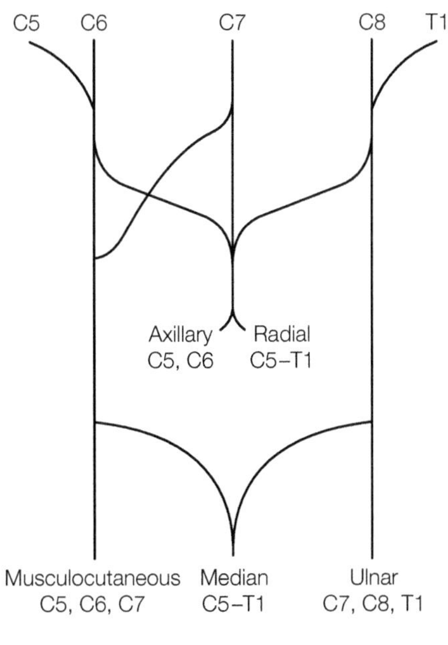


*Fig. 2. Schematic diagram of the distribution of fibers from spinal nerves C5–T1 in the brachial plexus (right side in this example).*

This is known as the **lumbosacral plexus**. It is located against the posterior wall of the abdomen and the pelvis. The major nerves arising from this plexus are the **ilioinguinal** and **iliohypogastric nerves**, the **genitofemoral nerve**, the **lateral cutaneous nerve of the thigh**, the **femoral nerve**, the **obturator nerve**, the **sciatic nerve**, the **pudendal nerve**, and the **pelvic splanchnic nerves**. The approximate distribution of nerve fibers from the various spinal nerves (L1–S4) in these named nerves is as shown in *Fig. 3*. Note that the sciatic nerve is in fact a large combined nerve that gives rise to many other named nerves with specific innervation territories. However, these are too numerous to mention here. Information presented here on the sciatic nerve refers to the whole nerve trunk, where it emerges from the lumbosacral plexus.

*Table 2* lists the major structures that are innervated by the various major nerves emerging from the lumbosacral plexus. In this case, the muscle groups innervated by the various nerves are more closely related to the skin and visceral organs that the same nerve supplies.

## Dermatomes and myotomes

The portion of skin supplied by a single spinal nerve is known as the dermatome of that spinal nerve. This may be innervated by several different individual (named) nerves because of the distribution of fibers from spinal nerves that takes place in the various nerve plexuses as described above. A dermatome map for the human body is shown in *Fig. 4*. Note the complicated pattern of sensory innervation of the skin of the limbs and the pelvic region. Contrast this to the simple pattern of the sensory innervation of the skin on the trunk; in fact because thoracic spinal nerves follow the underside of the ribs, the innervation pattern of spinal nerves T2–T12 resembles a ribcage to some extent. Damage to one named nerve will lead to sensory loss only in the area of skin that the nerve in question innervates, but damage to an entire **spinal cord**

*Table 2. Functions of the major nerves arising from the lumbosacral plexus*

| 'Named' nerve | Spinal nerve(s) | Sensory innervation contributing | Motor innervation |
|---|---|---|---|
| Ilioinguinal and iliohypogastric | L1 | Lower anterior abdominal wall | Lowest abdominal wall muscles |
| Genitofemoral | L1, L2 | Anterior scrotum (males), labia (females) | Cremaster muscle (males, raises testes) |
| Lateral cutaneous of thigh | L2, L3 | Skin over lateral thigh | None |
| Femoral | L2, L3, L4 | Hip and knee joints, skin over anterior thigh | Knee extensors |
| Obturator | L2, L3, L4 | Hip and knee joints, skin over medial (inner) thigh | Hip adductors (close legs) |
| Sciatic | L4–S3 | Skin on foot and on lateral parts of leg | Hip extension, knee flexion, movements of foot and toes |
| Pudendal | S2, S3, S4 | Perineum, bladder, other pelvic viscera | Bladder and urinary sphincters |
| Pelvic splanchnic nerves | S2, S3, S4 | Sensation in the anal canal | Parasympathetic motor to hindgut |

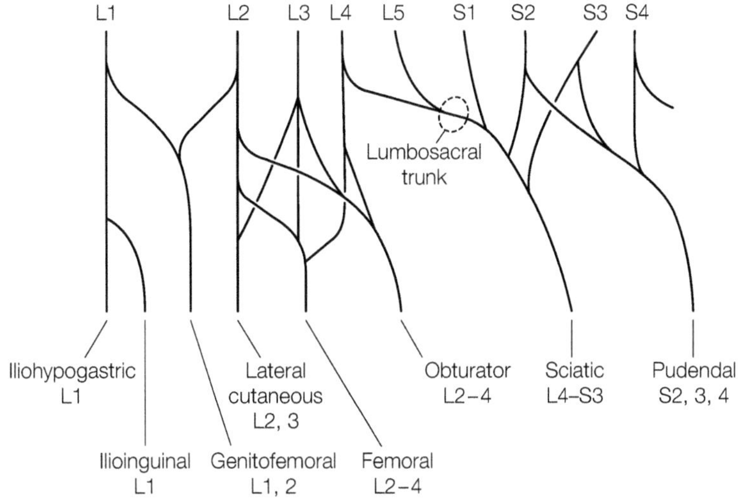


*Fig. 3.* *Schematic diagram of the distribution of fibers from spinal nerves L1–S4 in the lumbosacral plexus (right side in this example).*

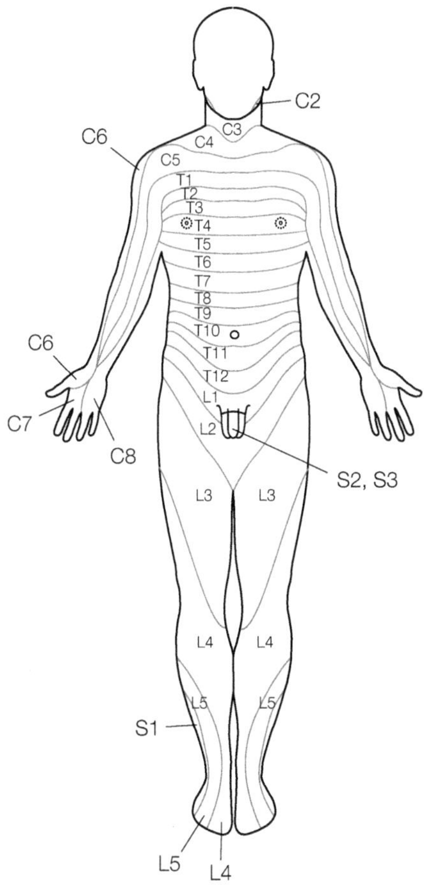


*Fig. 4.* *The human dermatome map, showing the anterior (front) surface only. C2, T1, L1, S1, etc. refer to areas innervated by the second cervical spinal nerve, first thoracic spinal nerve, first lumbar spinal nerve, first sacral spinal nerve, etc.*

segment or to a spinal nerve as it emerges from the vertebral column may have a more widespread effect.

The skeletal muscle (or muscle group) supplied by a given spinal nerve is known as the myotome of that spinal nerve. Damage to an entire spinal cord segment could result in a very different pattern of sensory loss from the pattern of weakness one might observe. Similarly, damage to the **ventral horn** (motor) of a given spinal cord segment would lead to a myotome deficit, but would probably result in little or no change in the function of the dermatome of that spinal cord segment.

# H3 STRUCTURE AND FUNCTION OF PERIPHERAL NERVE

## Key Notes

**Mixed peripheral nerves**

Peripheral nerves are composed of nerve fibers from different spinal nerves that subserve motor or sensory functions. Sensory fibers conduct nerve impulses from the periphery to the central nervous system, and motor fibers conduct nerve impulses in the opposite direction. Nerve fibers in peripheral nerves have a range of sizes, and some are myelinated whereas others are not; both these features determine the speed with which the nerve fiber can conduct nerve impulses.

**Motor functions of peripheral nerves**

Motor fibers in peripheral nerves innervate not only skeletal muscle (somatic motor), but also cardiac muscle, smooth muscle and glandular tissue (visceral motor). Visceral motor functions can be either facilitatory or inhibitory, but those of the somatic motor system are purely excitatory.

**Sensory modalities – touch, proprioception, and pain**

Different sensory modalities are picked up by specific specialized sensory receptors in the periphery, and all of these sensory neurons have their cell body in the dorsal root ganglia near the spinal cord. The specialized sensory receptors normally only respond to one type of stimulus (e.g. light touch), and there are several different types of sensory receptor. The different sensory modalities tend to be conducted by nerve fibers that have a particular appearance (e.g. fibers conducting information about joint position are normally large-diameter myelinated fibers). Some sensory neurons are involved in spinal reflexes, but all are involved in sending input to the central nervous system, usually to the somatosensory cortex in the parietal lobe.

**Related topics**

Action potentials (F2)
Spinal nerves and plexuses (H2)
Spinal reflexes (H4)
The neuromuscular junction (I2)
Innervation of smooth muscle (I7)

## Mixed peripheral nerves

When considering the nervous system, it is important to use certain terms correctly and precisely. For example, the word ***neuron*** is used to name the cells that (together with **glial cells**) make up the nervous system. The term ***nerve cell*** can be used to describe the same thing, but to use the word ***nerve*** is totally incorrect. This is because *nerves* are in fact composed of the axonal processes of neurons and their associated glial cells. **Peripheral nerves** are usually composed of the axons of neurons that subserve both sensory and motor functions, so they are mixed in that respect. Also, since many peripheral nerves supplying skeletal muscle, skin, and joints are composed of fibers from several different spinal nerves (see Topic H2), they can be considered mixed in that respect too. Although **action potentials** (APs) have the potential to be conducted in either direction along an axon, a combination of anatomical and biophysical factors (see Topic F2) mean that sensory axons conduct APs from the periphery to the central nervous system (CNS), and motor axons conduct APs from the CNS to the periphery.

As illustrated in *Fig. 1*, each axon is surrounded by its own connective tissue **endoneurium**. Axons are bundled together in peripheral nerves into **fascicles**, each of which is covered by **perineurium**. Blood vessels are present both within the fascicle and between fascicles. Fascicles are bundled together inside a nerve and surrounded by a connective tissue **epineurium**.

As described in topic F2, axon diameter and **myelination** of the axon affect the **conduction velocity** of nerve fibers. Peripheral nerve is composed of axons with a range of diameters (from 0.3 to 20 μm), some of which are myelinated and some of which are not. In general, larger axon diameter means a greater conduction velocity in that nerve fiber. Myelination also increases nerve fiber conduction velocity by allowing APs to be conducted in a saltatory manner, rather than relying on local circuits (see Topic F2). The factors of axon diameter and (non-) myelination combine in such a way that peripheral nerves are composed of different classes of nerve fiber, each with different conduction velocities (*Table 1*). These fibers are organized in such a way that individual classes of axon subserve different roles in the nervous system.

This mixed composition of peripheral nerve can be demonstrated if a peripheral nerve is stimulated at one point and the compound AP is recorded on the surface of the nerve some distance away. Several different peaks will be seen in the compound AP (*Fig. 2*), reflecting the different conduction velocities of the various types of fiber. Generally, one can distinguish three main groups: Aα/Aβ; Aγ/Aδ; and B/C.

## Motor functions of peripheral nerves

Peripheral nerves contain motor nerve fibers. The cell bodies of these motor neurons reside in the **ventral horn** of gray matter of the spinal cord in the case of somatic (voluntary) motor neurons, and in the cranial nerve motor nuclei and

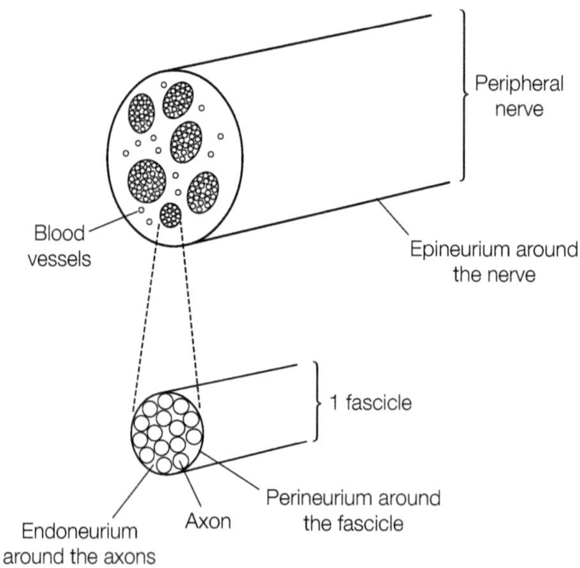


*Fig. 1. Cross-sectional diagram of fascicles of axons in a peripheral nerve.*

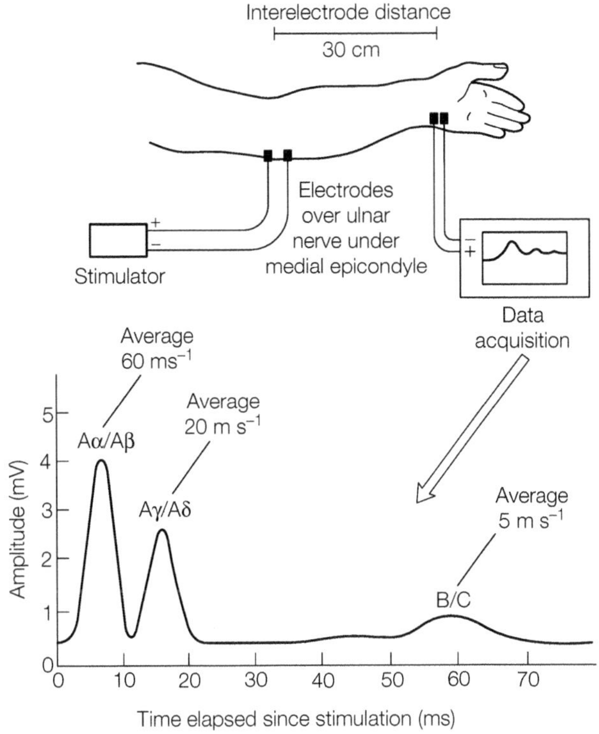


*Fig. 2. The compound action potential as recorded in peripheral nerve (e.g. ulnar nerve).*

*Table 1. Properties of nerve fibers in peripheral nerve*

| Fiber type | Diameter (μm) | Conduction velocity (m s$^{-1}$) | Function |
|---|---|---|---|
| Aα | 12–20 | 70–90 | Skeletal muscle motor, proprioception |
| Aβ | 5–12 | 30–70 | Touch, pressure |
| Aγ | 3–6 | 15–30 | Muscle spindle motor |
| Aδ | 2–5 | 12–30 | Pain, temperature, touch |
| B | <3 | 3–15 | Preganglionic autonomic (see Topic H5) |
| C | 0.3–1.3 | 0.5–2.3 | Pain, reflex, postganglionic sympathetic (see Topic H5) |

the **lateral horn** of gray matter in the thoracolumbar and sacral segments of spinal cord in the case of visceral (involuntary) motor neurons (see Topic H5). In this context, the use of the word ***motor*** is often misunderstood. This term refers not only to effects on muscle (skeletal, cardiac and smooth), but also to effects on glandular tissue. Similarly, it is erroneous to think that the term *motor* necessarily implies excitation; in the case of the somatic motor system this is the case, but in the visceral motor system the term is used for output from the CNS that may either excite or inhibit muscle fibers or glandular tissue. *Table 2* gives some examples of the actions of somatic and visceral motor innervation on various tissues.

*Table 2. Motor functions of various neurons in the peripheral nervous system*

| Motor innervation | Origin | Tissue | Effect |
|---|---|---|---|
| α-motor neurons | Cranial nerve motor nuclei and ventral horn of spinal cord gray matter | Skeletal muscle | Contraction of muscle cells outside muscle spindle only |
| γ-motor neurons | Cranial nerve motor nuclei and ventral horn of spinal cord | Muscle spindles in skeletal muscle | Contraction of cells within muscle spindle only |
| Parasympathetic | Sacral segments of spinal cord (lateral horn) | Smooth muscle and nerves in the colon | Increased contraction of smooth muscle |
| Sympathetic | Thoracic segments of spinal cord (lateral horn) | Cardiac muscle and nodal tissue in the heart | Increased heart rate and increased force of contraction |
| Parasympathetic | Sacral segments of spinal cord (lateral horn) | Smooth muscle in walls of blood vessels of the external genitalia (penis in males) | Relaxation of blood vessels, engorgement and erection |
| Sympathetic | Lumbar segments of spinal cord (lateral horn) | Smooth muscle and nerves in the colon | Relaxation |
| Parasympathetic | Cranial nerve VII and IX motor nuclei (pons and medulla) | Salivary glands | Increased secretion |
| Sympathetic | Thoracolumbar segments of spinal cord (lateral horn) | Sweat glands | Increased secretion |

## Sensory modalities – touch, proprioception and pain

As outlined in topic H1, **primary sensory neurons** with their axons in spinal nerves have their cell bodies in the **dorsal root ganglion (DRG)** of that spinal nerve. These neurons have two projections from the cell body; one ends in the **sensory receptor** in the innervated tissue (skin, joints, muscle, visceral organs), and the other synapses with the dendrites of second-order sensory neurons in the CNS (either in the spinal cord or the medulla). Most peripheral nerves contain a mixture of **Aβ-**, **Aδ-** and **C**-type sensory fibers (*Table 1*) for the purposes of receiving information about the environment in which we find ourselves. As is evident from *Table 1*, the various types of sensory fibers have different properties and conduct APs at different rates. Some sensory fibers also participate in **spinal reflexes** (e.g. the knee jerk reflex) by having direct interactions with motor neurons in the spinal cord; because such reflexes are designed to maintain balance and posture or to protect muscles from being stretched too much, the sensory component of the reflex is performed by fast-conducting myelinated nerve fibers, as is the motor component.

Another special quality of these neurons is the fact they normally only conduct information about a particular **sensory modality**. This is because the nerve fibers have specialized sensory receptors at their peripheral margins that are designed to detect a particular type of stimulus. The various types of specialized sensory receptors that exist in the skin for touch, pressure, temperature, and pain are shown in *Fig. 3*. Each of these is designed to detect a particular sensory modality and, when activated by a stimulus of a given amplitude, will send a signal to the CNS. To some extent, the brain then recognizes any input

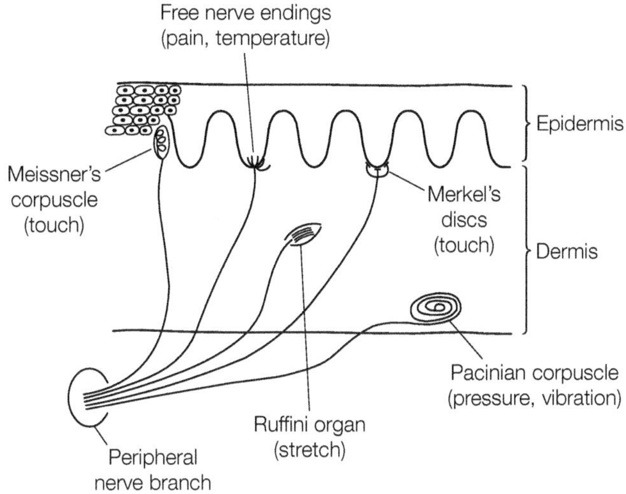


*Fig. 3. Sensory receptors in non-hairy skin. In hairy skin, nerve root plexuses detect movement of hairs.*

from that primary sensory neuron (and the cells with which it synapses along the way) as being related to a particular sensory modality (e.g. touch). For this reason, even if the sensory receptor is activated by a large stimulus of another type, or if it is destroyed and the pathway is activated in some other way, the brain recognizes this as touch. The same is true for pain, and for this reason, pain can be 'felt' in a limb that is either disconnected from the CNS or has been amputated. This is known as **phantom limb pain** and proves that pain is a central phenomenon. More information on sensory physiology is given in *Instant Notes in Neuroscience*, Sections F and G.

# H4 Spinal reflexes

## Key Notes

| | |
|---|---|
| **Components of a reflex arc** | Reflex responses involve the interaction of sensory neurons (responding to a stimulus) with motor neurons (producing a motor response). Normally these refer to spinal reflexes eliciting a somatic motor response, but brainstem reflexes and autonomic reflexes function along similar lines. The nature of the interaction may be monosynaptic (no interneuron), disynaptic (one interneuron), or polysynaptic (with several interneurons involved). Reflex responses cannot be eliminated completely by voluntary control, but can be suppressed to some extent. |
| **Myotatic reflex** | The simplest form of reflex involves a monosynaptic interaction between a sensory neuron and a motor neuron. An example of such a monosynaptic reflex is the muscle stretch reflex (also known as the tendon reflex or myotatic reflex). Stretching of a skeletal muscle by tapping the tendon that attaches it to a bone results in activation of muscle stretch afferents (sensory neurons innervating muscle spindles), which synapse on α-motor neurons in the ventral horn of spinal cord gray matter to elicit a contractile response in the muscle being stretched. |
| **Inverse myotatic reflex** | The inverse myotatic reflex is more complicated, and involves activation of sensory afferents from Golgi tendon organs by powerful contraction of the muscle to which the tendon is attached. In this case, interneurons in spinal cord gray matter are used to produce reflex inhibition of some of the α-motor neurons that innervate the muscle fibers in that muscle and its synergistic muscles to prevent tearing of the tendon. |
| **Related topics** | Spinal cord structure (H1)<br>Structure and function of peripheral nerve (H3)<br>Skeletal muscle function and properties (I1) |

## Components of a reflex arc

The simplest responses that the nervous system can make are known as **reflexes**. Unlike other responses of the nervous system, reflexes need little in the way of integrated nervous activity in order to take place. The components of a reflex arc mediated by the spinal cord (a spinal reflex) are as follows: a **primary sensory neuron** with its cell body in the **dorsal root ganglion (DRG)** and axonal processes to the periphery and to the spinal cord; a motor neuron residing in the **ventral horn** of spinal cord gray matter; and (in almost all cases) one or more **interneurons** between the primary sensory neuron and motor neuron (*Fig. 1*). There are no spinal interneurons between a primary sensory neuron and motor neuron in only one type of spinal reflex, known as the stretch reflex or **myotatic reflex** (see below). This reflex is therefore monosynaptic in that the only synapse occurs between the primary sensory neuron and the motor neuron in the ventral horn of spinal cord gray matter. The majority of reflexes involve one or more

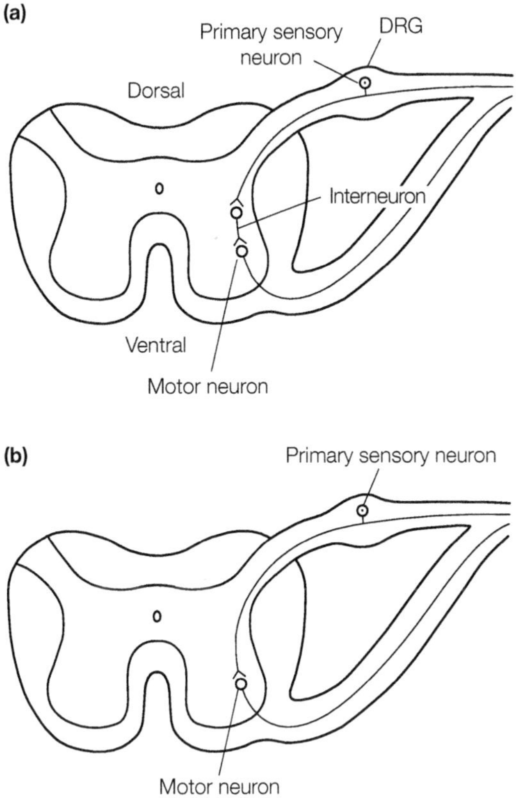


*Fig. 1. Arrangements of primary sensory neurons, interneurons and motor neurons in a spinal reflex pathway: (a) disynaptic reflex; (b) monosynaptic reflex. DRG, dorsal root ganglion.*

interneurons and thus are disynaptic or polysynaptic. The role of interneurons in spinal reflexes is generally to allow differential activation of different sets of muscle fibers in the same muscle, or to inhibit the contraction of antagonistic muscles. For example, flexion of a joint will be facilitated by simultaneous activation of flexor muscles and inhibition of extensor muscles.

Spinal reflexes do not require the involvement of higher centers in order to function, but that does not mean that they cannot be influenced *in vivo* by cortical input. We can all override reflexes to some extent by voluntary control.

**Myotatic reflex**

The myotatic reflex (synonymous with the **stretch reflex** or muscle spindle reflex) is a monosynaptic reflex arc designed to maintain muscle length within tight boundaries. For example, rapid loading of a muscle will elicit a stretch-induced contraction, keeping muscle length relatively constant. Such reflexes can be elicited by gently tapping the muscle or the tendon that connects the muscle to bone. Four major reflexes commonly tested in humans are the Achilles tendon (**ankle-jerk**) reflex, the patellar (**knee-jerk**) reflex, the biceps (**elbow flexor**) reflex, and the triceps (**elbow extensor**) reflex. All of these reflexes are mediated by discrete spinal cord segments (*Table 1*); this means that they can be tested in a clinical situation if damage to the spinal cord is suspected.

As mentioned above, these reflexes are monosynaptic, since their only neural components are a primary sensory neuron and a motor neuron (*Fig. 1b*). Stretch

*Table 1. Spinal cord segments mediating the most commonly tested stretch reflexes*

| Reflex | Spinal cord segments |
|---|---|
| Achilles tendon reflex | S1, S2 |
| Patellar reflex | L3, L4 |
| Biceps reflex | C5, C6 |
| Triceps reflex | C7, C8 |

of the muscle is sensed by specialized sensory nerve endings around **muscle spindles**. Muscle spindles (see Topic I1) are small collections of **intrafusal muscle fibers** that are separate from the muscle fibers that contribute to the generation of tension in skeletal muscle. The degree of stretch of the intrafusal muscle fibers is sensed by nerve endings of sensory neurons; as the intrafusal fibers are stretched, discharge of the sensory neuron increases. This increased activity of sensory neurons activates the motor neurons supplying that muscle group, leading to contraction of the muscle and reduction in the degree of stretch of the intrafusal fibers.

The patellar (knee-jerk) reflex is one such reflex, and most people are familiar with it. Tapping the patellar tendon results in a small degree of stretch of the intrafusal fibers within the muscle spindles of the **quadriceps femoris** (knee extensor) muscle group. This activates primary sensory neurons, which conduct nerve impulses towards the spinal cord and synapse with motor neurons in the ventral horn of spinal cord gray matter in spinal cord segments L3 and L4. Activation of these motor neurons results in contraction of the muscles of the quadriceps femoris via the **femoral nerve**, and knee extension, hence the term knee-jerk reflex.

## Inverse myotatic reflex

It is unfortunate that the myotatic reflexes are often called the tendon reflexes in a clinical setting, because other reflexes are mediated by sensory nerve endings that reside in the tendons called **Golgi tendon organs (GTOs)**. GTOs sense

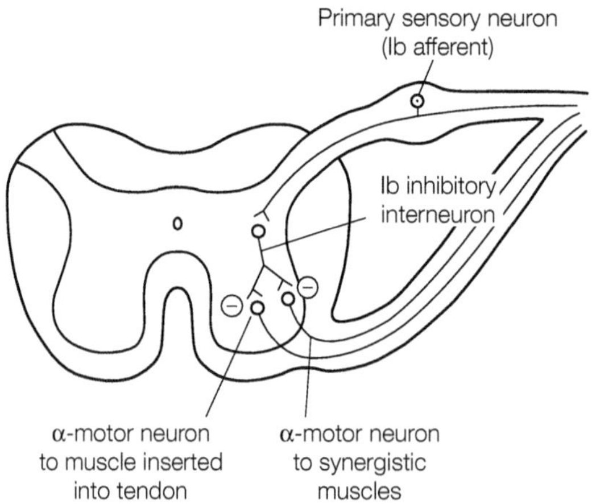


*Fig. 2. Arrangement of primary sensory neuron, interneuron and motor neurons in the inverse myotatic reflex from Golgi tendon organs.*

tension in the tendons exerted by the small proportion of the muscle fibers attached to tendons; as the muscle contracts, GTOs are activated. These GTOs are specialized nerve endings of Ib primary sensory neurons that synapse with inhibitory interneurons in spinal cord gray matter (*Fig. 2*). These inhibitory interneurons then synapse with a very small proportion (around 1%) of the motor neurons supplying that muscle and its synergistic muscles, leading to small decrease in overall contraction of the muscle and a small reduction in tension in the tendon. The **inverse myotatic reflexes** are subtle in comparison with myotatic reflexes and probably serve to prevent damage to tendons by excessively powerful muscle contractions.

# H5 Structure and function of the autonomic nervous system

## Key Notes

**Introduction**

The autonomic nervous system (ANS) is composed of two branches: the parasympathetic branch and the sympathetic branch. The ANS controls functions of which we are generally unaware and over which we have no control. Therefore the output of the ANS can be described as visceral or involuntary motor. The output of the ANS uses a two-neuron relay; preganglionic and postganglionic. The cell bodies of preganglionic neurons exist in the brainstem and in thoracolumbar and sacral segments of the spinal cord. Cell bodies of postganglionic neurons exist in autonomic ganglia in the head, neck, chest, and abdomen, in the sympathetic chain and in the adrenal medulla.

**The parasympathetic branch**

The cell bodies of preganglionic neurons exist in the brainstem and in the lateral horn of gray matter in the sacral segments of the spinal cord. This branch of the ANS is often said to have craniosacral output. Cell bodies of postganglionic neurons exist in autonomic ganglia in the head and neck and in the wall of the target organ in some cases. Preganglionic fibers are long, and postganglionic fibers are short.

**The sympathetic branch**

The cell bodies of preganglionic neurons exist in the lateral horn of gray matter in spinal cord segments T1–L2. This branch of the ANS is often said to have thoracolumbar output. Cell bodies of postganglionic neurons exist in autonomic ganglia in the neck, in the chest and abdomen, and in the adrenal medulla. Preganglionic fibers are relatively short, and postganglionic fibers are longer.

**Functions of the ANS**

The ANS comprises motor output to visceral organs. This includes structures in the head and neck, the heart, the airways, the gut and the urogenital tract. Visceral motor output therefore influences the function of smooth muscle, cardiac muscle, and glandular tissue. Visceral motor output, in contrast to somatic motor output to skeletal muscle, can be either excitatory or inhibitory.

**Visceral afferents**

Some visceral sensation is conveyed by the same nerves that the ANS uses to distribute its output; such afferents should never be termed 'sympathetic' or 'parasympathetic' because those terms refer to visceral motor output only. 'Visceral afferent' is a better term.

**Related topics**

Structure of cardiac muscle (D2)
Electrical activity of the heart – the ECG (D4)
The cardiac cycle, blood pressure, and its maintenance (D6)
Cranial nerves (G4)
Spinal cord structure (H1)
Autonomic pharmacology (H6)
Innervation of smooth muscle (I7)
Adrenal hormones (K8)

**Introduction**

The **autonomic nervous system (ANS)** is a division of the **peripheral nervous system** that controls processes that we are generally unable to control at a voluntary or conscious level. This includes the function of the cardiovascular system, the gastrointestinal tract, the airways and urinary system. The autonomic nervous system is divided into two branches: **parasympathetic** and **sympathetic**, which generally have opposing effects. Autonomic control of the function of an organ is normally determined by the balance of activation of these two branches of the ANS. In contrast to the **somatic motor** system for skeletal muscle, where upper motor neurons are known to reside in the motor cortex, the forebrain location of the upper motor neurons of the autonomic nervous system is not known, although many functions are believed to be controlled by the hypothalamus (see Topics G1 and G3).

The parasympathetic and sympathetic branches of the ANS are distinct from all other motor elements of the peripheral nervous system in that they utilize two neurons in their path from the central nervous system (CNS) to their target organs. The first neurons in this pathway have their cell bodies in cranial nerve motor nuclei in the brainstem (see Topic G4), or in the lateral horn of gray matter in the thoracic, upper lumbar and sacral sections of spinal cord (see Topic G5); these neurons are termed **preganglionic** (*Fig. 1*). The second neurons in the pathway have their cell bodies and dendrites in **autonomic ganglia** and send axonal projections to the target tissue. These neurons are termed **postganglionic** (*Fig. 1*). Autonomic ganglia are distinct from **sensory ganglia** (see Topics G4 and H2). Throughout this topic and Topic H6, the terms preganglionic and postganglionic will be used exclusively to refer to the ANS neurons, but the terms **presynaptic** and **postsynaptic** will be used to refer to either side of synapses both in the autonomic ganglia and in the target organ (*Fig. 1*).

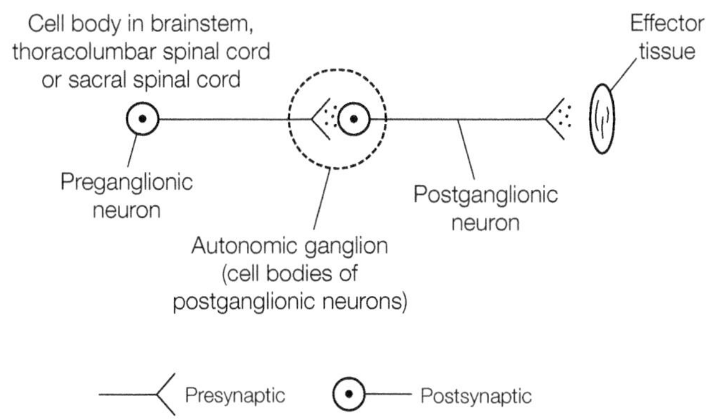


*Fig. 1. Arrangement of the two-neuron pathway in the autonomic nervous system.*

## The parasympathetic branch

The parasympathetic branch of the ANS has cell bodies of preganglionic neurons in the motor nuclei of **cranial nerves** III, VII, IX, and X, and in the **lateral horn** of spinal cord gray matter of spinal segments S2–S4. Thus, the term **craniosacral** is used to refer to parasympathetic ANS outflow. The axonal processes of these preganglionic neurons are generally longer than those of the sympathetic branch of the ANS, because the parasympathetic ganglia are closer to the target organ; in some cases they may even reside in the wall of the target organ (*Fig. 2*). For example, parasympathetic preganglionic fibers to the heart are carried by cranial nerve X (**vagus nerve**, preganglionic cell bodies in the **dorsal motor nucleus of vagus** in the brainstem) down to autonomic ganglia in the heart, where postganglionic neurons have very short projections to the **sinoatrial node** and the **atrioventricular node** (see Topic D4).

Structures in the head and neck receive their parasympathetic innervation from cranial nerves III, VII, IX, and X, whereas those in the trunk receive their parasympathetic innervation from cranial nerve X. The stomach, small intestine, ascending and transverse colon are also innervated by cranial nerve X, but parasympathetic outflow to the descending colon, bladder, and genitalia comes from spinal nerves S2–S4.

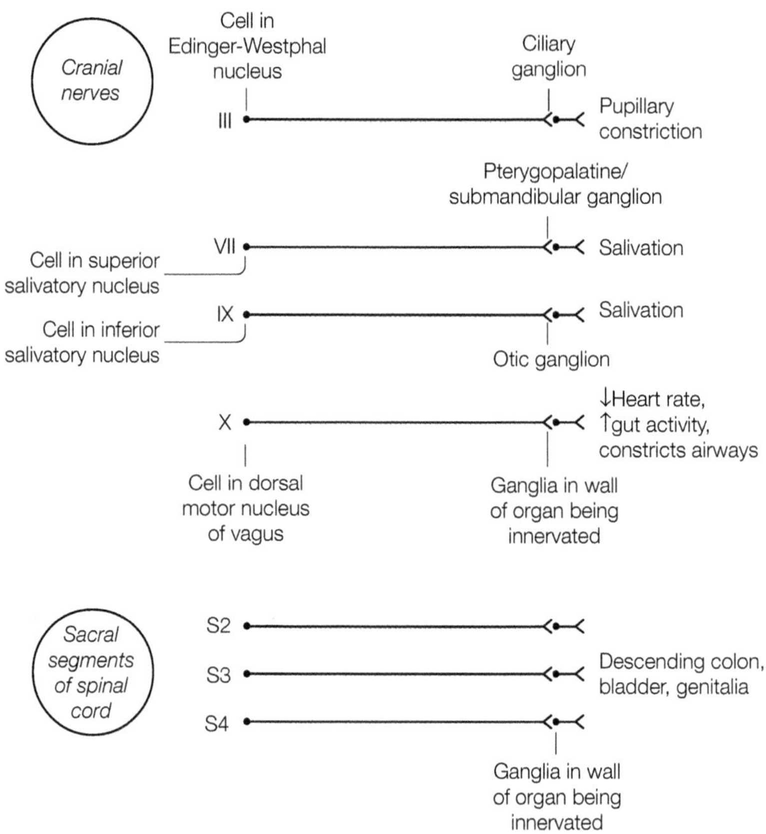


*Fig. 2. Specific arrangement between the preganglionic and postganglionic neurons in the parasympathetic branch of the autonomic nervous system. Note the relatively short post-ganglionic neurons.*

## The sympathetic branch

Distribution of the sympathetic outflow (preganglionic neuronal cell bodies in the lateral horn of spinal cord gray matter of spinal cord segments T1–T12, L1–L2) is more complex. The axonal processes of preganglionic neurons in the sympathetic branch of the ANS can be either long or short, depending upon where they make synaptic connections with postganglionic neurons. The majority of sympathetic ganglia exist in the **sympathetic chain**, a paired collection of ganglia on either side of the vertebral column; this acts as a distribution center for preganglionic fibers. An axon of a preganglionic sympathetic neuron emerging from spinal cord segments T1 to L2 may take a number of routes (*Fig. 3*). Firstly, such axons may synapse with a postganglionic neuron in the sympathetic chain at the same vertebral level, or ascend or descend the sympathetic chain before making synaptic contact with postganglionic neurons above or below the vertebral level from which they originally emerged. A third possibility is that the fibers may pass through the sympathetic chain without making

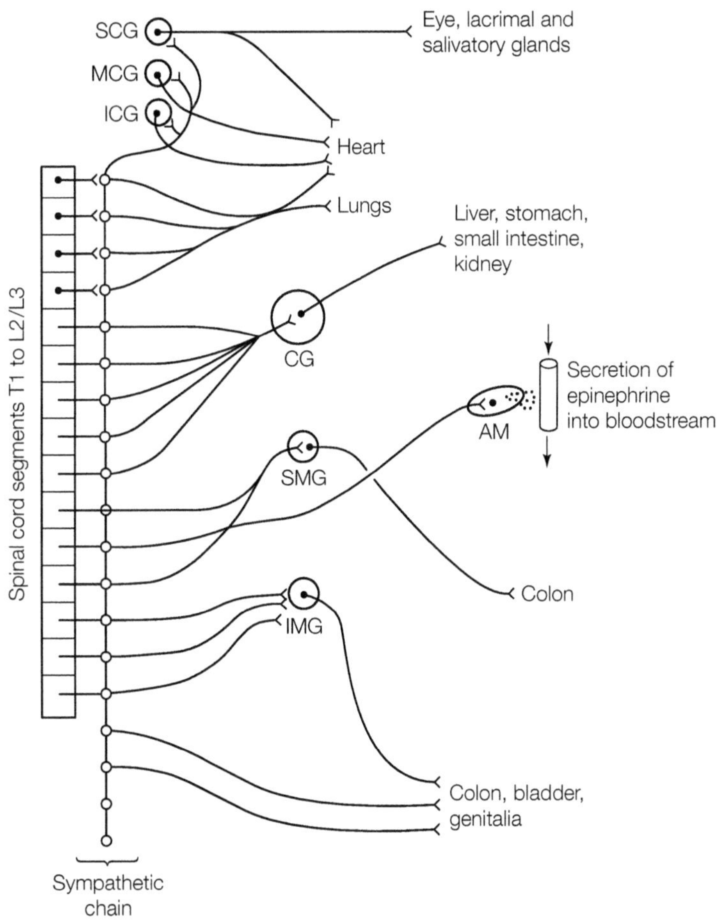


*Fig. 3. Specific arrangement between the preganglionic and postganglionic neurons in the sympathetic branch of the autonomic nervous system. Note the relatively long postganglionic neurons. AM, adrenal medulla; CG, celiac ganglion; ICG, inferior cervical ganglion; IMG, inferior mesenteric ganglion; MCG, middle cervical ganglion; SCG, superior cervical ganglion; SMG, superior mesenteric ganglion.*

a synaptic contact with a postganglionic neuron, and pass on to other sympathetic ganglia in the neck (**superior**, **middle** and **inferior cervical ganglia**) or in the trunk (the **celiac ganglion** and the **superior and inferior mesenteric ganglia**), before making synaptic contact with postganglionic neurons.

A fourth and final possibility is that the fibers may pass through both the sympathetic chain and the celiac ganglion before making contact with postganglionic cell bodies in the **adrenal medulla**. These cells are known as chromaffin cells, and in an embryological sense are neurons that did not develop axonal processes, but when activated release a mixture of **epinephrine** (80–90%) and **norepinephrine** (10–20%) into the bloodstream. Thus the sympathetic branch of the ANS can influence target organs through both neural and endocrine (hormonal) outflow.

## Functions of the ANS

The two branches of the ANS are often said to have opposing effects, but this implies that when one is active, the other is not. In the majority of organs, the net effect of the ANS on function of the organ at any one time is determined by the degree of activation of both branches. For example, it is true that the sympathetic branch of the ANS causes increased heart rate, and that the parasympathetic branch causes reduced heart rate, but at rest both branches of the ANS are active to some extent; cutting or blocking the sympathetic nerves to the heart and leaving the vagus nerve intact leads to a reduction in resting heart rate of 10–15 beats $min^{-1}$. Resting heart rate rises by 30–40 beats $min^{-1}$ when vagal influence is removed. The heart is therefore said to be under **vagal tone** at rest, but the data above show that both branches of the ANS are active.

The two branches of the ANS are also said to have the generalized effects of '**rest-and-digest**' (parasympathetic), and '**fight-or-flight**' (sympathetic), and one can see why this is so when the various effects of the two branches of the ANS on various organs are set out as in *Table 1* (this list is not exhaustive).

## Visceral afferents

As outlined in Topic H2, some cranial nerves and virtually all **spinal nerves** are mixed sensory and motor nerves. Sensation from visceral organs such as the heart, gut, and pelvic organs is conveyed back to the CNS in cranial nerves and branches of spinal nerves that also happen to carry the parasympathetic or sympathetic outflow to these organs. Hence, such sensory fibers are often termed '(para)sympathetic afferent fibers'. This term is erroneous and should be

*Table 1. Major effects of autonomic innervation of various tissues/organs*

| Tissue/organ | Effect of innervation | |
|---|---|---|
| | Parasympathetic | Sympathetic |
| Eye | Pupillary constriction | Pupillary dilation |
| Sweat glands | None | Secretion |
| Heart | Decreased heart rate | Increased heart rate |
| Heart | No direct effect on force of contraction | Increased force of contraction |
| Airways | Constriction | Dilatation |
| Gut smooth muscle | Promotes activity | Inhibits activity |
| Gut glands | Secretion increased | Secretion decreased |
| Gut sphincters | Relaxes | Contracts |
| Bladder smooth muscle | Contracts | Relaxes |
| Urinary sphincters | Relaxes | Contracts |

avoided; it almost implies that the pathway is a two-neuron arrangement like that of the autonomic visceral motor pathways described above. This is not so. It can also lead to confusion of cranial nerve sensory ganglia like the **trigeminal ganglion** with parasympathetic ganglia in the head and neck like the **ciliary ganglion**. The term 'sympathetic afferent', although widely used, is incorrect. The term **'visceral afferent'** is less ambiguous and is preferable. In the case of visceral afferent fibers that run in spinal nerves, the cell bodies of these primary sensory neurons reside alongside those of somatic sensory fibers in the **dorsal root ganglion** of that spinal nerve.

# H6 AUTONOMIC PHARMACOLOGY

## Key Notes

**Transmitters**

Preganglionic neurons of the parasympathetic and sympathetic branches of the autonomic nervous system (ANS) utilize acetylcholine (ACh) as their transmitter. Postganglionic neurons of the parasympathetic branch also utilize ACh. Postganglionic neurons of the sympathetic branch, with a few specific exceptions, use norepinephrine (NE). Adrenal medulla chromaffin cells (essentially sympathetic postganglionic neuronal cell bodies) release mostly epinephrine.

**Receptors**

The receptors for ACh are either nicotinic (ionotropic receptors) or muscarinic (G-protein-linked metabotropic receptors). The receptors for NE and epinephrine are adrenoceptors (G-protein-linked metabotropic receptors, of the α- or β-type). At parasympathetic and sympathetic ganglia, ACh acts on nicotinic ACh receptors. At the synapse between parasympathetic postganglionic neurons and the effector cell (e.g. smooth muscle), ACh acts on muscarinic ACh receptors. At the synapse between the majority of sympathetic postganglionic neurons and the effector cell (e.g. in cardiac muscle), NE acts on adrenoceptors (e.g $\beta_1$-adrenoceptors in cardiac muscle). Sympathetic postganglionic neurons innervating sweat glands release ACh that acts on muscarinic ACh receptors.

**Common ANS drugs**

The therapeutic action of several prescription drugs is predicated by their effects on neurotransmission in the ANS. For example, $\beta_1$-adrenoceptor antagonists such as atenolol act to block β-adrenoceptors in the heart, thereby limiting the effect of the sympathetic branch of the ANS in increasing heart rate and stroke volume; this means that such drugs can be used to treat hypertension. Muscarinic receptor antagonists such as ipratropium block the bronchoconstrictor action of ACh in the airways, thereby promoting bronchodilation and relief of the symptoms of asthma.

**Related topics**

Structure and function of the autonomic nervous system (H5)

## Transmitters

It is unusual for the two branches of the **autonomic nervous system (ANS)** to utilize neurotransmitters other than **acetylcholine (ACh)** or **norepinephrine (NE)** at their various synapses, although in some specific locations, other neurotransmitters may be used.

ACh (*Fig. 1a*) is a small-molecule neurotransmitter synthesized from acetyl coenzyme A and choline by the action of the enzyme **choline acetyl transferase** near its site of release. In ANS neurons, it is stored within vesicles and is released upon excitation of the cholinergic cell synapse by an action potential (AP). ACh is rapidly degraded at cholinergic synapses by the action of

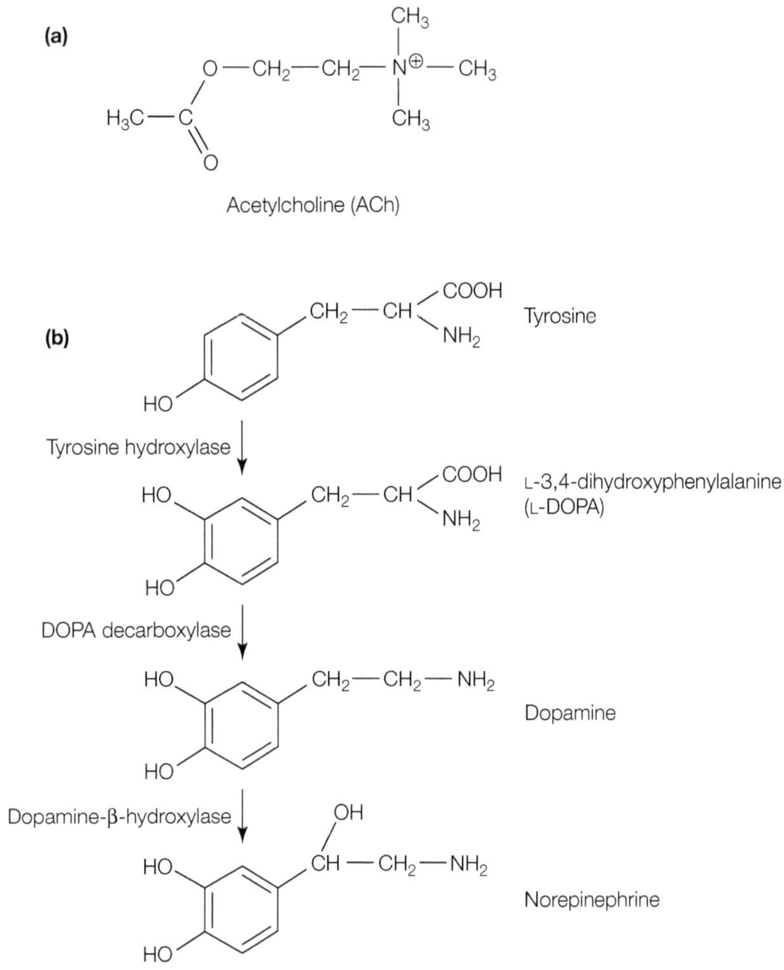


*Fig. 1. (a) Structure of acetylcholine; (b) biosynthetic pathway for norepinephrine.*

**acetylcholinesterase**, and choline is taken back up into cholinergic neurons by specific transport mechanisms, for use in synthesis of new ACh molecules.

NE is a **catecholamine** produced by the metabolism of the amino acid tyrosine. The biochemical pathway for its synthesis (*Fig. 1b*) involves three steps; the hydroxylation of tyrosine by **tyrosine hydroxylase** to form **L-DOPA (L-3,4-dihydroxyphenylalanine)**, the decarboxylation of L-DOPA by L-aromatic amino acid decarboxylase (**DOPA decarboxylase**) to form **dopamine**, and the hydroxylation of dopamine by **dopamine-β-hydroxylase** to form NE. NE can be converted to **epinephrine** by the action of **phenylethanolamine-*N*-methyl transferase** in cells that express this enzyme, such as the chromaffin cells of the **adrenal medulla**.

## Receptors

ACh acts on two classes of ACh receptors in the body, called nicotinic and muscarinic ACh receptors on the basis of the differential efficacy of nicotine and muscarine as agonists at the two classes of receptor (*Fig. 2*).

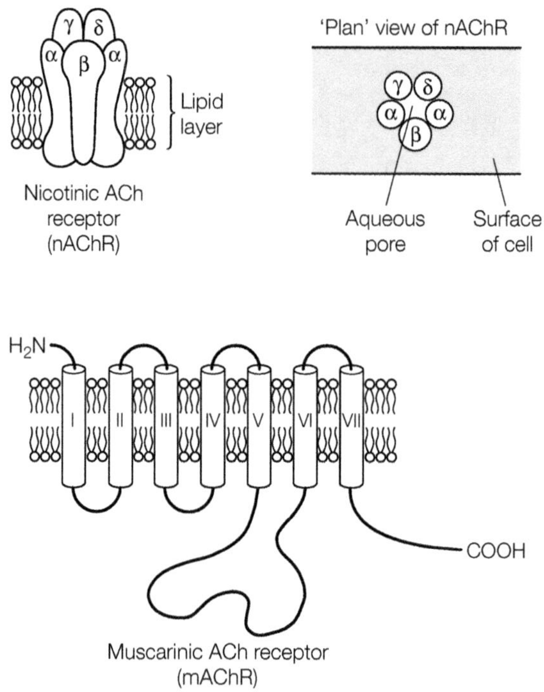


*Fig. 2. Diagrammatic representations of the structure of the two classes of acetylcholine receptor (AChR). Upper panel: nicotinic AChR (nAChR); lower panel: muscarinic AChR (mAChR).*

**Nicotinic ACh receptors (nAChRs)** are composed of five protein subunits, each of which spans the postsynaptic cell membrane. The five subunits form an aqueous pore through the membrane that can conduct small cations like $Na^+$ and $K^+$ when the receptor is activated by two molecules of ACh. Thus these receptors are known as **ionotropic receptors**. The net effect of activation of the receptor and subsequent opening of the integral ion channel is to produce an **excitatory postsynaptic potential (EPSP)** in the postsynaptic cell and make it more likely that an AP will be set up in the postsynaptic cell; more $Na^+$ ions enter the cell than $K^+$ ions diffuse out.

**Muscarinic ACh receptors (mAChRs)** are protein chains folded in such a way that they have an extracellular N-terminal portion, seven transmembrane domains, three intracellular and three extracellular loops, and an intracellular C-terminal portion. They belong to the superfamily of **G-protein-linked receptors**. There are several different types of mAChR, each with slightly different amino acid structure and encoded by different genes. On binding of ACh, mAChRs can elicit many different responses in postsynaptic cells. They may be linked by different G-proteins to different cellular enzymes and alter the level of different intracellular messengers (known as second messengers). Thus they alter metabolic processes in postsynaptic cells and are therefore known as **metabotropic receptors**. The most common effects of activation of mAChRs are changes in the level of **cyclic adenosine monophosphate (cAMP)** or **inositol trisphosphate ($IP_3$)** and **diacylglycerol (DAG)**.

NE acts on two main classes of receptor on postsynaptic cells, known as **α- and β-adrenoceptors**. Like mAChRs, these receptors are metabotropic G-protein-linked, seven-transmembrane domain receptors that, when activated, bring about alterations in the biochemistry (and thereby the activity) of the postsynaptic cell. There are many different types and subtypes of adrenoceptors that can be distinguished pharmacologically, but the most important types to be aware of are $\alpha_1$-, $\alpha_2$-, $\beta_1$-, $\beta_2$-, and $\beta_3$-adrenoceptors.

All **preganglionic neurons** of the ANS utilize ACh at the synapses between preganglionic and **postganglionic neurons** in the **autonomic ganglia** (*Fig. 3*). nAChRs are used as receptors on the dendrites of postganglionic neurons, so the effect of activation of preganglionic neurons is to activate postganglionic neurons. A very small proportion of the ACh receptors on postganglionic neurons are mAChRs, but these subserve a minor modulatory role on ganglionic neurotransmission, and can largely be ignored.

Parasympathetic postganglionic neurons also utilize ACh at their synapses with target cells (*Fig. 3*). In this case, the released ACh acts on mAChRs, and because their potential biochemical effects of activation of mAChRs are so diverse (see above), either excitation or inhibition of the target cell can take place. For example, activation of different classes of mAChRs in the heart and in glands results in inhibition (heart) and activation (glands). The action of ACh on mAChRs can be antagonized (blocked) by the drug **atropine**. Thus, administration of atropine or atropine poisoning results in tachycardia (increased heart rate), pupillary dilatation, urinary retention, and a dry mouth (see Topic H5).

Almost all sympathetic postganglionic neurons utilize NE at their synapses with target cells (*Fig. 3*). As with ACh acting on mAChRs, the effect of NE on adrenoceptors is diverse, and can result in excitation or inhibition of the various target cells. For example, NE released from sympathetic postganglionic neurons in the heart and the airways will act on β-adrenoceptors in both tissues ($\beta_1$ in the heart and $\beta_2$ in the airways) to produce activation (tachycardia and increased force of contraction) in the heart or inhibition (bronchodilatation) in the airways. Circulating epinephrine released from the adrenal medulla on sympathetic activation will have the same effect as NE released from sympathetic postganglionic neurons, but its effects will be widespread rather than local, and it is a more potent agonist at β-adrenoceptors than at α-adrenoceptors.

Some sympathetic postganglionic neurons release transmitters other than NE or epinephrine, and their effects are therefore mediated by receptors that are

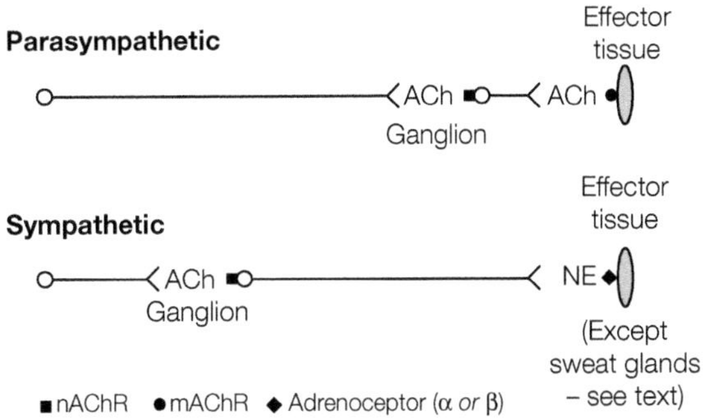


*Fig. 3. Organization of transmitters and receptor types in the two branches of the autonomic nervous system. ACh, acetylcholine; NE, norepinephrine.*

not adrenoceptors. Adenosine triphosphate (ATP), neuropeptide Y and other peptides are all utilized by a small subset of sympathetic postganglionic neurons. ACh is also released from a very specific set of sympathetic postganglionic neurons – those innervating **sweat glands**. Here, ACh acts on mAChRs and results in increased production of sweat. Thus, poisoning with the mAChR antagonist atropine (see above) can also cause **anhydrosis** (lack of sweating), while simultaneously interrupting *parasympathetic* influence on other tissues.

## Common ANS drugs

Many commonly used drugs in clinical practice are specifically designed to target the autonomic nervous system. These drugs are used in the treatment of cardiovascular, respiratory and gastrointestinal problems. With a few notable exceptions (**antimuscarinics** in the treatment of **asthma**), these drugs act upon adrenoceptors as either **agonists** (activating the receptors) or **antagonists** (blocking the action of NE at the receptors). One useful 'rule' to bear in mind is that:

- the majority of peripheral blood vessels have $\alpha$-adrenoceptors that, when activated, cause constriction
- cardiac muscle has $\beta_1$-adrenoceptors that, when activated, cause tachycardia and increased force of contraction
- airway smooth muscle has $\beta_2$-adrenoceptors that, when activated, cause bronchodilation.

$\beta_1$-adrenoceptor antagonists such as **atenolol** are therefore useful in the treatment of **hypertension** (high blood pressure), in view of their effect of lowering heart rate and the force of contraction of the ventricles (see Topics D1 and D5).

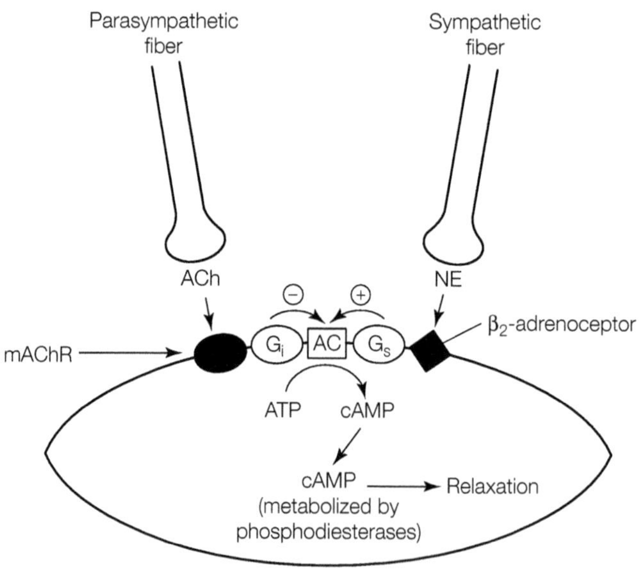


*Fig. 4. Schematic diagram of the interaction between parasympathetic and sympathetic nerve fibers innervating smooth muscle in the airways. AC, adenylyl cyclase; ACh, acetylcholine; ATP, adenosine triphosphate; cAMP, cyclic adenosine monophosphate; $G_i$, inhibitory G-protein; $G_s$, stimulatory G-protein; mAChR, muscarinic acetylcholine receptor; NE, norepinephrine.*

$\alpha$-adrenoceptor antagonists such as **doxazosin** are also useful in the treatment of hypertension because they reduce the degree of constriction of arterioles that is brought about by sympathetic activation. **Vasoconstriction** increases peripheral resistance (see Topics D1 and D5), so by administering an $\alpha$-adrenoceptor antagonist, blood pressure can be lowered. $\alpha$-adrenoceptor antagonists are not used as commonly as other classes of antihypertensive drugs, but can be useful in specific cases where other drugs do not work well.

$\beta_2$-adrenoceptor agonists such as **salbutamol** are useful in the treatment of the **bronchospasm** associated with an asthma attack. Inhalation of salbutamol or other $\beta_2$-adrenoceptor agonists produces bronchodilation and allows improved ventilation of the lungs. Activation of $\beta_2$-adrenoceptors increases levels of cAMP in bronchial smooth muscle cells, and this cAMP promotes relaxation (*Fig. 4*). The activation of mAChRs on bronchial smooth muscle cells by ACh released from parasympathetic postganglionic neurons in the airways reduces cAMP in these smooth muscle cells, thereby leading to contraction. Hence, the use of mAChR antagonists such as **ipratropium** can help in some cases of asthma. cAMP is broken down by **phosphodiesterase** enzymes, so administration of a drug that inhibits phosphodiesterase, such as **aminophylline**, can have a bronchodilator effect.

# I1 SKELETAL MUSCLE FUNCTION AND PROPERTIES

## Key Notes

**Flexion and extension**

Skeletal muscle acts principally on the joints of the skeleton. Muscle fiber contraction leads to shortening of the muscle and movement of the bones on either side of the joint. Thus contraction of flexor group muscles flexes a joint and of extensor group muscles extends a joint.

**Motor units**

Skeletal muscles are composed of motor units. A motor unit is comprised of all of the skeletal muscle fibers that are innervated by a single $\alpha$-motor neuron in the spinal cord. Graded contractions of muscle are produced by altering the number of motor units activated at any one time.

**Muscle spindles**

Muscle spindles are stretch receptors in skeletal muscle and are composed of small intrafusal muscle fibers. Muscle spindles have both motor and sensory innervation and are important in regulating and maintaining the length of skeletal muscles; they are important in stretch reflexes such as the knee jerk reflex.

**Striation and sarcomeres**

Skeletal muscle is striated (striped). The pattern of striation is caused by regular repeating arrangements of cytoskeletal proteins in units known as sarcomeres. The major cytoskeletal proteins that give rise to the striations of skeletal muscle are myosin and actin. As the filaments of these proteins in the skeletal muscle fiber slide over each other, contraction takes place.

**Related topics**

Bone and muscle (C5)
The locomotor system (G2)
Spinal reflexes (H4)
Excitation–contraction coupling (I3)
Properties of skeletal muscle fiber types (I4)
Exercising muscle (I5)

## Flexion and extension

**Skeletal muscle** most often connects bones together and acts on these bones to move joints in the body. For example, the skeletal muscles of the **biceps** group of the arm are elbow **flexors**, meaning that when they shorten, the elbow bends. This is the muscle group that is most often used when someone wants to show off his or her muscles. The largest muscle of this group, the **biceps brachii**, is connected by tendons at either ends to the radius (the more lateral of the two bones of the forearm, on the same side as the thumb) and to bones of the shoulder joint. When the muscle cells (fibers) in the biceps brachii contract, the muscle shortens, the belly of the muscle swells, and the elbow is flexed. Antagonistic muscle groups also act to extend the various joints of the body, so that rhythmic movement is determined by alternate contraction/relaxation of joint flexor and joint **extensor** muscles.

## Motor units

Skeletal muscle is voluntary muscle, meaning that skeletal muscle fibers do not normally contract unless instructed to do so by a nervous impulse from the locomotor system. As outlined in sections G and H, the final pathway for neural output to skeletal muscle is from **α-motor neurons** with their cell bodies in the **ventral horn** of spinal cord gray matter. These nerve fibers have myelinated axons and fast conduction velocities. An important fact that explains some of the ways in which skeletal muscles function is that each of these α-motor neurons makes synaptic connections with several muscle fibers in a muscle (a **motor unit**), but that each muscle fiber receives input from only one α-motor neuron (*Fig. 1*). This means that as more α-motor neurons are recruited by the locomotor system, more motor units become activated and the strength of contraction of a given muscle increases. **Acetylcholine (ACh)** released at the nerve endings of somatic motor neurons acts on **nicotinic ACh receptors** and always results in some degree of excitation of the muscle fiber (see Topic H6), so the motor unit is the smallest functional unit of any skeletal muscle.

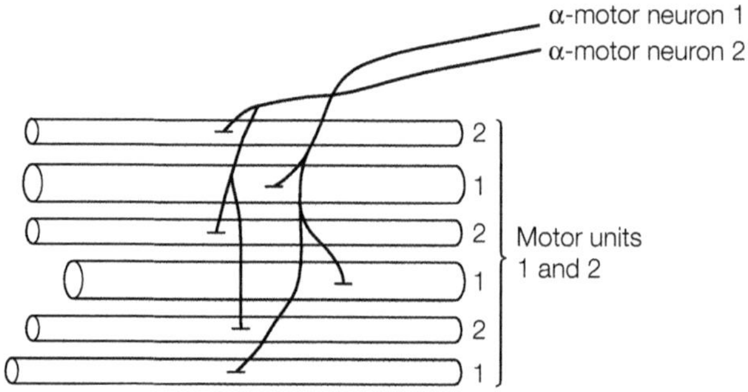


*Fig. 1. Arrangement of innervation of skeletal muscle motor units. A given skeletal muscle fiber is innervated by only one motor neuron, but a motor unit may consist of several muscle fibers.*

## Muscle spindles

Skeletal muscles contain specialized stretch receptors known as **muscle spindles** (*Fig. 2a*). These are fusiform (cigar-shaped) collections of small muscle fibers, afferent nerve endings and efferent nerve endings that are around 4–10 mm in length and around 0.1 mm in diameter. The muscle fibers within the spindles are known as **intrafusal muscle fibers**, and those outside the spindle, which do the work of the muscle, are known as **extrafusal muscle fibers**. Whereas extrafusal muscle fibers are innervated by α-motor neurons, intrafusal muscle fibers are innervated by **γ-motor neurons**, which have a slightly smaller diameter. These γ-motor neurons innervate the ends of the intrafusal muscle fibers (which can contract). The sensory nerve endings, which measure the degree of stretch of the spindle (length of the fibers at different points in time), are wrapped around the center of the intrafusal muscle fibers (which cannot contract). There are two main classes of sensory nerve endings in muscle spindles, known as **type Ia** and **type II** afferents.

The two types of sensory nerve fibers respond in slightly different ways to stretch (*Fig. 2b*). Type Ia afferents respond rapidly while the muscle is being stretched, whereas type II afferents respond more slowly. As the stretch is held, both types of afferents fire action potentials a little more frequently than they would have done at rest. When the stretch is released, type Ia afferents cease

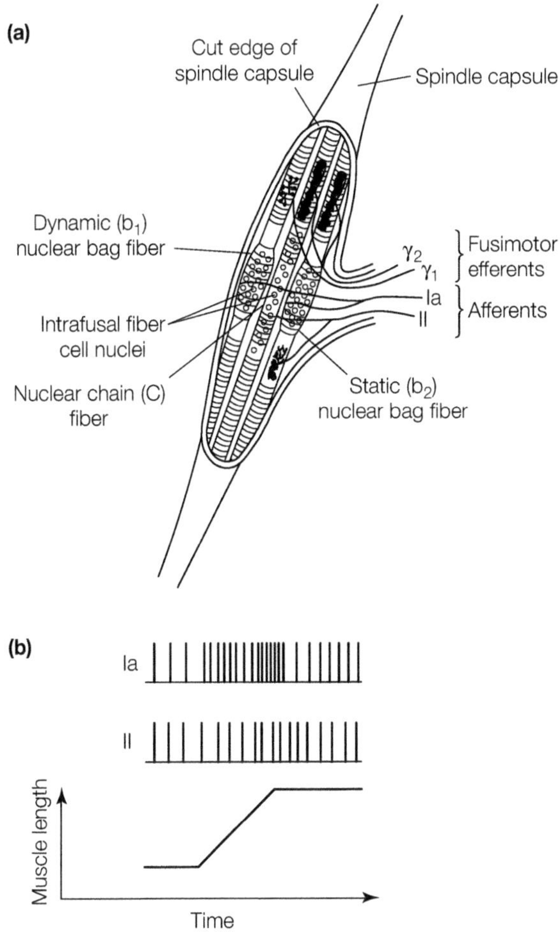


*Fig. 2. Muscle spindles: (a) a spindle opened to show intrafusal fibers and their innervation. A spindle normally contains one $b_1$, one $b_2$ and several c fibers; (b) responses of Ia and II afferents to muscle stretch.*

firing completely, whereas the type II afferents continue firing. Type Ia afferents therefore measure both speed of stretch and length of the muscle, whereas type II afferents measure only muscle length.

The purpose of the γ-motor neuron innervation of the muscle spindles is to allow the afferents to maintain the capacity to respond to stretch when the muscle is shortened. If α-motor neurons fired alone and contracted extrafusal fibers, the shortening of the muscle would unload the spindle afferents, particularly the Ia afferents, so that they would no longer be able to indicate muscle length. By activating both γ- and α-motor neurons, as extrafusal muscle fibers shorten, the intrafusal fibers are stretched slightly, maintaining the ability of the afferents to measure muscle length.

## Striation and sarcomeres

Skeletal muscle is a form of **striated muscle** (see Topic C5), meaning that when viewed using light microscopy it has a striped appearance. Each skeletal muscle is surrounded by **epimysium** and is composed of **fasciculi**, bundles of muscle

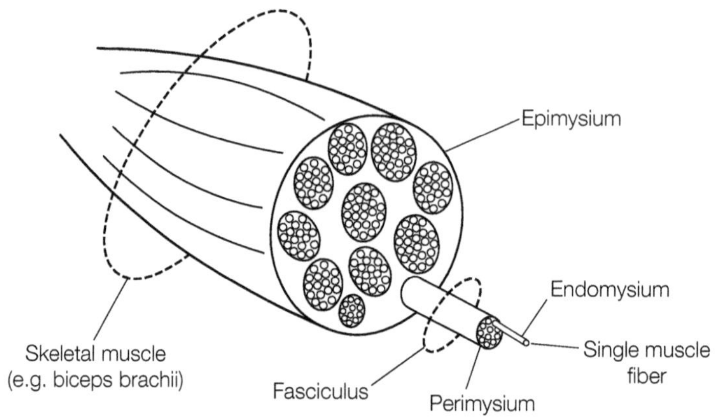


*Fig. 3. Arrangement of fasciculi and individual muscle fibers in a skeletal muscle.*

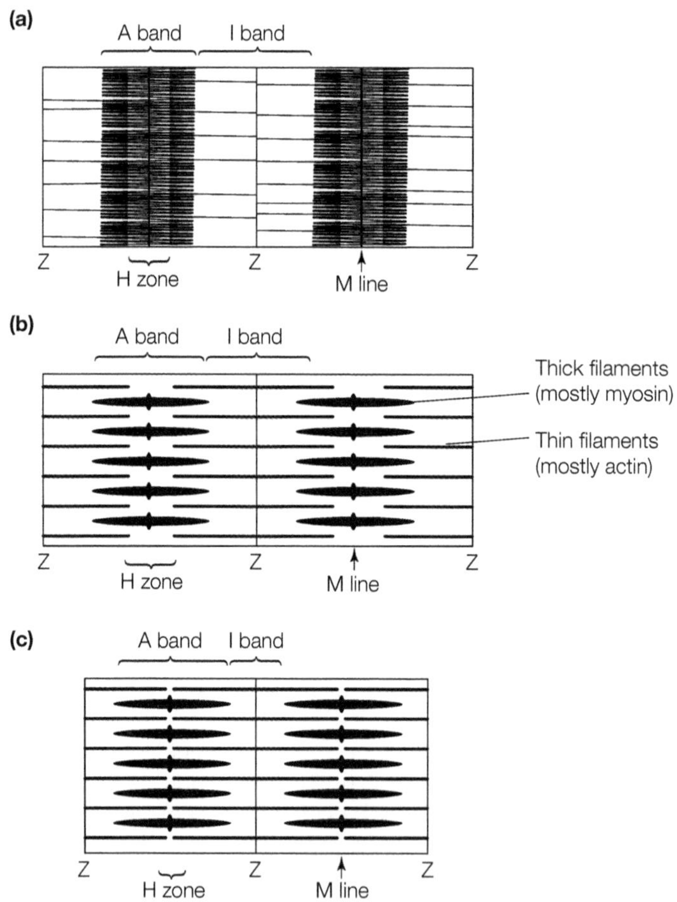


*Fig. 4. Organization of protein filaments in a myofibril sarcomere: (a) striation pattern of the sarcomere as viewed by light microscopy; (b) relationship of thick and thin filaments in relaxed muscle; (c) relationship of thick and thin filaments in contracted muscle.*

fibers that are surrounded by another connective tissue coat known as the **perimysium** (*Fig. 3*). Within the fasciculi, individual muscle fibers are surrounded by **endomysium** outside the muscle cell membrane (**sarcolemma**).

Within each muscle fiber, there are several myofibrils, bundles of specific cytoskeletal proteins that are known as **myofilaments**. Thus skeletal muscle is highly organized at a cellular level. However, skeletal muscle is also highly organized at a subcellular level; the proteins of the myofilaments are put together in such a way that it is the myofilaments that produce the striations. The unit of striation is known as the **sarcomere**. The myofilaments are made up of thick and thin filaments that are composed of polymers of **myosin** and **actin**, respectively. During contraction of the muscle fiber (see Topic I3), the thick filaments (myosin) interact with the thin filaments (actin) so that these two types of filaments pass over each other and the sarcomere shortens (*Fig. 4*). As the sarcomere shortens, the individual filaments do not change in length, but the **I band** (*isotropic*, light) becomes shorter, and the distance between **Z lines** decreases. This sarcomere pattern is repeated many times along the length of the muscle fiber, so that as individual sarcomeres shorten, the entire muscle fiber shortens, generating tension and moving the bone to which the muscle is attached.

# I2 THE NEUROMUSCULAR JUNCTION

## Key Notes

**Structure**

The neuromuscular junction (NMJ) is a specialized synaptic contact between an α-motor neuron and a skeletal muscle fiber. The axons of α-motor neurons are myelinated and have fast conduction velocities. When a nerve impulse arrives at the nerve ending, this causes release of acetylcholine (ACh) from synaptic vesicles. The ACh molecules diffuse across the synaptic cleft, where they bind to ACh receptors on the muscle cell membrane. A hydrolytic enzyme, acetylcholinesterase, is present on the muscle cell membrane and in the synaptic cleft, and this enzyme acts to break down ACh into acetate and choline.

**Action of acetylcholine**

When two ACh molecules bind to a muscle fiber nicotinic ACh receptor, this opens an integral ion channel in the receptor molecule. This ion channel conducts cations (principally $Na^+$ and $K^+$), and the net effect of opening of the channel is depolarization of the muscle fiber membrane to produce a miniature endplate potential (mEPP). If sufficient nicotinic ACh receptors are activated, these miniature endplate potentials summate to form an endplate potential that usually has sufficient amplitude to raise membrane potential to threshold, activate voltage-sensitive $Na^+$ channels on the muscle fiber membrane and thereby cause a muscle action potential.

**Pharmacology**

Numerous therapeutic drugs that target the NMJ are used in clinical medicine. For example, the bacterial poison botulinum toxin (Botox) can be administered to prevent ACh release and paralyze muscle fibers. Another example is succinylcholine, an analog of ACh, that is used to provide a depolarizing block of muscle contraction, which is useful during surgery.

**Related topics**

Synaptic transmission (F3)
The panoply of transmitters (F4)
Structure and function of peripheral nerve (H3)
Excitation–contraction coupling (I3)

## Structure

The specialized synaptic connection between **α-motor neurons** and skeletal muscle fibers is known as the **neuromuscular junction (NMJ)** or **motor endplate**. As indicated in topic I1, an individual skeletal muscle fiber receives motor input from only one α-motor neuron, but a single α-motor neuron innervates several skeletal muscle fibers in what is known as a **motor unit**. The NMJ is a specialized synapse between two excitable cells, the α-motor neuron and the skeletal muscle fiber (*Fig. 1*).

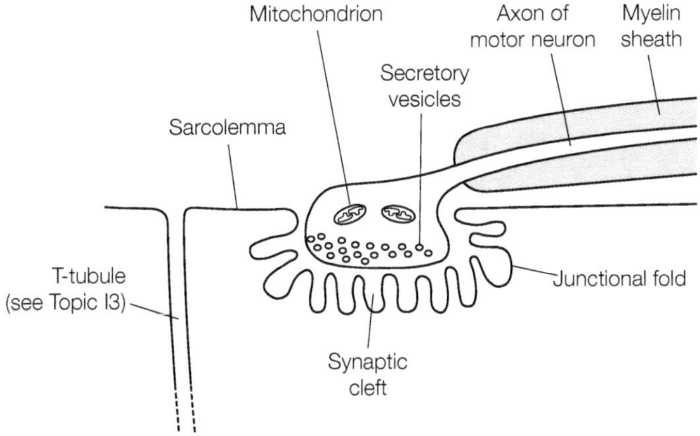


*Fig. 1. Schematic diagram of the neuromuscular junction, or NMJ. Note that not all elements can always be visualized, even using electron microscopy.*

The nerve ending of the NMJ is packed with **mitochondria** and with **secretory vesicles**. Each of the secretory vesicles contains a few thousand molecules of **acetylcholine (ACh)**. A single vesicle is said to hold a quantum of ACh. On arrival of an **action potential (AP)** at the nerve terminal, some of these vesicles fuse with the presynaptic membrane and discharge their contents into the **synaptic cleft**. The synaptic cleft measures between 30 and 50 nm across, and ACh molecules must diffuse across this gap in order to act on the muscle fiber. The density of **nicotinic ACh receptors (nAChRs)** is extremely high on the muscle cell membrane (**sarcolemma**) in the region of the motor endplate, although such protein elements cannot themselves be viewed by simple light microscopy or electron microscopy. The skeletal muscle fiber has numerous invaginations of its sarcolemma, known as **junctional folds**. The enzyme **acetylcholinesterase (AChE)** is expressed at the surface of the muscle fiber and can also be secreted into the synaptic cleft. Using labeled antibodies that recognize AChE, one can visualize motor endplates on skeletal muscle fibers. This enzyme serves to break ACh down into acetate and **choline**, thereby limiting the action of the transmitter at the nAChRs. Choline is actively transported back into the α-motor neuron end-bulb by a specific choline transporter protein.

## Action of acetylcholine

When ACh is released from the nerve terminal of α-motor neurons and diffuses across the synaptic cleft of the NMJ, it is then able to bind to the nAChRs on the sarcolemma. As outlined in Topic H6, nAChRs are composed of five protein subunits, arranged in such a way as to form an aqueous pore (channel) that is opened by binding of ACh to the receptor (*Fig. 2a*). In skeletal muscle, the nAChR subunits used are: two α-subunits, one β-, one δ- and one γ-subunit (all encoded by different genes). Binding of two molecules of ACh (one to each of the α-subunits) leads to opening of the channel.

When gated by ACh, the channels in nAChRs conduct small cations (predominantly $Na^+$ and $K^+$). The net effect of opening of the channel is to cause **depolarization** of the muscle fiber. A single quantum (see above) will produce a depolarization of the order of 0.5 mV, known as a **miniature endplate potential (mEPP)**. However, a single AP in an α-motor neuron causes a much larger

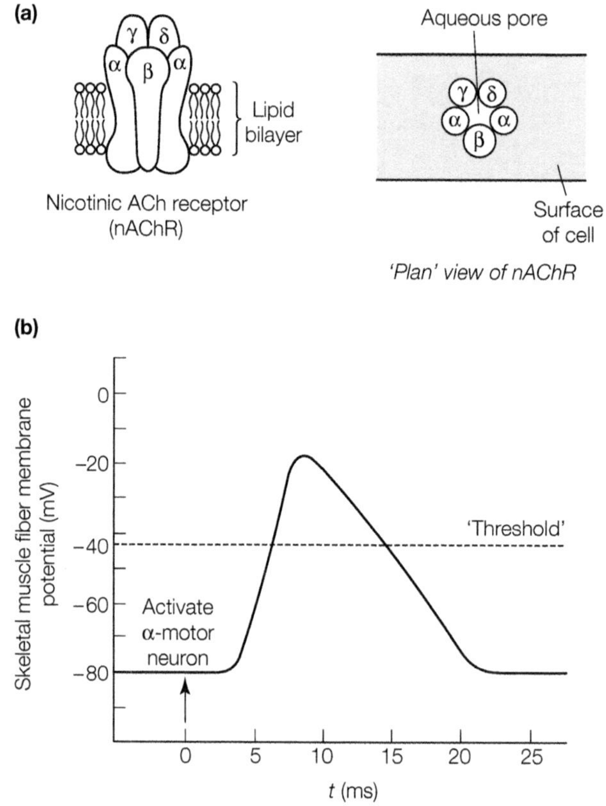


*Fig. 2. Action of ACh at the neuromuscular junction: (a) the nicotinic ACh receptor; (b) an endplate potential (EPP), recorded using an intracellular electrode in the skeletal muscle fiber near the endplate.*

change in muscle fiber membrane potential, of around 60 mV, known as the **endplate potential** (*Fig. 2b*). This magnitude of depolarizing potential is sufficient to achieve threshold and sets up a muscle AP, which is propagated over the sarcolemma by local circuits, in much the same way as an AP in an unmyelinated nerve fiber (see Topic F2). The muscle AP lasts only around 10 ms (compare this with 300 ms or so in cardiac myocytes, Topic D4) and leads to a twitch contraction of the muscle fiber lasting 50 ms or so (see Topic I3). Thus, the short **refractory period** of skeletal muscle fibers (the period during which no new action potential can be generated) means that twitches of skeletal muscle fibers can summate if a new nerve AP arrives at the NMJ within the period of contraction of the muscle fiber.

## Pharmacology

The NMJ is a site of action of both toxins and therapeutic drugs (*Fig. 3*). For example, a toxin produced by the bacterium *Clostridium botulinum* blocks the release of ACh by interfering with the mechanism by which vesicles dock with the presynaptic membrane. **Botulinum toxin** can cause death when produced by the bacteria themselves in cases of botulism, but this toxin can be also be used clinically to relieve the symptoms of dystonia and spasticity associated with spinal cord damage. In recent years, botulinum toxin injections (Botox)

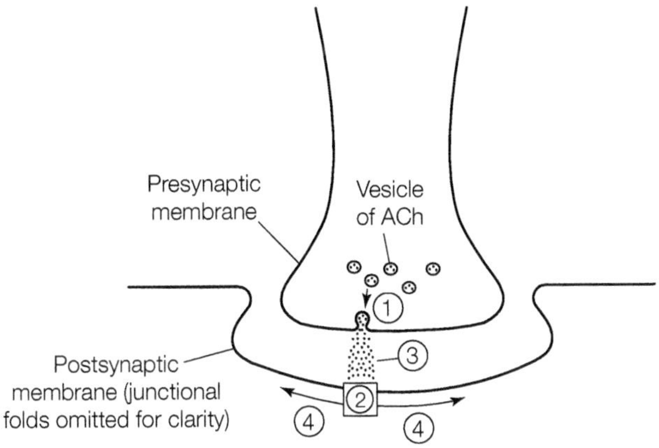


*Fig. 3. Sites of action of drugs that affect the neuromuscular junction. (1) Release of ACh (blocked by botulinum toxin); (2) action of ACh (blocked by curare); (3) degradation of ACh (blocked by neostigmine); (4) depolarization by ACh (mimicked by succinylcholine).*

have also found their use in cosmetic surgery, reducing wrinkles by effectively paralyzing underlying muscles.

Other drugs act postsynaptically at the NMJ. **Curare** is an antagonist of nAChRs at the NMJ and is used in hunting by native South Americans as a poison that causes paralysis of muscles. Drugs that are similar to curare, such as **pancuronium**, are commonly used in surgical procedures to bring about muscle relaxation. Drugs that are similar to acetylcholine, but less prone to breakdown by AChE, such as **succinylcholine**, can be administered to produce depolarizing muscle blockade as they remain on the nAChRs, preventing access of endogenously released ACh.

AChE inhibitors such as **neostigmine** can be administered to reverse the action of curare after surgery, by making more ACh available in the synaptic cleft. This drug can also be given to patients with **myasthenia gravis**, an autoimmune disease that causes a low density of nAChRs at the NMJ. These drugs result in a higher availability of ACh at the remaining functional nAChRs, and attenuation of the symptoms of the disease.

# 13 Excitation–contraction coupling

## Key Notes

**Propagation of the muscle action potential**

The muscle fiber action potential that is set up by α-motor neuron innervation is propagated by local circuits along the surface of the muscle fiber membrane. Voltage-sensitive $Na^+$ channels are involved in this action potential, in the same way that they are involved in the neuronal action potential. The depolarization associated with the muscle action potential must be transmitted to the heart of the muscle fiber, which may be up to 40 μm in diameter.

**T-tubules and triads**

Deep invaginations of the muscle fiber membrane called transverse tubules or T-tubules allow the muscle action potential to be conducted to the heart of the muscle fiber. When the T-tubule membrane is depolarized, a junctional complex of voltage-sensing proteins (dihydropyridine receptors) and calcium ion channels (ryanodine receptors) transduce the depolarization of the T-tubule into calcium ion release from the intracellular membrane system known as the sarcoplasmic reticulum (SR). Calcium ions are continually being pumped into the SR by the action of calcium-ATPases, so that when the T-tubules are repolarized, calcium ions are sequestered from the sarcoplasm.

**Structure of the sarcomere**

Skeletal muscle is striated because of a repeating pattern of sarcomeres. Each sarcomere is composed of a specific arrangement of cytoskeletal proteins known as myofilaments. Sarcomeres have a Z line at each end, an A band in the middle, and an I band spans two sarcomeres with the Z line in the middle. In the middle of the sarcomere, the H zone and M line are sometimes apparent. Thin filaments in skeletal muscle are composed mainly of actin, whereas thick filaments are composed mainly of myosin. Other regulatory proteins including troponins and tropomyosin are also present.

**The crossbridge cycle**

Sarcomeres and thus muscle fibers shorten by thin filaments being 'pulled' over thick filaments towards the center of the sarcomere. In order for this to occur, $Ca^{2+}$ ions released from the SR bind to troponin C, causing a conformational change in the troponin–tropomyosin complex, such that myosin binding sites on actin are revealed. The myosin head then binds to the thin filament, pulls toward the center of the sarcomere and then detatches. By undergoing cycling of this process (crossbridge cycling), the sarcomere becomes progressively shorter, using one molecule of adenosine triphosphate (ATP) per myosin head per cycle.

**Length–tension relationship**

The degree of tension generated by muscle fibers is not constant across all initial muscle lengths; optimal sarcomere length for tension development in human skeletal muscle is around 2 μm, and developed tension falls markedly below 1.95 μm and above 2.25 μm. This is known as the length–tension relationship of skeletal muscle and is determined by the arrangement and overlap of thick and thin myofilaments.

**Related topics**

Bone and muscle (C5)
Action potentials (F2)
Skeletal muscle function and properties (I1)
The neuromuscular junction (I2)

## Propagation of the muscle action potential

Topic I2 describes how release of **acetylcholine (ACh)** by **α-motor neurons** results in a change in the **membrane potential** of the postsynaptic cell (a skeletal muscle fiber). This topic describes the mechanisms that result in contraction of the muscle fibers in a motor unit – how nervous input is translated into shortening of the muscle. This process is known as **excitation-contraction coupling** and is now extremely well understood. The first step in excitation–contraction coupling is propagation of the muscle fiber action potential from the area near the **neuromuscular junction (NMJ)** to the rest of the muscle fiber membrane, so that the entire muscle fiber contracts, not just the area near the NMJ.

As outlined in topic I2, **endplate potentials** produced by the action of ACh on ligand-gated **nicotinic ACh receptors** are generally large enough in amplitude to bring the muscle fiber membrane to threshold and cause opening of **voltage-sensitive $Na^+$** channels in the muscle membrane. The flow of ions through these sodium channels leads to a muscle fiber action potential, and this action potential is propagated over the entire surface of the fiber by a mechanism that is similar to the propagation of action potentials in non-myelinated nerve fibers (see Topic F2). The process of propagation of the action potential leads to depolarization of the muscle fiber membrane on the surface of the cell, but in skeletal muscle fibers packed with **myofibrils** (see Topic I1), a structural feature of the muscle fiber membrane ensures propagation of this change in membrane potential to the very heart of the fiber. These structural features are deep invaginations of the muscle fiber membrane that are known as **transverse tubules** or **T-tubules**.

## T-tubules and triads

The presence of T-tubules in skeletal muscle means that changes in membrane potential near the NMJ can be translated into changes in membrane potential in portions of the muscle fiber membrane that extend deep into the muscle fiber, which can be up to 40 μm in diameter. The mechanism by which changes in membrane potential in the T-tubules leads to contraction of the muscle fiber hinges on another structural feature called **triads** (*Fig. 1*). Where T-tubules penetrate deep into the muscle fiber, they come into close apposition to an internal membrane system, the **sarcoplasmic reticulum (SR)**, which is the muscle fiber's more extensive equivalent of the **endoplasmic reticulum** in other cells. If one were to cut a skeletal muscle fiber along its length, one would see that the T-tubules come into apposition with SR on either side. It is this three-component structure (SR/T-tubule/SR) that is known as a triad.

The SR is an internal membrane system that one can think of as a set of large intracellar vesicles. As with the endoplasmic reticulum in other cells, which contains newly translated proteins before they are 'polished' for final use by the Golgi apparatus (see Topic B1), the SR is used as a store. However, in the case of the SR, the important substance stored is calcium ions. The protein **calsequestrin**

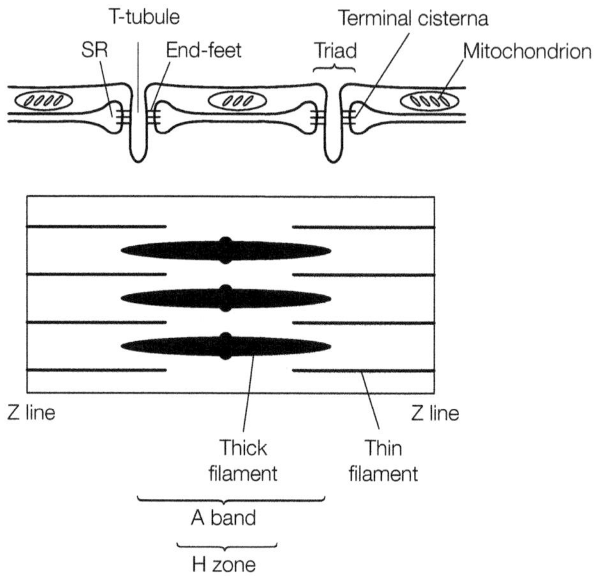


*Fig. 1. Organization of the T (transverse)-tubule and sarcoplasmic reticulum (SR) systems in skeletal muscle. Note the location of the triads near the sites of interaction between thick and thin filaments.*

is abundant in the SR, especially in the terminal cisternae (where the SR comes into close apposition to the T-tubules), and this is where the majority of calcium ions are stored, loosely bound to calsequestrin. There are many **calcium-ATPase** molecules in the SR membrane, which pump calcium ions from the sarcoplasm into the SR.

When the T-tubule membrane is depolarized by a muscle action potential, voltage-sensing proteins known as **dihydropyridine receptors** (that are embedded in the T-tubule membrane in close apposition to the SR) are activated. A chemical or mechanical interaction between the dihydropyridine receptors and **ryanodine receptors** (calcium channels present in the SR membrane) allows the flow of calcium ions from the SR into the **sarcoplasm** via the ryanodine receptor channels, although the exact mechanism for this has yet to be fully elucidated. The muscle action potential that induces this release of calcium ions is short-lived, and after a few milliseconds, the ryanodine receptor channels close and calcium-ATPases mop up the excess calcium ions from the sarcoplasm. The concentrations of calcium ions in the sarcoplasm are >10 000-fold higher during depolarization of the T-tubule (>10 μM) than when it is normally polarized (<1 nM). It is this change in calcium ion concentration in the sarcoplasm that induces contraction of the muscle fiber (see 'The crossbridge cycle' below).

## Structure of the sarcomere

When one looks at a skeletal muscle fiber using a light microscope, it is clear that the cell is striated (*Fig. 2a*). The basis of this striation is the arrangement of muscle proteins (*Fig. 2b*). An individual muscle fiber is packed with numerous myofibrils, which are highly ordered arrangements of protein filaments. These filaments consist of two types: thick filaments composed predominantly of

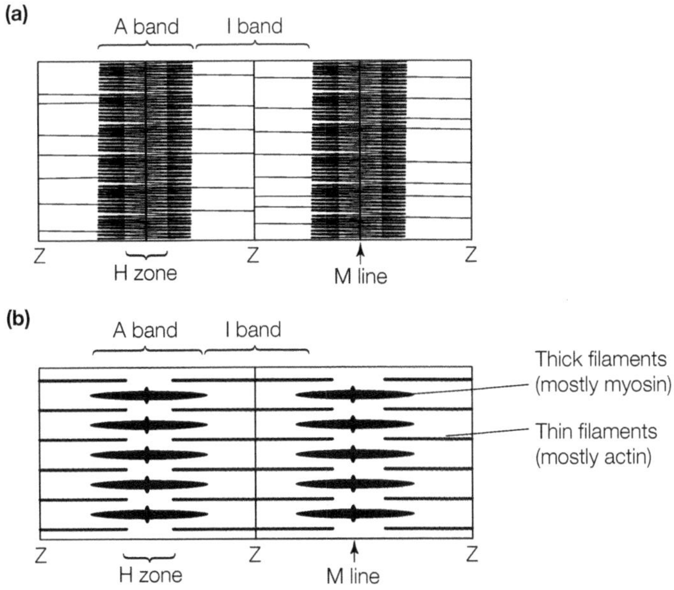


*Fig. 2. Organization of protein filaments in a myofibril sarcomere: (a) striation pattern of the sarcomere as viewed by light microscopy; (b) relationship of thick and thin filaments in relaxed muscle.*

**myosin**, and thin filaments composed predominantly of **actin**, but with other accessory proteins attached. The two most important accessory protein types in thin filaments are **troponins** (troponin-T, troponin-I and **troponin-C**) and **tropomyosin**.

The smallest functional unit of a skeletal muscle fiber is known as the **sarcomere** (*Fig. 2b*). This describes the arrangement of thin filaments around a set of thick filaments. The sarcomere is defined as a length of muscle fiber between two **Z lines**. At the end of sarcomeres in relaxed muscle fibers, there are only thin filaments. The region that spans two separate sarcomeres and contains only thin filaments is known as the **I band** (*isotropic*, light). In the center of the sarcomere in muscle fibers that are relaxed, there are only thick filaments. This region is known as the **H zone** (*hell*, from the German for 'bright'). Even in relaxed muscle fibers, there is some overlap of thick and thin fibers. This overlap zone, together with the H zone, makes up the **A band** (*anisotropic*, dark). At the part of the A band that does not include the H zone, the T-tubules project downward to come into close apposition with the SR, as outlined above.

The actin polymers that make up the thin filaments are a twisted arrangement of two chains of **F actin**, each of which is a chain of **G actin** monomers (*Fig. 3a*). The troponins and tropomyosin are arranged in a regulatory protein complex, and this complex is placed at distinct places along the length of the actin filament. When sarcoplasmic calcium ion concentrations are low, tropomyosin covers over the binding sites for myosin on the actin polymer. When sarcoplasmic calcium ion concentrations are high, calcium ions bind to troponin-C; this results in a conformational change in the troponin–tropomyosin complex so that the binding site for myosin on actin is uncovered.

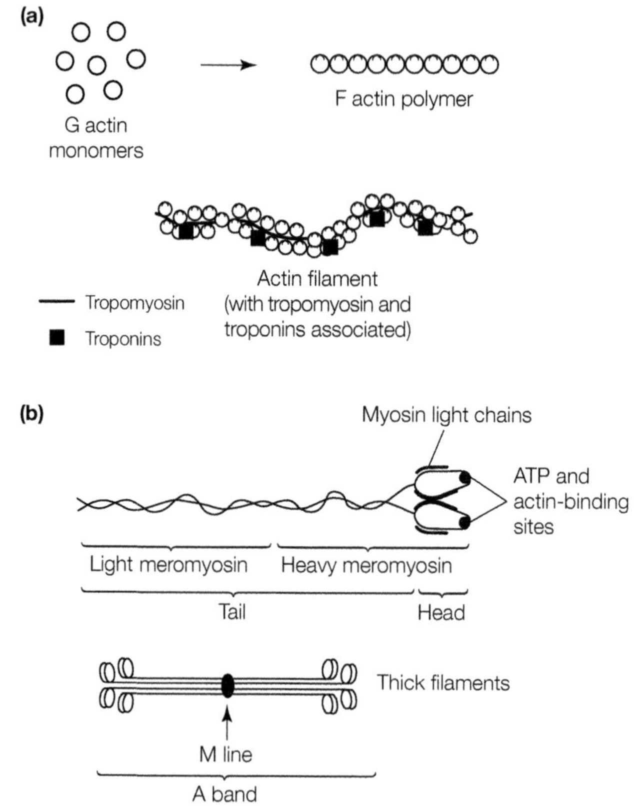


*Fig. 3. Protein composition of the thin and thick filaments in skeletal muscle. The troponin complex is composed of troponin-T, troponin-I and troponin-C.*

The myosin polymers that make up the thick filaments have a more complex structure (*Fig. 3b*). They are composed of several different types of light and heavy chain **meromyosin**, arranged so that there is a head portion and a tail portion. Several of these protein assemblies wrap around each other so that their tail portions make up the linear part of the thick filament, and their head portions project outwards. This arrangement is often said to resemble a bundle of golf clubs gathered together by the grips. Two such bundles of myosin molecules are brought together end to end (grip to grip) to form the thick filament. The heads of the myosin proteins are then in the correct orientation to interact with the thin filaments at either end of the sarcomere, and the tails of the myosin molecules come together to form the **M line** (*mittel*, from the German for 'middle').

## The crossbridge cycle

When a muscle fiber contracts, no appreciable change in the length of individual muscle fiber filaments takes place. Contracted sarcomeres have shorter I bands and shorter H zones than relaxed sarcomeres, but the A band remains the same length. The molecular basis for this sliding filament theory is known as the crossbridge cycle (*Fig. 4*). The heads of the myosin filaments have two binding sites, one for **adenosine triphosphate (ATP)** that also acts as an ATPase, and one for actin that allows interaction of the thick and thin filaments in the

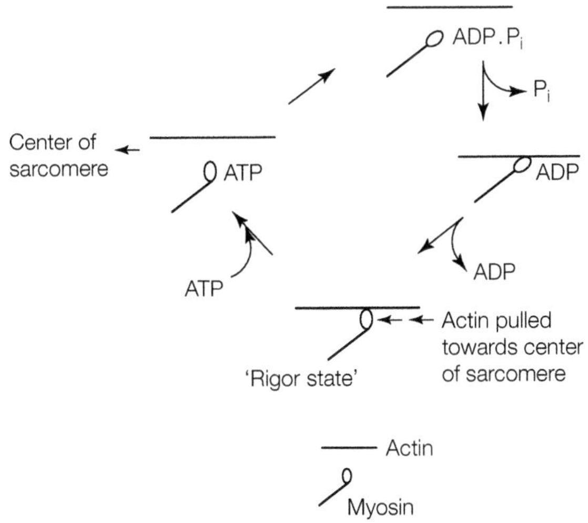


*Fig. 4. Events of the cross-bridge cycle. ADP, adenosine diphosphate; ATP, adenosine triphosphate; $P_i$, inorganic phosphate.*

sarcomere. At rest, the thick and thin filaments are detached from one another, and sarcomere length is around 2 μm. The myosin heads are 'charged'; that is to say they have bound ATP and split the molecule, but have not yet released the resultant **adenosine diphosphate (ADP)** and **inorganic phosphate ($P_i$)**. In this state, they can interact with the actin filaments, but only if the binding site for myosin on actin is free. However, a **troponin–tropomyosin regulatory complex** (see above) covers this site when calcium ion concentrations are low. When a muscle action potential is propagated down the T-tubule, calcium ions are released into the sarcoplasm from the SR, near the myosin heads. These calcium ions bind to troponin-C in the troponin–tropomyosin regulatory complex, and the tropomyosin is moved aside to uncover the myosin-binding site on actin.

Attachment of the myosin head to the actin filament results in the release of ADP and $P_i$ from the myosin head, and the myosin head then moves towards the center of the sarcomere (the **power stroke**). The crossbridge is then said to be in the '**rigor**' state. If ATP is available, this will then bind to the myosin ATPase, briefly detaching the myosin head from the actin filament before the ATP is split again and the myosin head is 'charged' again with ADP and $P_i$. If calcium ion concentrations remain high, further interactions between myosin and actin will again take place, and a further power stroke will result. Hundreds such interactions between myosin and actin take place along the length of the sarcomere, meaning that shortening of the sarcomere occurs even though some myosin heads are detached from actin filaments at all points during contraction. There is enough redundancy in the system for shortening of the sarcomere to take place in a ratchet-like fashion; when some myosin heads are detached from actin, others will be in the 'rigor' state and will be sufficient to hold the sarcomere at a given length. Only when the muscle fiber action potential is over and the T-tubule has become repolarized will calcium ion concentrations around the myosin heads begin to fall. This prevents any interaction between myosin heads and actin filaments as the troponin–tropomyosin regulatory complexes cover over the myosin-binding sites on actin. Thus, muscle contraction requires both

ATP and calcium ions. In **rigor mortis**, the stiffening of the muscles that occurs shortly after death, a lack of cellular ATP means that calcium ions cannot be sequestered by calcium-ATPases in the SR, and so calcium ion concentrations near the myosin heads begin to rise. This allows a tiny degree of muscle contraction to take place, but the lack of ATP then means that the interaction between myosin heads and actin filaments cannot be broken, leading to stiffening of the muscles.

## Length–tension relationship

The interaction between myosin heads and actin filaments also explains another phenomenon of muscle function, known as the **length–tension relationship** (*Fig. 5*). The degree of tension that can be generated by a muscle is usually highest around its resting length (sarcomere length between 1.95 and 2.25 μm). Above and below this resting length, the force that can be generated by a muscle falls off steeply. This can be explained by the degree of overlap between myosin heads and actin filaments. This is greatest at resting length, and as the muscle is lengthened the possibilities for interaction between the two types of filament become fewer, until eventually there is no overlap at sarcomere lengths above around 3.5 μm. At shorter initial muscle lengths, compression of the ends of the actin filaments and then the ends of the myosin filaments are detrimental to the development of tension. This phenomenon is very important in explaining the types of contraction that both skeletal muscle and cardiac muscle can carry out under different initial lengths (see Topic D8 and the **Frank–Starling law of the heart**).

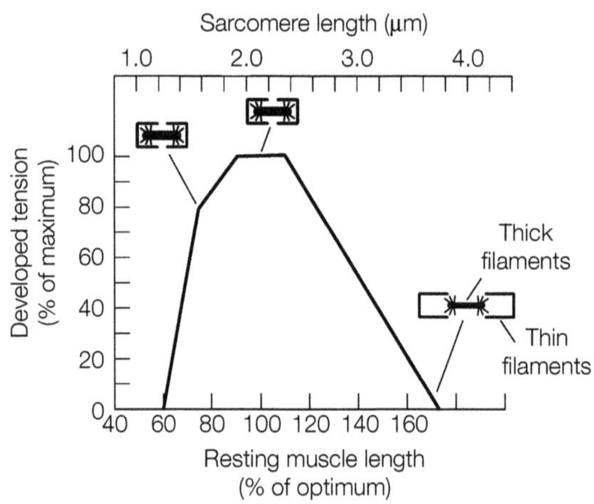


*Fig. 5. The length–tension relationship of skeletal muscle. Note the degree of potential interaction of thick and thin filaments at various sarcomere lengths.*

# I4 Properties of skeletal muscle fiber types

## Key Notes

**Morphology and twitch character**

Skeletal muscles are composed of muscle fibers with different characteristics. In muscles designed to hold contractions at a given length for a long period of time, small-diameter, slow-twitch fibers predominate. In muscles designed to contract quickly and powerfully, but which require significant period of rest between periods of activity, larger diameter, fast-twitch fibers predominate. The speed of contraction is determined by the kinetics of the ATPase site on the myosin heads.

**Biochemical characteristics**

In addition to morphology, the different types of skeletal muscle fiber have different biochemical characteristics. For example, slow-twitch fibers contain high levels of myoglobin to encourage oxygen diffusion into the muscle cell. Some fast-twitch fibers contain high amounts of glycogen so that they can generate sufficient ATP from the relatively inefficient glycolytic pathway. $Ca^{2+}$-ATPase activity of the sarcoplasmic reticulum is high in fast-twitch fibers so that relaxation is rapid also.

**Effects of training**

Although the number and type of fibers of the various types is relatively fixed in a given individual, some alterations in fiber type and fiber size can be achieved by appropriate training for various activities. An obvious example is the larger size of muscle in people who train with weights, as more fibers become fast-twitch glycolytic and increase in size.

**Related topics**

Metabolic processes (B3)
Skeletal muscle function and properties (I1)
Excitation–contraction coupling (I3)
Exercising muscle (I5)

## Morphology and twitch character

Skeletal muscles are composed of fibers that have the same basic structure as outlined in Topics I1 and I3. However, the jobs that different skeletal muscles in the body have to perform are very different. Take, for example, muscles that are involved in postural control during standing, as compared with muscles whose job is to move joints quickly against load, or to run fast for short periods of time. The postural control muscles must be able to hold joints in position for long periods of time, whereas the 'sprinting' muscles need not have this property. This difference in function is reflected in the composition of the muscles themselves.

The muscle fibers of the postural muscles tend to be smaller in diameter than those being used for sprinting; this is because if such muscle fibers must contract for long periods of time, they must be able to receive a good supply of oxygen. By having a smaller diameter, oxygen has only a short distance to diffuse to reach the mitochondria in the very center of the fibers, which supply **adenosine**

**triphosphate (ATP)** for use in muscle contraction. Thus, such muscle fiber types do not generally build up an **oxygen debt**, and can work for longer periods of time before tiring. In the case of muscles used for sprinting, the muscles will be active rhythmically over a short period of time. Thus, it is not essential that they have a constant oxygen supply. This relatively high diameter (and thus high distance for oxygen diffusion) is traded off against the need for such muscle fibers to have more **myofibrils** than those involved in postural control (more myofibrils = more power).

If one compares the physical properties of the types of muscle fibers in postural muscles with those used to lift loads quickly, other differences become apparent. The muscle fibers of postural muscles have a **slow-twitch** character, whereas those in muscles used for sprinting tend to have a **fast-twitch** character (*Fig. 1*). This fits with their roles in the body; explosive sprinting is better suited to fast muscle fiber twitches than slow ones.

## Biochemical characteristics

If one examines the biochemistry of different muscle fiber types, there are many differences that can be explained by the function required of that muscle (*Table 1*). Muscle fibers can first be grouped into **red fibers** and **white fibers**. The red

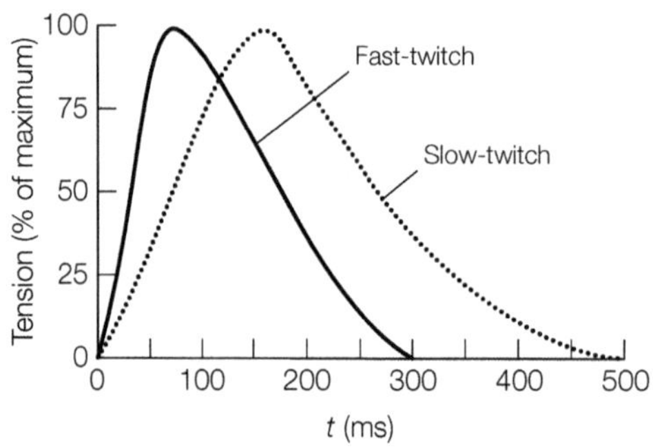


*Fig. 1. Characteristics of evoked contractions in fast-twitch and slow-twitch muscle fibers.*

*Table 1. Properties of the various types of skeletal muscle fibers found in man*

| | Slow twitch | Fast twitch | Fast twitch |
|---|---|---|---|
| Metabolic type | Slow oxidative | Fast oxidative-glycolytic | Fast glycolytic |
| 'Color' | Red | Red | White |
| Myosin ATPase activity | Low | High | High |
| Source of ATP | Oxidative phosphorylation | Oxidative phosphorylation/glycolysis | Glycolysis |
| Sarcoplsmic reticulum calcium-ATPase activity | Moderate | High | High |
| Myoglobin content | High | High | Low |
| Glycogen content | Low | Moderate | High |
| Fiber diameter | Low | Moderate | High |
| Resistance to fatigue | High | Moderate | Low |
| Functional role | Posture/endurance | Medium endurance | Rapid powerful movements |

fibers have a higher **myoglobin** content than the white fibers. Myoglobin is a low-affinity oxygen storage protein than can be used by muscle fibers to store oxygen and to keep the levels of dissolved oxygen inside the fiber relatively low, thus encouraging diffusion of oxygen from the extracellular fluid. Red fibers tend to use **oxidative phosphorylation** as the major source of cellular ATP production. In contrast, white fibers have lower levels of myoglobin and tend to use the glycolytic, anaerobic pathway to produce ATP. These fibers also contain a large amount of **glycogen** that can be broken down to produce fuel to run **glycolysis** (see Topic B3).

Some red fibers are slow-twitch and some are fast-twitch, whereas all white fibers are generally fast-twitch (*Table 1*). Whether a given fiber contracts quickly or slowly is dependent upon the rate of crossbridge cycling (see Topic I3). The faster the ATPase portion of the myosin head (**myosin ATPase**) can split ATP and then interact with **actin** (a process repeated many times during contraction), the faster the rate of shortening of the sarcomere. Thus, fibers that have fast-twitch characteristics express forms of myosin ATPase that have fast ATPase kinetics and those that have slow-twitch characteristics express forms of myosin ATPase that have slower kinetics. Histochemical staining with antibodies against the different forms of myosin can be used to distinguish fast- and slow-twitch fibers. Fast-twitch muscle fibers must also relax quickly, and since this is dependent on the sequestration of calcium ions into the **sarcoplasmic reticulum** by **calcium-ATPases** (see Topic I3), this enzymatic activity is high in fast-twitch muscle fibers.

## Effects of training

If all humans carry out the same range of movement and have the same muscles, how is it that some of us are clearly more able sprinters/throwers/jumpers than others, who might be ideal distance runners or swimmers? The answer is twofold: nature *and* nurture. In the untrained state, a small proportion of people simply have more slow-twitch fibers in their muscles than others, who may well have more fast-twitch fibers. The person with the preponderance of slow-twitch fibers is the more 'natural' distance athlete, whereas the person with the preponderance of fast-twitch fibers is the more 'natural' sprinter. The average person, with the appropriate training, can fairly easily become a distance athlete or a sprinter, but it is those people at the more extreme ends of the normal spectrum who are likely to become, with training, the international-class athletes of either description.

The effects of training on muscle composition are twofold. In both cases, there is usually no change in the number of muscle fibers in a given muscle. Firstly, training with a particular regimen of activity naturally suits the appropriate type of muscle fibers in a given muscle. Activity produces **hypertrophy** of those muscle fibers, and in the case of fast, powerful movements, this usually results in an obvious increase in muscle bulk. Of course, such changes can occur in fast-twitch fibers because they rely less upon oxygen supply. In the case of endurance training, hypertrophy of slow-twitch fibers also occurs, but is limited by the need for the fibers to have lower diameter and thus receive enough oxygen. The second type of effect involves a degree of class-switching by muscle fibers from one type to another. This is more often seen in endurance training where fibers that were once **fast glycolytic** become **fast oxidative-glycolytic** or **slow oxidative** fibers over a prolonged period of time.

# I5 Exercising muscle

## Key Notes

**Isometric versus dynamic exercise**

Muscle performs work, irrespective of whether shortening takes place or not. It is sometimes more difficult to hold a heavy weight in a constant position than to move it up and down. Isometric exercise describes a form of exercise in which muscles are held at a relatively constant length. Dynamic exercise describes a form of exercise in which muscles actually shorten and relax during the exercise period.

**Active hyperemia**

Muscle blood flow increases as demand for oxygen increases during exercise. This is facilitated by metabolic control of the diameter of the blood vessels supplying skeletal muscle. In muscles undertaking isometric exercise, these blood vessels can become squeezed, and oxygen supply quickly becomes insufficient. Accumulation of adenosine diphosphate (ADP) is an important factor in causing muscle fatigue.

**Related topics**

Metabolic processes (B3)
Layout and function of the vasculature and lymphatics (D5)
Cardiovascular response to exercise (D8)
Utilization of $O_2$ and production of $CO_2$ in tissues (E5)
Properties of skeletal muscle fiber types (I4)

## Isometric versus dynamic exercise

Irrespective of the type and severity of exercise being carried out, all exercise involves increased activity of skeletal muscle. There are two main forms of work done by muscle. In **isometric exercise**, muscles produce tension while being held at a relatively constant length; a good example is holding a heavy object for a long period of time. In **dynamic exercise**, rhythmic and regular contraction and relaxation of muscle groups allows work to be performed; a good example is cycling on an exercise bike where movement of the pedals and flywheel is accomplished predominantly by the muscle groups of the lower limb, with some contribution from arm and back muscles to gain more 'purchase'.

As outlined in topic I3, tension is developed in **muscle fibers** by **crossbridge cycling** between **myosin** and **actin** filaments. In the case of dynamic exercise, bursts of nerve impulses in the **α-motor neurons** that innervate a given muscle to produce contraction will be interspersed with periods of quiescence of those α-motor neurons. In the case of isometric exercise, maintenance of tension over a long period of time will require the **tetanic contraction** of some or all of the motor units in a given muscle until the work being done can stop, or the muscle becomes fatigued. In both cases, the muscle fibers must use **adenosine triphosphate (ATP)** in order to facilitate the crossbridge process. In the case of dynamic exercise, the act of detachment of myosin from the actin filament (that is necessary for the ratchet mechanism that progressively shortens the sarcomere) and for relaxation requires ATP to break the bond between the two types of filament.

In the case of isometric exercise, a high rate of crossbridge cycling is also necessary in order to hold the muscle at a given length against load, so this uses ATP also. Measures of vertebrate **myosin ATPase** activity suggest that each myosin head uses 5–6 ATP molecules per second during fast contraction.

The source of the ATP used for contraction of muscle fibers is **cellular respiration** (see Topics B3 and E5), although an additional store of high-energy phosphate that can be used to regenerate ATP is **creatine phosphate**. This creatine phosphate itself is synthesized from ATP during situations when there is more ATP than is necessary for work (i.e. when muscle is resting), and an adequate supply of fuel and oxygen to rephosphorylate the **adenosine diphosphate (ADP)** is produced as a result (*Fig. 1*). Fatigue and cessation of activity in motivated individuals is not necessarily linked to complete depletion of muscle ATP stores; the most obvious marker is an increase in the ADP/ATP ratio above a certain level.

More details of muscle mechanics during exercise are given in *Instant Notes in Sport and Exercise Physiology*.

## Active hyperemia

As mentioned in Topic D8, blood flow to skeletal muscles increased markedly during exercise, by up to 17-fold. Some of this increase in blood flow results from increased **cardiac output** during exercise, but a further contributory factor is the decrease in the level of tone of arterioles supplying active muscles. This **active hyperemia** occurs as a response to the high metabolic demands of active skeletal muscle and by the vasodilatory action of circulating epinephrine on $\beta_2$-adrenoceptors in skeletal muscle arterioles. By-products of cellular metabolism such as **carbon dioxide** and **lactate**, when elevated, produce a vasodilator response in the arterioles supplying skeletal muscle, which are extremely sensitive to these substances. In effect, the vasodilator response not only washes out the metabolites, but helps to supply more oxygen and nutrients to the tissue in order to keep the levels of these metabolites low. As blood flow increases, muscle tissue carbon dioxide levels and lactate levels will fall, leading to a reduced level of vasodilation, and a subsequent build-up of the metabolites again, leading to vasodilation. This is presumably why, in skeletal muscle, cellular ATP levels never fall to zero. Note in *Fig. 1* that the ultimate source of

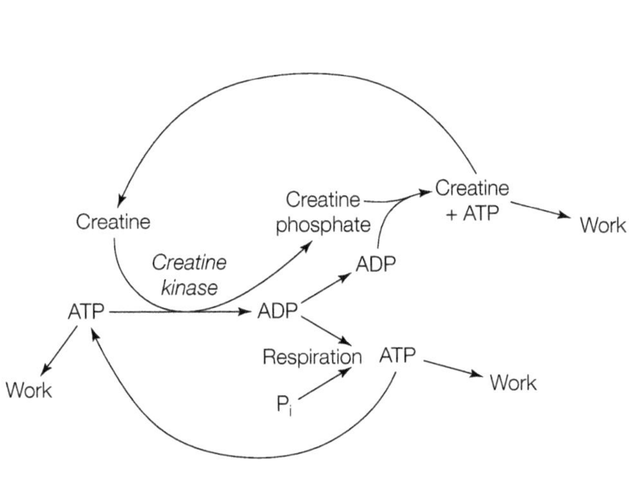


*Fig. 1. Energy substrates for muscle contraction.*

ATP and creatine phosphate is still cellular respiration. If oxygen or nutrient supply is inadequate, ADP will begin to accumulate. Near the end of heavy exercise, cellular ADP levels rise as respiration cannot keep pace with ATP utilization (arterioles are maximally dilated) and fatigue occurs.

# 16 SMOOTH MUSCLE FUNCTION AND PROPERTIES

## Key Notes

**Structure of smooth muscle cells**

Smooth muscle cells contain many of the same contractile proteins as skeletal and cardiac muscle fibers, but the proteins are not arranged in such a rigid fashion. Thus smooth muscle cells are not striated. Contraction of smooth muscle cells takes place in different orientations so that force is applied to many neighboring cells. The number of electrically leaky gap junctions between smooth muscle cells in various organs determines some of the characteristics of smooth muscle function. Multi-unit smooth muscle has few gap junctions, whereas single-unit smooth muscle has many.

**Contractile events in smooth muscle**

Smooth muscle cells do not generate as much force as either skeletal or cardiac muscle fibres. Some smooth muscle cells are spontaneously active. Calcium ions are essential for smooth muscle contraction, but most of these calcium ions enter the cell from the extracellular fluid via calcium channels. Calcium ions induce contraction in smooth muscle by altering the phosphorylation state of myosin via a calcium–calmodulin–myosin light chain kinase complex. Raised intracellular calcium ion concentrations cause phosphorylation of myosin light chains, allowing interactions between myosin and actin to take place. This leads to contraction of the smooth muscle cell.

**External influences**

Smooth muscle contraction can be influenced by neural, hormonal, and local factors. Some of the hormonal and local factors include circulating epinephrine and locally released nitric oxide, which influence the action of myosin light chain kinase or the opening or closure of ion channels on the smooth muscle cell membrane.

**Related topics**

Layout and function of the vasculature and lymphatics (D5)
Layout and structure of the respiratory tract (E1)
Excitation–contraction coupling (I3)
Innervation of smooth muscle (I7)
Adrenal hormones (K8)

## Structure of smooth muscle cells

**Smooth muscle cells,** in contrast to those of skeletal muscle, are very small in both length and diameter (*Fig. 1* and see Topic C5, *Table 1*). Smooth muscle cells are also not striated, although they contain much of the same contractile machinery as skeletal muscle fibers. The **myofilaments** are less regularly organized than in striated muscle like skeletal and cardiac muscle. When viewed under the electron microscope, smooth muscle cells have **dense bodies**

(cytoplasmic and membrane-associated), where thin and intermediate filaments come together, rather like the Z lines of skeletal muscle sarcomeres (see Topic I3). Connective tissue strands attach individual smooth muscle cells to their neighbors, transmitting contractile force throughout the tissue. The interaction between cytoplasmic and membrane-associated dense bodies means that contraction does not take place along the axis of the cell but obliquely, allowing contractile force to be transmitted to other cells both in series and in parallel.

Smooth muscle cells are electrically coupled to each other by **gap junctions**, which allow ions to pass from one cell to the next, meaning that depolarization of one cell can lead to depolarization of its neighbors. This is similar to the situation in cardiac muscle (see Topic D2), but in smooth muscle, the presence of gap junctions between cells varies according to the role of the smooth muscle in the organ in which it resides, or according to certain physiological stimuli that influence the degree of interaction between individual smooth muscle cells. For example, hormonal changes during pregnancy and then near birth alter the electrical activity of **uterine smooth muscle** to maintain quiescence during pregnancy and facilitate coordinated contractions during labor. There are two major classes of smooth muscle, defined by the degree of electrical connectivity between the cells. The first is **multi-unit smooth muscle,** in which gap junctions between cells are relatively few. This type of smooth muscle usually requires nervous input for contraction to take place. **Single-unit smooth muscle**, on the other hand, has many gap junction connections between cells so that the mass of smooth muscle acts like a **functional syncytium**.

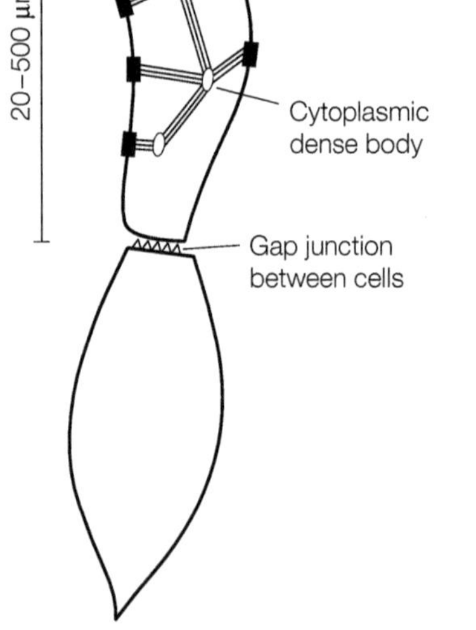


*Fig. 1. Schematic diagram of the structure of smooth muscle cells. Lack of striation reflects the less highly organized nature of the contractile proteins.*

## Contractile events in smooth muscle

Smooth muscle plays a very different role from either skeletal muscle or cardiac muscle. In general, it contracts with relatively low force, but contractions are maintained for long periods of time. This fits with the role of smooth muscle in maintaining the diameter of blood vessels, the degree of constriction of the airways, and slow propulsive movements in the gut. Smooth muscle cells have a resting **membrane potential** of around –50 mV, which is slightly less negative than that of skeletal or cardiac muscle. The membrane potential rises and falls around this average value of –50 mV in a pattern of electrical activity that is known as **slow wave activity**. This slow wave activity brings about partial contractions, which are converted to twitch contractions of the smooth muscle cell if membrane potential reaches a level whereby a smooth muscle action potential is set off. Such activity is independent of hormonal or neural influences, so smooth muscle contraction is often said to be **myogenic** in nature.

As with skeletal and cardiac muscle, the trigger for contraction of smooth muscle is an increase in **intracellular free calcium ion concentrations**. However, in the case of smooth muscle, voltage-gated **calcium ion channels** allow calcium to flow into the cell, raising intracellular calcium *per se*, but also triggering calcium-induced calcium release from the **sarcoplasmic reticulum**. In smooth muscle, this calcium plays a different role from that performed by calcium in skeletal and cardiac muscle contraction (*Fig. 2*). In smooth muscle cells, calcium ions activate a complex of **calmodulin** and **myosin light chain kinase (MLCK)**, which allows splitting of ATP to ADP and inorganic phosphate ($P_i$), and phosphorylation of **myosin**. This MLCK-catalyzed phosphorylation of myosin is akin to the 'charging' of the myosin head in the crossbridge cycle of skeletal muscle (see Topic I3); the phosphorylated myosin can then interact with **actin** because smooth muscle thin filaments lack troponin. This is known as **myosin-linked regulation**. So long as calcium ion concentrations are high enough, the MLCK will be activated and will allow **crossbridge cycling** to occur (the ATP molecule utilized in crossbridge cycling being distinct from that used to phosphorylate the myosin light chains). As calcium ion concentrations fall (by active pumping of calcium ions into the sarcoplasmic reticulum and extracellular fluid, and by sodium–calcium exchange between the extracellular and intracellular compartments, MLCK will become inactivated and more of the myosin will be in a dephosphorylated state (there being constant levels of dephosphorylation of myosin light chains by the enzyme **myosin light chain phosphorylase**). The dephosphorylation of myosin light chains results in lower affinity of myosin for actin and, ultimately, the relaxation of the smooth muscle cell.

## External influences

As outlined above, some smooth muscle requires nervous input in order to contract (e.g. in the muscles controlling pupil diameter in the eye), whereas others are spontaneously active (e.g. in the gut). This is dependent on the type of electrical coupling between smooth muscle cells. However, even single-unit smooth muscle, which is spontaneously active, can be influenced by external factors. Neural input, levels of circulating hormones, and factors released by other cells in the tissue can all alter smooth muscle contractions by either facilitating or inhibiting activity. Neural influences on smooth muscle are covered in Topic I7, so will not be discussed further here. Hormonal and local influences are so numerous that it would be inappropriate to discuss them all here, but some examples are useful in describing how these elicit their effects.

Circulating **epinephrine** released from the adrenal medulla acts on the majority of smooth muscle cells in the body. For example, epinephrine acts to

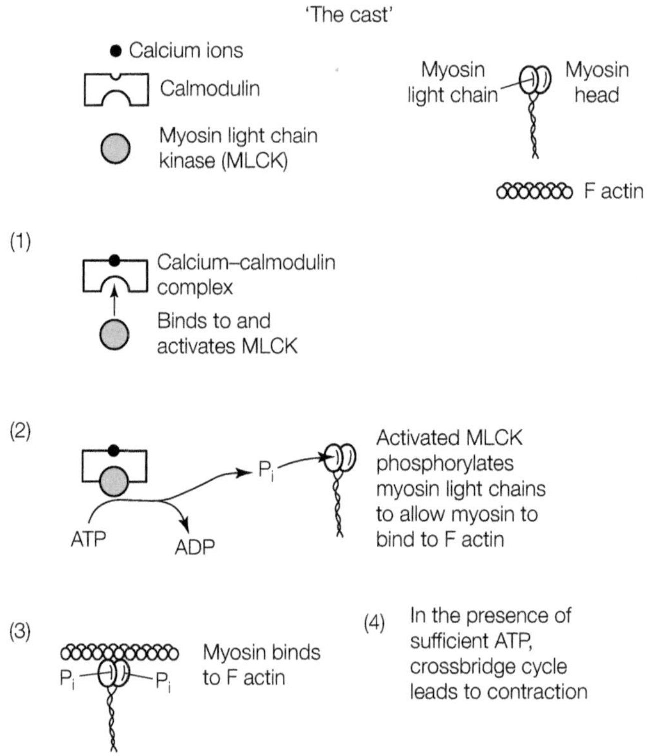


*Fig. 2. Role of calcium ions in contraction of smooth muscle. (1) When calcium ions concentrations are great enough, calcium binds to calmodulin. This calcium–calmodulin complex can bind to and activate the enzyme myosin light chain kinase (MLCK); (2) activated MLCK phosphorylates the light chains of myosin, allowing interaction between the myosin heads and F actin filaments; (3) and (4) the interaction of myosin and F actin and crossbridge cycling causes contraction of the smooth muscle cell.*

relax gut smooth muscle by acting on both **α-** and **β-adrenoceptors** (see Topic H6). Gut smooth muscle α-adrenoceptors are coupled to production of **inositol trisphosphate ($IP_3$)**, and opening of **$IP_3$-dependent potassium channels** (*Fig. 3a*). Activation of these α-adrenoceptors is likely, therefore, to result in potassium efflux from the smooth muscle cells and repolarization or hyperpolarization of the cell membrane (see Topic B4). This effect would reduce calcium ion entry through voltage-sensitive calcium channels and lead to relaxation or reduced contractility of the cells.

In contrast, gut smooth muscle β-adrenoceptors are coupled to **adenylyl cyclase**, an enzyme which catalyzes the formation of **cyclic adenosine monophosphate (cAMP)** from ATP. When epinephrine binds to these β-adrenoceptors, activity of the adenylyl cyclase is increased and intracellular cAMP levels rise (*Fig. 3b*). cAMP causes smooth muscle relaxation by two mechanisms. Firstly, cAMP increases the activity of **cAMP-dependent protein kinases**, which help increase calcium ion sequestration by the sarcoplasmic reticulum, lowering intracellular calcium. Secondly, cAMP-dependent protein kinase phosphorylates MLCK, resulting in a lowered activity of MLCK, a reduction in myosin phosphorylation, and relaxation.

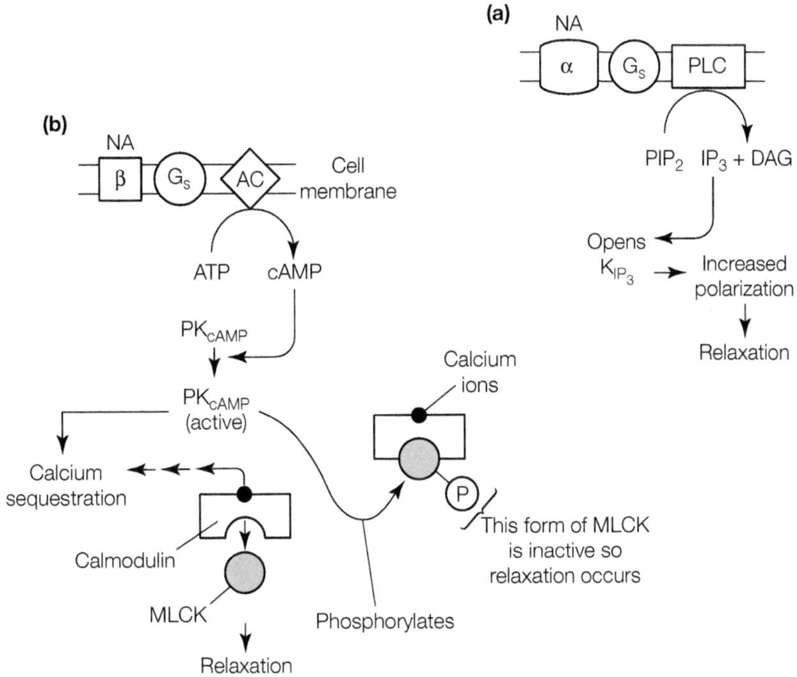


*Fig. 3. Relaxation of gut smooth muscle through activation of (a) α- and (b) β-adrenoceptors is brought about by different second-messenger systems. AC, adenylyl cyclase; cAMP, cyclic adenosine monophosphate; DAG, diacyl glycerol; $G_s$, stimulatory guanine nucleotide binding protein; $IP_3$, inositol trisphosphate; $K_{IP3}$, $IP_3$-dependent $K^+$ channels; MLCK, myosin light chain kinase; NE, norepinephrine; $PIP_2$, phosphatidyl inositol bisphosphate; $PK_{cAMP}$, cyclic adenosine monophosphate-dependent protein kinase; PLC, phospholipase C.*

Locally released factors also influence smooth muscle contraction. One such molecule is **nitric oxide (NO)**, which is released from vascular endothelial cells and elicits a relaxatory response in vascular smooth muscle cells. Before its chemical nature was elucidated, NO was originally known as **endothelium-derived relaxant factor (EDRF)**. Shearing forces on endothelial cells lining blood vessels cause the release of NO from endothelial cells (see Topic D8, *Fig. 3*). NO is a small, diffusible, lipophilic gas molecule that is synthesized from **L-arginine** by the action of **nitric oxide synthase**. Due to its lipophilic nature, it is able to enter vascular smooth muscle cells, where it stimulates **guanylyl cyclase**, causing production of **cyclic guanosine monophosphate (cGMP)**. cGMP activates **cGMP-dependent protein kinase** and, in a fashion similar to that for cAMP above, this lowers intracellular calcium, resulting in smooth muscle relaxation.

# I7 INNERVATION OF SMOOTH MUSCLE

## Key Notes

**Autonomic innervation**

Smooth muscle is usually spontaneously active to some degree. Multi-unit smooth muscle has few gap junctions between cells, whereas single-unit smooth muscle has many. Neural influence on smooth muscle activity comes from two sources: the autonomic nervous system, and the enteric nervous system in the gut. The effects of the two branches of the autonomic nervous system on smooth muscle activity vary depending on the role of the smooth muscle. For example, sympathetic activity causes increased vasoconstriction but decreased activity of gut smooth muscle. These effects are mediated by different types of adrenoceptors in the two tissues. In general, parasympathetic actions on smooth muscle can be described as 'rest-and-digest', whereas those of the sympathetic branch can be described as 'fight-or-flight'.

**Enteric neurons**

Enteric neurons exist in plexuses in the gut and constitute the 'gut brain'. The neurochemistry of the enteric nervous system is extremely complex. Intrinsic enteric reflexes and autonomic influences are important in mediating the degree of activation of gut smooth muscle and gut secretions.

**Related topics**

Structure and function of the autonomic nervous system (H5)
Autonomic pharmacology (H6)
Smooth muscle function and properties (I6)
Introduction to the gastrointestinal tract (J1)
Birth and lactation (L6)

## Autonomic innervation

The predominant source of neural input to smooth muscle in the body comes from the **autonomic nervous system (ANS)**. The contractile state of smooth muscle is under the influence of both the **parasympathetic** and **sympathetic** branches of the ANS. In **multi-unit smooth muscle** (see Topic I6), neural input is normally required for muscle contraction to take place, and **gap junctions** between cells are relatively few. A good example of multi-unit smooth muscle is that controlling the diameter of the pupil in response to light. In order for the pupil to become smaller, parasympathetic outflow from the preganglionic neurons in the **Edinger–Westphal nucleus** in the midbrain must activate post-ganglionic neurons in the ciliary ganglion that, in turn, cause contraction of the smooth muscle of the iris in order to bring about **pupillary constriction**. Sympathetic outflow to the smooth muscle of the iris (via preganglionic neurons in the spinal cord, and postganglionic neurons in the superior cervical ganglion) produces relaxation of iris smooth muscle and pupillary dilation. Under normal resting circumstances, the effects of the parasympathetic branch of the ANS on pupil diameter predominate, which means that application of **atropine** (a mAChR antagonist, see Topic H6) leads to pupillary dilation.

In the case of **single-unit smooth muscle**, which has an abundance of gap junction contacts between cells, the effect of ANS is to modulate a pre-existing level of activity. For example, smooth muscle cells in the wall of the gut are spontaneously active because of **slow-wave activity** (see Topic I6). This means that gut smooth muscle is active even in the absence of neural or hormonal influences. If one takes some smooth muscle from the gut, one finds that it is spontaneously active *in vitro*. However, the two branches of the ANS act to alter the pattern of smooth muscle activity on the gut. In general, parasympathetic outflow from the vagus nerve and the sacral segments of spinal cord increase the level of activity of gut smooth muscle, whereas sympathetic outflow from the thoracolumbar portions of spinal cord decrease the level of activity of gut smooth muscle.

The effects of ANS innervation of different types of smooth muscle are diverse (see Topic H5) but, in general, these effects fit broadly within the simple notion that the parasympathetic branch of the ANS governs **'rest-and-digest'** functions, whereas the sympathetic branch governs **'fight-or-flight'** functions (*Table 1*). For example, during a stressful of frightening situation, it would be beneficial if the sympathetic outflow associated with such events led to an elevation of blood pressure (through increased peripheral resistance via constricted arterioles), opened airways (to let more air in for adequate oxygen delivery), relaxed the gut and bladder, and constricted sphincters, in the event that one must run away from the threat (real or imagined).

*Table 1. Actions of the autonomic nervous system on smooth muscle*

| Smooth muscle of: | Effect of: | |
|---|---|---|
| | Parasympathetic system ('rest-and-digest') | Sympathetic system ('fight-or-flight') |
| Iris | Pupillary constriction | Pupillary dilation |
| Arterioles | None | Constriction |
| Airways | Constriction | Dilation |
| Gut wall | Promotes activity | Inhibits activity |
| Gut sphincters | Relaxes | Contracts |
| Bladder | Contracts | Relaxes |
| Urinary sphincters | Relaxes | Contracts |
| Uterus | Varies | Contracts |

## Enteric neurons

Between the circular and longitudinal smooth muscle layers of the gut wall (see Topic J1), there exists a plexus of neurons known as the **myenteric plexus**. Together with the **submucous plexus** (between the submucosa and the circular layer of smooth muscle), this makes up the **enteric nervous system**. The enteric nervous system is often described as the third branch of the ANS, or as the **'gut brain'**. The myenteric plexus is most concerned with modulating smooth muscle activity and the submucous plexus with altering gut secretion. Stretch-mediated changes in activation of the two different smooth muscle layers underlie the process of **peristalsis** by which food is moved from the oral end to the anal end of the gut (see Topic J2); in portions of gut that have been removed from the body, such peristaltic activity can still be elicited. The neurochemistry of enteric neurons is complex. Probably all known neurotransmitter substances are represented, from gases such as nitric oxide up to peptides like substance P. It is therefore unlikely that you will be expected to know a great deal about the neurochemistry of the enteric nervous system. The parasympathetic and sympathetic branches of the ANS can have an effect on the activity of smooth muscle fibers themselves, or on neurons of the myenteric or submucous plexuses in order to alter the activity of the gut. For example, **norepinephrine** can cause relaxation of gut smooth muscle by acting on **adrenoceptors** on the smooth muscle cells themselves, or by acting on adrenoceptors on excitatory neurons to inhibit transmitter release (**presynaptic inhibition**).

# J1 INTRODUCTION TO THE GASTROINTESTINAL TRACT

## Key Notes

**Parts of the gastrointestinal tract**

The gastrointestinal (GI) tract is composed of the mouth and oropharynx, the esophagus, the stomach, the small intestine (duodenum, jejunum and ileum), the large intestine (cecum, ascending, transverse and descending colon, sigmoid colon, rectum), and the anal canal and anus. Other accessory organs such as the liver and pancreas are important in the function of the GI tract.

**Major functions of the GI tract**

The major functions of the GI tract are to digest food and allow nutrients to be absorbed into the bloodstream. The GI tract also has a barrier function to prevent infection and poisoning, and some parts of the GI tract produce hormones that act on other parts of the body.

**Common features of the gut wall**

The gut wall, when viewed in cross-section, has four main layers. The innermost layer is the mucosa, and deep to this lies the submucosa. Outside the submucosa lies the muscularis externa layer, composed of smooth muscle. Two intrinsic plexuses of neurons (the submucous and myenteric plexuses) exist in the gut, and these have a role in the control of secretion and gut motility. The outermost connective tissue coat is known as the adventitia, and a mesentery may also be present to allow a particular part of the gut to move in the abdominal cavity. The features of these four layers vary depending upon the function of the part of the gut in question.

**Related topics**

Epithelia and connective tissue (C2)
Structure and function of the autonomic nervous system (H5)
Absorption of nutrients (J6)

### Parts of the gastrointestinal tract

The **gastrointestinal (GI) tract** is a set of organs whose main functions are the digestion and absorption of food and the excretion of solid waste matter in the form of **feces**. In thinking about the GI tract, a useful analogy is an oil refinery, where crude oil is broken down into its constituent hydrocarbons for use as fuels, and in manufacturing industries such as plastics. An oil refinery uses different methods to extract these substances from the crude oil, in much the same way that the GI tract uses different chemical methods to break down food and to ensure the constituents of a meal are absorbed properly. The end-product of oil refining is bitumen that can be used in road building, but this is where the analogy falls down, as we have yet to discover such a utilitarian end-point for human feces. The important thing to realize is that the formation of all these products is done by several different processes, just like in the GI tract. Thus, different parts of the GI tract look different and have different functions.

From oral to anal end, the parts of the GI tract are (*Fig. 1*): mouth and oropharynx, esophagus, stomach, small intestine (duodenum, jejunum and ileum), large intestine (cecum, ascending colon, transverse colon, descending colon, sigmoid colon, rectum), anal canal, and anus. Each of these regions has a specific role to play in digestion and absorption of food, as shown in *Table 1*.

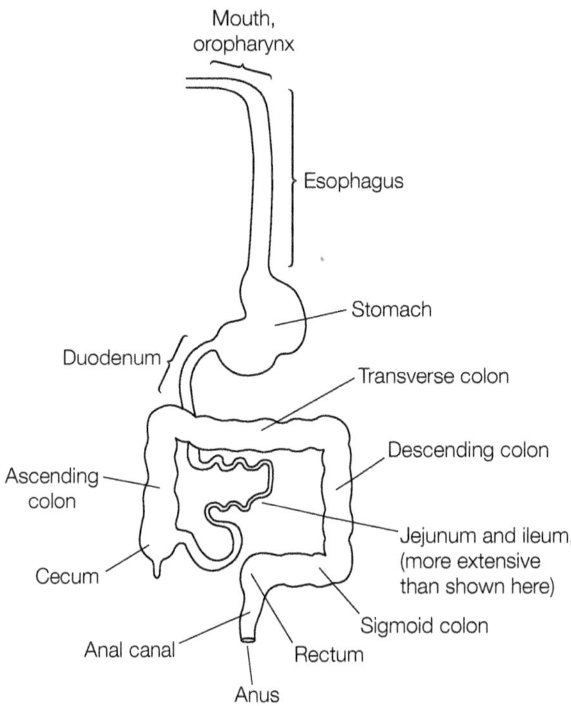


*Fig. 1. Parts of the gastrointestinal (GI) tract.*

*Table 1. Broad functions of the parts of the gastrointestinal (GI) tract*

| Part of GI tract | Functions |
|---|---|
| Mouth and oropharynx | Lubrication, chewing, swallowing |
| Esophagus | Coordinated propulsion of food to the stomach |
| Stomach | Mechanical digestion, chemical digestion |
| Small intestine | Chemical digestion and absorption of nutrients |
| Large intestine | Final absorption of water and electrolytes, storage, compaction |
| Rectum | Storage of feces |
| Anal canal and anus | Continence/defecation |

## Major functions of the GI tract

The GI tract is the means by which we obtain **nutrients** from our environment. Food is consumed and is broken down mechanically by chewing and by churning movements of the GI tract, during which it is mixed with digestive juices that help to break it down into its chemical constituents. The digestive juices are produced by glands in the wall of the GI tract, or by accessory structures such as the pancreas and liver. The chemical constituents of food, comprising sugars, fats, amino acids, vitamins, and minerals, are absorbed across the wall of the GI tract into the body before being used in metabolic processes, or else stored for future use, in the form of complex carbohydrates (e.g. glycogen) or as adipose (fat) tissue.

However, the GI tract has other functions in addition to dealing with the food that we consume. It is also a set of endocrine organs that produce a number of hormones that are released into the bloodstream. Some of these hormones are designed to alter the activity of other parts of the GI tract or its accessory glands, while others have more widespread roles in the body. For example, **secretin** is a hormone produced by the small intestine that alters stomach and small intestinal function as well as increasing the alkalinity of the secretions being produced by the pancreas. Its role is therefore relatively local and it serves to create the correct environment for digestion of food in the small intestine. The hormone **cholecystokinin**, on the other hand, as well as having local effects on the GI tract, acts on the central nervous system (certain brainstem and hypothalamic nuclei) to cause feelings of satiety. Thus its effects are both local and distant.

A third major function of the GI tract is as a barrier to infection and poisoning. As outlined in Topic C2, the GI tract is lined with epithelial cells, and the lumen of the GI tract can therefore be considered to be an outside surface of the body. Therefore, the GI tract could be colonized by pathogens ingested from the environment, and we could be easily poisoned (deliberately or otherwise). There are three main ways in which the GI tract performs this barrier function. Firstly, once food has passed down the esophagus and entered the stomach, it is immersed in the highly acidic environment of the stomach (see Topic J3). The pH of the stomach contents after a meal is usually just above 3 (i.e. the $H^+$ ion concentration is somewhere between $10^{-4}$ and $10^{-3}$ M), which is said to be sufficiently acidic to kill many bacteria. Secondly, a large amount of **mucosa-associated lymphoid tissue (MALT)** exists in the wall of the GI tract, with the greatest amount in the small intestine. Like lymphoid tissue elsewhere, the function of these patches of white blood cells is to act as sentinels and facilitate the destruction of foreign organisms that may have broken through the epithelial cell barrier in the GI tract. Thirdly, the large number of chemoreceptive cells in the gut can signal the presence of noxious substances (such as alcohol or bacterial toxins) to the central nervous system, and induce vomiting. **Vomiting**, a highly complex and coordinated form of gut motility, can be considered to be part of the barrier function of the GI tract

It should be mentioned that not all micro-organisms that find their way into the GI tract are necessarily harmful to us. A large number of commensal bacteria live in the GI tract. The purpose of these bacteria is to help us to digest food and excrete substances such as **bilirubin** (see Topics J5 and J6). One only has to take a course of oral antibiotics, which can kill some of these bacteria, in order to discover how important they are for gut function; diarrhea is one of the most common side-effects of oral antibiotic therapy.

## Common features of the gut wall

Although the various parts of the gut have different roles to play in the process of digestion and absorption of food, their walls all have the same four-layer structure (*Fig. 2*). The gut wall consists of an inner layer nearest to the lumen

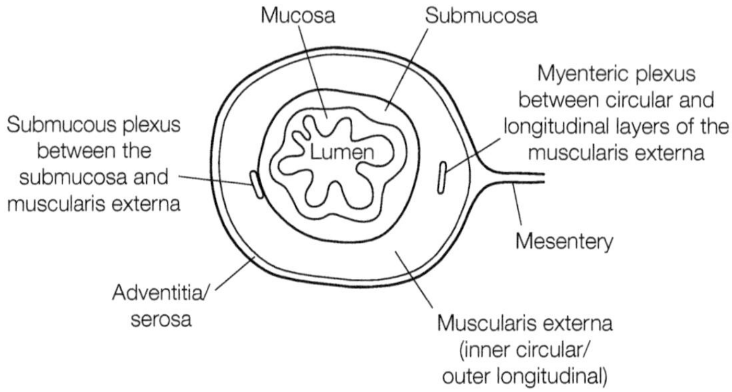


*Fig. 2. General cross-sectional structure of the gut wall.*

known as the **mucosa**. Moving out from the mucosa, one encounters the **submucosa**, the **muscularis externa**, and the **serosa/adventitia**. Where the gut tube is free to move in the abdomen and surrounded by **mesentery**, this outer layer is known as the serosa, and where it is more rigidly fixed to the body wall without mesentery, it is known as the adventitia. Regional variations within this basic structure are explained by the functional requirements of that part of the GI tract, as explained below.

The mucosa is composed of epithelial cells lining the lumen of the gut, loose connective tissue known as the **lamina propria**, and a thin layer of smooth muscle known as **muscularis mucosae** (distinct from muscularis externa). The type of epithelial arrangement differs depending on the function of that part of the gut. In the case of the esophagus, where the epithelium must resist damage caused by friction during swallowing, it is stratified squamous, whereas in the small intestine, where a large surface area is necessary for absorption of nutrients, it is simple columnar, but thrown into many folds, with each cell having finger-like projections known as **microvilli**. These microvilli serve to increase the surface area available for absorption of nutrients in the small intestine (see Topic J6). The lamina propria contains blood vessels, lymph vessels and in some cases may also contain glands and collections of MALT. Some parts of the GI tract have more MALT than others; the small intestine and the appendix are especially well endowed in this regard. In the ileum, the patches of MALT even have a specific name: **Peyer's patches**. The role of the muscularis mucosae is to confer an additional layer of support for the mucosa as a whole, and contraction of the smooth muscle therein can change the folding of this entire layer.

The submucosa is composed of connective tissue fibers such as collagen and elastin, and may contain glandular tissue. These cells are epithelial and although on histological section they may appear disconnected from the epithelial surface of the gut tube, this is usually an artefact of the way in which the tissue has been sectioned. These glands are connected to the surface of the gut tube by ducts. An example of a part of the gut where such glands in the submucosa are numerous is the duodenum, with its extensive network of **Brunner's glands**. Deep to the connective tissue of the submucosa is a nerve plexus known as the **submucous plexus** (sometimes called **Meissner's plexus**), the predominant role of which is to control secretory capacity of the glands in the submucosa (see Topic I7).

The muscularis externa is always composed of at least three constituent parts: an inner, circularly arranged layer of smooth muscle fibers, a **myenteric plexus** (sometimes called **Auerbach's plexus**, to control gut motility), and an outer longitudinally arranged smooth muscle layer. Contraction of the circular layer of smooth muscle narrows the lumen of the gut tube, whereas contraction of the longitudinal layer of smooth muscle shortens that portion of the gut tube. The purpose of the muscularis externa is to move the gut tube in a particular way that fits the function of that part of the gut, so the smooth muscle therein varies in terms of its spontaneous electrical activity and contractile function. Further specialization in the muscularis externa is found in the esophagus, the stomach, and the large intestine. In the case of the esophagus, the upper third is composed primarily of skeletal muscle under voluntary control, important in swallowing. This skeletal muscle component gradually runs out, so that in the lower third there is only smooth muscle. In the case of some parts of the stomach, there are three muscular layers: an inner oblique layer, a middle circular layer, and an outer longitudinal layer. This is important in generating the various types of churning motion that the stomach must perform in order to break up larger lumps of food effectively. In the case of the large intestine, there are only two layers again, but the outer longitudinal layer is arranged into three thickened bands known as the **taeniae coli** (singular, taenia coli), with only very thin longitudinal smooth muscle between. These taeniae coli are shorter than the colon itself, so contraction causes bunching of the colon into **haustrae**, helping to squeeze the feces to conserve fluid.

The accessory organs of digestion, such as the liver and pancreas, have ducts that connect them to the gut tube. Secretions produced by these organs (**bile** in the case of the liver, and bicarbonate- and enzyme-rich secretion in the case of the pancreas) pass down these ducts and enter the lumen of the gut, where they are able to exert their effects. Knowledge and understanding of these regional variations in the structure of the gut wall is important to pathologists, who may be called upon to diagnose everything from cancerous change in the gut epithelium to appendicitis. From the perspective of physiology, such knowledge makes it easier to understand the function of the GI tract as a whole.

# J2 THE MOUTH, SALIVARY SECRETION, AND THE ESOPHAGUS

## Key Notes

| | |
|---|---|
| **The oral cavity – features** | The oral cavity is lined with stratified squamous epithelium to resist the friction of biting, chewing, and swallowing. The teeth and tongue are important features of the oral cavity; the teeth are involved in mechanical breakdown of food, and the tongue helps us to taste food and to swallow. |
| **The salivary glands and secretion** | There are three sets of salivary glands in the oral cavity. These are the sublingual, submandibular, and parotid glands, which produce pH neutral mucous saliva, rich in enzymes involved in chemical digestion of food, and some proteins involved in antimicrobial action. |
| **Swallowing and peristalsis** | Swallowing is a process that is under the control of lower cranial nerves that emerge from the medulla of the brainstem; it involves relaxation of the upper esophageal sphincter, and co-ordinated movements of the tongue and pharynx. Food is propelled down the esophagus by peristalsis, an intrinsic reflex pattern of gut motility. |
| **The gastro-esophageal junction** | The esophagus is lined with stratified squamous epithelium, but at the gastro-esophageal junction, the epithelium abruptly becomes glandular to help protect the lining of the stomach from acid secretions. When persistent reflux of stomach contents into the esophagus occurs, this can lead to metaplasia of the esophageal epithelium so that it changes from stratified squamous to glandular. This metaplasia (Barrett's esophagus) is associated with future cancerous change. |

**Related topics**

Epithelia and connective tissue (C2)
Cranial nerves (G4)
Taste and olfaction (G8)
The stomach and gastric secretion (J3)
Chemical digestion of food (J4)

## The oral cavity – features

The mouth is the first part of the **gastrointestinal (GI) tract** and has all the features that one might expect in order that it fulfils its purpose. The epithelial covering of the mouth and the subsequent parts of the GI tract down to the junction between the stomach and esophagus (**gastro-esophageal junction**) is of the non-keratinized, **stratified squamous** type. This type of epithelium is designed to resist damage caused by friction of relatively solid food before it passes into the stomach. The function of the teeth, as we all know, is to allow us to mechanically break down food by chewing, making it easier for us to swallow. However, this chewing action also increases the surface area to volume ratio of the food, preparing it for chemical digestion by secretions in the mouth and, subsequently, the stomach. The 'milk teeth' are formed from 20 buds of epithelial cells in the first few months of infant life, and these are eventually replaced by 32 adult teeth over the course of childhood and adolescence. They can be considered similar to bones; they have a calcified outer layer (**enamel** and **dentine**) and a spongier center (the **pulp**) that is richly endowed with blood vessels and nerves. The teeth are held in place by **periodontal ligaments**, in the same way that bones are held together by ligaments. This allows the teeth to move somewhat during biting and chewing, without becoming too loose. Like bones, they can also undergo remodeling and repair processes, and are as prone to damage by mineral deficiency and hormonal imbalance (see Topics C5 and K7).

Another important feature of the oral cavity is the tongue. This is a modified skeletal muscle mass under the neural control of the **hypoglossal nerve** (cranial nerve XII). Together with the muscles of the pharynx, jaw, palate, and face, it is important in modifying sounds made by the vocal cords so that we produce intelligible speech. Together with the pharynx, the tongue is also important in swallowing (see below). The dorsal surface of the tongue is covered in stratified squamous epithelium, thrown into different types of folds, with **taste buds** (see Topic G8). Taste buds contain, amongst other cell types, chemoreceptor cells that are in contact with sensory nerve endings that have their axons in the **facial nerve** (cranial nerve VII) and the **glossopharyngeal nerve** (cranial nerve IX). Individual chemoreceptor cells in taste buds recognize the four classic taste 'modalities'; salty, bitter, sweet, and sour. Recent research has indicated that a fifth taste modality, **umami** (from the Japanese word for savory) also exists. Serous and mucous glands help to hydrate and lubricate the tongue.

## The salivary glands and secretion

Once food has been broken down into smaller pieces by chewing, it can then begin to be digested by **enzymes** present in **saliva**. Saliva is a hypo-osmotic secretion that contains proteins and mucus, and has a neutral pH. It is produced by three sets of ducted glands present in the oral cavity; the **sublingual**, **submandibular**, and **parotid** glands. The sublingual and submandibular glands produce a secretion that is rich in **mucins**, which help to lubricate the mouth. The sublingual glands have more mucus-producing cells than the submandibular glands. The parotid glands produce a secretion that is rich in enzymes such as **amylase**, and other proteins that have an antimicrobial action.

Salivary glands have a classical tubuloacinar shape (*Fig. 1*). A primary secretion rich in sodium chloride and proteins is produced by cells in the end-piece (**acinus**), and this secretion is modified by the duct cells, which reabsorb some sodium chloride and add some bicarbonate ions to produce a final secretion that is hypo-osmotic and pH neutral. This is known as the **two-stage hypothesis** of exocrine secretion, but the actual mechanisms are slightly different in the wide range of ducted exocrine glands in the body. Salivary secretion is under the control of the **autonomic nervous system**, which can alter the rate of secretion from around 50 ml $h^{-1}$ to 400 ml $h^{-1}$. As secretory rate increases, sodium and chloride ion content of the saliva increases; this is as a direct result of the two-stage nature of the process of secretion, as the duct cells have less time in which to reabsorb sodium and chloride ions from the primary secretion at high flow rates.

The purpose of the enzymes secreted by the parotid glands has been the focus of recent debate. In a test tube, and in a pH neutral environment, it is true that salivary amylase can catalyse the breakdown of starch into disaccharides such as

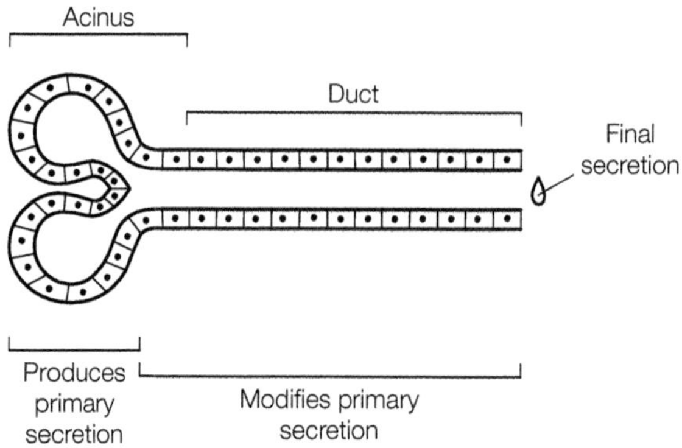


*Fig. 1. Tubuloacinar arrangement of salivary glands.*

maltose and other longer-chain oligosaccharides. However, it is not clear that this is the role of amylase *in vivo*; whether enough amylase penetrates the relatively large bolus of food in the mouth or whether there is enough time for the amylase to have its effect is uncertain. An alternative theory suggests that the role of amylase is to help keep the mouth clean by limiting any build-up of starchy foodstuffs, thereby limiting the likelihood that the growth of microorganisms in the mouth may cause infection. The role of bicarbonate ions may be to help neutralize acids that are present in foods and produced by bacteria. If the function of salivary glands is impaired, this can lead to tooth decay, yeast infections and inflammation in the mouth.

## Swallowing and peristalsis

The act of swallowing is a complex response to the introduction of food or saliva to the posterior surface of the tongue. Muscles supplied by the glossopharyngeal and **vagus nerves** (cranial nerves IX and X) propel the food backwards, the **cricopharyngeal sphincter** (upper esophageal sphincter, UES) opens, and the nasopharynx closes. Food can then pass into the esophagus and hopefully not into the larynx, although sometimes it nearly does, resulting in the coughing and spluttering associated with choking. The majority of these actions are mediated (in both sensory and motor terms) by the vagus nerve and, to a lesser extent, the glossopharyngeal nerve. Thus, brain injuries (e.g. a stroke) or neurodegeneration in the medulla (see Topic G4) can lead to potentially life-threatening deficits in this process.

Once the bolus of food has been introduced into the esophagus, it is propelled towards the stomach by a process known as **peristalsis**, coordinated waves of smooth muscle relaxation and contraction that help propel food along the gut tube. Although the force of gravity has some role to play, one can still eat standing on one's head; peristalsis was presumably necessary when our primitive ancestors resided flat on the ocean floor and could not rely on the force of the water to get food all the way into the stomach. Less primitive ancestors may have had to bolt down food before they themselves were eaten, and thus may have had to retain peristalsis in order to get food into the stomach more quickly than is strictly necessary.

The complex process of peristalsis is an intrinsic mechanism of the gut tube; it persists even when parts of the gut are removed and thereby isolated from neural and hormonal influences. The trigger for this coordinated sequence of events is input to the intrinsic nerve plexuses of the GI tract from sensory nerves in the **submucous plexus**. In its simplest terms, peristalsis involves **receptive relaxation** of the segment ahead of the bolus of food, and then contraction of the segment behind the bolus (propulsion) in order to move the food along (*Fig. 2*). In fact, receptive relaxation involves relaxation of the circular smooth muscle layer (see Topic J1), and contraction of the longitudinal smooth muscle layer and propulsion involves contraction of the circular smooth muscle and relaxation of the longitudinal muscle layer. Once the bolus of food has been moved on, the same process starts over again in the next segment of GI tract, although when viewed in real time, this can look like a smooth, coordinated squeezing motion. It is only by experimental analysis of neurotransmission and electrophysiology that it has been possible to distinguish the basic components of this intrinsic reflex.

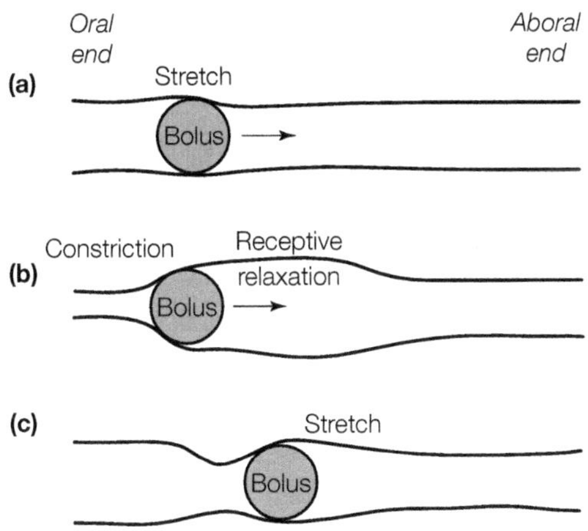


*Fig. 2. The process of peristalsis. Stretch of the gut wall aboral to the bolus of food (a) induces aboral receptive relaxation and constriction on the oral side of the bolus (b). This causes the bolus to be propelled (c) from the oral to aboral end of the gut tube, inducing stretch in the next aboral segment.*

## The gastro-esophageal junction

The esophagus is lined with stratified squamous epithelium and contains many mucus-producing glands in the **mucosa** and **submucosa**. The mucus produced by these glands serves to lubricate the passage of the bolus of food down the esophagus, and the stratified squamous epithelium is designed to be friction resisting. At the gastro-esophageal junction, the epithelium changes form to become simple columnar, and noticeably more glandular in the stomach, in order to protect the wall of the stomach from the **hydrochloric acid** that is produced to activate stomach enzymes (see Topic J3). Reflux of stomach contents into the esophagus (e.g. in **gastro-esophageal reflux disease, GERD**) can lead to the development of **Barrett's esophagus**, a form of metaplasia where the esophageal epithelium becomes simple and glandular in an effort to protect the esophagus from the acidic stomach contents. When this change is seen in patients with GERD, it can be indicative of future cancerous change in the esophageal epithelium. However, in most people, gastro-esophageal reflux is prevented by the functionally defined **lower esophageal sphincter (LES)**. No anatomical sphincter can be seen here (unlike at the upper esophageal sphincter (UES)); the 'LES' is formed in part by the folded arrangement of the gastro-esophageal junction and in part by the constricting action of the diaphragm where the esophagus passes between the thoracic and abdominal cavities.

# J3 THE STOMACH AND GASTRIC SECRETION

## Key Notes

**Parts of the stomach**

The stomach is composed of the cardia, fundus, body, and antrum. Upper parts of the stomach store food, and lower parts of the stomach mix food with digestive juices.

**Cardiac and gastric glands**

The epithelium of the stomach has different specializations depending upon the function of that region. In the cardia, the epithelial glands are much simpler and produce a mucus-rich secretion to protect the lining of this region from stomach acid. In the rest of the stomach, the glands are much more complex and are composed of cells producing acid, enzymes, mucus, and hormones. Pepsinogen is an enzyme precursor that is important in the digestion of proteins, and intrinsic factor is important in absorption of vitamin $B_{12}$ and thereby in the production of mature erythrocytes.

**Gastric acid secretion**

Highly specialized parietal cells in the stomach epithelium produce a secretion that is rich in hydrochloric acid (HCl). Drugs designed to prevent excessive acid secretion and dyspepsia target various processes in parietal cells. HCl secretion is influenced by neural input from the vagus nerve, histamine released from enterochromaffin cells, and gastrin produced by G cells.

**Phases of gastric acid secretion**

HCl secretion by the stomach is stimulated by three main actions: the thought, sight, smell, and taste of food (the cephalic phase); entry of food into the stomach (the gastric phase); and entry of the liquid component of the meal into the duodenum (the intestinal phase). Later entry of calorifically rich food into the duodenum results in a reduction in gastric acid secretion.

**Control of gastric emptying**

Mashed-up food leaves the stomach very slowly so that the duodenum is not overwhelmed. In part, this is caused by the shape of the pyloric region of the stomach, but gastric emptying is also under the control of hormones such as cholecystokinin (CCK). CCK is released from the small intestine when calorifically rich food enters the duodenum. CCK slows down stomach motility and increases secretion of pancreatic juices and release of bile from the gall bladder to aid digestion of foods such as fats, sugars, and proteins.

**Related topics**

Cranial nerves (G4)
Structure and function of the autonomic nervous system (H5)
Introduction to the gastrointestinal tract (J1)
Chemical digestion of food (J4)
Absorption of nutrients (J6)
Defining 'hormones' (K1)

## Parts of the stomach

The stomach is important in the first phases of digestion of food. Glands in the stomach produce many different secretions that facilitate the process of digestion. However, the stomach also has an important role in the storage of food, and regulates the amount of food that enters the small intestine in a given period of time, so that small intestinal digestive and absorptive processes are not overwhelmed. **Gastric emptying** is controlled by neural input and by the release of hormones from the small intestine.

The stomach consists of four main anatomical parts: the **cardia**, the **fundus**, the **body**, and the **antrum** (*Fig. 1*). These various parts of the stomach are said to be analogous to the 'hopper and mill' components of the process of grinding cereals like wheat and corn; the cardia, fundus, and upper parts of the body act as storage areas (the 'hopper'), whereas the lower parts of the body and the antrum participate more in the process of mechanical digestion of food (the 'mill'). The cardia and fundus regions are important in the process of **receptive relaxation**: an active relaxatory response that allows more food to enter the stomach after a meal. The fundus and body are important in producing gastric secretions that are designed to chemically digest food (particularly protein), and the antrum plays an important role in retention of substances in the stomach, or allowing food to enter the **duodenum**.

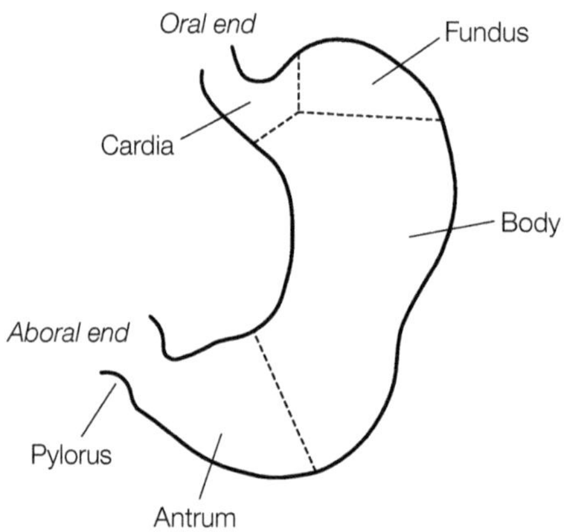


*Fig. 1. Parts of the stomach.*

## Cardiac and gastric glands

As with other parts of the **gastrointestinal (GI) tract**, the stomach is lined with epithelium. At the gastro-esophageal border, the epithelium changes abruptly from **stratified squamous** epithelium to **simple columnar** epithelium (see Topic J2). In the cardia (*Fig. 2a*), some of the epithelial cells are arranged into coiled glands and secrete large amounts of mucus; this is important not only in lubricating the bolus of food but also in protecting the epithelium from damage that might by caused by the acid secretion of other stomach glands.

In the fundus and body, the epithelium is arranged into **gastric glands** (*Fig. 2b*), tubular structures that are composed of epithelial cells with a number of different phenotypes. Four main epithelial cell types exist in gastric glands: mucous cells in the neck region of the gastric gland, **parietal cells** (producing **hydrochloric acid** and **intrinsic factor**), and in the lower region of the gastric glands, **chief cells** (producing the enzyme precursor **pepsinogen**), and a few **G**

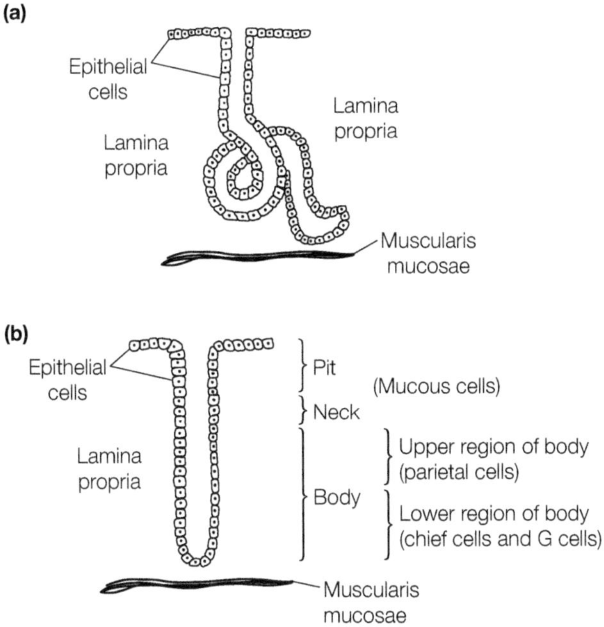


*Fig. 2. Types of secretory glands in the stomach. (a) Cardiac glands; (b) gastric glands.*

**cells** (enteroendocrine cells that release the hormone **gastrin** into the bloodstream). A few other enteroendocrine cells, known as **D cells**, are also present and release **somatostatin** into the bloodstream. In the antrum, the glands are similar in appearance, but they contain fewer parietal cells and more chief cells and G cells.

In the case of both cardiac and gastric glands, the epithelial tissue does not extend past the **mucularis mucosae** so in cross-section, all of the epithelial tissue is clearly confined to the mucosa (see Topic J1).

The digestive properties of gastric secretion are determined by the composition of the substance being produced, which is a mixture of all these secretions. Pepsinogen, released from chief cells, is converted to an active peptidase (**pepsin**), by the action of hydrochloric acid, released from parietal cells. The pepsin can then act at low pH to break proteins down into smaller peptides. Intrinsic factor, released from parietal cells, facilitates the later absorption of **vitamin $B_{12}$** in the ileum. Vitamin $B_{12}$ is essential for the production of mature erythrocytes, and vitamin $B_{12}$ deficiency can cause **pernicious anemia**. The mucus secreted by cells near the surface of the stomach lumen is also slightly alkaline, which helps in protecting the stomach lining from damage by hydrochloric acid produced by parietal cells. The G cells, which are endocrine and therefore release their secretions into the bloodstream, produce the hormone gastrin, which is an important factor in stimulating the parietal cells. Other influences on gastric secretion include **acetylcholine** released from nerves, and **histamine** released from resident **enterochromaffin**-like cells in the stomach.

## Gastric acid secretion

The secretion of hydrochloric acid by parietal cells (also known as **oxyntic cells**) does not, at first glance, seem compatible with life. After a meal, the pH of stomach contents is around 3, meaning that the $H^+$ ion concentration is around 1 mM. Few people would be prepared to drink a solution of hydrochloric acid (HCl) as concentrated as this. Nevertheless, this low pH is a requirement for both the activation of pepsinogen and the action of pepsin itself, with the added advantage of being bactericidal. The stomach lining is protected by mucus secretion. The formation of HCl by parietal cells is an elegant example of many of the processes involved in exocrine secretion, involving active pumping of ions, enzymatic formation of intermediates, and electrically neutral exchange of ions by facilitated diffusion.

While actively producing HCl, parietal cells require large amounts of **adenosine triphosphate (ATP)** to run ion pumps, and the production of ATP by these metabolically active cells results in the production of carbon dioxide ($CO_2$). ATP and a ready source of $CO_2$ in the aqueous (water-rich) environment inside the parietal cell is all that the parietal cell needs to start producing HCl (*Fig. 3*). Like all cells, parietal cells contain ion transport proteins, ion channels, and ion pumps. Post-translational trafficking of proteins inside the cell means that various proteins are sent to various parts of the cell in order that the cell can do its job. Parietal cells have basolateral and apical surfaces, and trafficking of the various ion pumps, ion channels and ion transporters to either the basolateral or apical membranes is centrally important in the process of HCl secretion.

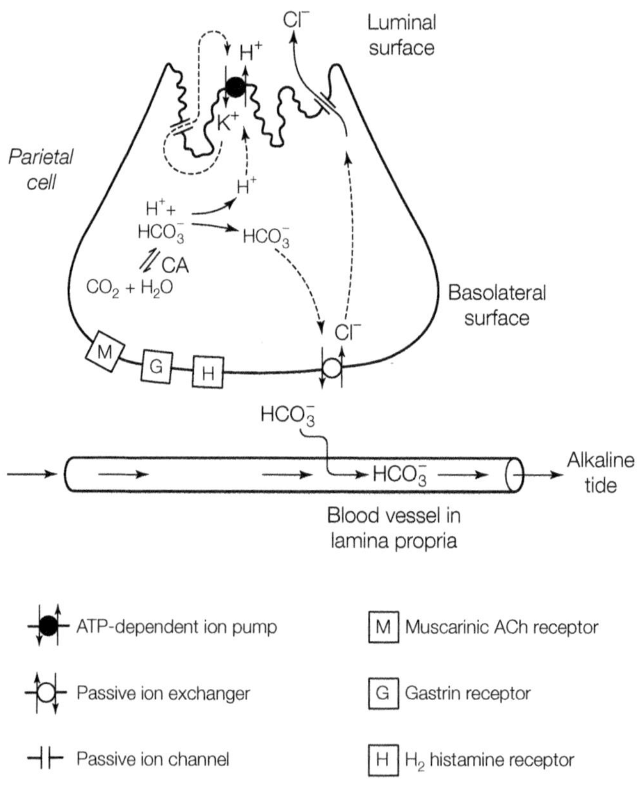


*Fig. 3. Cell model of hydrochloric acid (HCl) secretion by parietal cells in the stomach. CA, carbonic anhydrase.*

Parietal cells have **$H^+/K^+$ ATPase** pumps on their apical surfaces. This pump removes $H^+$ ions from the cell, and pumps $K^+$ ions into the cell. The $K^+$ ions are able to leak back out of the cell through $K^+$ channels (down their concentration gradient), and together with the availability of ATP, this keeps the pumps working. However, if $H^+$ concentrations inside the cell became low, this could lead to pump failure also. The parietal cells therefore have a useful mechanism for generating $H^+$ ions. Water ($H_2O$) is combined with the plentiful supply of $CO_2$ to form $H^+$ ions and $HCO_3^-$ ions, a reaction catalyzed by **carbonic anhydrase**. The $H^+$ ions are the source of $H^+$ to keep the apical $H^+/K^+$ ATPases running. However, a build-up of $HCO_3^-$ ions would result in a significant degree of alkalinization of the parietal cell intracellular fluid, so **$HCO_3^-/Cl^-$ exchangers,** placed on the basolateral membranes, allow $HCO_3^-$ ions to leave the cell and $Cl^-$ ions to enter the cell down their concentration gradients (facilitated diffusion). The $HCO_3^-$ ions are removed from the extracellular fluid by the high levels of blood flow during gastric acid secretion (this results in an **'alkaline tide'** in blood leaving the stomach after a meal), and the $Cl^-$ ions leave the cell through $Cl^-$ ion channels on the apical surface of the parietal cells. These $Cl^-$ ions combine with the $H^+$ ions that have been pumped out by the apical $H^+/K^+$ ATPases, to form HCl.

Parietal cells are stimulated to produce gastric acid by three main molecules: acetylcholine (ACh), gastrin, and histamine. ACh released from nerves, and gastrin released from G cells in the stomach act on muscarinic ACh receptors and gastrin receptors, respectively, to cause an increase in parietal cell intracellular $Ca^{2+}$ ions. In both cases, this is sufficient to cause an increase in HCl secretion. Histamine released from enterochromaffin-like cells in the stomach activates **histamine $H_2$ receptors,** switches on **adenylyl cyclase**, and leads to an increase in **cyclic adenosine monophosphate (cAMP)** levels in parietal cells. When ACh and histamine, or gastrin and histamine, or all three substances are released, the combination of high levels of cAMP and high levels of intracellular $Ca^{2+}$ ions produces an increase in HCl secretion that is significantly greater than the additive effect of the compounds when applied individually. This seems to be related to an increased ability of the parietal cells to pump $H^+$ and $K^+$ ions at the apical surface, but the mechanism is not well understood. However, this helps us understand why the two most effective pharmacological treatments for excess gastric acid secretion are used. **$H_2$ histamine receptor antagonists** like cimetidine were the mainstays of treatment for such conditions for many years. In recent years, drugs such as omeprazole, **proton pump inhibitors** which block the action of the $H^+/K^+$ ATPases on parietal cells, have become important.

## Phases of gastric acid secretion

Gastric secretions, quite sensibly, increase when we eat. However, the pattern of gastric secretion in response to a meal has several phases. These are known as the **cephalic phase**, the **gastric phase**, and the **intestinal phase** of gastric secretion. During the cephalic (head) phase, the mere sight and smell of food is enough to produce increases in gastric secretions. Tasting food, chewing, and swallowing also have a large part to play in this phase. The increase in gastric secretions that is observed during the cephalic phase is mediated by neural pathways, the final output of which is parasympathetic output in the **vagus nerve** (cranial nerve X). ACh released from vagal fibers activates both the parietal cells and the G cells, and the gastrin release from G cells further activates the parietal cells. The gastric (stomach) phase of gastric secretion is produced by distention of the stomach when food enters and, later, the presence of digested peptides. Local neural reflexes and longer loop reflexes with afferent and efferent arms in the vagus nerve (**vago-vagal reflexes**) produce ACh release and gastrin release as in the cephalic phase. Once some digestion of proteins has taken place, the presence of peptides in the stomach also increases gastrin release and increases parietal cell secretion. In the final phase of gastric secretion, the presence of the first, liquid, component of the meal in the duodenum provides a final top-up (10%) of parietal cell secretion, in an effort to digest the remaining solid matter that is still in the stomach. Towards the end of the intestinal phase, the presence of acid in the duodenum shuts down gastric secretion by releasing the hormone **secretin**. The presence of fatty acids and salts in the duodenum releases the enterogastrone **gastric inhibitory peptide**, which switches off parietal cell secretion.

## Control of gastric emptying

The pattern of smooth muscle motility in the stomach is complex. In simple terms, the contractile activity of the 'hopper' part of the stomach (cardia, fundus and upper body) consists mainly of tonic contractions that move the food around a little, and the activity of the 'mill' parts of the stomach (upper body and antrum) are more phasic, rapid and propulsive. This **antral pump** serves two purposes; it churns the food and gastric secretions around and propels the food towards the pyloric sphincter between the stomach and duodenum. The activity pattern that leads to the antral pump is myogenic, meaning that the basic pattern of activity is independent of neural and hormonal influences. However, vagal input and circulating hormones can modulate this basic pattern of activity.

The antral pump serves to churn food in the stomach and propel it towards the **pyloric sphincter**. This pattern of activity (*Fig. 4*) only allows a small amount of food to exit the stomach on each contraction. This is because it takes longer

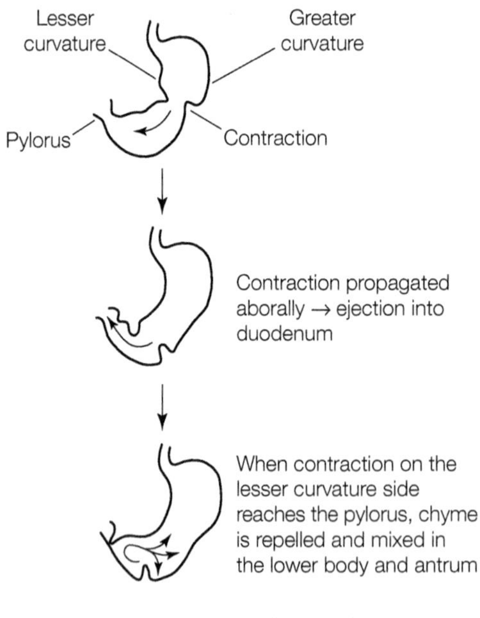


*Fig. 4. Mechanism of controlled gastric emptying and churning of chyme in the antrum.*

for the contraction on the greater curvature side of the stomach (trailing contraction) to reach the pylorus than the contraction on the lesser curvature side (leading contraction). By the time that the trailing contraction reaches the pylorus, the pyloric sphincter has been closed by the leading contraction. The majority of food is propelled back towards the antrum on each contraction, and only a small amount of food (liquid components of a meal first, followed by more solid matter) squirts through into the duodenum. The rate of gastric emptying is proportional to the frequency and strength of the contractions in the gastric antrum. Neural and hormonal input can alter the rate of gastric emptying by modulating the basic pattern of activity.

One of the major factors that determine gastric emptying is the composition of the liquid food (**chyme**) in the first part of the small intestine, the duodenum. When the calorific value of the chyme in the duodenum increases (fatty acids are the most calorific form of food that the duodenum encounters), this results in a reduction in gastric emptying. The most important hormone mediator of this effect is **cholecystokinin (CCK)**. CCK also causes contraction of the **gall bladder** to release bile (the name cholecystokinin suggests that it 'moves' the gall bladder), and stimulation of pancreatic secretion. Both of these effects exemplify the high degree of coordination of gastric emptying and digestion of food in the small intestine.

# J4 CHEMICAL DIGESTION OF FOOD

## Key Notes

**Digestion of carbohydrates**

The major dietary carbohydrate, starch, is a polysaccharide consisting entirely of glucose units. It is metabolized successively by salivary and pancreatic amylases predominantly to a disaccharide, maltose. Maltose and other dietary disaccharides, sucrose (cane sugar), and lactose (milk sugar) are hydrolyzed to the monosaccharides glucose, fructose, and galactose by disaccharidases on the microvillar membranes of the intestinal epithelial cells, with concomitant transport of the monosaccharide into the cell.

**Digestion of proteins**

After denaturation in the acid environment of the stomach, limited hydrolysis of protein to smaller peptides occurs under the action of pepsin (a protease), secreted in an inactive precursor form (pepsinogen) by the chief cells. The major site of proteolysis is the small intestine where a battery of proteases secreted in precursor form from the pancreas is activated and hydrolyzes the peptides to amino acids and di- and tripeptides. Di- and tripeptides are hydrolyzed by brush border peptidases during transport.

**Pancreatic secretion**

The exocrine pancreas synthesizes and secretes a broad spectrum of hydrolytic enzymes to facilitate the breakdown of the major dietary carbohydrates, proteins, and fats into molecules that can be absorbed by the small intestine. These enzymes are secreted as inactive precursors, becoming active in the small intestine under the action of trypsin. Trypsin itself is secreted from the pancreas as trypsinogen, and is activated by enterokinase, a protease released from dying intestinal epithelial cells.

**Digestion of lipids**

Although a lingual lipase may hydrolyze a little dietary triglyceride in the mouth, most is broken down in the small intestine under the action of pancreatic lipase. This activity requires the presence of bile salts. The pancreatic lipase hydrolyzes triglycerides to monoglycerides and fatty acids. Other pancreatic enzymes hydrolyze cholesterol esters and phospholipids to free cholesterol and lysophospholipids respectively, with the release of free fatty acids. Monoglycerides, fatty acids, cholesterol, lysophospholipids, and fat-soluble vitamins are absorbed from mixed micelles formed with bile salts in the intestinal lumen.

**Related topics**

Epithelia and connective tissue (C2)
The mouth, salivary secretion, and the esophagus (J2)
The stomach and gastric secretion (J3)
The liver, gall bladder, and spleen (J5)
Absorption of nutrients (J6)

## Digestion of carbohydrates

Nutritional experts (e.g. The British Nutrition Foundation) recommend that around 50% of the calories consumed in the diet should come from carbohydrates. This includes sugars (mono- and disaccharides), and complex (polymeric) carbohydrates like starch. Some dietary carbohydrates are not absorbed; this is known as **fiber**, which helps to keep the gastrointestinal tract healthy. The daily dietary energy requirements of a healthy 70 kg man who is moderately active is around 2500 kiloCalories (kCal), and therefore 1250 kCal should ideally come from carbohydrates. Carbohydrates have a **calorific value** around 4 kCal $g^{-1}$, and so a healthy 70 kg man should consume 313 g carbohydrate per day.

Carbohydrates are absorbed in the small intestine as monosaccharide sugars (see Topic J6), and so complex carbohydrates like starch must be chemically digested by the action of several enzymes; there are three main digestive enzymes (*Table 1*). The first of these is **salivary amylase**, which is secreted by the cells of the **parotid gland** (see Topic J2). Salivary amylase is mixed with food entering the mouth during chewing, and its action is facilitated by the neutral pH of salivary secretions. Salivary amylase catalyzes the conversion of polysaccharides like starch to maltose (a disaccharide), maltrotriose (a trisaccharide), and α-limit dextrins (polysaccharides). There is debate about whether a significant amount of digestion of carbohydrates is produced by salivary amylase, because ingested food is rapidly swallowed and enters the acidic environment of the stomach. However, it is possible that the central portion of pieces of food that have entered the stomach are protected from this acidic environment, giving the salivary amylase more time to act. The second enzyme involved in the digestion of carbohydrates is **pancreatic amylase**. This is secreted as an integral part of the alkaline pancreatic secretion (see below), which helps to neutralize the acidic chyme from the stomach. Pancreatic amylase is therefore provided with a pH neutral environment in which to act, and produces the same digestion products as salivary amylase. The third set of enzymes involved in the digestion of carbohydrates to monosaccharides are **brush border disaccharidases**. These are proteins associated with the highly folded plasma membrane of the cells in the small intestine that actually absorb monosaccharides (see Topic J6). The products of carbohydrate digestion by amylases, together with sucrose and lactose, are further digested to the monosaccharides **glucose**, **fructose**, and **galactose**, before being absorbed by cells in the small intestine.

*Table 1. Major enzymes involved in carbohydrate digestion in the gut*

| Enzyme | Source | Reactions catalyzed |
|---|---|---|
| Salivary amylase | Secretions from the parotid gland | Starch to maltose, maltotriose and α-limit dextrins |
| Pancreatic amylase | Pancreatic secretion | Starch to maltose, maltotriose and α-limit dextrins |
| Brush border enzymes | Part of the luminal plasma membrane of epithelial cells in the small intestine involved in absorption of carbohydrates | Maltose, maltotriose, α-limit dextrins, sucrose, and lactose to monosaccharides |

## Digestion of proteins

Around 15% of the calories consumed in the diet should ordinarily come from proteins. This is low compared to carbohydrates, because amino acids (the monomers that make up proteins) are used primarily for rebuilding muscle in the body, rather than for energy production. Proteins have a calorific value around 4 kCal $g^{-1}$, and so a healthy 70 kg man should consume around 94 g protein per day.

Proteins are absorbed in the small intestine as short peptides or amino acids (see Topic J6) and so, as with carbohydrates, they must be chemically digested by the action of several enzymes. The digestive enzymes come from three main sources (*Table 2*). The first of these is **pepsin**, produced in the stomach. Chief cells in the stomach epithelium produce an inactive precursor of pepsin known as **pepsinogen**. The **hydrochloric acid** in the stomach has two main roles in the digestion of proteins; it acts to convert pepsinogen to pepsin, and helps maintain an acidic pH, ideal for the action of pepsin. Pepsin digests proteins into smaller polypeptides by acting on specific bonds within protein molecules, and is thus known as an **endopeptidase**. Although important, a lack of pepsin activity in the stomach can be compensated for by the action of another major group of peptidases, which are produced by the pancreas. These are produced as precursor molecules (pro-enzymes) that are activated to become **trypsin**, **chymotrypsin**, **elastase** and **carboxypeptidase**, each of which have a specific role in the digestion of proteins and polypeptides to peptides and amino acids. Trypsin is formed from **trypsinogen** by the action of **enterokinase**, which is an enzyme produced by small intestinal cells. Trypsin then activates the various other pancreatically-secreted pro-enzymes. **Pancreatic peptidases** are very important, and diseases that affect the function of the pancreas (such as cystic fibrosis) can result in malabsorption and malnutrition. The third set of enzymes involved in the digestion of proteins are **brush border peptidases**. Amino acids and peptides are absorbed by specific small intestinal transport processes, described in Topic J6.

*Table 2. Major enzymes involved in protein digestion in the gut*

| Enzyme | Source | Reactions catalyzed |
|---|---|---|
| Pepsin | Chief cells in stomach epithelium; pepsinogen hydrolyzed to pepsin | Protein to smaller polypeptides |
| Pancreatic peptidases | Pancreatic secretion | Protein and polypeptides to peptides and amino acids |
| Brush border peptidases | Part of the luminal plasma membrane of epithelial cells in the small intestine involved in absorption of tripeptides, dipeptides, and amino acids | Larger peptides to tripeptides, dipeptides, and amino acids |

## Pancreatic secretion

The small intestine represents the major site of digestion and absorption of food, and the microvillar surface of the epithelial cells lining the tissue presents a huge surface area for this absorption. The epithelium is rapidly turned over as it is replaced by cells emanating from the crypts; loss of these surface cells releases enzymes into the lumen of the small intestine, one of which, enterokinase, activates trypsinogen to trypsin. This initiates a hydrolytic cascade in which trypsin activates further molecules of trypsinogen to trypsin (autocatalysis), which activate the other inactive precursor hydrolases secreted by the exocrine pancreas (see above). The overall activity of these hydrolases is to digest the macromolecules of food to much smaller molecules that can be absorbed by specific transfer mechanisms across the brush border and into the enterocytes. Thus, proteins are hydrolyzed to tripeptides, dipeptides and amino acids; carbohydrate, principally starch, to the disaccharide maltose; triglyceride to monoglyceride and fatty acids.

## Digestion of lipids

Around 35% of the calories consumed in the diet should ordinarily come from fats. Fat and carbohydrates are the major sources of energy production in the body. However, fat has a calorific content of 9 kCal $g^{-1}$, so a healthy 70 kg man should consume only 90–100 g fat per day. Under circumstances where we consume more calories than required for our activities, fat is stored in the body as adipose tissue. Each kilogram of fat tissue has a calorific value of around 9000 kCal; this means that if a person who wishes to lose weight cuts his or her calorie intake from the recommended 2500 kCal $day^{-1}$ to 2250 kCal $day^{-1}$, it would theoretically take at least 36 days for them to lose 1 kg of fat.

**Triglycerides**, **phospholipids** and **cholesterol** (the major forms of fat intake in the diet) are absorbed in the small intestine as monoglycerides, fatty acids, lysophosphatidylcholine, and free cholesterol. All of these components are absorbed by a process known as **micellar solubilization**, which depends on **bile** secretion (see Topics J5 and J6). Three main pancreatic enzyme groups are involved in fat digestion: **pancreatic lipases**, **phospholipase $A_2$**, and **cholesterol esterase**. Triglycerides (an assembly of glycerol and fatty acids) are digested to 2-monoglyceride and free fatty acid by the combined action of pancreatic lipase and co-lipase. Phosphatidylcholine (also known as **lecithin**, a combination of glycerol phosphate, choline, and two fatty acids) is digested by phospholipase $A_2$ to free fatty acid and lysophosphatidylcholine. Cholesterol esterase converts the major dietary form of cholesterol, cholesterol ester (a combination of cholesterol and a fatty acid) to free cholesterol, so that it can be absorbed in the small intestine.

# J5 THE LIVER, GALL BLADDER, AND SPLEEN

## Key Notes

**Liver structure**

The cells of the liver are known as hepatocytes. Hepatocytes are arranged into columns of cells that, when cut in cross-section, have a roughly hexagonal shape. Each column (a liver lobule) is drained of blood by a central vein, and supplied with blood by six branches of the hepatic portal vein and the hepatic artery. Each liver lobule is drained of bile by six biliary ducts. Thus from a secretory standpoint, the hepatic lobule is not the functional unit of the liver. The liver is supplied with blood with low oxygen content because 80% of blood flow comes from the hepatic portal vein. The liver is perfused by specialized capillaries known as sinusoids, which are extremely leaky; this allows the liver to secrete proteins into the blood and to destroy abnormal blood cells.

**Liver functions**

The liver is important in storage of nutrients such as sugars and fats, and of fat-soluble vitamins such as vitamin A. It is also important in the production of blood plasma proteins and clotting factors. The liver is also a major site of metabolism and excretion of many substances, including therapeutic drugs.

**Bile**

Bile is produced by hepatocytes and flows towards the biliary ducts at the end of the hepatic acini. The biliary ducts combine to conduct the bile to the gall bladder, where it is concentrated and stored. Upon entry of food into the duodenum, a number of hormones cause the gall bladder to contract and discharge some bile into the duodenum. Bile also contains bilirubin, the end-product of heme metabolism. When bile flow is obstructed, excretion of bilirubin is affected, leading to jaundice.

**Spleen functions**

The spleen has a number of immune and hematological functions. Large numbers of white blood cells reside in the white pulp of the spleen, and these cells check the blood for signs of infection. Old and damaged erythrocytes are phagocytosed in the red pulp of the spleen. The red pulp is also a major site of platelet storage.

**Related topics**

Blood and immunity (C6)
The stomach and gastric secretion (J3)
Chemical digestion of food (J4)
Absorption of nutrients (J6)

## Liver structure

The liver is one of the body's largest organs, weighing around 1.8 kg. It is made up of specialized liver cells known as **hepatocytes** that are unique in many respects. The liver is supplied with blood from two sources: the **hepatic artery** (oxygenated) and the **hepatic portal vein** (deoxygenated, see Topic J6). Approximately 80% of the liver's blood supply of around 1.2 l $min^{-1}$ comes from the hepatic portal vein. Blood leaves the liver via **hepatic veins** that drain into the **inferior vena cava** in the upper abdomen.

When sectioned, stained and viewed with the aid of a microscope, the liver is seen to have a regular structure. Columns of hepatocytes are arranged in hexagonal **liver lobules**, each with a **central vein** that passes on to a branch of a hepatic vein (*Fig. 1a*). At the edge of these hexagonal liver lobules there are '**portal triads**', consisting of a branch of the hepatic portal vein, a branch of the hepatic artery, and a **biliary duct** (*Fig. 1b*). In some specimens, it may be possible to discern a fourth structure, a lymphatic vessel, sitting alongside this 'triad'.

Hepatic lobules are, however, structural units that do not tell us much about the secretory function of the liver. Blood flows from the portal triad (supplied by branches of the hepatic artery and the portal vein), past columns of hepatocytes, before draining into the central veins. Liver secretions such as **bile**, however, are produced by hepatocytes and flow in the opposite direction to the blood, into biliary ducts. Thus one can consider the functional unit of the liver in a secretory sense to consist of all of the cells contained within a prism with a biliary duct at its center (*Fig. 1c*). This is known as the **portal lobule** or **hepatic acinus**, conceptually similar to the acinus of other exocrine glands such as salivary glands (see Topic J2). Within a hepatic acinus, blood flows in one direction and bile flows in

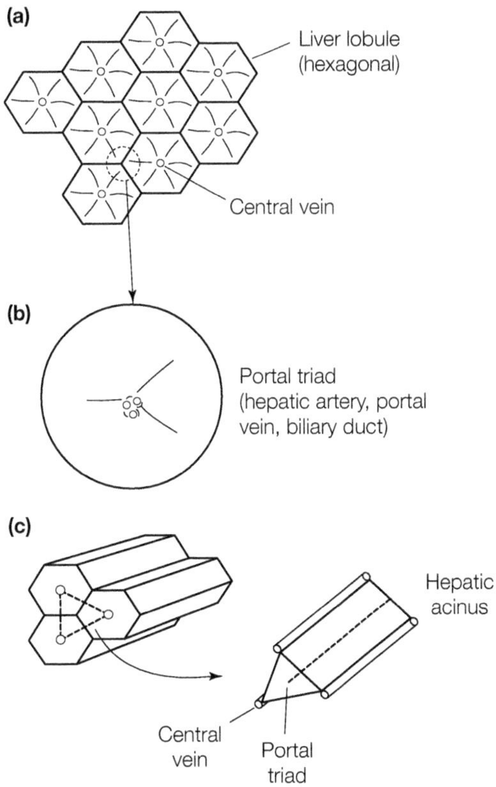


*Fig. 1. Liver structure. (a) The structural unit of the liver is the liver lobule; (b) blood supply to the liver and biliary drainage takes place through vessels in the region where three liver lobules abut one another, in the portal triad; (c) the secretory unit of the liver is a hepatic acinus, drained by the biliary duct within the portal triad.*

the opposite direction. In view of the fact that the liver receives a blood supply that is relatively low in oxygen when compared to other tissues (see above), one can appreciate that the cells nearer to the center of a hepatic lobule are more prone to hypoxia than others. On the other hand, cells nearer to the center of a hepatic acinus would be more prone to damage by blood-borne toxins, because these cells would encounter the highest concentrations of such chemicals.

Blood supply to the liver is unusual in one other way; the capillaries in the liver are extremely 'leaky'. In fact, these capillaries are known as liver **sinusoids** (*Fig. 2*). The gaps between endothelial cells constituting the sinusoids are so large that proteins and cell fragments are able to pass between the blood and the hepatocytes, an arrangement that exists nowhere else in the body. This means that the protein concentration in the perisinusoidal space (the **space of Disse**) is much higher than in the interstitial space of other tissues; in fact the protein content of the space of Disse is around 90% that of the plasma. This is important for liver function, as described below. Other cells are important in liver function; **Kupffer cells** are the resident macrophage-like cells of the liver, and **stellate cells** store fat droplets and fat-soluble vitamins such as **vitamin A**.

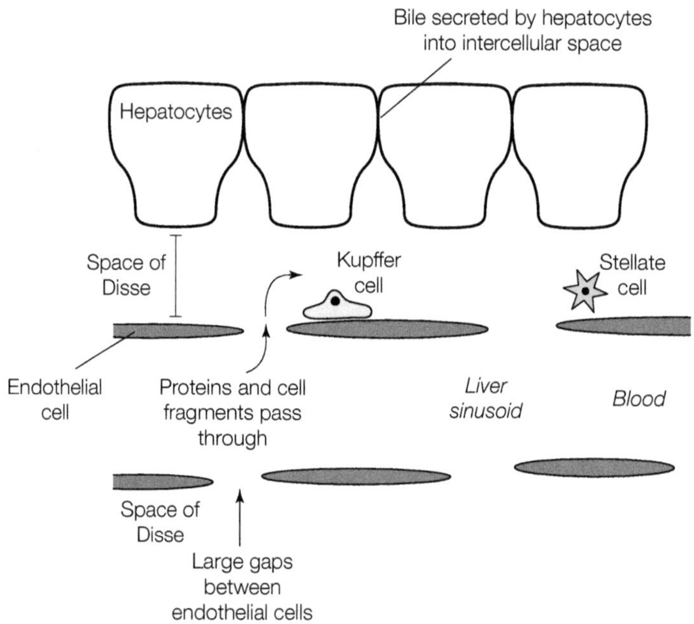


*Fig. 2. The relationship between liver sinusoids, the space of Disse, and hepatocytes. Bile is secreted by hepatocytes into channels that would run through the plane of the page.*

## Liver functions

The liver has an important storage function (see Topic J6). Nutrients that have been absorbed from the gastrointestinal tract are transported to the liver, where they are stored. In the case of glucose, the sugar is stored as **glycogen**. When blood glucose rises above a particular level, the endocrine portion of the **pancreas** releases **insulin**, facilitating the transport of glucose into hepatocytes, where it is converted to glycogen. This process is described in more detail in Topic K9.

The liver also synthesizes triglycerides which are exported into the blood as part of **very low-density lipoproteins (VLDL)** and **high-density lipoprotein (HDL)** precursor particles, which are involved in reverse cholesterol transport. Several water-soluble vitamins are stored in the tissue. It is also important in modulating endocrine processes in the body; it is important in the metabolism of many hormones, and it produces some factors that are important in activating some hormones. For example, **angiotensinogen** is released from the liver and is converted to **angiotensin I** by the action of the enzyme **renin**, produced by the kidney (see Topic M5 for more details).

The liver is also the major site of production of **plasma proteins**; the concentration of albumin in the blood is maintained at approximately 45 g $l^{-1}$ by release of this protein from liver cells. Other plasma proteins such as **fibrinogen** and **prothrombin** (important in blood clotting) and **complement proteins** (important in the immune response) are also produced by the liver, and find their way into the plasma by passing into the space of Disse (see above), which is almost continuous with the blood.

The liver is an important site of protein metabolism in the body. Not only do hepatocytes synthesize new proteins, but they also help to remove nitrogenous waste products from the body in the form of **urea**, which can then be renally excreted. In keeping with its role as a metabolic organ, the liver is also important in the detoxification and excretion of many different substances. In some instances, however, the intermediate metabolites can be toxic to hepatocytes. One such substance is **paracetamol**, which when taken in moderate doses, i.e. as little as 8 × 500 mg/day (4 g/day), can be converted to ***N*-acetyl-*p*-benzoquinone imine (NAPQI)**. NAPQI is toxic to liver cells, and so paracetamol overdose (deliberate or accidental) can result in liver failure.

**Bile**

The major secretory product of the liver is bile. Bile contains bicarbonate ions, sodium ions, **bile salts**, phospholipids, cholesterol and **bilirubin**, the end-product of the metabolism of **heme** groups in **hemoglobin**. Bile salts are derived from cholesterol, and act as detergent molecules since they have both hydrophilic and hydrophobic domains. Biliary secretion produced by hepatocytes flows towards the center of a hepatic acinus (see above), where it is conducted towards the intestinal tract. Bile is stored in the **gall bladder** until required, and it is also concentrated there by secondary active transport of sodium ions and thence osmotic movement of water. One of the triggers for gall bladder contraction is release of **cholecystokinin (CCK)** from the intestinal epithelium into the bloodstream in response to the introduction of calorifically rich food (including fats) into the duodenum. CCK also causes relaxation of the **sphincter of Oddi** (between the common bile duct and the duodenum). Once introduced into the intestinal tract, the bile salts are able to emulsify 2-monoglycerides and fatty acids prior to absorption (see Topic J6).

Bilirubin is formed from heme metabolism in macrophages that have digested old or malformed erythrocytes. This **unconjugated bilirubin** is transported in blood bound to albumin and absorbed by hepatocytes from the space of Disse. The bilirubin in bile that is secreted by hepatocytes has been conjugated to glucuronic acid and is introduced into the intestine when bile is released from the gall bladder. In the small intestine, this **conjugated bilirubin** is virtually unabsorbed, but near the end of the ileum it is deconjugated and some is converted to **urobilinogen**. Urobilinogen can then be: (a) converted to **stercobilin** and excreted (stercobilin is responsible for the brown color of feces); or (b)

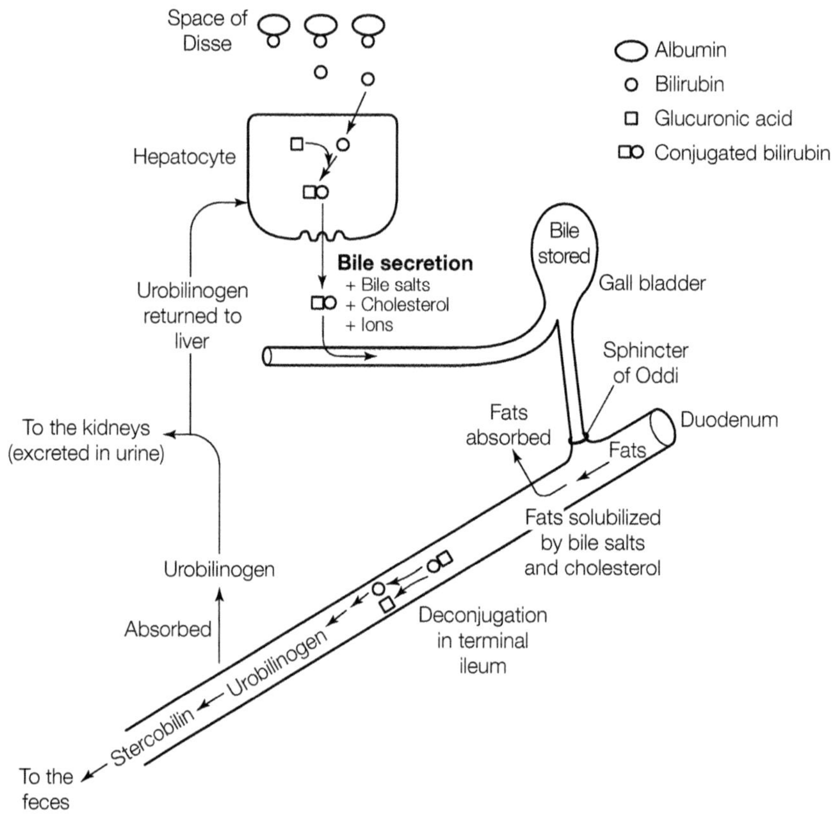


*Fig. 3. The role of bile secretion in the digestion and absorption of lipids, and in bilirubin excretion. Note that the anatomical features of the figure are not drawn to scale.*

absorbed in the small intestine and excreted in either urine or bile. Liver diseases can adversely affect this process of bilirubin elimination, leading to a number of dangerous forms of **jaundice**, as levels of plasma bilirubin rise. Jaundice in newborns is relatively common because of the high turnover from erythrocytes containing **fetal hemoglobin** (see Topic L4), and because the liver does not work at full capacity for a number of days after birth.

## Spleen functions

The spleen sits in the upper left quadrant of the abdomen and is tucked up underneath the diaphragm against the posterior abdominal wall. It is bounded posteriorly and on the left by ribs 9–11, on the right by the left kidney, and inferiorly by the colon. The spleen is the largest mass of lymphatic tissue in the body and is responsible for a number of hematological functions. The spleen has a profuse blood supply and is made up of **white pulp** and **red pulp**. The white pulp is composed of white blood cells, predominantly lymphocytes and macrophages, and is important in acting as part of the body's defense against infection. The red pulp is composed of venous sinuses (filled with blood, of course). Blood entering the spleen via the **splenic artery** first perfuses the white pulp and is checked for signs of infection. Blood then drains into the red pulp, where old or damaged erythrocytes are phagocytosed by **macrophages**. Hemoglobin from these erythrocytes is metabolized to form bilirubin, dealt with by the liver (see above). The red pulp is also a site of storage of platelets, which are important in blood clotting.

Damage to the spleen can cause a number of problems. When a blow to the ribcage is forceful enough to fracture ribs, these can then rupture the spleen, leading to abdominal bleeding. If the ruptured spleen cannot be repaired, it is often removed, and this then causes other problems; someone who has had a **splenectomy** will have compromised immune function, and will probably have to take prophylactic antibiotics at various times (e.g. before even minor surgical procedures and venepuncture) in order to protect against infections.

# J6 ABSORPTION OF NUTRIENTS

## Key Notes

| | |
|---|---|
| **Absorption of carbohydrates** | Glucose and galactose are transported into the enterocyte via a membrane transporter (sodium–glucose ligand transporter-1, SGLT1) which requires cotransport of sodium. From the enterocyte, glucose and galactose enter the blood via a sodium-independent transporter (glucose transporter-2, GLUT2). Fructose is absorbed at the luminal membrane via separate sodium-independent transport system (GLUT5) but uses the same system as glucose and galactose to traverse the basolateral membrane into the blood. |
| **Absorption of amino acids and peptides** | A number of separate transport systems with somewhat broad specificity for neutral, aromatic, basic, and acidic amino acids are present in the brush border membrane. These require cotransport of sodium for amino acid uptake. Some di- and tripeptides are transported intact and undergo hydrolysis in the enterocyte. |
| **Absorption of lipids** | Lipids are absorbed from mixed micelles formed from bile salts and the products of hydrolysis of dietary lipids in the intestine: monoglycerides, free fatty acids, cholesterol, lysophospholipids, and fat-soluble vitamins. Bile salts are reabsorbed in the terminal ileum. |
| **Absorption of vitamins** | Specific transport systems exist in the jejunum for the absorption of all of the water-soluble vitamins apart from vitamin $B_{12}$, which is absorbed from its complex with intrinsic factor in the ileum. |
| **Related topics** | Epithelia and connective tissue (C2)<br>Chemical digestion of food (J4)<br>The liver, gall bladder, and spleen (J5) |

## Absorption of carbohydrates

The major dietary carbohydrate is starch, a polymer of glucose derived mainly from cereals (e.g. wheat), and root vegetables (e.g. potatoes). The disaccharide sucrose (from cane sugar), containing **glucose** and **fructose**, and lactose (from milk), containing **galactose** and glucose, may also be significant components of the diet. Other monosaccharides including glucose and fructose in fruit may also be present in the diet. Sugars are absorbed as monosaccharides, and the terminal step in the hydrolysis of disaccharides occurs on the **brush border** membrane of the small intestine. The **villi** of the small intestine are lined by epithelial cells, each of which has a highly folded luminal membrane. These folds are known as **microvilli** and are apparent as the 'brush border' of the intestinal epithelium; their purpose is to maximize the surface area available for absorption of nutrients (*Fig. 1a*). Final hydrolysis of carbohydrates to monosaccharides is followed by transport of the monosaccharide products into the **enterocyte** (*Fig. 1b*). Intestinal uptake of glucose and galactose requires the cotransport of sodium and water by the **sodium-glucose ligand transporter (SGLT1)**. The sodium gradient across the cell membrane is maintained through the action of an energy-dependent **$Na^+/K^+$-ATPase** (sodium pump), and so the luminal transport of glucose and galactose takes place via secondary active transport. This permits the facilitated diffusion of glucose and galactose from a region of low concentration (the intestinal lumen) to a region of high concentration (the cytoplasm of the enterocyte). Galactose and glucose then leave the enterocyte via the sodium-independent glucose transporter **GLUT2** (down their concentration gradients) and enter the portal circulation.

Fructose, derived mainly from the hydrolysis of sucrose by **sucrase** on the brush border membrane, is transported by another sodium-independent facilitative glucose transporter (**GLUT5**) that is present on both the apical and basement membranes. GLUT5 is energy independent, and has a low affinity for glucose. Also, like glucose and galactose, fructose can be transported across the basement membrane into the blood via GLUT2. These sugars are then transported to the liver via the **hepatic portal vein**.

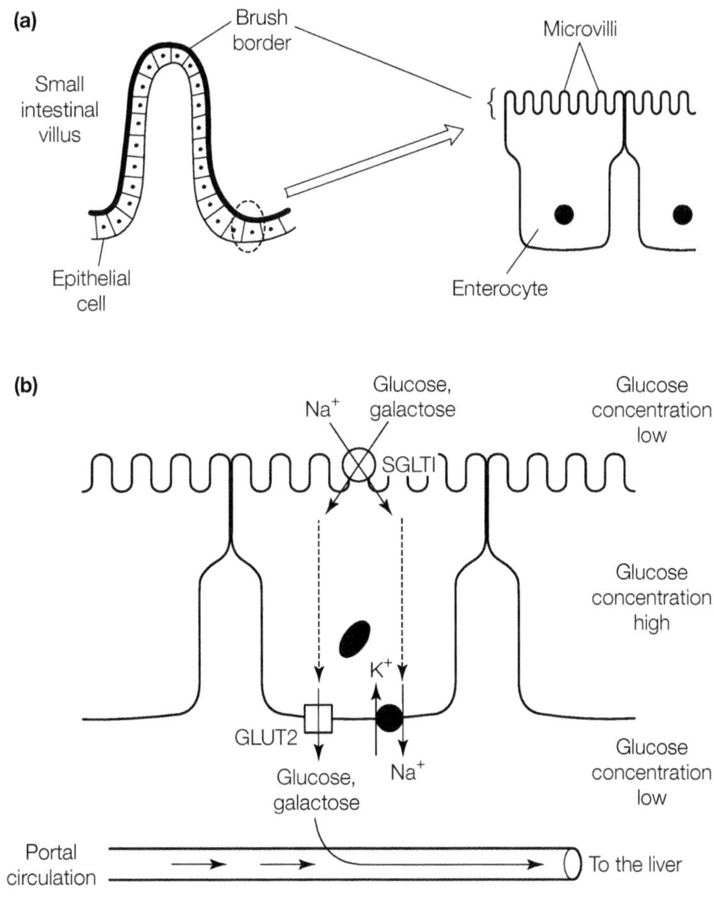


*Fig. 1. Absorption of glucose and galactose in the small intestine, showing (a) the arrangement of enterocytes into villi and (b) a cell model for absorption. GLUT2, sodium-independent glucose transporter 2; SGLT1, sodium-glucose ligand transporter 1.*

## Absorption of amino acids and peptides

There are several different sodium-dependent amino acid transporters, some specific but others more general, with overlapping substrate specificities. For example, there exist separate transporters for acidic amino acids, basic amino acids, aromatic amino acids, neutral aliphatic amino acids, and imino acids (proline and hydroxyl proline), and these co-transport sodium ions and the amino acids into the villus cell.

Some oligopeptides derived from hydrolysis of proteins in the lumen of the small intestine by **pancreatic peptidases** undergo further hydrolysis to amino acids and di- and tripeptides by the action of **endopeptidases** on the brush border surface. Free amino acids produced in this way are transported on the sodium-dependent transporters described above, while the di- and tripeptides are absorbed across the enterocyte membrane by a **proton-coupled oligopeptide transporter (PepT1)**. Once inside the cell, these di- and tripeptides are hydrolyzed to their constituent amino acids by intracellular peptidases.

Amino acids are then transported across the basement membrane into the portal blood, where they are transported to the liver.

## Absorption of lipids

The major lipid component of the diet is **triglyceride** (triacylglycerol). The extremely hydrophobic nature of the molecule presents problems in terms of both its hydrolysis and absorption. Apart from a little hydrolysis of triglycerides containing short-chain fatty acids by the lingual lipase, most of the dietary lipid arrives in the duodenum intact. Peristalsis helps in breaking down the lipid mass into smaller particles, but it is only when the food bolus is mixed with bile salts, released by contraction of the **gall bladder** in response to cholecystokinin (see Topic J5), that any appreciable emulsification of the dietary lipid occurs (**micellar solubilization**). Emulsification of the lipid serves two purposes: it makes the lipid compatible with the aqueous environment of the lumen, and presents the lipid in a form that can be hydrolyzed by the pancreatic lipase. The presence of a smaller protein, **colipase** (also secreted by the pancreas) is essential for this latter action, by anchoring the lipase to the lipid surface. The products of hydrolysis of the triglyceride are **2-monoglyceride** (2-monoacylglycerol) and **free fatty acids**.

Other lipid components of the diet are free **cholesterol**, **cholesterol esters**, and **phospholipids**. Cholesterol esters are hydrolyzed by a pancreatic **cholesterol esterase** to free cholesterol and fatty acids, while phospholipids are hydrolyzed to lysophospholipids and fatty acids by a pancreatic **phospholipase $A_2$**. The hydrolyzed lipids then form mixed micelles with bile salts, and absorption across the brush border membrane takes place from these micelles. Fatty acids and monoglyceride are thought to dissolve in the membrane, and are thus transported into the cell down a concentration gradient. A specific transporter that regulates the movement of cholesterol into the enterocyte is thought to exist, and this transporter can be inhibited by **ezetimibe**, a drug used to decrease the absorption of dietary cholesterol in hypercholesterolemic patients. Bile salts undergo enterohepatic circulation, and are reabsorbed not from the mixed micelle in the jejunum but in the terminal ileum.

Once inside the enterocyte, the absorbed lipid is reconverted to the form in which it occurred in the diet. Thus triglyceride is synthesized from monoglyceride and fatty acids. Cholesterol ester and phospholipids are also resynthesized and packaged with the triglyceride, and some newly synthesized hepatic protein into **chylomicrons**. These **lipoprotein** particles are transported not into the portal circulation but into the lymphatic system, from which they exit into the peripheral blood via the thoracic duct.

## Absorption of vitamins

The fat-soluble vitamins (A, D, E, K) in the diet are incorporated into the mixed micelles described above, and absorbed in the jejunum. Inside the enterocyte, they are incorporated into the chylomicrons and eventually reach the liver as part of the chylomicron remnant, the product of chylomicron hydrolysis in the periphery.

Most of the water-soluble vitamins are absorbed by facilitated transport on specific transporters in the jejunum. The one exception is **vitamin $B_{12}$ (cobalamin)**. Vitamin $B_{12}$ binds to **intrinsic factor** in the jejunum, and the complex passes into the ileum where it binds to a specific receptor. The vitamin is transported into the enterocyte and from there into the blood where it circulates bound to a specific transport protein, **transcobalamin II**.

# J7 The large intestine, fluid, and electrolyte reabsorption

## Key Notes

**Motility of the large intestine**

As with the rest of the gut, the large intestine has muscular walls. In the large intestine, the longitudinal smooth muscle layer of muscularis externa is gathered together into three bands known as the taeniae coli. When these contract, the large intestine becomes bunched up. When the circular smooth muscle layer of muscularis externa contracts, this produces haustrations. Large intestinal motility takes two main forms: phasic contractions like those mentioned above that do not propel feces along the colon, but squeeze the colon contents; and mass propulsive movements that are designed to move feces toward the rectum for defecation. Control of motility in the large intestine is not particularly well understood, but mass propulsive movements tend to follow a meal; neural and hormonal reflexes such as the gastroileal reflex and the gastocolic reflex are important.

**Absorption of fluid and electrolytes**

The colon is the site of a significant amount of fluid reabsorption from the gut contents. Squeezing of the contents of the colon in the haustrations helps preserve fluid, as does the slow transit time of feces in the colon, especially in the transverse section. Water reabsorption by epithelial cells in the colonic mucosa is a secondary active process, with water following the active transport of solutes. Mucus is secreted into the colon to ease the passage of stools during defecation.

**Related topics**

Introduction to the gastrointestinal tract (J1)

## Motility of the large intestine

The large intestine extends from the **ileo-cecal valve** at the end of the ileum to the anal canal. The major functions of the large intestine are storage of **feces**, drying and compaction of feces, and propulsion of feces toward the rectum and anus for defecation. The large intestine is composed of the **cecum** and attached **appendix**, the **ascending colon**, the **transverse colon**, the **descending colon**, the **sigmoid colon**, the **rectum**, and the **anal canal** (see Topic J1). Unlike the transverse and sigmoid parts of the colon, the ascending and descending colon are retroperitoneal. They are fixed to the posterior abdominal wall, whereas the transverse and sigmoid colon have a **mesentery** and are more mobile. The smooth muscle coat of the colon can be distinguished from all other parts of the gastrointestinal tract by the presence of three bands of longitudinal smooth muscle known as the **taeniae coli** (*Fig. 1a*). These taeniae coli are shorter than the colon itself, so the colon is often bunched. This is made more prominent when the taeniae coli and the circular smooth muscle contract (as they do frequently); when this happens, the colon has **haustrations** (*Fig. 1b*). The purpose of this activity is not to propel feces along the colon, but to compress and compact the fecal material.

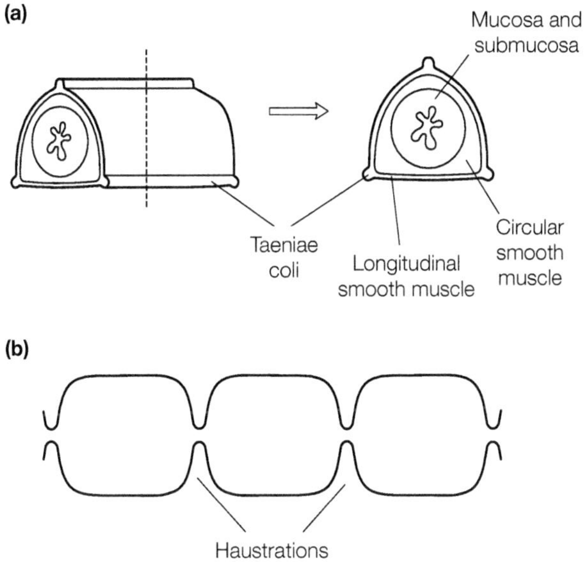


*Fig. 1. Colonic longitudinal smooth muscle is thickened into three bands known as taeniae coli (a), which, together with circular smooth muscle, help to produce non-propulsive contractions (b) that lead to the formation of haustrations in the colon. Such contractions are readily visualized by radiographic methods.*

The smooth muscle in the wall of the large intestine is active almost constantly, although this background level of activity can be modified markedly by both neural and hormonal influences. The large intestine receives input from the **sympathetic** branch of the **autonomic nervous system (ANS)** via thoracolumbar segments of the spinal cord. The **parasympathetic** input to the large intestine up to the splenic flexure comes from the **vagus nerve** (cranial nerve X) and from the splenic flexure onwards from sacral segments of the spinal cord. The transit time through the large intestine for material introduced into the cecum is 2–3 days. The part of the colon in which food remains longest is the transverse colon; this has been shown by instillation of radio-opaque dyes and imaging of the large intestine over a number of days. While the fecal material is in the colon, the last small amounts of water and electrolytes are absorbed from it (see below), and its passage through the colon, anal canal and the anus is eased by the secretion of mucus from numerous mucous cells in the wall of the large intestine. Major propulsive motions of the large intestine normally follow a meal, but the pathways involved in such reflexes are not well understood. The **gastroileal reflex** leads to opening of the ileo-cecal valve to allow more chyme to enter the lower parts of the ascending colon, whereas the **gastrocolic reflex** elicits powerful propulsive contractions of the colonic smooth muscle to propel feces towards the rectum. Distention of the rectum then leads to the urge to defecate. These reflexes are most noticeable in babies, who will regularly defecate following feeding. The reflexes persist into adulthood, but are less noticeable since voluntary control over defecation (via contraction of the **external anal sphincter**, composed of skeletal muscle) allows us to delay defecation until a socially appropriate moment.

## Absorption of fluid and electrolytes

Pumping of ions to maintain the correct balance of various ions on either side of the cell membrane is central to all cellular activities. It is therefore crucially important to maintain normal fluid and electrolyte balance in the face of variable intake of fluid and electrolytes on a daily basis. *Table 1* shows the average amounts of fluid consumed, secreted, absorbed, and lost in the gastrointestinal tract during a normal day in a healthy, 70 kg male subject. It is evident that the amount of fluid and electrolytes absorbed from the large intestine, although relatively small, is important. If the large intestine did not absorb these amounts of fluid and electrolytes and diarrhea resulted, it would not take long before bodily functions could be compromised.

*Table 1. Sources and fate of fluid in the GI tract*

| Part of gut | Amount ingested (l) | Amount secreted (l) | Amount reabsorbed (l) | Net excretion (lost to feces) (l) |
|---|---|---|---|---|
| Mouth | 2 | | | |
| Saliva | | 1.4 | | |
| Stomach | | 2.3 | | |
| Biliary tree | | 0.6 | | |
| Pancreas | | 1.6 | | |
| Duodenum | | 1.1 | | |
| Jejunum | | | 5.4 | |
| Ileum | | | 2.2 | |
| Colon | | | 1.2 | |
| | | | | 0.2 |

As in the small intestine, large intestinal fluid and electrolyte absorption is carried out by epithelial cells lining the wall of the gut. In view of the fact that the amount of fluid and electrolytes to be absorbed is relatively small, and because fecal material is present in the large intestine for a long period of time, the lining of the large intestine is folded, but is not thrown into villi in the same way as the small intestine (see Topics J1 and J6). *Fig. 2* shows a schematic view of the processes involved in electrolyte absorption by large intestinal epithelial cells. **$Na^+/K^+$ ATPase** pumps are placed on the basolateral membranes of colonic epithelial cells, creating a concentration gradient for $Na^+$ ions between the luminal fluid and the interior of the epithelial cell. $Na^+$ channels are placed at the apical membranes of the epithelial cells, facilitating the diffusion of $Na^+$ ions into the cell from the lumen of the colon. $K^+$ ions pass through channels on both the basolateral and apical membranes of colonic epithelial cells, and are also secreted as a component of colonic mucus, but this secretion of $K^+$ can be limited by the action of a **$H^+/K^+$ ATPase** on the luminal membrane that actively transports both $H^+$ ions and $K^+$ ions against their concentration gradients, and helps keep $K^+$ ions in the cell. However, long-lasting diarrhea can be problematic

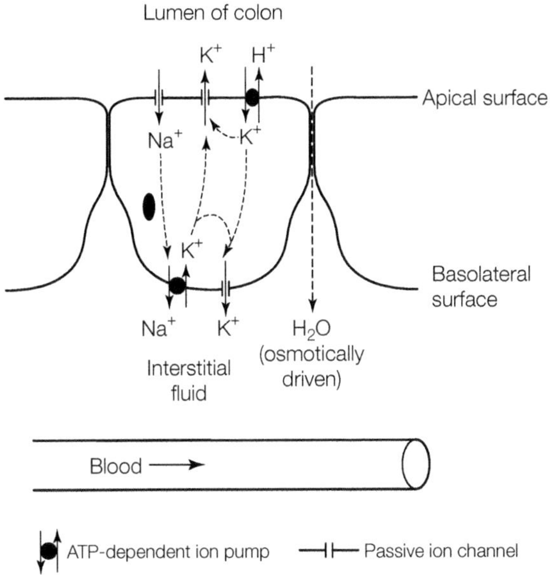


*Fig. 2. Cell model for electrolyte and water reabsorption in the colon.*

because significant numbers of $K^+$ ions can still be lost in the feces, resulting in reduced plasma $K^+$ (**hypokalemia**). This can have potentially fatal effects on the excitability of the heart and conduction of action potentials in heart tissue (see Topic D4). Water moves from the lumen of the gut to the interstitial fluid of the mucosa or vice versa, driven by osmotic forces; when osmotic pressure on either side of the intestinal epithelial cell is balanced, there should be no net movement of water.

# K1 DEFINING 'HORMONES'

## Key Notes

**The endocrine system** — The endocrine system controls metabolic processes through a series of molecules (hormones) that are synthesized in endocrine tissues and travel in the blood to a target tissue on which they exert an effect. Such effects can be acute or chronic, and under normal circumstances serve to maintain homeostasis throughout the day.

**Hormone or enzyme?** — The term 'hormone' has broadened somewhat recently to accommodate molecules that do not fit the strictly classical definition. For example, renin activates a plasma peptide, and some locally acting cytokines are not secreted into the bloodstream.

**Hormone or neurotransmitter?** — Some molecules appear to act as a hormone and as a neurotransmitter in different situations. For example, the classification of hormones becomes a moot point with a molecule such as epinephrine, which is not only synthesized and secreted from the adrenal medulla, but also released from neurons. Other molecules performing such dual roles are cholecystokinin and vasoactive intestinal peptide.

**Related topics**

Homeostasis and integration of body systems (A2)
The panoply of transmitters (F4)
The stomach and gastric secretion (J3)
Hormone action (K3)
Adrenal hormones (K8)
Hormonal control of fluid balance (M5)

## The endocrine system

The **endocrine system** represents one of the two major, interconnected, control systems in the body, the other being the **nervous system**. The distribution of endocrine tissues in the body is shown in *Fig. 1*. The system involves the regulation of organ activity by chemicals (**hormones**) synthesized elsewhere in the body, and a hormone may be defined as a molecule which is synthesized in a tissue (endocrine tissue), secreted into the bloodstream, and acts upon a target tissue to regulate metabolic processes (e.g. the rate of metabolic reactions, transmembrane transport including secretion, protein synthesis, cell growth and division). Such effects can be short lived, lasting for just a few seconds, or more chronic, lasting for days or more. The magnitude of the response is concentration related, and the sensitivity of a tissue to a given hormone concentration is governed by the number of receptors in the target tissue. The precise mode of action of the hormone varies with its structure, but will invariably involve the interaction of the hormone with a specific receptor, often a large polypeptide, located on the target cell surface or in the cytoplasm. This interaction initiates an intracellular response in the target cell. The release of a hormone itself usually represents the last step in a cascade of reactions initiated in the brain (**hypothalamus**) in response to external stimuli such as pain, depression, smell, visual stimuli, etc., or to metabolic stimuli such as metabolite, electrolyte and hormone concentration in the blood. In this way, the hypothalamus acts as a collector of information regarding a number of processes that control whole-body **homeostasis** (see Topic A2), and will act to restore homeostasis when this is disturbed. The hypothalamus signals to the endocrine tissue via the **pituitary gland**, and gives rise to the concept of the **hypothalamic–pituitary–end-organ axes** responsible for the control of **thyroid**, **adrenal**, sex gland, and probably also alimentary functions.

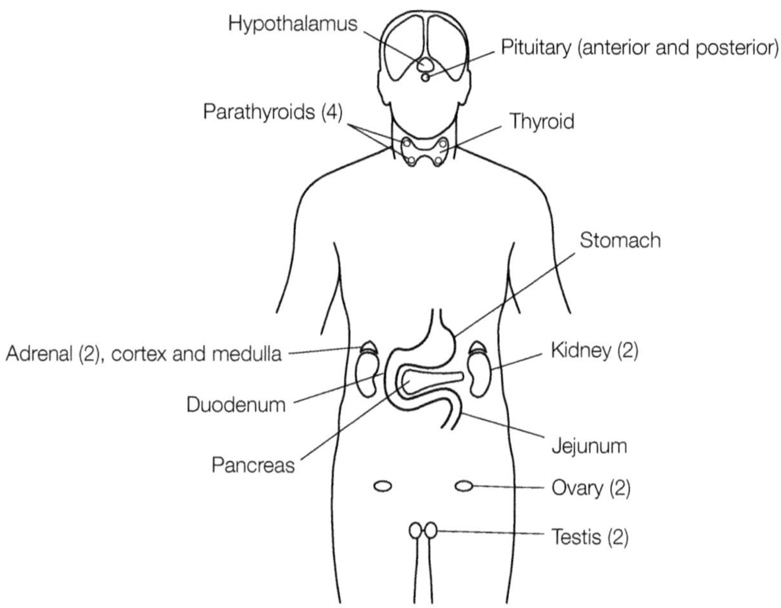


*Fig. 1. The major endocrine organs of the body.*

## Hormone or enzyme?

Endocrine hormones were classically described as intercellular messenger molecules that are secreted from ductless glands into the bloodstream. These ductless glands include the hypothalamus, the pituitary gland, the **pineal gland**, the **thyroid gland**, the **parathyroid glands**, the **pancreas**, the **adrenal gland**, the **testes**, and the **ovaries**. A number of substances that are called hormones have markedly different functions from those classically described. One such example is **renin**, secreted from the cells of the **juxtaglomerular apparatus** in the kidney in response to lowered kidney perfusion pressure and lowered **glomerular filtration rate** (see Topics D6 and M5). Renin is released into the bloodstream, where it acts not on a cellular receptor but on a pro-hormone known as **angiotensinogen**, produced by the liver, to form **angiotensin I**. Renin is a peptidase that cleaves certain amino acids from angiotensinogen to form angiotensin I and thus, renin does not fit easily within the historical description of a hormone. Angiotensin I is inactive and does not bind to **angiotensin receptors** (**AT receptors**), but is converted by **angiotensin-converting enzyme (ACE)** on the surface of vascular endothelial cells to **angiotensin II**. Angiotensin II binds to AT receptors to cause contraction of vascular smooth muscle, and to release **aldosterone** from the **adrenal cortex**. Which of these molecules is the hormone in this system? Angiotensinogen is released from the liver (not endocrine glands) and does not act on cellular receptors. Angiotensin I is converted in the blood to angiotensin II, which then binds to cellular receptors, so perhaps angiotensin II is the hormone? Or perhaps angiotensin II is merely the stimulus for the production of the 'true' hormone in the pathway, aldosterone, which after all is released from a ductless endocrine gland and acts upon receptors on target cells?

## Hormone or neurotransmitter?

Another example is **epinephrine** released from the **adrenal medulla** (*Fig. 2*). The cells of the adrenal medulla are in fact postganglionic **sympathetic** neurons (see Topic H5) that do not have axonal processes. When activated by preganglionic fibers, they release a mixture of epinephrine (80%) and **norepinephrine** (20%) into the bloodstream, which bind to **adrenoceptors** on the surface of target cells, and cause a multitude of physiological effects like tachycardia and bronchodilation, as well as many metabolic effects (see Topic K8). Epinephrine is often described as a hormone, but many neuroscientists cannot help but think of it as a neurotransmitter because it is released from neurons that just happen to reside in a ductless endocrine gland. Then there are all of the substances once described as hormones that we now know also to be neurotransmitters. Interestingly, many of these molecules were identified as hormones released from the gastrointestinal tract such as **cholecystokinin (CCK)** and **vasoactive intestinal peptide (VIP)**; perhaps the gut–brain and the head–brain share more common features than we first thought. When CCK and VIP are released from presynaptic neurons to act on postsynaptic neurons, should we still think of them as hormones?

The purpose of these last few paragraphs has been to encourage thought. In recent times, the list of 'hormones' has grown to include intercellular signaling molecules that do not fit the classical criteria for endocrine hormones. Perhaps the term is outmoded; we should certainly think more about the situations in which we use the word.

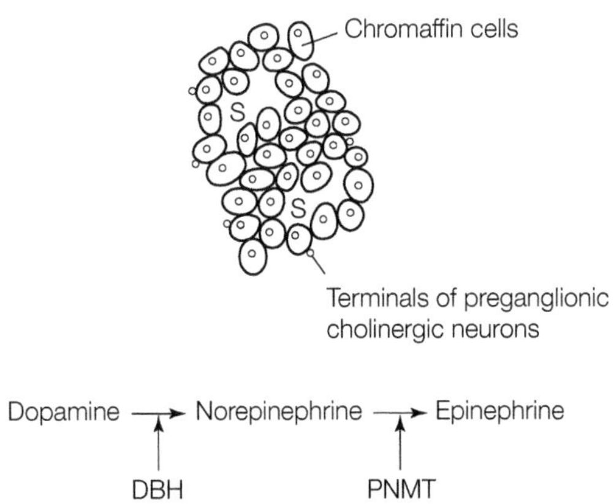


*Fig. 2. Cell bodies of postganglionic sympathetic neurons make up the adrenal medulla, which secretes epinephrine and norepinephrine. These cells are known as chromaffin cells. DBH, dopamine-β-hydroxylase; PNMT, phenylethanolamine-N-methyltransferase; S, venous sinusoid.*

# K2 SYNTHESIS OF HORMONES

## Key Notes

| | |
|---|---|
| **Hormone classes** | Hormones may be subdivided broadly into peptides, phenylalanine derivatives, steroid hormones, and eicosanoids, which reflect their chemical nature and often their mechanism of action. |
| **Biosynthesis of peptide hormones** | Peptide hormones are synthesized initially with extra N-terminal amino acid sequences that do not appear in the mature hormone. These extra sequences direct the peptide through the endoplasmic reticulum into storage vesicles ready for export. |
| **Biosynthesis of steroid hormones** | All steroid hormones are synthesized from cholesterol. The initial common step, the formation of pregnenolone, involves the loss of the aliphatic side-chain from cholesterol, in a reaction stimulated by adrenocorticotropic hormone. Further metabolism requires the action of a number of NADPH-linked hydroxylases. |
| **Arachidonic acid derivatives** | Arachidonic acid, released from membrane phospholipids by the action of phospholipase $A_2$, is the precursor for the synthesis of prostaglandins, prostacyclin $I_2$ and thromboxane $A_2$ via the cyclo-oxygenase pathway, and of leukotrienes via the 5-lipoxygenase pathway. The particular product formed is tissue specific. |
| **Related topic** | Hormone action (K3) |

## Hormone classes

Hormones may be divided into classes that reflect their chemical nature and biosynthetic route. With one or two exceptions, the mechanism of action for members of any group is similar:

- peptide hormones (e.g. insulin, glucagon, growth hormone, follicle-stimulating hormone)
- derivatives of phenylalanine metabolism (e.g. epinephrine, norepinephrine, thyroid hormones)
- steroid hormones, derivatives of cholesterol (e.g. cortisol, aldosterone, sex hormones)
- derivatives of arachidonic acid (e.g. prostaglandins, thromboxanes, leukotrienes).

## Biosynthesis of peptide hormones

**Peptide hormones** are translated from their **messenger ribonucleic acid (mRNA)** on the rough endoplasmic reticulum, and translocation of the nascent polypeptide into the lumen of the endoplasmic reticulum is directed through the **signal peptide** encoded by the 5′ end of the mRNA. The signal sequence is not normally present in the mature hormone, being cleaved from the primary translation product cotranslationally. Further post-translational processing of the hormone may occur during passage of the nascent peptide through the smooth endoplasmic reticulum, Golgi apparatus, and into the **secretory vesicle** where it is stored (*Fig. 1*).

| | |
|---|---|
| Initial translation product: | Pre-prohormone |
| | ↓ Peptidase |
| Product of cotranslational cleavage: | Prohormone |
| | ↓ Peptidase |
| Product of post-translational cleavage: (storage form) | Parent hormone |

*Fig. 1. Scheme for the synthesis and storage of peptide hormones.*

In some cases, for example **parathyroid hormone** (Topic K7), the mature hormone may undergo further proteolysis after release into the circulation. In the absence of stimulated release of the stored peptide hormone, much of the hormone is degraded in the tissue in which it is synthesized.

## Biosynthesis of steroid hormones

The major steroid-secreting tissues, the **adrenal cortex**, **gonads**, and **placenta**, synthesize **steroid hormones** from cholesterol which is stored as cholesterol ester in the tissue. This cholesterol may be synthesized *de novo* in the tissue, but is more likely to have come from cholesterol ester derived from circulating low-density lipoprotein (LDL). LDL-borne cholesterol ester undergoes hydrolysis to free cholesterol and fatty acid in the tissue before being re-esterified for storage. Steroid hormone synthesis requires activation of a cholesterol ester hydrolase to release free cholesterol, which is transported into the mitochondrion for further metabolism. The common intermediate in the biosynthesis of all steroid hormones is **pregnenolone**, formed under the action of the rate-limiting enzyme, desmolase. This step is stimulated by **adrenocorticotropic hormone (ACTH)**. Further metabolism of pregnenolone to individual steroid hormones is discussed in later sections dealing with specific endocrine tissues.

## Arachidonic acid derivatives

A number of local hormones that are important in the regulation and control of inflammatory responses are produced from **arachidonic acid**. Arachidonic acid is a polyunsaturated fatty acid ($C_{20:4}$; containing 20 carbons and 4 double bonds). It is present in cell membranes, esterified to the glycerol backbone of phospholipids (mainly phosphatidylcholine). A number of biologically active peptides (e.g. cytokines) stimulate an intracellular **phospholipase $A_2$** to release arachidonic acid from the phospholipids, and for its subsequent metabolism. The possible fates of arachidonic acid are shown in *Fig. 2*.

The **prostanoids** (e.g. prostaglandin, prostacyclins, and thromboxanes) are formed via the **cyclo-oxygenase pathway** under the initial action of a prostaglandin endoperoxide synthase which exhibits both cyclo-oxygenase and peroxidase activity to form initially **prostaglandin $G_2$ ($PGG_2$)** which is subsequently reduced to **prostaglandin $H_2$ ($PGH_2$)**. The cyclo-oxygenase reaction is inhibited by non-steroidal anti-inflammatory drugs such as **ibuprofen**. **Aspirin** (acetylsalicylic acid) causes irreversible inhibition by acetylating the enzyme.

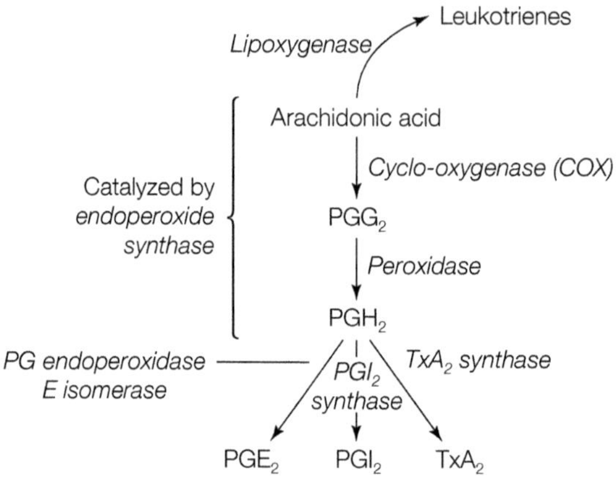


*Fig. 2. Derivatives of arachidonic acid. PG, prostaglandin; Tx, thromboxane.*

Two isoforms of the cyclo-oxygenase have been described; a constitutively expressed **cyclo-oxygenase 1 (COX1)** and the cytokine-inducible **cyclo-oxygenase 2 (COX2)**.

$PGH_2$ is the common precursor for **prostaglandin $E_2$ ($PGE_2$), prostacyclin ($PGI_2$)** and **thromboxane $A_2$ ($TxA_2$)**; the particular product formed is dependent on the enzymes found in an individual tissue. For example, the major metabolite in vascular endothelial cells is $PGI_2$, whereas in kidney cells it is $PGE_2$, and in the platelet, $TxA_2$. In each case, the capital latter denotes the structure of the ring formed during cyclization, and the numerical subscript the number of double bonds in the product. Since two double bonds of the original arachidonic acid substrate are lost during metabolism, all prostanoids derived from arachidonic acid will bear the subscript 2. The prostanoids synthesized from the fish oil fatty acids, eicosapentaenoic ($C_{20:5}$) and docosahexaenoic ($C_{22:6}$) acids have subscripts of 3 and 4 respectively. Another inflammatory mediator, **leukotriene $B_4$**, is synthesized in mast cells from arachidonic acid via the **5-lipoxygenase pathway**.

# K3 Hormone action

## Key Notes

**Mechanisms of hormone action**

Hormones activate metabolic processes in the target cell by one of two mechanisms. They may bind to a specific receptor on the target cell surface and stimulate synthesis of an intracellular messenger molecule, which in turn elicits a response, or they may pass across the cell membrane and bind to an intracellular receptor prior to being translocated to the nucleus, where they modulate the expression of specific genes.

**Cyclic AMP**

The binding of many peptide hormones to their membrane receptors activates the synthesis of cyclic adenosine monophosphate (cAMP) from adenosine triphosphate (ATP) by the enzyme adenylyl cyclase. This action is mediated through the action of a guanine-nucleotide binding protein (G-protein). An increased intracellular concentration of cAMP, a second messenger, leads to a rapid change in cellular function through activation of a protein phosphorylation cascade initiated by protein kinase A.

**$PIP_2$, $IP_3$ and diacylglycerol**

Inositol trisphosphate ($IP_3$) and diacylglycerol (DAG; diglyceride) are the products of hydrolysis of a minor plasma membrane phospholipid, phosphatidylinositol bisphosphate ($PIP_2$), by the enzyme phospholipase C, in response to activation of G-protein-linked receptors. Both $IP_3$ and DAG act as second messengers. $IP_3$ releases calcium from intracellular stores, causing activation of calcium-dependent processes. DAG activates a membrane-bound protein kinase C, which initiates a cascade of protein phosphorylation to phosphorylate nuclear transcription factors and modify gene expression.

**Hormone response elements**

Steroid hormones are translocated to the cell nucleus by specific intracellular protein receptors, and this complex recognizes specific sequences (hormone response elements) in the promoter region of hormone-responsive genes. Binding of the hormone–receptor complex to these sites can modify gene expression.

**Related topics**

Gene transcription and protein translation (B2)

Synthesis of hormones (K2)

## Mechanisms of hormone action

As mentioned in Topic K1, hormones affect tissues other than that in which they are synthesized, and by implication must be recognized in some way by the target tissue. Furthermore, in order to trigger intracellular responses, some method of signaling across the plasma membrane must exist. Two quite distinct signaling pathways have been identified. The response to **peptide hormones** is generally rapid, involving a change in the conformation of existing cell proteins, usually enzymes, with a consequent change in the activity of these proteins. **Steroid hormones**, on the other hand, act through structurally related receptors that bind to the promoter region of specific target genes to regulate their transcription. In this way they alter the amount of specific cell proteins, thereby changing the metabolic activity of the cell and producing the associated physiological response.

The plasma membrane poses a barrier to water-soluble hormones such as peptide hormones, epinephrine, growth factors, and neurotransmitters and these hormones exert their effect from outside the cell. The interaction of these hormones with specific **plasma membrane receptors** stimulates the intracellular synthesis of distinct molecules which activate **protein kinases**, which in turn modify the metabolic response of the cell. Such molecules are known as **second messengers**, as they transmit the extracellular signal from the hormone, first messenger, inside the cell. The binding of a single hormone molecule to its specific receptor on the plasma membrane can give rise to the synthesis of many second messenger molecules inside the cell, and such an amplification of the original signal is an essential feature of the pathway.

Steroid hormones and **thyroid hormone**, on the other hand, are lipid soluble and can traverse the plasma membrane of target cells and bind to **intracellular receptors**. These then translocate to the nucleus and alter rates of transcription of specific genes.

## Cyclic AMP

Many peptide hormones (*Table 1*) signal intracellularly via **cyclic adenosine monophosphate (cAMP)**, formed by the action of **adenylyl cyclase** on adenosine triphosphate (ATP). Binding of the hormone to its specific receptor causes activation or inhibition of adenylyl cyclase via **guanine nucleotide regulatory proteins (G-proteins)** thereby altering the intracellular concentration of cAMP. G-proteins are heterotrimeric proteins consisting of an $\alpha$-subunit (39–52 kDa), a $\beta$-subunit (35–36 kDa), and a small $\gamma$-subunit (8 kDa), and can be stimulatory ($G_s$) or inhibitory ($G_i$). In the unstimulated state, guanosine diphosphate (GDP) binds to the $\alpha$-subunit. Binding of the hormone to the receptor induces a conformational change in the receptor such that it interacts with the G-protein and

*Table 1. Some hormone-induced cell responses mediated by cyclic AMP*

| Target tissue | Hormone | Major response |
|---|---|---|
| Thyroid | TSH | Thyroid hormone synthesis and secretion |
| Adrenal cortex | ACTH | Cortisol synthesis and secretion |
| Adipose | Epinephrine, ACTH, TSH Glucagon | Triglyceride breakdown |
| Liver | Glucagon | Glycogen breakdown |
| Muscle | Epinephrine | Glycogen breakdown |
| Kidney | Parathyroid hormone | Phosphate excretion, calcium reabsorption, 1$\alpha$-hydroxylation of (25-OH) vitamin D |
| Ovary | LH | Progesterone secretion |
| Heart | Epinephrine | Increased heart rate and force of contraction |

ACTH, adrenocorticotropic hormone; LH, luteinizing hormone; TSH, thyroid-stimulating hormone.

promotes the exchange of GDP for guanosine triphosphate (GTP) on the α-subunit. Binding of GTP to the α-subunit causes it to dissociate from the β- and γ-subunits, allowing it to bind to and activate ($G_s$) or inhibit ($G_i$) adenylyl cyclase on the internal surface of the plasma membrane. The intrinsic GTPase activity ($GTP \rightarrow GDP + P_i$) of the α-subunit limits the activation of the adenylyl cyclase, and promotes the reformation of the inactive trimeric G-protein. The importance of this termination of response is demonstrated by the action of cholera toxin. **Cholera toxin** causes a covalent modification (ADP-ribosylation) of the α-subunit of the G-protein, thereby inhibiting its GTPase activity and irreversibly activating adenylyl cyclase. This action in the intestine results in the massive secretion of water and sodium ions, and is responsible for the dehydration and salt-depletion characteristic of cholera.

A similar chain of events occurs for the inhibitory receptor where the $G_i$-protein has identical β- and γ-subunits to the $G_s$-protein. However, in this instance, adenylyl cyclase is inhibited when GTP binds to the α-subunit. **Pertussis toxin** promotes ADP-ribosylation of the α-subunit of the inhibitory G-protein, preventing its interaction with the receptor and subsequent inhibition of adenylyl cyclase. As in the case of cholera toxin, adenylyl cyclase acts in an unregulated manner.

The physiological effects of cAMP are virtually all mediated via the action of cytoplasmic cAMP-dependent kinases (**protein kinase A, PKA**). These protein kinases are tetrameric proteins consisting of two regulatory and two catalytic subunits. Binding of cAMP to the regulatory subunit induces a conformational change and releases the catalytic subunits, allowing them to phosphorylate serine and threonine residues on target proteins, and initiate a series of protein phosphorylation reactions.

In some neuroendocrine cells, the catalytic subunits of PKA may translocate to the nucleus and catalyze the phosphorylation and activation of regulatory transcription factors. In this way cAMP may also stimulate gene transcription.

The binding of cAMP to the regulatory subunits of PKA induces a conformational change, causing these subunits to dissociate from the catalytic subunits. This activates the kinase activity of the catalytic subunits. The release of the catalytic subunits requires the binding of more than two cAMP molecules to the regulatory subunits in the tetramer. This increases the sensitivity of the response of the kinase to changes in the cAMP concentration. There are at least two types of PKA:

- *type I*: mainly in the cytosol
- *type II*: bound by its regulatory subunit to the plasma membrane, nuclear membrane, mitochondrial outer membrane, or microtubules.

However, once catalytic subunits are freed and active they can migrate into the nucleus where they can regulate gene regulatory proteins, while the regulatory subunits remain in the cytoplasm. The action of protein phosphatases limits the action of the PKAs.

## $PIP_2$, $IP_3$ and diacylglycerol

**Phosphatidylinositol-bisphosphate ($PIP_2$)** is a minor phospholipid (0.4% of total) of the inner surface of the plasma membrane. On hydrolysis by **phospholipase C (PLC)**, it yields two intracellular second messengers, **diacylglycerol (DAG)** and **inositol trisphosphate ($IP_3$)**. PLC may be activated via the cAMP pathway or via a receptor with intrinsic **tyrosine kinase** activity. In the latter case, binding of a hormone (e.g. epidermal growth factor) to its receptor causes the receptor to dimerize and undergo autophosphorylation. This allows it to bind PLC, and in turn phosphorylate and activate it. The activated PLC can then hydrolyze $PIP_2$. PLC is deactivated by a tyrosine phosphatase.

$IP_3$, as described above, is derived from hydrolysis of $PIP_2$. It is a water-soluble molecule and mobilizes calcium from intracellular stores in the endoplasmic reticulum where the cation is bound to proteins which exhibit low-affinity, high-capacity binding sites. $IP_3$ binds to specific glycoprotein receptors in the endoplasmic reticulum, opening calcium gates to allow a rise in cytosolic calcium concentration which, in turn, activates calcium-dependent processes.

The other product of PLC-catalyzed hydrolysis of $PIP_2$ is DAG. This molecule can activate a calcium-dependent, membrane-bound protein kinase, **protein kinase C (PKC)**, which can phosphorylate cellular proteins including transcriptional regulators and thereby lead to altered rates of gene expression.

## Hormone response elements

The response to steroid hormones and thyroid hormone involves regulation of transcription of specific genes in the nucleus, and is mediated by intracellular receptors. In the non-stimulated state, these receptors are bound to other proteins in inactive complexes. Binding of the hormone facilitates dissociation of the receptor protein from the complex, and dimerization of the receptor with bound hormone so that the receptor can now interact with DNA in the nucleus and modify the rate of gene transcription. Targeting of particular genes for up- or downregulation of transcription is done by specific **hormone response elements** in the region of the gene responsible for control of transcription. As an example, **cortisol**, when bound to receptors in target cells, alters the transcription of several genes by binding of the hormone–receptor complex to **glucocorticoid response elements (GREs)** on DNA. An example of the action of cortisol is the upregulation of transcription of genes encoding adrenoceptors in vascular smooth muscle cells.

# K4 HYPOTHALAMIC HORMONES

## Key Notes

**The hypothalamus**

The hypothalamus lies in the wall and floor of the third ventricle of the brain, and communicates both directly and indirectly with the pituitary. Axons from the supraoptic and paraventricular nuclei release oxytocin and vasopressin in the posterior pituitary, and other axons secrete releasing factors into the hypothalamic-pituitary portal system to act on the anterior pituitary cells to stimulate trophic hormone production. The release of hypothalamic hormones is under tight control via short and long feedback loops from the pituitary and target organs, respectively. In this way the hypothalamus and pituitary exercise control over the five major endocrine axes, controlling the biosynthesis and secretion of thyroid hormone, glucocorticoids (cortisol), sex hormones, growth hormone, and prolactin. Another important hormone, leptin, acts on the hypothalamus to affect feeding behavior and thermoregulation.

**TRH and PRH**

Thyrotropin-releasing hormone (TRH) is a tripeptide which stimulates synthesis and secretion of thyroid-stimulating hormone (TSH, thryotropin) from the anterior pituitary thyrotrophs. It can also stimulate prolactin release from the anterior pituitary lactotrophs. Prolactin-releasing hormone (PRH) is a small peptide that may also stimulate prolactin release.

**GHRH and somatostatin**

Growth hormone (GH) release by the anterior pituitary somatotrophs is subject to stimulatory (growth hormone-releasing hormone, GHRH) and inhibitory (somatostatin) signals from the hypothalamus. The actions of these two small peptides allow for fine control of GH release.

**CRH and GnRH**

Release of the small peptide, corticotropin-releasing hormone (CRH), by the hypothalamus, stimulates the production in the anterior pituitary corticotrophs of a protein, pro-opiomelanocortin, which is the precursor of a number of peptide hormones including adrenocorticotropic hormone. The decapeptide gonadotropin-releasing hormone (GnRH) stimulates the synthesis and secretion of the gonadotropins, follicle-stimulating hormone and luteinizing hormone, in the anterior pituitary gonadotrophs.

**Vasopressin and oxytocin**

Although vasopressin (antidiuretic hormone) is synthesized in the hypothalamus, it is stored in neuronal terminals in the posterior pituitary. Its primary action is on the kidney where it facilitates reabsorption of water. Oxytocin stimulates milk ejection from the breast by inducing contraction of the myoepithelial cells around the alveolae and ductules. It is released from the posterior pituitary in response to afferent signals arriving in the hypothalamus from the tactile receptors in the nipple during suckling. It also has a role with other hormones in the induction of contraction of the uterus during birth.

| **Related topics** | The limbic system (G3)<br>Pituitary hormones (K5)<br>Thyroid hormones (K6)<br>Adrenal hormones (K8) | The gonads and sex hormones (K10)<br>Hormonal control of the menstrual cycle (L3)<br>Birth and lactation (L6) |
|---|---|---|

## The hypothalamus

The **hypothalamus**, which forms the wall and floor of the third ventricle of the brain, lies above the **pituitary stalk**. This stalk leads down to the **pituitary gland** and carries the hypophyseal–pituitary blood supply. The hypothalamus contains the vital centers that control appetite and thirst, and autonomic activity such as the regulation of body temperature. It is actively involved in circadian rhythm, menstrual cycle activity, and response to stress, exercise and mood, through a cascade of processes controlled initially through the synthesis and release of **releasing hormones**. These in turn regulate the synthesis and release of **trophic hormones** in the pituitary gland, which have effects on target endocrine tissues. The hypothalamic releasing hormones are synthesized by neural cell bodies and are stored, prior to release, in the axonal terminals of the neurons located in capillary networks in the **median eminence** of the hypothalamus and lower infundibular stem.

Synthesis and secretion of hormones from endocrine tissues are controlled from higher centers in the brain, the hypothalamus, via the pituitary. Such a pathway enables tight regulation of hormone secretion via a series of **feedback loops** involving target hormone actions at both the hypothalamic and pituitary levels. These feedback loops can be disturbed acutely or chronically by a number of factors including stress, nutritional status, and systemic illness.

Hypothalamic hormones regulate hormone secretion by the pituitary. The mechanism of regulation can be somewhat complex since many hormones are produced by the hypothalamus, and a single hypothalamic hormone may affect more than one pituitary hormone. The action of the hormones is mediated by binding to specific hormone receptors in the target cell membrane followed by modulation of intracellular events via transmembrane signal transduction systems. The hypothalamic hormones may be split into two groups:

1. those that travel to the anterior pituitary from the median hypothalamic eminence via a unique system of portal veins (*Fig. 1*):
   - thyrotropin-releasing hormone (TRH)
   - gonadotropin-releasing hormone (GnRH)
   - corticotropin-releasing hormone (CRH)
   - growth hormone-releasing hormone (GHRH)
   - somatostatin (somatotropin release-inhibiting factor; SRIF)
   - prolactin-releasing hormone (PRH)
   - dopamine
2. those that travel down neurons to the posterior pituitary (*Fig. 1*). These hormones are synthesized in specialized neurons in the supraoptic and paraventricular nuclei of the hypothalamus. They are bound to glycoproteins and pass down the axons of the hypothalamopituitary tract through the pituitary stalk (**infundibulum**), and are stored in the distended parts of the axons in the posterior pituitary:
   - vasopressin (antidiuretic hormone; ADH)
   - oxytocin.

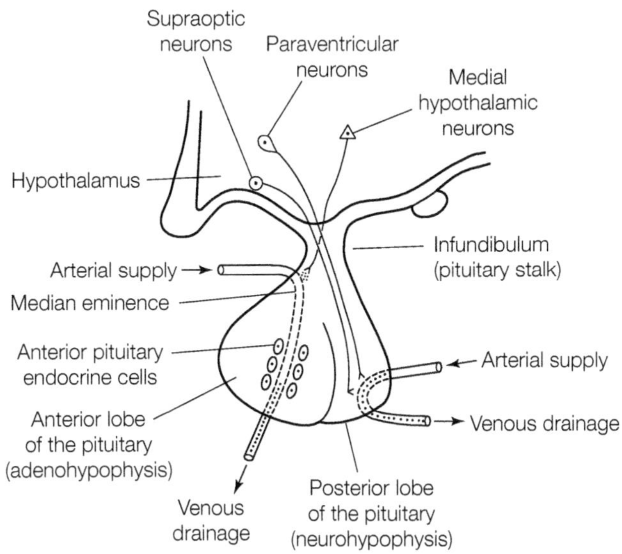


*Fig. 1. Arrangement of hypothalamic input to the two lobes of the pituitary. Neurons with cell bodies in the supraoptic and paraventricular nuclei of the hypothalamus have projections to the posterior pituitary, where they release their hormones directly into the venous drainage of the gland. Neurons with cell bodies in medial hypothalamic nuclei project to the median eminence, where they release their hormones into the pituitary portal veins. These hormones influence secretion of hormones from cells in the anterior lobe of the pituitary, into the venous drainage of the gland.*

**Leptin**, a hormone produced mainly by fat cells, circulates to the hypothalamus and regulates the satiety center and body weight. Circulating leptin levels are correlated with the amount of adipose tissue in the body. Elevated blood leptin concentrations lead to decreased food intake and increased energy expenditure, and low blood leptin concentrations, in contrast, lead to increased food intake and decreased energy expenditure. Leptin synthesis is increased by hyperglycemia and hyperlipidemia, and is decreased during starvation. The low blood leptin concentration during starvation may be the driving force for refeeding.

## TRH and PRH

Thyrotropin-releasing hormone (TRH) is a tripeptide (GluHisPro) released from a 26 kDa prohormone secreted in response to **norepinephrine** and dopamine. Its major roles are to stimulate the synthesis and secretion of **thyroid-stimulating hormone (TSH)** by the thyrotropic cells, and **prolactin** by the lactotropic cells of the **anterior pituitary**. This is the first step in the amplification process through which secretion of a small number of TRH molecules by the hypothalamus results eventually in the release of many thousands of molecules of **thyroxine** by the thyroid (see Topic K6).

A number of hypothalamic peptides other than TRH appear to have the ability to stimulate prolactin release. One of these is a 31-amino acid peptide given the trivial name prolactin-releasing hormone (PRH). Its importance in humans is unclear. Dopamine represents the predominant prolactin inhibitory factor secreted by the hypothalamus.

### GHRH and somatostatin

Growth hormone releasing-hormone (GHRH) is a 44-amino acid peptide which stimulates the secretion of **growth hormone** (GH) by the somatotropic cells of the anterior pituitary. Again, release of the hormone by the hypothalamus is feedback-inhibited by growth hormone, and this is accompanied by an increase in somatostatin release by the tissue.

Somatostatin is a 14-amino acid peptide derived from a 28-amino acid precursor. It antagonizes the stimulation of growth hormone secretion by GHRH in a concentration-dependent manner. The release of both stimulatory and inhibitory signals by the hypothalamus allows for a very fine control on GH release by the pituitary. It has been suggested that somatostatin also inhibits the release of TSH by the pituitary, but this is not very sensitive and its physiological relevance is unclear.

### CRH and GnRH

Corticotropin-releasing hormone (CRH) is a 41-amino acid peptide which stimulates the release of peptides derived from **pro-opiomelanocortin (POMC)** in the corticotropic cells of the anterior pituitary. These peptides, which are released in equimolar amounts, include **adrenocorticotropic hormone (ACTH)**, β-endorphin, γ-lipotropin and **α-melanocyte-stimulating hormone (α-MSH)**. Vasopressin is present in about half of the CRH-containing paraventricular neurons, and acts synergistically with it on ACTH release by the anterior pituitary. During stress, CRH is responsible for approximately 75% of the release of ACTH, while the remainder is due to vasopressin. In this condition it appears that the vasopressin-containing CRH neurons are selectively activated. Negative feedback on ACTH is exhibited by both cortisol, synthesized in the adrenal cortex, and ACTH and β-endorphin from the pituitary.

Gonadotropin-releasing hormone (GnRH) is a decapeptide, synthesized from a 92-amino acid precursor, whose major role is to stimulate the synthesis and secretion of both **luteinizing hormone (LH)** and **follicle-stimulating hormone (FSH)** by the gonadotropic cells of the anterior pituitary. LH and FSH are the **gonadotropins**. Differential rates of synthesis and secretion of the two hormones are mediated by differences in sensitivity to feedback control by a number of molecules including steroid and peptide hormones. GnRH is secreted in a pulsatile manner, allowing the number of GnRH receptors on the pituitary gonadotrophs to remain upregulated. The feedback control of the **hypothalamic–pituitary–gonadal axis** is described later in Topic K10. The feedback loops in this axis are found primarily in adults since gonadotropin and sex steroid concentrations are very low in children. This situation changes at puberty when gonadotropin and steroid hormone levels start to rise and trigger sexual development. After puberty, intense physical training or **anorexia nervosa** can lead to decreased secretion of GnRH and amenorrhea.

### Vasopressin and oxytocin

Although, as described above, vasopressin is synthesized in the hypothalamus, it is generally considered to be a hormone of the **posterior pituitary**. It acts on the collecting ducts in the kidney, facilitating water reabsorption and reducing urinary volume. It is for this property that it is also known as antidiuretic hormone (ADH). The sensing system (osmostat) that controls vasopressin release is located in the hypothalamus, anterior to the third ventricle. This center controls the release of vasopressin stored in the posterior pituitary. Inability to secrete vasopressin or a non-functioning receptor is accompanied by an excessive loss of urine (diabetes insipidus).

Vasopressin, synthesized in some neurons that terminate on the pituitary portal system, may be released into the capillaries and carried in the blood to the anterior pituitary. Here it can stimulate the release of growth hormone by the somatotrophs.

Although present in both men and women, a physiological role for oxytocin has only been described in women. It stimulates contraction of the uterus at birth facilitating delivery of the fetus. It is also essential for the secretion of milk from the **mammary gland**, being released from the posterior pituitary in response to suckling via afferent neural signals to the hypothalamus from the nipple. The hormone has a specific effect on myoepithelial cells around the alveolae and ductules, causing them to contract and eject milk (see Topic L6).

# K5 PITUITARY HORMONES

## Key Notes

**The pituitary gland**

The pituitary gland, which is attached to the hypothalamus by the pituitary stalk, represents the fusion of two embryologically distinct glands. The two parts, anterior and posterior, synthesize and secrete different hormone classes in response to quite separate stimuli. The anterior pituitary synthesizes a number of trophic hormones in response to stimulation by releasing factors from the hypothalamus. The posterior pituitary secretes only vasopressin (antidiuretic hormone) and oxytocin.

**TSH (thyrotropin)**

Thyroid-stimulating hormone (TSH) is a peptide hormone released in response to thyrotropin-releasing hormone from the hypothalamus. Its synthesis is under negative feedback control by thyroid hormones. It acts on the thyroid to stimulate the synthesis and secretion of thyroxine and triiodothyronine.

**FSH and LH (gonadotropins)**

Follicle-stimulating hormone (FSH) and luteinizing hormone (LH) are responsible for the secretion of estrogens and androgens from their target organs. Cyclical changes in the secretion of FSH and LH in the female cause the ovarian and uterine cycles that form the reproductive cycle. FSH initiates development of the ova and stimulates the ovaries to produce estrogens, while LH stimulates ovulation and secretion of progesterone by the corpus luteum. In males, FSH initiates sperm production in Sertoli cells, and LH stimulates testosterone production in the testes. Both FSH and LH are subject to long- and short-loop feedback control.

**ACTH (corticotropin)**

Adrenocorticotropic hormone (ACTH) is produced along with other small peptide hormones by proteolytic cleavage from a precursor protein, pro-opiomelanocortin, in response to corticotropin-releasing hormone. ACTH stimulates cortisol (glucocorticoid) and, to a lesser extent, aldosterone (mineralocorticoid) and androgen production in the adrenal cortex.

**Growth hormone (GH)**

Growth hormone secretion is controlled by stimulatory (growth hormone-releasing hormone) and inhibitory (somatostatin) signals from the hypothalamus. As its name implies, under anabolic conditions (raised insulin), growth hormone stimulates the release of insulin-like growth factor (IGF-1) from the liver, and acts either alone or with IGF-1 to promote protein synthesis in muscle and lipogenesis in adipose tissue (i.e. growth). However, when insulin levels are decreased (catabolic conditions) it induces insulin resistance and promotes lipolysis in adipose tissue.

**Prolactin**

The major role of prolactin is the induction and maintenance of milk production in late pregnancy and during lactation. It is produced by the lactotroph cells which, stimulated by estrogens, form the major pituitary cell type during pregnancy. Prolactin release is inhibited by dopamine and stimulated by TRH and vasoactive intestinal peptide from the hypothalamus. Suckling represents a physiological stimulus for prolactin release.

**Related topics**

Hypothalamic hormones (K4)
Thyroid hormones (K6)
Adrenal hormones (K8)
The gonads and sex hormones (K10)
Overview of the male reproductive system (L1)
Overview of the female reproductive system (L2)
Hormonal control of the menstrual cycle (L3)
Pregnancy (L5)
Birth and lactation (L6)

## The pituitary gland

The **pituitary gland** is a small oval-shaped organ weighing about 0.6 g and lying in a bony cavity (sella turcica, from the Latin meaning *Turkish saddle*) immediately below the third ventricle in the brain. It has direct neural and vascular communication with the **hypothalamus** via the **pituitary stalk** (see Topic K4, *Fig. 1*). The gland consists of two parts, anterior and posterior, which have quite

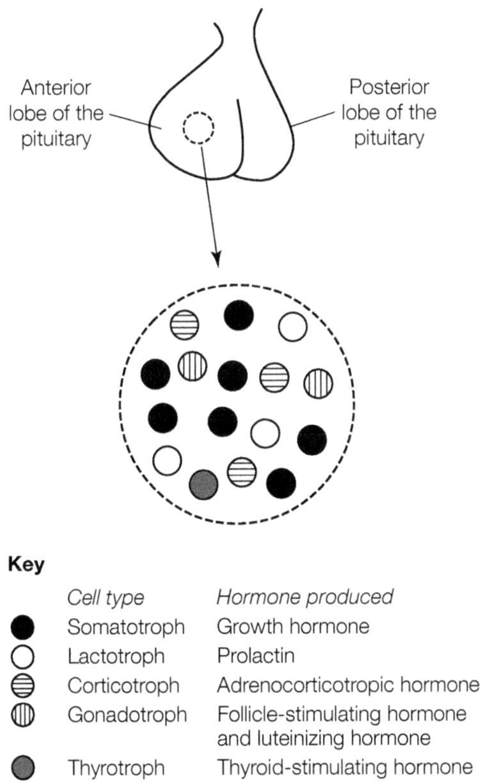


*Fig. 1. Cells of the anterior pituitary (adenohypophysis); the endocrine cells in the adenohypophysis have diverse phenotypes and influence several different hormone systems in the body.*

distinct embryological origins and functions. As described in topic K4, the posterior pituitary is in effect a component of the hypothalamus with direct neural connections from the supraoptic and paraventricular nuclei of that tissue, **vasopressin** and **oxytocin** being synthesized in the cell bodies of the neurons in these nuclei and traveling down the axons to the **posterior pituitary** from where they may be secreted into the bloodstream.

The **anterior pituitary**, which constitutes about 80% of the gland, is under the control of releasing hormones from the hypothalamus from where they are transported via the hypophyseal portal venous system. There are six major cell types in the anterior pituitary, and these are components of the five major endocrine axes associated with the pituitary (*Fig. 1*). These cells are responsible for the synthesis and secretion of trophic hormones: **thyroid-stimulating hormone (TSH)** by thyrotrophs (5%), **follicle-stimulating hormone (FSH)** and **luteinizing hormone (LH)** by gonadotrophs (10%), **adrenocorticotropic hormone (ACTH)** by corticotrophs (10–20%), **growth hormone (GH)** by somatotrophs (40–50%), and **prolactin** by lactotrophs (15–25%). The sixth cell type, the **folliculostellate cells**, do not contain secretory granules or secrete hormones in the same way as the other five cell types. They appear to produce growth factors that have paracrine effects on neighboring cells.

The glycoprotein hormones (TSH, LH, and FSH) of the anterior pituitary are heterodimers, consisting of common $\alpha$-subunits and similar but unique $\beta$-subunits. The $\alpha$- and $\beta$-subunits are encoded on different genes. Both subunits are glycosylated, and the specificity of action for each hormone thus lies in its $\beta$-subunit.

## TSH (thyrotropin)

TSH is secreted in a pulsatile fashion from the thyrotrophs in response to **thyrotropin-releasing hormone (TRH)**, and the synthesis and secretion of TSH is extremely sensitive to feedback inhibition by thyroid hormone. There is also a small diurnal rhythm in TSH secretion with a maximum at night. TSH stimulates growth of the **thyroid gland**, and also regulates the synthesis and secretion of thyroid hormones, **thyroxine ($T_4$)** and **triiodothyronine ($T_3$)**, in the thyroid gland. It acts by binding to its specific receptor, a member of the G-protein-coupled seven-transmembrane family, and stimulating the synthesis of cAMP. This rise in intracellular cAMP activates cAMP-dependent kinases, which regulate many steps in the synthesis and secretion of thyroid hormone:

- increased iodide uptake by the thyroid
- increased iodination of **thyroglobulin**
- increased lysosomal proteolysis of thyroglobulin
- increased thyroidal deiodination of $T_4$ to $T_3$
- increased secretion of $T_4$ and $T_3$ into the circulation.

## FSH and LH (gonadotropins)

The gonadotropins, released from the gonadotrophs in response to **gonadotropin-releasing hormone (GnRH)**, bind to their specific receptors (members of the G-protein seven-transmembrane family), and stimulate cyclic adenosine monophosphate (cAMP) production in the target cells. They are involved in the control of sexual differentiation, steroid hormone synthesis, and **gametogenesis**. Their roles in males and females are quite different. Thus in males, FSH and LH act in concert to induce **spermatogenesis**, FSH being involved in sperm maturation in the Sertoli cells and seminiferous tubules, and LH stimulating androgen production in the Leydig cells in the testes.

In females, FSH receptors are found on the **granulosa cells** of the ovary, and binding of FSH stimulates the biosynthesis of **estrogens**. LH stimulates the **thecal cells** in the ovary to produce **androgens** and steroid precursors, which are then transported to the granulosa cells for aromatization which is the final step in estrogen biosynthesis. The specific secretory pattern of FSH and LH during the **menstrual cycle** is described in Topic L3.

## ACTH (corticotropin)

ACTH is a 39-amino acid peptide derived from a 241-amino acid precursor, **pro-opiomelanocortin (POMC)**. As mentioned in Topic K4, hydrolysis of the precursor peptide gives rise to equimolar amounts of several other peptides including γ-lipotropin, β-endorphin and **α-melanocyte stimulating hormone (α-MSH)**. Initial cleavages of the POMC precursor are made by proteases at positively charged amino acid residues, and the peptides are trimmed to produce the final products. Different corticotropic cells express different proteases to produce equimolar amounts of different peptide hormones. Those in the anterior lobe of the pituitary produce only ACTH and β-lipotropin, while γ-lipotropin, β-endorphin and α-MSH are produced in cells of the intermediate lobe. These peptides are released from the pituitary in response to stimulation by **corticotropin-releasing hormone (CRH)** from the hypothalamus. ACTH secretion exhibits a diurnal variation with a peak at about 0500 h, but is also increased by stress, both psychological and physical. Factors affecting its secretion are listed in *Table 1*.

*Table 1. Factors regulating CRH and ACTH secretion*

| Factor | Effect on CRH and ACTH secretion |
|---|---|
| Cortisol | ↓CRH and ↓ACTH secretion |
| ACTH and β-endorphin | ↓CRH secretion |
| Morphine | Suppresses ACTH response to CRH in man |
| ACh, DOPA, epinephrine, norepinephrine | ↑CRH secretion |
| GABA | ↓CRH secretion |
| Epinephrine, norepinephrine | ↑ACTH secretion |

ACh, acetylcholine; ACTH, adrenocorticotropic hormone; CRH, corticotropin-releasing hormone; DOPA, 3,4-dihydroxyphenylacetic acid; GABA, γ-aminobutyric acid.

The biological activity of ACTH resides within amino acids 1–18 of the molecule, and it exerts its effects on the target cells of the adrenal cortex through binding to a receptor of the G-protein-coupled seven-transmembrane-spanning class. Its primary acute effect is to stimulate the hydrolysis of stored cholesterol ester to provide cholesterol for the synthesis of steroid hormones. It also stimulates the action of the 20,22-desmolase, the rate-limiting enzyme responsible for the cleavage of a six-carbon aliphatic chain from cholesterol. This generates **pregnenolone**, the common precursor for the synthesis of all the steroid hormones (see Topic K2). Chronically, ACTH induces the transcription of genes encoding the enzymes of the steroidogenic pathway, thereby increasing the capacity of the adrenal cortex for **cortisol** production. It also has some stimulatory action on the secretion of adrenal **androgens** and mineralocorticoids. Regulation of ACTH synthesis and release is via the **hypothalamic–**

**pituitary–adrenal axis** with short- and long-loop feedback. ACTH is secreted from the pituitary in discrete pulses, and the response of ACTH to stimulation by CRH depends on the circulating concentrations of cortisol. This hormone, the end-product of the hypothalamic–pituitary–adrenal axis, inhibits CRH production in the hypothalamus as well as inhibiting both *POMC* gene expression and ACTH secretion in the pituitary. ACTH itself induces a diurnal variation in cortisol secretion with the highest plasma levels of cortisol occurring in the early morning, and lowest levels in the late afternoon to early evening. Trauma, either psychological or physical, is a major effector of ACTH secretion, as is **hypoglycemia**. In this latter case, increased ACTH leads to increased cortisol release by the adrenal cortex, which in turn contributes to an increase in **gluconeogenesis** (see Topic K8). Chronic stimulation by CRH causes adrenal hyperplasia, and excessive ACTH production, from **pituitary adenomas**, for example, results in **Cushing's disease**.

## Growth hormone (GH)

Large amounts (5–10 mg) of GH are stored in the pituitary, and about 5% of this is released in a pulsatile manner each day in response to a number of hypothalamic peptides, neurotransmitters, steroid hormones, glucose, and growth factors. The major factors controlling GH secretion are **growth hormone-releasing hormone (GHRH)**, which is stimulatory, and **somatostatin**, which is inhibitory, as mentioned in Topic K4. Although GH is secreted in bursts every 3–4 h, the major burst in secretion occurs at night, and is associated with slow-wave sleep.

GH exists in blood in a number of variant forms, the predominant one being a 191-amino acid (22 kDa) protein, which circulates bound to a 29 kDa protein having identical sequence homology with the extracellular domain of the GH receptor. Binding to this protein helps to prolong the plasma half-life of the hormone. The action of insulin-like growth factor is discussed below.

The GH peptide has two receptor-binding domains that allow it to bind to two separate receptor molecules on the target cell surface, thereby inducing dimerization of the receptor. Dimerization of the receptors causes activation of an intracellular signaling cascade via **tyrosine kinase** activation.

The actions of GH on a target tissue may be direct, through binding to its receptor, or indirect through the action of **insulin-like growth factor-1 (IGF-1)**, whose production it stimulates in many tissues. In some cases it is difficult to distinguish between the actions of the two hormones. Most of the circulating IGF-1 derives from the liver.

As the name implies, GH is an anabolic hormone, promoting amino acid uptake into muscle, and protein synthesis. However its anabolic properties are manifest only in the presence of raised insulin concentrations. Under these conditions it also induces **lipogenesis** in adipose tissue, such that lean body mass is increased and nitrogen balance is positive. However, in the presence of **hypoglycemia**, when insulin levels are decreased, GH stimulates **lipolysis** in adipose tissue and induces peripheral **insulin resistance**. This has the effect of providing fatty acids as a respiratory substrate for most tissues, and sparing blood glucose for the brain and erythrocytes which have an obligatory requirement for it. Furthermore, GH increases the rate of hepatic gluconeogenesis by increasing the expression of key gluconeogenic enzymes, thereby helping to relieve the hypoglycemia.

GH also has indirect metabolic actions through its stimulation of IGF-1 synthesis. IGF-1 is a 70-amino acid peptide bearing much sequence homology with **proinsulin**. Stimulation of its synthesis by GH in a tissue causes an increase in growth of that tissue. Its major function appears to be to control GH secretion by feedback inhibition at the pituitary. IGF-1 also stimulates linear growth by increasing the proliferation of chondrocytes, and the synthesis of cartilage matrix in skeletal tissues.

## Prolactin

Prolactin is a 23 kDa protein that shares sequence homology with GH, suggesting that they are derived from a common ancestral gene. It is secreted from the lactotropic cells of the pituitary in response to TRH and **vasoactive intestinal peptide (VIP)**, but the major controlling influence is **dopamine**, which is inhibitory. Estrogen has a proliferative effect on the lactotrophs, and thus females have a higher number of these cells than men, and indeed, during pregnancy they constitute almost 70% of the cells of the pituitary. A prolactin-releasing peptide (**prolactin-releasing hormone, PRH**) from the hypothalamus also stimulates prolactin release but its role *in vivo* has not yet been established. Like many of the other pituitary hormones described above, prolactin is released at low levels in a pulsatile manner in both males and females, with increases in plasma concentration occurring during sleep and stress. Its primary role, however, is during pregnancy, where it binds to its receptor in the mammary gland and, in concert with estrogens, insulin and cortisol, it induces the synthesis of milk proteins and milk fat.

Prolactin can also inhibit the release of GnRH from the hypothalamus, thereby reducing the synthesis and secretion of FSH from the pituitary, and suppressing **ovulation**.

# K6 THYROID HORMONES

## Key Notes

**$T_3$ and $T_4$**

The two thyroid hormones, thyroxine ($T_4$) and triiodothyronine ($T_3$) are released in response to the action of thyroid-stimulating hormone (TSH) on the thyroid. Although $T_3$ is the active hormone, $T_4$ is the major circulating form, and both are transported bound to plasma proteins, thyroid-binding globulin and transthyretin. $T_4$ undergoes deiodination to $T_3$ in peripheral tissues.

**Synthesis of thyroid hormones**

$T_4$ and $T_3$ are synthesized by condensation of two iodinated tyrosine residues of thyroglobulin (TG), a precursor protein synthesized in the thyroid cells and stored as colloid in the follicular lumen. Tyrosine residues in the protein are iodinated post-translationally. Formation of $T_4$ requires two di-iodotyrosine residues. Formation of $T_3$ requires one di-iodotyrosine residue, and one mono-iodotyrosine residue. Proteolysis of TG in response to TSH releases $T_4$ and some $T_3$ into the circulation.

**Actions of thyroid hormones**

$T_3$ is required for growth and development; absence in the neonate gives rise to cretinism. $T_3$ also increases basal metabolic rate and thermogenesis. This involves stimulation of lipolysis, glycogenolysis, and gluconeogenesis to provide substrates for oxidation in mitochondria. $T_4$, the major circulating form of the hormone, exerts precise negative feedback on the release of TSH by the pituitary.

**Calcitonin**

Calcitonin is a peptide hormone produced by the parafollicular C cells of the thyroid in response to an acute increase in plasma calcium. In pharmacological doses it acts to reduce plasma calcium levels by inhibiting resorption of bone, and by promoting uptake of calcium and phosphate into bone. However, calcitonin supplementation is not required in patients who have undergone total thyroidectomy, and its physiological role in calcium homeostasis is unclear.

**Related topics**

Metabolic processes (B3)
Bone and muscle (C5)
Hypothalamic hormones (K4)
Pituitary hormones (K5)
Parathyroid hormone (K7)

## $T_3$ and $T_4$

The **thyroid hormones** (*Fig. 1a*), **thyroxine ($T_4$)** and **triiodothyronine ($T_3$)**, are released from the cells of the gland in response to signaling from the **hypothalamus** and **pituitary gland**. The biologically active form of the hormone is $T_3$. Control of thyroid hormone secretion is exerted via a feedback loop involving the hypothalamus, **anterior pituitary**, and thyroid gland. **Thyrotropin-releasing hormone (TRH)** from the hypothalamus triggers the release of thyrotropin (**thyroid-stimulating hormone, TSH**) from the pituitary which, in turn, stimulates the formation and release of $T_4$ and $T_3$ from the thyroid (*Fig. 1b*). Thyroid

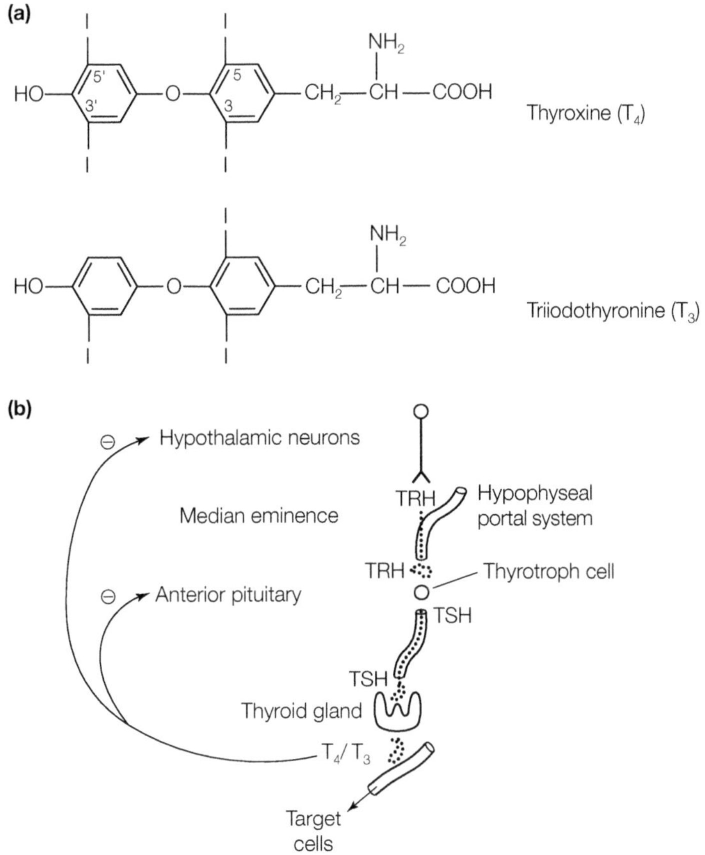


*Fig. 1. (a) Structures of thyroxine ($T_4$) and triiodothyronine ($T_3$); (b) the hypothalamic–pituitary–thyroid axis. TRH, thyrotropin-releasing hormone; TSH, thyroid-stimulating hormone.*

hormone feeds back on both the pituitary and hypothalamus, serum TSH levels being inversely log-linearly related to those of $T_4$. Thus, small increases/decreases in free $T_4$ produce marked decreases/increases in TSH levels.

## Synthesis of thyroid hormones

The synthesis of thyroid hormones by the **follicular cells** requires that the gland be capable of concentrating iodide from the circulation. Iodine, which constitutes 64% of the weight of $T_4$, is a scarce element in the earth's crust, and although optimal daily intake is 150–300 μg $day^{-1}$, dietary intake varies widely throughout the world (20–700 μg $day^{-1}$). Inadequate intake of iodine gives rise to endemic **goiter**, and is the rationale for the iodination of salt in the UK and USA. Dietary iodine is reduced in the gut to iodide and is rapidly absorbed in this form. It is cleared from the circulation either through the urine or by concentration by the thyroid. The efficiency of clearance can be seen by the fact that of the 50 mg of iodine found in the body, about 15 mg is present in the thyroid, and the thyroid cell to plasma concentration ratio is around 30:1. Thyroid clearance is inversely proportional to plasma concentration. The iodide concentrated in the thyroid is rapidly incorporated into **thyroglobulin**, a 660 kDa dimeric protein stored extracellularly in the follicular lumen of the tissue as **thyroid colloid**. The N-terminal of thyroglobulin is enriched in tyrosine residues and, under the action of a peroxidase at the apical border, iodide is oxidized and incorporated into tyrosine residues, the process of organification of iodine. This produces a protein in which some tyrosine residues have one or two iodines (**mono-iodotyrosine (MIT)** or **di-iodotyrosine (DIT)**). Coupling of two of these residues during degradation of thyroglobulin, in response to TSH, yields $T_4$ and $T_3$, a process inhibited by thioureas (*Fig. 2*).

When there is a demand for thyroid hormone, the thyroglobulin is brought back into the epithelial cells by endocytosis and degraded by lysosomal enzymes to release $T_4$ and $T_3$. Approximately 25 mg $day^{-1}$ of thyroglobulin undergoes proteolysis, yielding about 110 nmol (90 μg) of thyroid hormones. Any MIT or DIT not conjugated to produce thyroid hormones is deiodinated to conserve and recycle the iodine. $T_4$ is the major secretory product from the thyroid (approximately 80–100 μg $day^{-1}$), constituting about 80% of the secretion of thyroid hormones from the tissue. Only 20% of the $T_4$ is deiodinated in the thyroid gland itself.

Thyroid hormones are transported in the circulation bound to specific plasma proteins; $T_4$ is bound to **thyroid-binding globulin (TBG)** and **transthyretin** (also called thyroid binding prealbumin), while $T_3$ binds only to TBG. $T_3$ is less avidly bound to TBG than $T_4$. The protein-bound forms of the two hormones are inert, but binding increases the plasma half life of the hormone ($t_{1/2}$: $T_4$ ~6–7 days; $T_3$ ~1 day). The biologically important fraction of $T_4$ and $T_3$ is the free, non-protein-bound fraction which usually represents <1% of the plasma total, and the active form of the hormone is $T_3$. Thus $T_4$ has to be deiodinated to $T_3$, and while this can occur to some extent in the thyroid, 80% of $T_4$ is deiodinated in peripheral

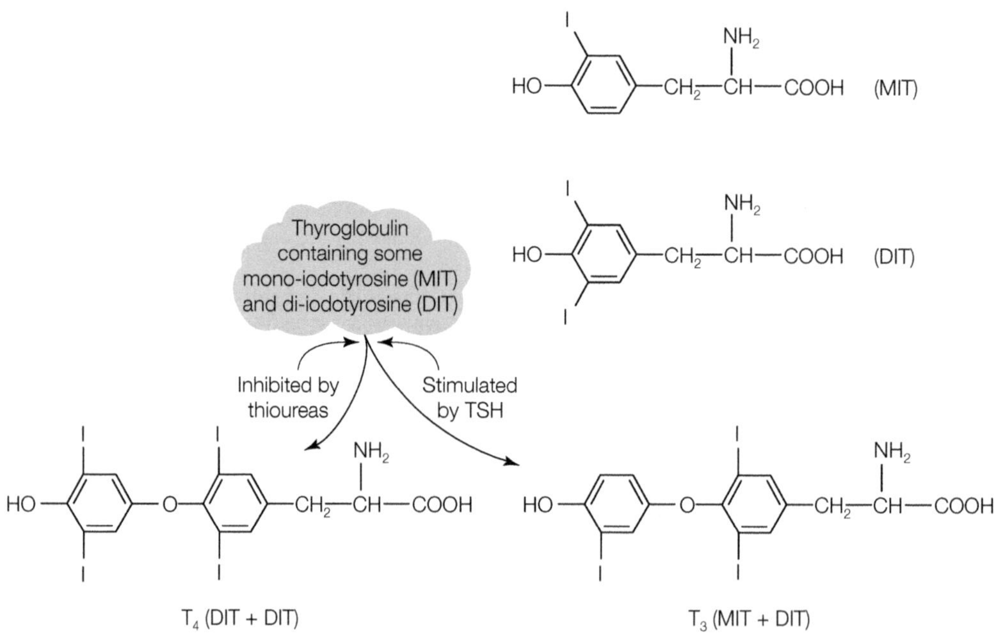


*Fig. 2. Thyroid hormones are synthesized from the hydrolysis of thyroglobulin to produce thyroxine ($T_4$) and triiodothyronine ($T_3$). DIT, di-iodotyrosine; MIT, mono-iodotyrosine.*

tissues, principally liver and kidney. Less than 0.02% of the plasma $T_4$ is in the free form, while 0.3% of the $T_3$ is free. 5′-deiodination of $T_4$ yields $T_3$ (*Fig. 1a*), while 5-deiodination yields inactive, **reverse-$T_3$ ($rT_3$)**. Under normal circumstances, $T_4$ is metabolized to equal amounts of $T_3$ and $rT_3$. However, preferential deiodination can occur. For example, exposure to cold leads to increased production of $T_3$ and decreased $rT_3$, while hepatic production of $rT_3$ may be elevated in patients with systemic illnesses or undergoing major surgery. Also, during pregnancy, the fetus is intrinsically hypothyroid, obtaining a little $T_4$ from the maternal circulation. The fetal **hypothalamic–pituitary–thyroid** axis does not become active until the second trimester. This does not present a problem unless the mother too is hypothyroid. The percentage conversion of fetal $T_4$ to inactive $rT_3$ is high until birth, when there is an immediate switch to the production of $T_3$ and this rises during week one after birth and steadily through infancy. $T_3$ is inactivated by deiodination or by conjugation and biliary excretion.

## Actions of thyroid hormone

At the target cell, $T_3$ dimerizes with a retinoid prior to translocation to the nucleus. Here it binds to specific DNA sequences adjacent to genes regulated by $T_3$, and influences the expression of tissue-specific genes. The biochemical responses to $T_3$ are:

- raised **basal metabolic rate**
- increased **thermogenesis** through increased mitochondrial oxidative metabolism to supply sufficient adenosine triphosphate to meet the demand for increased sodium pump activity
- higher rates of **lipolysis** through activation of hormone-sensitive lipase to supply free fatty acids for mitochondrial oxidation
- increased rates of **glycogenolysis** and **gluconeogenesis** to supply glucose for thermogenesis.

Thyroid hormone is also essential for growth and development, especially post partum growth and brain development. Absence of thyroid hormones in the first few months of life leads to **cretinism**, and tests for hypothyroidism are made in neonatal screening procedures. It also influences cardiac function through interactions with the sympathetic nervous system.

## Calcitonin

**Calcitonin** is a 32-amino acid peptide (3.5 kDa) secreted by the parafollicular 'C' cells of the **thyroid gland**. The initial translation product (17 kDa) undergoes co- and post-translational hydrolysis (see Topic K2) to yield the parent hormone. Secretion of calcitonin is stimulated by a rise in plasma calcium and the gut hormone, **gastrin**. The physiological role of calcitonin is unclear, since patients with either a very high or zero plasma calcitonin appear to have no obvious skeletal abnormality or altered **calcium homeostasis**.

However, calcitonin does have a transient inhibitory effect on osteoclastic **bone resorption** (Section C5). This action has led to the suggestion that the arrival of food in the stomach stimulates the release of gastrin from the G cells of the gastric mucosa (see Topic J3), and that this in turn stimulates calcitonin release from the thyroid. A rise in plasma calcitonin leads to a transient hypocalcemia through inhibition of calcium efflux from bone. This transient hypocalcemia will in turn stimulate secretion of **parathyroid hormone (PTH)** from the **parathyroid glands**. The overall effect will be the retention within the body of dietary calcium absorbed through the intestine, since raised plasma PTH will increase renal calcium reabsorption and prevent loss of calcium in the urine (see Topic K7). Such a series of events allows the body to achieve calcium homeostasis. Calcium absorbed will balance that retained within the bone, due to the action of calcitonin, and extracellular fluid calcium homeostasis will be restored.

Pharmacologically, calcitonin is used to inhibit osteoclastic bone resorption in conditions such as **Paget's disease** and **osteoporosis**. At pharmacological concentrations, calcitonin inhibits calcium and phosphate reabsorption in the kidney. This action of calcitonin is also made use of in the treatment of **hypercalcemia** of malignancy.

# K7 Parathyroid hormone

## Key Notes

**Parathyroid hormone**

Parathyroid hormone (PTH) is released from the parathyroid glands in response to a fall in plasma calcium. Its principal acute action is to restore calcium homeostasis by increasing calcium reabsorption in the distal tubules of the kidney, and the movement of calcium from bone extracellular fluid into bulk extracellular fluid. The former effect is accompanied by increased urinary excretion of phosphate. PTH also upregulates renal 25-hydroxy vitamin D 1α-hydroxylase, the enzyme responsible for synthesis of the active form of vitamin D (1,25-dihydroxy vitamin D), which leads to increased absorption of dietary calcium. Synthesis and secretion of PTH is subject to feedback control by raised plasma calcium and 1,25-dihydroxy vitamin D.

**Parathyroid hormone-related protein**

Structurally similar to PTH, parathyroid hormone-related protein (PTHrP) is thought to be involved in the differentiation of chondrocytes and skin cells in the embryo. PTHrP binds to PTH receptors on the osteoblast, and thereby increases osteoclast differentiation, bone resorption and hypercalcemia. Ectopic production of PTHrP, associated with squamous cell carcinomas, can lead to hypercalcemia of malignancy.

**Related topics**

Homeostasis and integration of body systems (A2)
Bone and muscle (C5)
Thyroid hormones (K6)
Reabsorption of electrolytes and glucose (M3)

## Parathyroid hormone

**Parathyroid hormone (PTH)** is important in **calcium homeostasis**, and is synthesized and secreted by the **parathyroid glands** in response to a fall in plasma calcium. The initial translation product undergoes cotranslational cleavage of a 25-amino acid **signal peptide** during translocation of the protein into the lumen of the endoplasmic reticulum. Further cleavage of a 6-amino acid fragment during packaging into secretory vesicles yields the parent hormone (84 amino acids; 10 kDa). Interestingly it is the N-terminal 1–34-amino acid fragment that is responsible for the biological activity of the hormone, and it is likely that this fragment is formed during circulation.

Release of PTH is exquisitely sensitive to an individual's set-point calcium, generally in the range 2.2–2.5 mM. Its actions serve to increase plasma calcium concentration both acutely, through actions on the kidney and bone, and chronically, through indirect effects on the gut, mediated via **1,25-dihydroxy vitamin D (1,25$(OH)_2$ vitD)**. Greater than 95% of the non-protein-bound calcium filtered at the glomerulus is subsequently reabsorbed by a hormone-independent mechanism by the renal tubules, predominantly in the proximal nephron (see Topic M3). PTH critically promotes **calcium reabsorption** in the distal tubules of the kidney. This is accompanied by a **phosphaturia** due to increased phosphate excretion proximally. PTH acting with $1,25(OH)_2$ vit D will also increase the movement of calcium from bone extracellular fluid into bulk extracellular fluid through actions on the bone-lining **osteoblast** cells (see Topic C5). These two actions will cause an acute rise in plasma calcium.

A further renal action of PTH is the stimulation of **25-hydroxy vitamin D 1α-hydroxylase**, probably through depletion of intracellular phosphate. This enzyme converts the **25-hydroxy vitamin D**, formed by hydroxylation of vitamin D in the liver, to $1,25(OH)_2$ vit D. This molecule acts as a steroid hormone and is carried in the blood bound to **vitamin D-binding protein**. Its major action is in the small intestine where it increases the synthesis of a brush border calcium transport protein leading to increased absorption of dietary calcium. A return to a plasma calcium in the normal range will inhibit PTH secretion.

Interrelationships of various hormones in the control of serum calcium are shown in *Table 1*.

*Table 1. Hormones affecting calcium metabolism*

| Hormone | Target organ | Action |
|---|---|---|
| Parathyroid hormone | Kidney, proximal and distal tubules | ↓Reabsorption of phosphate |
| | Kidney, distal tubule | ↑Reabsorption of calcium |
| | Kidney, proximal tubule | Activates 1α-hydroxylase (↑plasma1,$25(OH)_2$ vitD) |
| | Bone | ↑Calcium and phosphate resorption |
| $1,25(OH)_2$ vitD | Small intestine | ↑Calcium absorption |
| | Bone (with PTH) | ↑Calcium and phosphate resorption |
| Calcitonin | Bone | ↓Calcium and phosphate resorption[a] |

[a]At pharmacological dose; physiological action unclear.

## Parathyroid hormone-related protein

**Parathyroid hormone-related protein (PTHrP)** is synthesized in three isoforms (139, 141 and 173 amino acids), all of which share sequence homology with PTH, particularly in the biologically active N-terminal 1–34 amino acids (e.g. 8 of the first 13 amino acids are identical). This structural similarity allows PTHrP to bind to the PTH receptor and elicit biological actions similar to PTH. However, secretion of PTHrP is not subject to feedback control by a rise in plasma calcium. Although it has been suggested that PTHrP might be involved in ensuring calcium supply to the fetus, the placenta being a source of PTHrP, there is no evidence as yet for its involvement in adult calcium homeostasis. However, it does appear to play a role in dermatologic and chondrocytic differentiation in the embryo. Certain types of cancer (e.g. squamous cell tumors of the head and neck and those derived from kidney, breast and lymphoid tissue) secrete PTHrP and, by stimulating bone resorption by **osteoclast** cells, contribute to the hypercalcemia of malignancy.

# K8 ADRENAL HORMONES

## Key Notes

**Adrenal gland structure**

The adrenal glands are two organs lying above each kidney and consist of two quite separate parts, an outer cortex and an inner medulla. The cortex, which represents the major site of steroid hormone biosynthesis, is in turn divided into three zones; the outer zona glomerulosa secretes mineralocorticoids, principally aldosterone; the intermediate zona fasciculata secretes corticosteroids, principally cortisol; and the inner zona reticularis secretes small amounts of androgens, principally dehydroepiandrosterone and androstenedione. The adrenal medulla is the major site of synthesis of catecholamines, principally epinephrine and norepinephrine, which are stored in vesicles (chromaffin granules) and released upon sympathetic stimulation.

**Biosynthesis of steroid hormones**

All steroid hormones are synthesized from cholesterol, stored in the tissue as cholesterol ester. This cholesterol comes from low-density lipoprotein delivered to the tissue or from *de novo* synthesis in the tissue itself. The common intermediate for steroid hormone synthesis is pregnenolone, formed by loss of part of the aliphatic side chain of cholesterol, a reaction stimulated by adrenocorticotropic hormone. Further metabolism to the final steroid product involves hydroxylations catalyzed by cytochrome $P_{450}$-linked hydroxylases, and modifications to the aliphatic rings.

**Actions of adrenal cortical hormones**

Glucorticoids are transported in the blood bound to a cortisol-binding globulin (transcortin) synthesized in the liver. They are released as part of the stress response (physical and/or psychological) and exert anti-insulin effects, promoting the availability of energy-providing substrates – glucose, fatty acids, and amino acids. This is done by stimulation of the breakdown of glycogen (liver), triglyceride (adipose tissue) and protein (muscle). Cortisol also promotes gluconeogenesis and the centripetal redistribution of fat, the pear-shaped physique characteristic of Cushing's syndrome. The major action of aldosterone is in the kidney, where it acts to control sodium and water balance. The androgens secreted by the zona reticularis are transported by a sex-hormone-binding globulin (produced in the liver) to the gonads, where they are further metabolized to testosterone. Only a little testosterone is formed in the adrenal cortex.

**Biosynthesis of epinephrine**

Phenylalanine is the precursor for the biosynthesis of both norepinephrine and epinephrine in the chromaffin cells of the adrenal medulla. The biosynthetic pathway involves two hydroxylations of the aromatic ring to form 3,4-dihydroxyphenylalanine (DOPA) and decarboxylation of DOPA to dopamine. A further hydroxylation forms norepinephrine, which undergoes methylation to epinephrine. The catecholamines are stored complexed with magnesium and adenosine triphosphate (ATP) in vesicles, prior to release.

**Actions of epinephrine**

Epinephrine is released in response to acute stress and increases blood pressure and heart rate. Metabolically, it stimulates the breakdown of body energy stores to supply glucose from hepatic glycogen, fatty acids from adipose tissue triglyceride, and lactate from muscle glycogen via glycolysis, for immediate use by tissues.

**Related topics**

The cardiac cycle, blood pressure, and its maintenance (D6)
Structure and function of the autonomic nervous system (H5)
Autonomic pharmacology (H6)
Hypothalamic hormones (K4)
Pituitary hormones (K5)
Reabsorption of electrolytes and glucose (M3)
Hormonal control of fluid balance (M5)

## Adrenal gland structure

The **adrenal glands** consist of two types of endocrine tissue, the outer **adrenal cortex** and inner **adrenal medulla**, which are quite distinct both developmentally and functionally, having different embryological origins. They can be considered as two separate endocrine glands. The adrenal cortex arises from mesoderm and, in response to stimulation by **adrenocorticotropic hormone (ACTH)** from the pituitary (Topic K5), synthesizes and secretes **steroid hormones**. It is thus part of the **hypothalamic–pituitary–adrenal** endocrine system. The common precursor for the steroid hormones is cholesterol. The adrenal medulla is part of the sympathetic nervous system, arising from neural crest ectoderm. It synthesizes and secretes **epinephrine**.

Three classes of steroid hormone are secreted by the adrenal cortex:

- **mineralocorticoids** (e.g. **aldosterone**) contribute to the control of electrolyte and fluid homeostasis
- **glucocorticoids** (e.g. **cortisol**) have effects on carbohydrate, fat and protein metabolism
- **androgens** (e.g. **dehydroepiandrosterone**): small quantities are secreted and supplement sex hormone secretion by the gonads.

Three zones of the adrenal cortex can be identified histologically (*Fig. 1*).

The **zona glomerulosa** secretes mineralocorticoids, particularly aldosterone, which regulates sodium and potassium levels through actions on the sodium pump in the kidney tubules (see Topic M5). Aldosterone is also involved in the regulation of blood pressure via the renin–angiotensin system (see Topic D6). Thus, aldosterone promotes sodium retention and urinary potassium loss, leading to an increase in total body sodium and blood pressure within a few hours. In contrast to the synthesis of the glucocorticoids and sex hormones, the major stimulator of aldosterone synthesis is angiotensin II, and synthesis of aldosterone is controlled by the renin–angiotensin system.

The **zona fasciculata**, the intermediate and broadest of the three zones of the cortex, secretes glucocorticoids, principally cortisol (the most important $C_{21}$ steroid). The major metabolic actions of cortisol are summarized in *Tables 1* and *2*. Released in response to **stress**, both physical and psychological, it serves to increase the supply of oxidizable substrates to tissues. Control of cortisol secretion is maintained through the hypothalamic–pituitary–adrenal axis, and immediately via ACTH and long- and short-feedback loops. ACTH secretion is

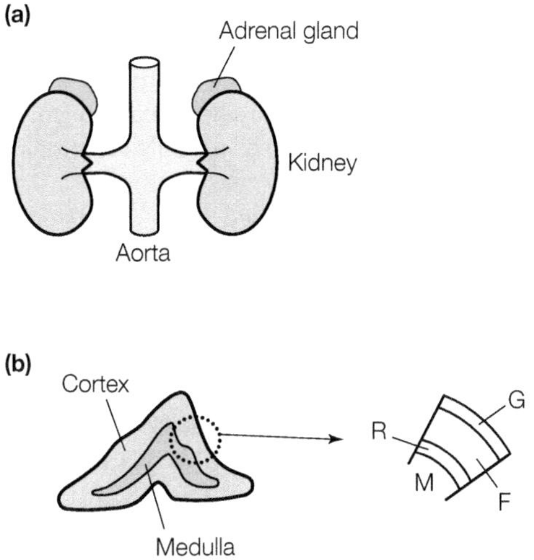


*Fig. 1. (a) Position of the adrenal glands; (b) subdivisions of the adrenal glands. F, zona fasciculata; G, zona glomerulosa; M, medulla; R, zona reticularis.*

*Table 1. Actions of glucocorticoids*

| Increased/stimulated | Decreased/inhibited |
|---|---|
| Gluconeogenesis | Protein synthesis |
| Glycogen degradation | Host response to infection |
| Protein catabolism | Lymphocyte transformation |
| Fat deposition | Delayed hypersensitivity |
| Sodium retention | Circulating lymphocytes |
| Potassium loss | Circulating eosinophils |
| Free water clearance | |
| Uric acid production | |
| Circulating neutrophils | |

*Table 2. Major metabolic actions of glucocorticoids*

| Tissue | Response |
|---|---|
| Liver | ↑GNG → liver glycogen; ↑amino acid breakdown (anti-anabolic) |
| Peripheral tissues | ↓Insulin response; ↓protein synthesis (anti-anabolic) |
| Lymphoid tissues | ↓Inflammatory response; ↓prostaglandin synthesis |
| General | ↓Stress response |

GNG, gluconeogenesis.

stimulated by **corticotropin-releasing hormone** from the hypothalamus (see Topic K4). There is also a circadian variation in the plasma cortisol concentration, with highest levels in the morning and lowest at around midnight. The zona fasciculata may also secrete small amounts of sex hormones.

The **zona reticularis**, the innermost zone of the adrenal cortex, may also secrete small amounts of sex hormones, chiefly the androgens **dehydro-epiandrosterone (DHA)** and **androstenedione**. This may, however, reflect storage of hormone synthesized in the zona fasciculata, and *in vivo* the two zones identified histologically may be a single functional unit.

**Biosynthesis of steroid hormones**

All steroid hormones are synthesized from **cholesterol** stored in the tissue as **cholesterol ester**. This cholesterol is derived predominantly from low-density lipoprotein-borne cholesterol ester or from *de novo* synthesis in the tissue itself. The first step in steroid hormone synthesis thus requires the stimulation of a **cholesterylesterase** to release free cholesterol. An outline of the biosynthetic pathway of steroid hormones is shown in *Fig. 2*. Free cholesterol is transported into the mitochondrion for further metabolism involving shortening of the aliphatic side-chain and subsequent hydroxylations. The common intermediate in steroid synthesis is **pregnenolone**, formed in three steps catalyzed by the mitochondrial enzyme, **desmolase**. This rate-limiting step, which leads to loss of

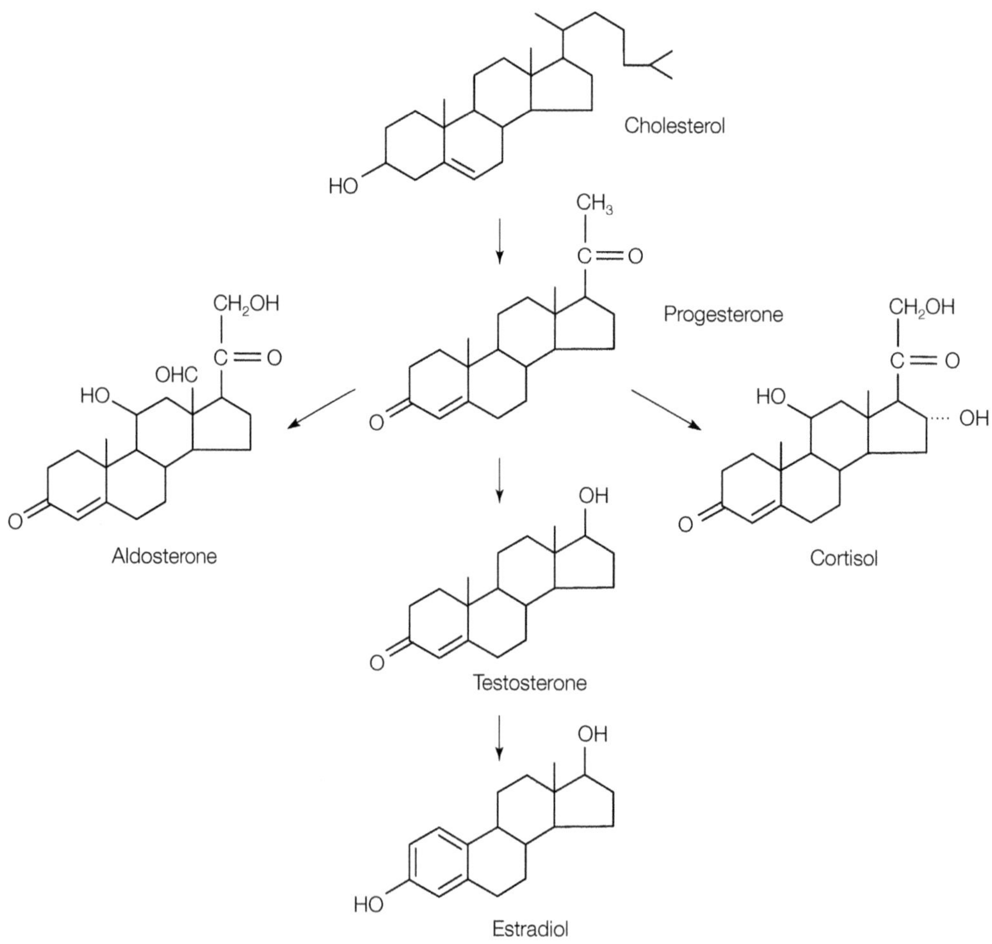


*Fig. 2. Biosynthesis of steroid hormones.*

a six-carbon fragment is stimulated by ACTH and involves **cytochrome $P_{450}$-linked mixed function hydroxylase** activity. Further shuttling of the intermediates of steroid synthesis in and out of the mitochondrion continues, because some of the hydroxylases (C-17, C-21 and C-11 for cortisol; C-18 for aldosterone) that catalyze subsequent metabolic steps are cytosolic, while others are mitochondrial. Androgens like testosterone are formed after removal of the side-chain to produce C-19 steroids. The zone-specific distribution of the enzymes required for the synthesis of the separate classes of steroid hormone allows for the independent control systems mentioned above: the **renin–angiotensin system** for aldosterone and ACTH for cortisol.

## Actions of adrenal cortical hormones

The major naturally occurring glucocorticoids are cortisol and **corticosterone**, and both exert anti-insulin effects on intermediary metabolism. Thus, they promote **gluconeogenesis** in the liver, **proteolysis** in muscle, and **lipolysis** in adipose tissue. In homeostasis, cortisol helps in the maintenance of extracellular fluid volume and normal blood pressure. In excess however, such as in **Cushing's disease** or when given as part of tissue transplantation therapy, cortisol leads to impaired glucose tolerance, redistribution of adipose tissue, and fluid retention. Cortisol is transported in the circulation bound to a specific, hepatically synthesized globulin, **cortisol-binding globulin (CBG, transcortin)**. This protein, a 59 kDa $\alpha_2$-globulin, protects circulating cortisol from hepatic clearance and gives cortisol a relatively long plasma half-life of 60–80 min. Under normal circumstances about 95% of cortisol is bound to the protein, leaving the unbound 5% as the physiologically active fraction. Under these conditions, the plasma CBG is almost fully saturated, and thus any increased secretion or therapeutically administered cortisol will cause a rise in the free, active fraction. This may become important, since there is some overlap in the actions of the C-21 steroids so that when the free fraction of cortisol is significantly increased, it may exert a significant mineralocorticoid effect. Glucocorticoids are inactivated by conjugation with glucuronate or sulfate in the liver, and excreted via the kidneys.

Unlike the other steroid hormones, the mineralocorticoid aldosterone is not transported in plasma bound to a specific protein. Its major actions are to stimulate the exchange of sodium for protons and potassium ions across cell membranes, and to control sodium and water balance in the kidney as discussed in Topic M5. Like the corticosteroids, aldosterone is inactivated by conjugation in the liver with glucuronate or sulphate, and excreted in the urine.

DHA and androstenedione promote protein synthesis, but have only a mild androgenic action at physiological concentrations. They are transported in the circulation bound to **sex-hormone-binding globulin (SHBG)** and albumin, and may be converted to testosterone in the testes or ovaries. The major metabolic end-products of androgen metabolism, androsterone and aetiocholanone, are conjugated with glucuronate or sulfate in the liver and excreted in the urine .

## Biosynthesis of epinephrine

Epinephrine is synthesized from phenylalanine by **chromaffin cells** of the medulla. These chromaffin cells are post-ganglionic neuronal cell bodies of the sympathetic branch of the **autonomic nervous system (ANS)** that did not produce axons during development (see Topic H5). The rate-limiting step in this synthetic pathway (*Fig. 3*) is the hydroxylation, by **tyrosine hydroxylase**, of tyrosine, itself the product of hydroxylation of phenylalanine. The formation of the product of the tyrosine hydroxylase reaction, **3,4-dihydroxyphenylalanine**

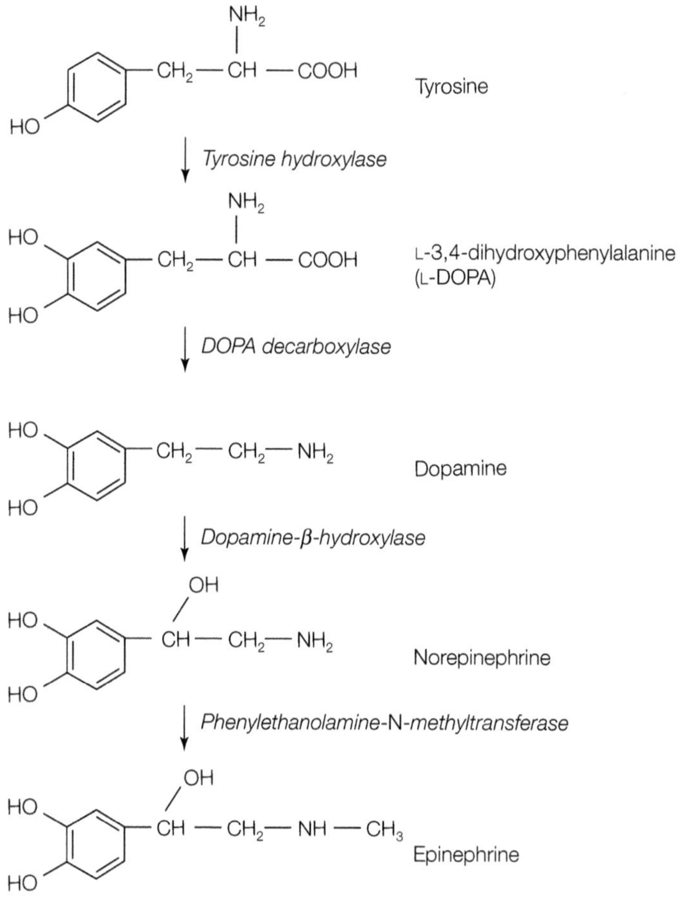


*Fig. 3. Biosynthesis of epinephrine.*

**(DOPA)**, is subject to feedback inhibition by both **dopamine** and **norepinephrine**. DOPA is decarboxylated by the action of **DOPA decarboxylase** to dopamine (dihydroxyphenylethylamine), which undergoes further hydroxylation to norepinephrine by **dopamine β-hydroxylase**, a copper-dependent mixed function oxidase. The final step, *N*-methylation of norepinephrine to epinephrine, is catalyzed by **phenylethanolamine-*N*-methyltransferase (PNMT)** using *S*-adenosylmethionine as methyl donor, an activity which is induced by glucocorticoids. Epinephrine is stored as $Mg^{2+}$-ATP complexes with membrane proteins (**chromogranins**), in discrete storage vesicles within the medullary cell, and released via exocytosis on stimulation.

The reason why adrenal medullary cells produce epinephrine rather than norepinephrine (as most other postganglionic sympathetic neurons do) is unclear. However, it may be related to the way in which steroid hormones being produced by the adrenal cortex exit the gland: via venous channels that run past the medullary cells, before draining into the adrenal vein that drains the medulla. Steroid hormones are know to induce high levels of expression of the *PNMT* gene, so it is possible that this steroid hormone influence converts the medullary cells from a noradrenergic to an adrenergic phenotype.

## Actions of epinephrine

In addition to its effects on cells innervated by the ANS, such as the heart, airways and blood vessels (see Topics H5 and H6), epinephrine has a number of major metabolic actions, as shown in *Table 3*. They are mainly catabolic, as seen in the initial stress response to trauma, where the immediate response is an increase in the plasma concentrations of glucose, lactate, and fatty acids as a result of increased hepatic **glycogenolysis** and **gluconeogenesis**, increased muscle glycogenolysis and **glycolysis**, and increased adipose tissue **lipolysis** respectively. Epinephrine is metabolized to an inactive product, **4-hydroxy-3-methoxymandelate (vanillyl mandelate)** that is excreted in the urine.

*Table 3. Some metabolic actions of epinephrine and other catecholamines*

| Tissue | Effect | Result |
|---|---|---|
| Whole body | ↑Thermogenesis | ↑Metabolic rate |
| Liver | ↑Glycogenolysis | ↑Glucose output |
| | ↑Gluconeogenesis | ↑Ketone body output |
| | ↑Ketogenesis | Maintains $K^+$ homeostasis |
| | ↑$K^+$ uptake | |
| Adipose tissue | | |
| white | ↑Lipolysis | ↑Blood free fatty acids |
| brown | ↑Lipolysis | ↑Local thermogenesis |
| Muscle | ↑Glycogenolysis | ↑Lactate output |
| | ↓Glucose uptake | ↑Blood glucose |
| | ↑$Ca^{2+}$ uptake | ↑Contractile strength |
| | Gluconeogenesis | ↑Glucose and ammonia output |
| | ↑$K^+$ uptake | |
| Kidney | ↑Free water clearance | ↑Urine output |
| | ↑$Na^+$ reabsorption | ↑ECF $Na^+$ |
| | ↑$Ca^{2+}$ excretion | Hypercalciuria |

ECF, extracellular fluid.

# K9 PANCREATIC HORMONES

## Key Notes

**Synthesis of insulin and glucagon**

Insulin, the major anabolic hormone, is synthesized and secreted by pancreatic β-cells in the islets of Langerhans, in response to a rise in plasma glucose concentration. The insulin gene codes for a pre-prohormone, which undergoes loss of short peptide sequences during processing to the mature hormone. This includes cleavage of a C-peptide (connecting peptide) during storage of the hormone in secretory vesicles prior to release. The mature hormone consists of two peptides, A and B, linked by two disulfide bridges. Glucagon, a major catabolic hormone, is synthesized in the pancreatic α-cells. It is synthesized from a large precursor peptide which yields two further peptides on hydrolysis, glucagon-like peptides-1 and -2, as the hormone is concentrated in secretory vesicles.

**Release of insulin and its effects**

An increase in plasma glucose concentration is sensed by the β-cells and leads to an increase in cytosolic calcium in the cell and secretion of insulin and its C-peptide. The measurement of plasma C-peptide is a good indicator of insulin secretion since, unlike insulin, it does not undergo a first-pass metabolic effect through the liver. Insulin acts peripherally to increase glucose uptake into muscle and adipose tissue. The major metabolic actions of insulin are anabolic, and serve to promote storage of glucose as glycogen in liver and muscle, by activating glycogen synthesis, and fatty acid as triglyceride in adipose tissue. Insulin also promotes amino acid uptake into muscle and muscle protein synthesis. Its action on increasing the uptake of potassium into cells has some use therapeutically in the treatment of hyperkalemia.

**Release of glucagon and its effects**

Glucagon is released in response to a fall in plasma glucose concentration and, by stimulating hepatic glycogenolysis, immediately provides glucose for the blood. Plasma glucagon concentrations are raised chronically during fasting, and this leads to an increase in hepatic gluconeogenesis to provide glucose for glucose-dependent tissues such as brain, kidney and erythrocytes. Glucagon also stimulates the hormone-sensitive lipase in adipose tissue to provide fatty acids for metabolism by tissues that are not glucose dependent, particularly skeletal muscle and liver.

**Related topics**

Metabolic processes (B3)
The liver, gall bladder, and spleen (J5)
Absorption of nutrients (J6)

## Synthesis of insulin and glucagon

The two peptide hormones secreted by the **pancreas** have essentially opposing actions in terms of the control of intermediary metabolism. **Insulin**, synthesized in the **β-cells** of the **islets of Langerhans**, is the major anabolic hormone while **glucagon**, synthesized in the **α-cells**, stimulates catabolic pathways. The genes for both code for a pre-prohormone which undergoes editing before storage in secretory vesicles, as described in Topic K2.

The gene for insulin is on the short arm of chromosome 11, and codes for a 105-amino acid **pre-proinsulin** (*Fig. 1a*), from which the signal sequence leader peptide of 24 amino acids is cleaved as it crosses the cisternal membrane of the endoplasmic reticulum (ER). The **proinsulin** at this stage has three intra-chain disulfide bonds. Following movement through the Golgi body in microvesicles, proinsulin is packaged in storage vesicles which mature into β-granules. During this process, proteases in the vesicle are activated, and these cleave a **connecting peptide (C-peptide)**, to leave two further peptides, A and B, which are linked by two disulfide bonds forming the mature hormone. In the granules, insulin binds zinc and the mature granule contains zinc-insulin crystals as well as the C-peptide. Stimulation of the β-cell leads to secretion of equimolar amounts of insulin and C-peptide.

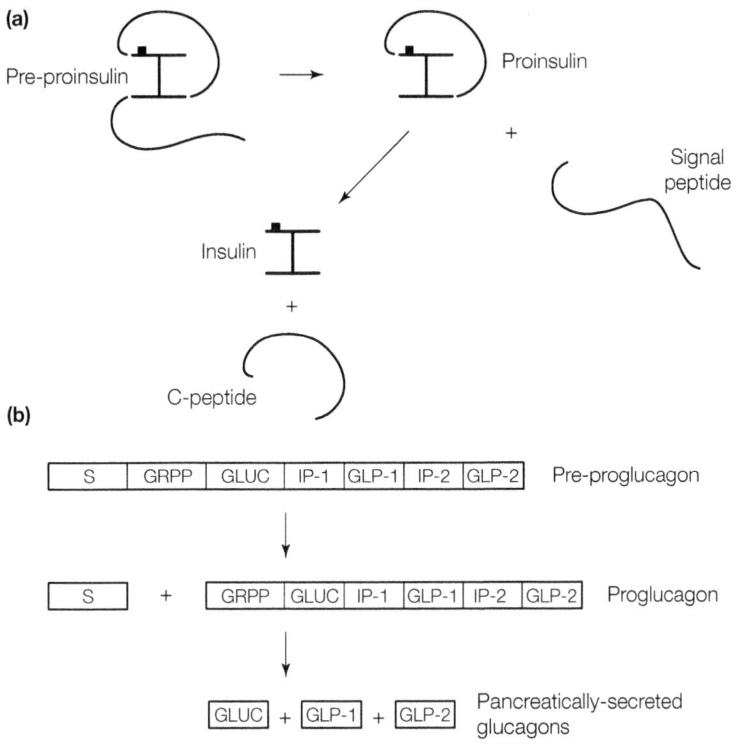


*Fig. 1. Structure of (a) insulin; and (b) glucagons. C-peptide, connecting peptide; GLP-1, glucagon-like peptide 1; GLP-2, glucagon-like peptide 2; GLUC, glucagon; GRPP, glucagon-related polypeptide; IP-1, intervening peptide 1; IP-2, intervening peptide 2; S, signal peptide.*

The gene for **glucagon** is located on the long arm of chromosome 2, and codes for a large precursor molecule, **pre-proglucagon** (*Fig. 1b*) which is expressed predominantly in the α-cells of the pancreas. It is also expressed to a much lesser extent in the intestine and brain. The glucagon gene is a member of a multigene superfamily which includes **secretin**, **vasoactive intestinal peptide** and **gastric**

**inhibitory peptide** amongst others. Although a peptide of only 29 amino acids (3.5 kDa), glucagon is synthesized as part of a much larger peptide which includes two further peptides, **glucagon-like peptide 1 (GLP-1)**, and **glucagon-like peptide 2 (GLP-2)**. The initial translation product, pre-proglucagon, contains an N-terminal signal sequence to allow it to be translocated across the membrane of the ER and into the ER lumen. The product of this hydrolysis, **proglucagon**, then undergoes further processing by a prohormone convertase during packaging, to release glucagon and the two glucagon-like peptides.

## Release of insulin and its effects

The major regulator of insulin secretion is plasma glucose concentration. As the plasma glucose rises above the homeostatic concentration of about 4.5 mM, glucose enters the pancreatic β-cell via the **GLUT2 transport protein** and, after being phosphorylated to glucose-6-phosphate, is metabolized, setting off a signal transduction system which in turn closes ATP-sensitive $K^+$ channels, depolarizing the cell and stimulating calcium entry into the cytoplasm. Calcium entry triggers fusion of **secretory vesicles**, leading to exocytosis of the contents of the vesicles into the bloodstream. The magnitude of insulin secretion is dependent on the plasma glucose concentration and the initial secretion is of prestored hormone, followed by secretion of hormone synthesized *de novo*. Insulin and C-peptide are released into the **portal circulation**, and approximately 50% of the parent hormone is metabolized in the first pass through the liver, such that the concentration of insulin in the portal vein is 2–4-fold greater than in the peripheral circulation. The C-peptide, on the other hand, does not undergo such a 'first-pass' effect, and measurement of C-peptide rather than of insulin itself is therefore a more precise marker of endogenous insulin secretion.

As mentioned above, insulin is secreted in response to an increase in plasma glucose. After a high glucose load, its primary actions in liver, muscle, and adipose tissue serve to oppose this rise by promoting uptake of glucose into muscle and adipose tissue, and storage of glucose in liver. Insulin acts on its target tissues through a specific heterodimeric **insulin receptor** consisting of two α- and two β-subunits linked via disulfide bonds. Insulin binds to the α-subunits, which are on the extracellular surface of the cell, and these subunits then interact with the membrane-spanning β-subunits. The β-subunits have **tyrosine kinase** activity, and can themselves be phosphorylated on serine, threonine, and tyrosine residues on the cytoplasmic surface. Binding of insulin to the α-subunits leads to phosphorylation of the β-subunits and the initiation of a phosphorylation cascade, which ultimately produces the metabolic effects associated with the hormone. One such signal transduction pathway involving 1-phosphatidylinositol 3′-kinase leads to the translocation of the **glucose transport protein** (**GLUT4**) to the cell surface, to facilitate uptake of glucose into muscle and adipose tissue. Approximately 30% of the glucose ingested during a meal rich in carbohydrate is stored in the liver as glycogen. The actions of insulin and glucagon on some metabolic pathways are shown in *Table 1*. Insulin has no effect of hepatic glucose uptake, but increases glycogen production from glucose entering the hepatocyte. It also stimulates triglyceride synthesis and thereby stimulates the secretion of **very low-density lipoprotein (VLDL)** by the liver. The insulin-stimulated glucose uptake by muscle leads to increased glycogen synthesis and storage by the tissue. In adipose tissue, insulin-stimulated glucose uptake leads to storage of triglyceride, after metabolism of the glucose to glycerol-3-phosphate.

*Table 1. Metabolic effects of insulin and glucagon*

| Tissue | Insulin | Glucagon |
|---|---|---|
| Liver | ↑Glycolysis<br>↑Glycogen synthesis<br>↓Lipolysis<br>↑Lipogenesis<br>↓Gluconeogenesis | ↓Glycolysis<br>↓Glycogen synthesis<br>↑Glycogenolysis<br>↓Lipogenesis<br>↑Gluconeogenesis<br>↑Ketogenesis |
| Adipose | ↑Triglyceride synthesis | ↑Lipolysis |
| Muscle | ↑Glucose transport<br>↑Glucose metabolism<br>↑Glycogen synthesis<br>↑Amino acid uptake<br>↑Protein synthesis | Only limited action |

Under normal dietary conditions, where glucose consumption is more moderate, the major hepatic action of insulin is to decrease glucose production by inhibiting glycogen breakdown (**glycogenolysis**) and glucose synthesis (**gluconeogenesis**), actions which are more sensitive to smaller increases in insulin concentration. Under these conditions, insulin acts peripherally to promote amino acid uptake and protein synthesis, and to inhibit protein breakdown in muscle. In adipose tissue it activates lipoprotein lipase, and inhibits the intracellular hormone-sensitive lipase, thereby increasing the hydrolysis of VLDL triglyceride and promoting the synthesis and storage of triglyceride in the tissue.

Insulin also promotes the uptake of $K^+$ ions by cells, by stimulation of the **$Na^+/K^+$ ATPase**. Excess insulin secretion can cause **hypokalemia** in persons with certain types of pancreatic tumors. This $K^+$-lowering action of insulin is sometimes used therapeutically to correct hyperkalemia; glucose is administered at the same time to prevent hypoglycemia.

### Release of glucagon and its effects

Glucagon secretion responds rapidly to a fall in blood glucose concentration, and acts to restore normoglycemia. The hormone has a very short half-life in plasma (5 min), and its concentration falls while eating, especially with a carbohydrate-rich diet. Blood glucagon concentration rises between meals as blood glucose concentration falls, and is chronically raised during fasting or on consumption of low carbohydrate diets. Direct sympathetic stimulation of the pancreas during times of acute psychological or physical stress also leads to increased secretion of glucagon.

As mentioned above, glucagon acts to mobilize readily metabolizable fuels, glucose, and fatty acids, at times when dietary supply of these fuels is reduced or non-existent. In general, glucagon will mobilize glucose from whatever source it can. Its major sites of action are liver and adipose tissue. It binds to its trimeric G-protein-linked **glucagon receptor** on these tissues, and signals via adenylyl cyclase and cyclic adenosine monophosphate (cAMP) to activate phosphorylase in liver, and **hormone-sensitive lipase** in adipose tissue by phosphorylation. In liver (*Fig. 2*), this results in hydrolysis of glycogen stores to glucose-1-phosphate, which is converted to glucose-6-phosphate and then to

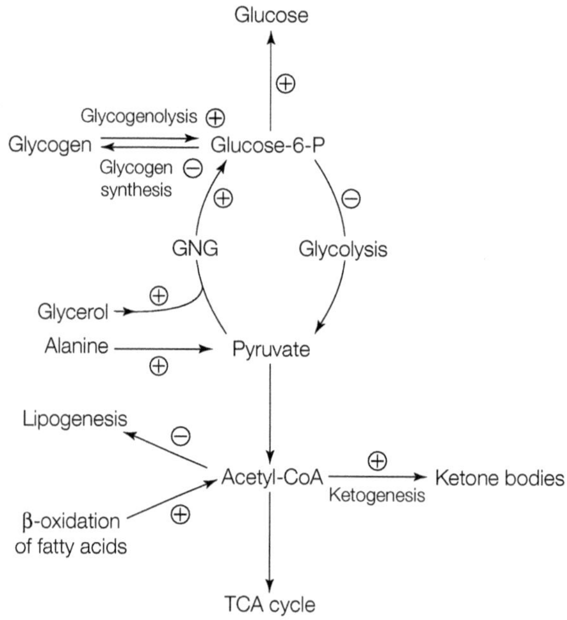


*Fig. 2. Hepatic effects of glucagon. GNG, gluconeogenesis; glucose-6P, glucose-6-phosphate; TCA, tricarboxylic acid.*

glucose. This glucose is released into the blood and used by glucose-dependent tissues such as brain, erythrocytes, nervous tissue, and the kidney. Simultaneously, the same phosphorylation system acts directly on **glycogen synthase** to inactivate the enzyme. Thus, the activation of the catabolic enzyme, phosphorylase, is coordinated with the inactivation of the anabolic enzyme, glycogen synthase, thereby preventing the futile cycling of glucose-6-phosphate, and allowing the liver to contribute to blood glucose.

Activation of the hormone-sensitive lipase in adipose tissue by glucagon results in hydrolysis of stored triglyceride to free fatty acids and glycerol. The free fatty acids are carried in the blood, bound to albumin, for use by tissues as fuel, and glycerol is returned to the liver to serve as a substrate for gluconeogenesis.

# K10 THE GONADS AND SEX HORMONES

## Key Notes

**Testicular hormones**

The major testicular hormone, testosterone, is synthesized in the Leydig cells in response to stimulation by luteinizing hormone (LH) from the pituitary gonadotrophs. It is transported in the blood by sex-hormone-binding globulin. Systemically, in conjunction with its metabolite, dihydrotestosterone, testosterone induces the development of the male reproductive organs, secondary sex characteristics such as facial hair and enlargement of the larynx at puberty, and growth. Locally, it stimulates spermatogenesis in the Sertoli cells, where maturation of the spermatozoa is facilitated by follicle-stimulating hormone (FSH).

**Ovarian hormones**

The ovaries produce a mature ovum each month for fertilization during the female reproductive cycle. They also synthesize a number of hormones including estrogens, progesterone, inhibin and relaxin, under the control of gonadotropins (LH and FSH) from the anterior pituitary, to regulate the uterine cycles. Estradiol from the developing follicle controls oocyte development and maintenance of female reproductive structures. Progesterone, from the corpus luteum, is responsible for the maintenance of the endometrium in preparation for the fertilized egg.

**Related topics**

Hypothalamic hormones (K4)
Pituitary hormones (K5)
Adrenal hormones (K8)
Overview of the male reproductive system (L1)
Overview of the female reproductive system (L2)
Hormonal control of the menstrual cycle (L3)

## Testicular hormones

The **testes** are a pair of organs lying in the scrotal sac, and specialize in the production of **spermatozoa (sperm cells)** by the **Sertoli cells** of the **seminiferous tubules**, and the synthesis and secretion of male sex hormones, mainly **testosterone** in the interstitial **Leydig cells**. The production of testosterone in the fetal Leydig cells is critical for the development of the male phenotype. The testes thus control sexuality and fertility in males. The role of the testes in producing sperm is covered in Topic L1.

Endocrine control of the testes is mediated via the gonadotropins **luteinizing hormone (LH)** and **follicle-stimulating hormone (FSH)**, synthesized in the pituitary gonadotropic cells, which in turn are under the control of pulses of **gonadotropin-releasing hormone (GnRH)** from the hypothalamus (**hypothalamic–pituitary–gonad axis**). Short- and long-feedback loops ensure tight control (*Fig. 1*). Both LH and FSH signal intracellularly via G-proteins and cyclic adenosine monophosphate (cAMP).

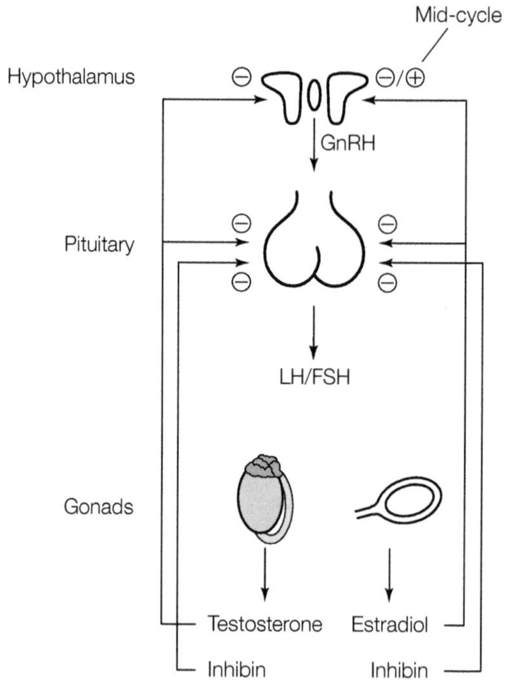


*Fig. 1. Hormones of the hypothalamic–pituitary–gonadal axis. Note that feedback of estradiol on release of hormones from the hypothalamus and pituitary switches from negative feedback to positive feedback at a critical point in the middle of the ovarian cycle, to induce ovulation. GnRH, gonadotropin-releasing hormone; FSH, follicle-stimulating hormone; LH, luteinizing hormone.*

LH activates the rate-limiting enzyme, **desmolase**, in the conversion of **cholesterol** to **pregnenolone**, the precursor for testosterone synthesis. This activity requires the action of the steroid acute regulatory protein. Leydig cells have the complete complement of enzymes required to convert pregnenolone to testosterone. Testosterone is transported in the blood bound to two plasma proteins, 44% to **sex-hormone-binding globulin (SHBG)**, and 54% to albumin, with 2–3% being present as free, unbound hormone. The SHBG fraction is tightly bound, and acts as a storage form of the hormone. The free and albumin-bound hormone represent the bioavailable fraction. Testosterone acts systemically to produce the secondary male characteristics such as enlargement of the larynx at puberty, and facial hair and maintenance of **libido** throughout life. It also has a general anabolic action. Locally, it stimulates spermatogenesis in the Sertoli cells. FSH stimulates the maturation of spermatozoa in the Sertoli cells.

**Ovarian hormones** The **ovaries** are a pair of organs lying in the peritoneal cavity, and are required for **oogenesis** and **ovulation** of a mature **secondary oocyte** into the female reproductive tract approximately mid-way through the normal **menstrual cycle**. Female sex hormones such as **estradiol**, produced by cells in the ovary, are essential in preparing the lining of the uterus in anticipation of fertilization of the ovum with a spermatozoon, and implantation of the fetus. The maturation of the secondary oocyte and ovulation take place under the influence of FSH and

LH, the release of which are controlled by GnRH, as outlined in Topics K4 and K5. FSH, as its name suggests, promotes development of primary oocytes and surrounding cells in the ovary into follicles. Each month, one of these follicles will go on to become a **Graafian follicle** and, under the influence of **estrogens** and LH, will release a secondary oocyte into the peritoneal cavity so that the oocyte can find its way into the uterus for fertilization. The other cells of the follicle (**granulosa cells**) remain in the ovary and become the **corpus luteum**. The corpus luteum is important in producing the hormone progesterone that maintains the thick, secretory, highly vascularized endometrial lining of the uterus that is necessary for implantation of the fetus if fertilization takes place. The role of the ovaries in ovulation is covered more extensively in Topic L2.

Estradiol is produced from the developing follicle and granulosa cells under the influence of FSH, via a mechanism that involves induction of the **aromatase** enzyme. The estradiol produced by the developing follicle then feeds back in a negative fashion on FSH release from the **anterior pituitary**, so that only the most developed follicles ever become dominant; those which by that stage have failed to develop regress as FSH levels fall. Estradiol and other hormones such as **inhibin** are important in negative feedback control of FSH and LH (*Fig. 1*).

At a critical point, normally around mid-way through the menstrual cycle, estrogens being produced by the developing follicle have a positive feedback

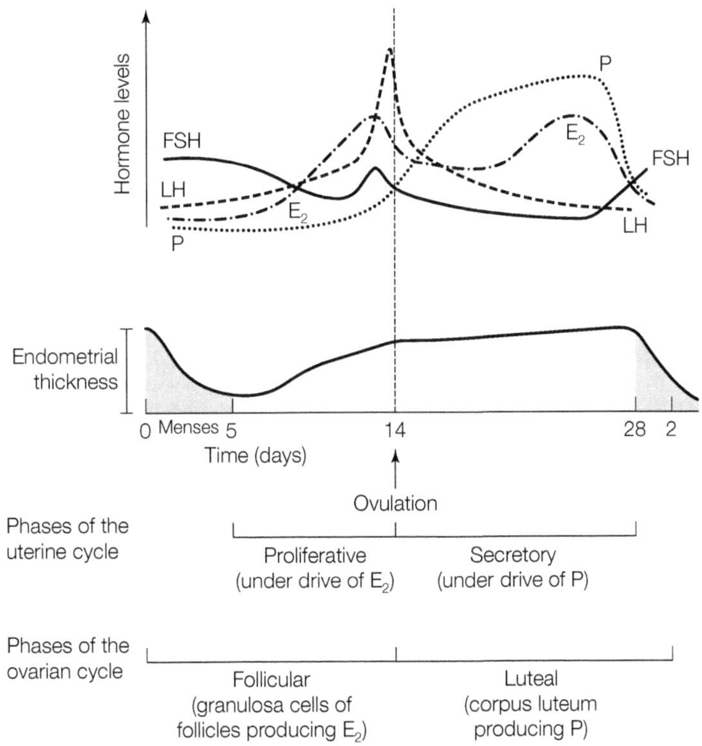


*Fig. 2. Circulating levels of various hormones and development of the endometrium during the menstrual cycle. $E_2$, estradiol; FSH, follicle-stimulating hormone; LH, luteinizing hormone; P, progesterone. FSH and LH are produced by the anterior pituitary, $E_2$ is produced by granulosa cells of the developing follicles in the ovary, and P is produced by the corpus luteum after ovulation.*

effect on GnRH and LH release. This mid-cycle **LH surge** is the trigger for ovulation to occur. The reasons why the normal negative feedback effect of estrogens on GnRH, FSH and LH release is converted at this point in the menstrual cycle to a positive feedback effect have yet to be elucidated. *Figure 2* shows a graph of the concentrations of FSH, LH, estradiol and progesterone during the menstrual cycle. Study this in tandem with the information presented in Topics L2 and L3.

# L1 Overview of the male reproductive system

## Key Notes

**Functional anatomy**

The male reproductive system is composed of the testes, the epididymis, the vasa deferentia, the penis, and several glandular structures. The testes and epididymis produce sperm cells and the vasa deferentia conduct the sperm cells to the urethra, where they are mixed with secretions from the seminal vesicles, the prostate gland, and the bulbourethral glands to form semen. The secretions from the various glands have particular functions.

**Spermatogenesis**

Sperm cells (spermatozoa) are produced in the testes from stem cells known as spermatogonia, a process that is under the control of follicle-stimulating hormone, testosterone, and luteinizing hormone. The sperm cells produced by spermatogenesis are haploid, having only one of each of the 23 chromosome pairs that other body cells have. Mature sperm cells have a structure that is ideally suited to their required motile function.

**Erectile function**

Erection of the penis is caused by dilation of the arteries supplying the corpora cavernosa. This causes engorgement of the organ, occluding its venous drainage. Erection is controlled by parasympathetic nervous output from the sacral section of the spinal cord. Ejaculation of semen from the penis is controlled by a sympathetic reflex. Psychological or physical factors may cause problems with erectile function. Some therapeutic measures are effective.

**Characteristics of semen**

Semen contains sperm cells in a mixture of secretions from the various glands of the male reproductive tract. Male infertility can be caused by such factors as low sperm count or a high proportion of damaged sperm in the semen. The World Health Organization publishes normal values for these various parameters.

**Related topics**

Structure and function of the autonomic nervous system (H5)

The gonads and sex hormones (K10)

## Functional anatomy

The male reproductive system comprises the **testes** and **epididymis**, the vasa deferentia (singular **vas deferens**), the **seminal vesicles**, the **prostate gland**, **bulbourethral glands**, and the **penis** (*Fig. 1*). The male gametes (**spermatozoa** or **sperm cells**) are made in the testes and mature in the epididymis (see below). This process is under the control of **testosterone** (produced by **Leydig cells** in the testis) and the anterior pituitary hormone **follicle-stimulating hormone (FSH**, see Topic K5). Blood supply to the testes comes from the extremely long testicular (gonadal) arteries that arise from the abdominal aorta. These arteries

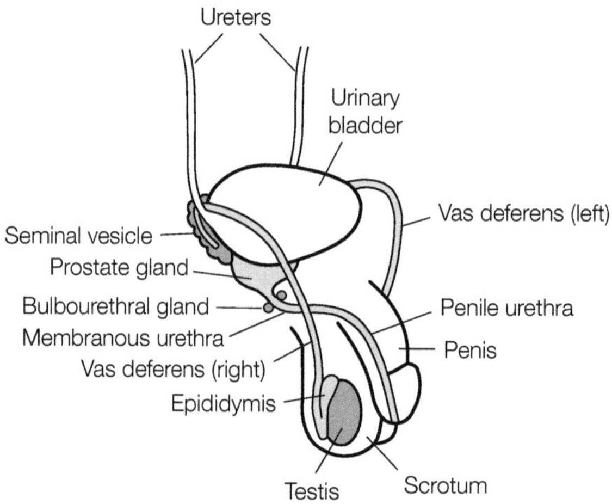


*Fig. 1. Layout of the male reproductive organs.*

are long because the testes began development in the abdominal cavity before descending into the scrotum, maintaining their blood supply as they migrated.

The vasa deferentia conduct mature spermatozoa upwards to the urethra, where they are expelled from the reproductive tract by ejaculation during coitus (sexual intercourse). The lumen of the vas deferens is lined with epithelial cells, and is continuous with that of the epididymis and the **seminiferous tubules** of the testis. The vasa deferentia are normally around 45 cm long, passing up through the spermatic cord, over and around various other organs in the pelvis, before joining the ejaculatory duct and **urethra** within the body of the prostate gland. The various glands, including the prostate gland, produce secretions that are designed to provide the spermatozoa with the nutrients that they require in order to make their way to an ovum in the female reproductive tract after coitus.

The combination of spermatozoa and secretions that is ejaculated is known as **semen**. The secretion of the seminal vesicles is a viscous, alkaline secretion that is rich in fructose. The alkaline nature of the secretion helps neutralize acid in the female reproductive tract, and the fructose can be used by spermatozoa to make adenosine triphosphate (ATP) for motility. Seminal vesicular secretion also contains proteins that help the semen coagulate after ejaculation, to help keep the semen in the female reproductive tract. Prostatic fluid is a slightly acidic, milky secretion that also helps the semen to coagulate after ejaculation, but which additionally contains enzymes that subsequently help the semen to liquify so that the spermatozoa can become motile again within the female reproductive tract. The bulbourethral glands (also known as **Cowper's glands**) lie within the muscle of the pelvic floor and produce an alkaline, lubricating fluid that helps to protect the surface of the penis during coitus.

## Spermatogenesis

The male reproductive tract is lined with epithelial cells, all the way down to the highly coiled seminiferous tubules of the testes. In the seminiferous tubules, the stratified epithelium is composed of two types of cells: **Sertoli cells** and **spermatogenic cells** (*Fig. 2a*). One of the roles of the Sertoli cells is to respond to FSH and produce **androgen-binding protein (ABP)**. ABP has high affinity for

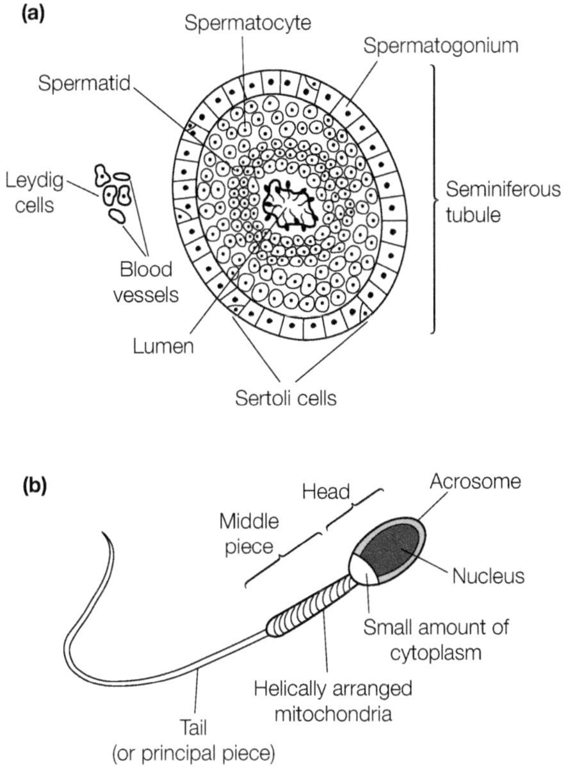


*Fig. 2. Spermatogenesis in the testes. (a) Structure of the seminiferous tubule; (b) structure of mature sperm cells (spermatozoa).*

testosterone and **dihydrotestosterone**, which are produced by the Leydig cells present in the intertubular space. ABP is important in the trafficking of testosterone to the epididymis, where it may have a role in maturation of spermatozoa. The action of testosterone in promoting the production of mature spermatozoa is therefore dependent upon a complex interaction of several hormones; **luteinizing hormone (LH)** and **prolactin** control testosterone secretion from Leydig cells. This testosterone promotes spermatogenesis in the seminiferous tubules. FSH stimulates Sertoli cells to 'capture' testosterone by secreting ABP.

Spermatogenesis consists of two discrete processes; **spermiogenesis** and **spermiation**. During spermiogenesis, spermatogenic cells undergo **meiosis**, changing from **spermatogonia**, through **spermatocytes** and **spermatids**, until they become spermatozoa (sperm cells). During meiosis, cells that are **diploid** (having the normal complement of chromosomes, 46 in 23 pairs) divide and become **haploid**, with only one set of 23 chromosomes. In post-pubertal males, this process continues throughout life, although the process is much reduced in elderly men. The haploid sperm cells, when fused with a haploid ovum at fertilization, form a diploid cell. This diploid cell is the **zygote**, and by undergoing subsequent mitotic divisions (with duplication and conservation of genetic material) the zygote gives rise to the embryo that can become implanted in the uterus. See Topics L2–L4 for more details of these processes.

Spermiation is the process by which sperm cells are released into the lumen of the seminiferous tubules. Haploid spermatids, which do not have the structure of sperm cells (*Fig. 2b*), are embedded into crypts in the plasma membrane of Sertoli cells. The spermatids undergo a number of changes so that they eventually have a **flagellum** (tail) and an **acrosome sac**. The purpose of the flagellum is to enable the spermatozoa to move in the semen and in the female reproductive tract. The acrosome contains hydrolytic enzymes that allow the head of the spermatozoa to enter the ovum at fertilization.

## Erectile function

The penis is composed of three columns of tissue: two dorsolateral corpora cavernosa (singular **corpus cavernosum**), and one ventral **corpus spongiosum**. The urethra runs through the corpus spongiosum, but the corpora cavernosa are composed almost entirely of blood sinuses, supplied by arteries. During sexual arousal, these arteries dilate so that the corpora cavernosa become engorged with blood, their venous drainage becomes compressed, and penile **erection** results. This engorgement is controlled by vascular effects of parasympathetic nerves from the sacral regions of the spinal cord and by release of **nitric oxide**, which is a vasodilator (see Topic D8). **Ejaculation** is a sympathetic reflex that involves contraction of the **internal urethral sphincter** (to ensure that urine does not enter the ejaculate and that semen does not enter the bladder), and coordinated contraction of the smooth muscle in the walls of the vasa deferentia so that semen is ejected from the penis into the female reproductive tract. Within a few minutes of ejaculation, the arteries that supply the corpora cavernosa constrict, and the penis returns to its non-erect state.

Erectile dysfunction is a common problem, the cause(s) of which can be either psychological or physical. Peripheral vascular disease can affect the arteries supplying the penis as well as other organs, so (without trivializing the seriousness of erectile dysfunction), patients with this problem should always be investigated for signs of more dangerous disease. Pharmacological treatments for erectile dysfunction such as **papaverine**, α-adrenoceptor antagonists or **prostaglandin $E_1$** have, in recent years, been superseded by the use of **sildenafil (Viagra)**. This has mainly been because of the better tolerability of the oral route of administration for sildenafil as compared with the other drugs (intracavernosal injection or intraurethral pessary), although none are devoid of the risk of adverse events.

## Characteristics of semen

As described above, semen is a mixture of sperm and the fluids produced by the various glands associated with the male reproductive tract. *Table 1* shows the

*Table 1. Characteristics of normal human semen*

| Parameter | Normal value |
|---|---|
| Volume | >2.0 ml |
| Spermatozoon concentration | $>2 \times 10^7$ $ml^{-1}$ |
| Total spermatozoa | $> 4 \times 10^7$ $ml^{-1}$ |
| Vitality | >75% |
| Spermatozoal morphology normal | >30% |
| Average number of defects per abnormal spermatozoa | <1.5 |
| pH | 7.2–8.0 |
| Leukocytes | $<1 \times 10^6$ $ml^{-1}$ |

normal values (World Health Organization statistics) for various parameters that might be tested in a sample of semen. The normal volume of semen ejaculated during orgasm is between 2 and 5 ml. If less than 2 ml are ejaculated, this is considered abnormal, and is likely to adversely affect the man's fertility. Equally, it is necessary that the concentration of spermatozoa in the semen sample is greater than 20 million $ml^{-1}$ (i.e. a total of greater than 40 million in the ejaculate) to increase the likelihood of an ovum (egg) being fertilized. This is because it is highly likely that a number of the spermatozoa in the sample will not be viable, either because they are not motile or because they have defects. However, as many as 70% of the spermatozoa may have abnormal morphology before fertility is compromised. The pH (normally between 7.2 and 8.0) of the semen is also important. The slightly alkaline nature of the semen, contributed by fluid from the seminal vesicles and from the bulbourethral glands, serves to neutralize the slightly acidic environment of the female reproductive tract to ensure that more spermatozoa survive deposition there. If a test for white blood cells is positive (greater than 1 million $ml^{-1}$), this may indicate an infection of the reproductive tract.

# L2 OVERVIEW OF THE FEMALE REPRODUCTIVE SYSTEM

## Key Notes

**Oogenesis**

Oocytes are produced from stem cells by the process of oogenesis. During fetal life, all of the stem cells (oogonia) undergo a few stages of meiosis before being arrested as primary oocytes. Following puberty, several primary oocytes and their surrounding cells are triggered to undergo follicular development, but usually only one oocyte is ovulated. The remaining stages of meiosis take place at this time. Oocytes that are ovulated later in life have a greater chance of having chromosomal abnormalities.

**Ovulation**

After puberty, follicle-stimulating hormone induces follicular development of some of the primary oocytes and their surrounding cells in both ovaries. Normally only one of these follicles will become a Graafian follicle and be ovulated in the middle of the ovarian cycle, a process brought about by a surge in luteinizing hormone secretion by the anterior pituitary. Once ovulated, the ovum (secondary oocyte, zona pellucida, and corona radiata) finds its way into the Fallopian tube, where fertilization by a sperm cell may take place. If fertilization does not take place, the corpus luteum (in the ovary) degenerates, causing a decline in progesterone levels and sloughing off of the uterine lining during menstruation.

**Functional anatomy**

The female reproductive tract comprises the ovaries, Fallopian tubes, uterus, cervix, and vagina. Semen is deposited in the vagina during coitus. Each month (after puberty) the ovaries produce an ovum that finds its way into the uterus via the Fallopian tube. If fertilization takes place, the zygote implants in the wall of the uterus and is protected during gestation by a mucus plug in the cervix. The muscle in the wall of the uterus is important during labor and birth; uterine contractions help to expel the baby.

**Related topics**

The gonads and sex hormones (K10)
Hormonal control of the menstrual cycle (L3)
From ovum to fetus (L4)
Pregnancy (L5)
Birth and lactation (L6)

## Oogenesis

In both males and females, the sex cells or gametes develop in the gonads by the process of **meiosis** (*Fig. 1*). In meiosis, **haploid** cells (with only half the complement of genetic material of other cells) develop from **diploid** cells. In the male, the diploid cells in the testes are known as **primary spermatocytes**, and these exist throughout life. In the female the diploid cells are known as **oogonia** and only exist *in utero*. Even before birth, these oogonia undergo the first few stages

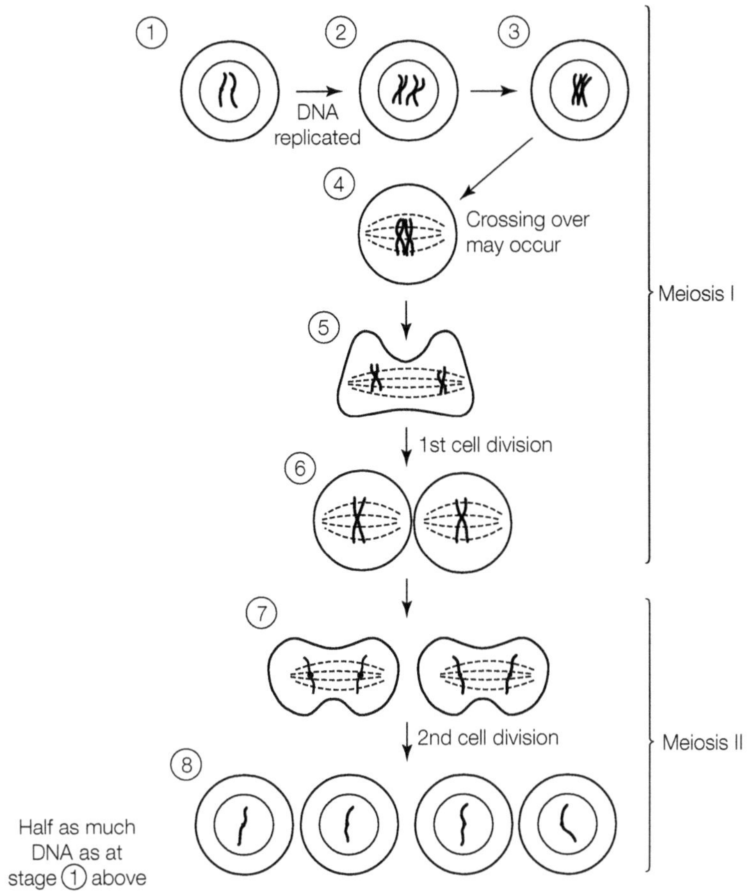


*Fig. 1. Schematic diagram of the events in meiosis. See text for details.*

of meiosis before being suspended as **primary oocytes**. Primary oocytes are acted upon by pituitary hormones after puberty (see Topics K5 and K10), which normally cause one to be ovulated each month as a **secondary oocyte**. However, it is not until the point of fertilization that these secondary oocytes actually complete meiosis to become truly haploid.

In stages 1–3 of the scheme presented in *Fig. 1* (corresponding approximately to **prophase I**), DNA is replicated and homologous **chromosomes** align with each other near the center of the cell (note that at stage 2 the cell has twice as much DNA as normal). Primary oocytes are arrested at this stage of meiosis I. At stage 4, the nuclear membrane has broken down, and a cytoskeletal element known as the **spindle apparatus** has formed to help the chromosomes to move apart. Some 'crossing over' of genetic material from one homologous chromosome to another may occur at this stage. During stages 5 and 6, the homologous chromosomes separate and move into two separate cells. It is important to note that the DNA content of these daughter cells is the same as that of most other cells in the body at this stage. In stages 7 and 8, a further cell division takes place, forming a total of four haploid cells, each with half as much DNA as normal. In spermatogenic meiosis, all daughter cells (sperm cells) are of equal size, but in oogenic meiosis, the size of these daughter cells is uneven such that the mature ovum retains most of the cytoplasm.

These differences in male and female **gametogenesis** can have important consequences; as a woman ages, her primary oocytes age also. This means that genetic defects are more common in the ova of older women (from their mid-30s onwards), which is one of the reasons why the incidence of genetic abnormalities rises as women get older. For example, **Down's syndrome** is a genetic disease in which children have three copies of chromosome 21 (trisomy 21), rather than the usual two copies, and the rate of birth of children with Down's syndrome in women aged 40 years (1:100) is ten times greater than that of children born to women aged 29 years (1:1000). This is caused largely by the fact that primary oocytes in older women are themselves 'older', so that more errors are made during the second stage of meiosis.

Loss of the stem cells (they are converted to primary oocytes) in females as opposed to males (where they persist as primary spermatocytes) also means that women have only 30–40 reproductive years, whereas men remain fertile more or less from puberty to (in some cases) very old age. This reduction in reproductive function in women, marked by the cessation of menstrual bleeding, is known as the **menopause**, and although a man's fertility wanes somewhat as he ages, the decline in circulating levels of sex hormones in men is not as rapidly abrupt as in women. Bear in mind, however, that the menopause in women affects only female sex hormones like the **estrogens**, because a major source of **androgens** in women is the adrenal cortex (see Topic K8). Thus, although post-menopausal women may have markedly reduced fertility, the *biochemical* source of their **libido** (sex drive) is normally not markedly altered.

## Ovulation

As outlined in Topic K10, ovulation takes place around 14 days after the first day of a woman's last menstrual period. A secondary oocyte is released from the **Graafian follicle** into the peritoneal cavity, from where it can find its way into the uterus, ready for **fertilization** and **implantation**.

Early in fetal development, oogonia undergo the first stages of meiosis, but become arrested at prophase of meiosis I as primary oocytes. Other epithelial cells surround the primary oocytes to form **primordial follicles** in the broad outer cortex of the ovary. After puberty, **follicle-stimulating hormone (FSH)**, produced by the anterior pituitary, induces development of several primary oocytes and their surrounding cells in the ovary into **primary follicles**. Each month, normally only one of these follicles will go on to become a Graafian follicle, ready for ovulation; the other primary and secondary follicles regress. *Figure 2* shows a diagrammatic representation of follicular development, together with hormonal changes and uterine changes over the normal 28-day menstrual period. Granulosa cells in the dominant follicle secrete **estradiol**, which 'drives' the **proliferative phase** of the uterine cycle and, through negative feedback on the hypothalamus and pituitary, reduces FSH secretion, thereby causing regression of the other recruited follicles.

A surge in **luteinizing hormone**, normally at around day 14 of the menstrual cycle, causes **ovulation** by a mechanism that is, as yet, not well understood. Once ovulated, what is commonly known as the **ovum** consists of a central secondary oocyte and its surrounding **zona pellucida**, and an outer layer of **corona radiata** cells. The cells that are left behind in the ovary undergo change to become the **corpus luteum**; this secretes **progesterone** to help to maintain the lining of the uterus and promote the secretory phase of the uterine cycle. If the

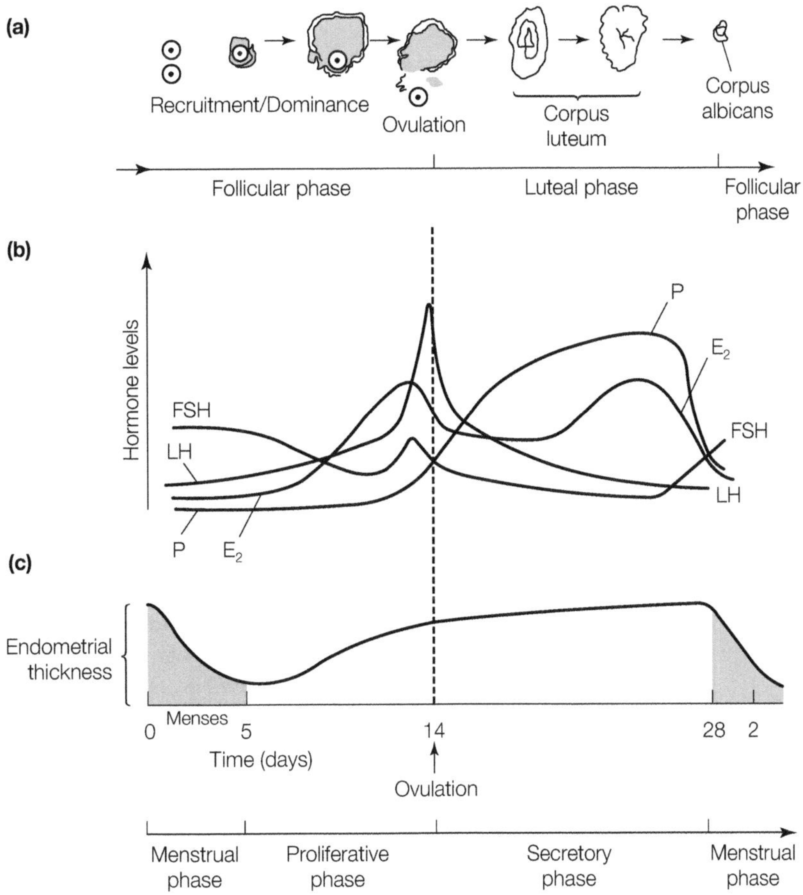


*Fig. 2. Folliculogenesis in the ovary (a) and its relationship to hormone levels during the menstrual cycle (b) and appearance of the endometrium during the uterine cycle (c). $E_2$, estradiol; FSH, follicle-stimulating hormone; LH, luteinizing hormone; P, progesterone.*

ovum remains unfertilized, the corpus luteum undergoes **apoptosis** (programmed cell death) around 12 days after ovulation, and stops secreting progesterone. This fall in progesterone levels leads to **menstruation**, sloughing off of the functional layer of the endometrial epithelium.

## Functional anatomy

The female reproductive tract is composed of the **ovaries**, the **Fallopian tubes**, the **uterus**, the **cervix** and the **vagina** (*Fig. 3*). The ovaries are the site of formation of eggs (ova) that, after adolescence, are released each month in response to cyclic changes in the circulating level of a number of hormones. These processes are known as oogenesis (production of ova) and ovulation (release of ova). Once released from the ovary, the ovum (on average one is released each month, around 14 days after the start of the woman's last menstrual period) finds its way to the Fallopian tube at the end of the uterus. Viability of the ovum is ensured by higher circulating levels of certain hormones, at least for a few days. If the ovum encounters a **spermatozoon (sperm cell)** within 2–3 days of ovulation, fertilization can occur. Otherwise, the ovum degenerates, and no pregnancy can result until another ovum is released the following month.

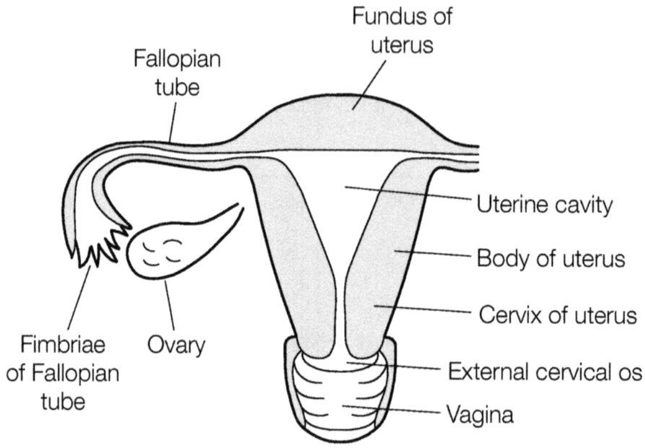


*Fig. 3. Layout of the female reproductive organs.*

If fertilization occurs, the cell made from fusion of ovum and spermatozoa (the **zygote**) may start dividing until a ball of cells exists and then implants in the lining of the uterus, which has become thickened under hormonal influence. Hormones produced by the fertilized ovum help maintain this thickened, highly vascular uterine lining until the placenta has developed. The sterile, enclosed environment of the uterus is maintained by a thick mucus plug that develops in the cervix. Around 40 weeks after the start of the woman's last menstrual period, labor contractions of the uterine muscle begin (under hormonal influence), and the cervix dilates and becomes effaced (thin). These labor contractions and 'ripening' of the cervix result in the birth of the baby, which emerges into the outside world via the cervix and vagina, with the placenta following shortly afterwards.

# L3 HORMONAL CONTROL OF THE MENSTRUAL CYCLE

## Key Notes

**The female reproductive cycle**

The female reproductive cycle is composed of the ovarian cycle and the uterine cycle. Events in the ovaries drive the secretion of hormones that are important in producing the changes to the uterus that happen during the uterine cycle. The menstrual cycle is driven by both these cycles, resulting in menstruation, normally every 28 days.

**The ovarian cycle**

The ovarian cycle can be broken up into the follicular phase and the luteal phase. In the follicular phase (driven by the secretion of follicle-stimulating hormone from the anterior pituitary), developing follicles produce estrogens that have effects on the uterine endometrium. After ovulation, the ovarian corpus luteum produces progesterone, which in turn causes the development of the secretory phase of the uterine cycle.

**The uterine cycle**

During the follicular phase of the ovarian cycle, estrogens being produced by the developing follicles result in proliferation of the uterine endometrium (the proliferative phase of the uterine cycle). After ovulation, progesterone secreted by the corpus luteum causes an increase in the secretory capacity of the glands in the endometrium (the secretory phase of the uterine cycle). Menstruation occurs if fertilization of the ovulated ovum does not occur, as the corpus luteum degenerates and progesterone levels decline.

**Related topics**

The gonads and sex hormones (K10)
Overview of the female reproductive system (L2)
From ovum to fetus (L4)

## The female reproductive cycle

The female reproductive cycle involves coordinated changes in both the **ovaries** and **uterus** which are responsible, respectively, for maturation of the **oocyte** and preparation of the **endometrium** for the reception of the fertilized ovum. Should fertilization not occur, the endometrial surface sloughs off and is lost along with the unfertilized ovum. Two separate, but linked, cycles can be described:

- the **ovarian cycle**
- the **uterine cycle**.

Control of the **menstrual cycle** in females is exerted via positive and negative signals from the **hypothalamic–pituitary–gonadal** axis. The cyclical release of **gonadotropin-releasing hormone (GnRH)** from the **hypothalamus** (see Topic K4) causes a cyclical release of **follicle-stimulating hormone (FSH)** and **luteinizing hormone (LH)** from the **pituitary**. This induces cyclical activity in the ovary, leading to the ovarian cycle and the production of **estrogen** and

**progesterone**, which in turn leads to cyclical activity in the uterus (uterine cycle) and monthly bleeding (menstrual cycle). Thus there are two major interrelated functions of the ovary: synthesis and secretion of sex steroids and the monthly release of a mature ovum. The changes in hormone secretion during the menstrual cycle are shown in *Fig. 2* of Topic K10.

A menstrual cycle begins on the first day (day 1) of bleeding (**menses**), and the whole cycle has a median length of 28 days (range 21–40 days). The function of the menstrual cycle is to present a **secondary oocyte** for fertilization around day 14 of the cycle. This involves selection of a single follicle from the approximately 400 000 follicles present in the human ovary at birth. Such primordial follicles, which contain encapsulated primary oocytes arrested at the first meiotic division during fetal development, remain dormant until puberty (see Topic L2).

## The ovarian cycle

The ovarian cycle consists of two phases. These are the **follicular phase** and the **luteal phase**, separated by **ovulation**. During each ovarian cycle, a few (up to 20) secondary follicles are activated to begin maturation under the influence of FSH. The initiation of follicular growth begins towards the end of the previous cycle, at about day 25, when a rise in FSH secretion occurs. This elevation in plasma FSH continues into the early follicular phase, when it initiates growth and development of a group of follicles, and ends with a slight surge at ovulation. Usually, only one follicle reaches maturity and undergoes ovulation, but exactly how a single preovulatory follicle is selected from this group is unknown. The dominant follicle suppresses the growth of the others through local actions. It is possible that the role of the non-selected follicles is to act as an independent endocrine gland. As the selected follicle matures, the layer of cells surrounding the follicle synthesizes and secretes estrogen and **inhibin**, and later (as the **corpus luteum**), progesterone. Further maturation involves changes in the oocyte, follicular cells and the surrounding stromal tissue. As the follicle grows, the layer of cells surrounding the follicle produces estrogens (**estradiol** and **estrone**) in response to FSH. Estrogen levels rise from about 7–8 days prior to ovulation, reaching a peak on about day 13 of the cycle. This increase in estrogens, in the absence of progesterone, has a positive feedback effect on the hypothalamus, to increase GnRH and also secretion by pituitary gonadotrophs, specifically those secreting LH, thereby initiating the LH surge which marks the end of the follicular phase. There is a concomitant increase in FSH secretion at the same time, but this is of a much lower magnitude than for LH.

Plasma **androgens** also increase slightly over this period with a peak concentration on the day of the LH surge. However, progesterone levels are not increased until immediately prior to the LH surge. During this phase, the follicle matures from a single layer of cells (**theca interna**) encircling the oocyte, to a multilayered state. The growing follicle acquires a cavity filled with liquid (**liquor folliculi**) secreted by the **granulosa cells**, and the primary oocyte moves to an eccentric position in the follicle. A rise in LH causes the developing oocyte to complete its first meiotic division (which had stopped at **prophase** during fetal development) and become a secondary oocyte. Meiosis II now starts and halts in **metaphase** (see Topic L2). Eventually, the follicle achieves a maximum size (**Graafian follicle**), and bulges on the antrum of the ovarian surface.

About 14 days after the start of the follicular phase, the Graafian follicle ruptures to release its secondary oocyte, still surrounded by its **zona pellucida** and **corona radiata**, into the peritoneal cavity, from where it travels to the

**Fallopian tube**. This process is known as ovulation and is stimulated by the LH surge mentioned above. At this stage, all other growing follicles cease to grow and begin to involute, a process known as **atresia**. The next 14 days of the ovarian cycle represent the luteal phase during which time the remaining tissue of the follicle, having shed the secondary oocyte and its surrounding cells, rearranges itself and acquires a yellowish tinge becoming the corpus luteum. Under the influence of LH, the corpus luteum secretes progesterone, estrogens, **relaxin** and inhibin, and plasma progesterone is maximal about 6–8 days after the LH surge of the follicular phase. The secretion of progesterone inhibits the development of new follicles and, in concert with inhibin, inhibits contraction of the uterus thereby helping to prepare the endometrium for possible pregnancy. If fertilization does not occur and there is no source of **human chorionic gonadotropin (HCG)**, derived from the **chorion** of the conceptus, the corpus luteum is programmed to undergo apoptosis after two weeks.

The raised concentrations of both progesterone and estrogen feed back to the hypothalamus to inhibit the release of GnRH, which in turn inhibits further LH secretion from the pituitary and eventually the synthesis of progesterone in the corpus luteum. As the corpus luteum degenerates, plasma progesterone then falls, and the endometrium starts to break down. This fall in progesterone and estrogen towards the end of the luteal phase releases the inhibition of GnRH synthesis and secretion, and thus plasma GnRH begins to rise again, promoting FSH secretion by the pituitary and consequently stimulating follicular growth to start a new ovarian cycle. The inhibition of uterine contraction is also removed, and menstruation begins, marked by detachment and shedding of the endometrial surface and loss of blood.

However, if fertilization occurs at the ovulatory stage, the secondary oocyte starts to divide and the lifespan of the corpus luteum is prolonged, supported by HCG from the chorion of the conceptus from about eight days post-fertilization.

## The uterine cycle

The uterine cycle begins on day 1 of the menstrual cycle with the loss of around 50–150 ml of blood, tissue fluid, and sloughed off epithelial cells from the endometrium (of this, only 30–50 ml is blood). The fall in progesterone towards the end of the luteal phase of the ovarian cycle results in the secretion of **prostaglandins**, which cause constriction of the **spiral arterioles**. This leads to the death of endometrial epithelial cells through oxygen starvation.

Two phases of the uterine cycle can be distinguished: the proliferative and secretory phases caused, respectively, by the follicular and luteal phases of the ovarian cycle. The preovulatory, **proliferative phase** defines the period of proliferation of the endometrium. Estrogens from the growing ovarian follicles stimulate both the repair of the endometrium and the division of cells from the basal layer (**stratum basalis**) to produce a new functional layer (**stratum functionalis**). During this time, estrogens from the theca interna around the growing follicle cause the endometrium to double in thickness to between 4–10 mm, with proliferation of the simple tubular glands. Vascularization of the new stratum functionalis arises through coiling and stretching of arterioles into this layer from the stratum basalis.

In the (postovulatory) **secretory phase** of the uterine cycle, progesterone and estrogens from the corpus luteum promote the new endometrial glands to assume an irregular corkscrew shape and, one week after ovulation, to secrete fluid rich in glycogen to support the arrival of the fertilized ovum in the uterus.

The endometrium approaches its maximum thickness at this time. If fertilization does not occur, plasma progesterone levels fall as described earlier, the endometrial blood vessels contract, and the stratum functionalis is shed, together with fresh blood. Another menstrual cycle begins.

# L4 FROM OVUM TO FETUS

## Key Notes

**Fertilization and division**

Fertilization is the fusion of male and female gametes (sperm cell and ovum) to form a zygote. Polyspermy (entry of more than one sperm cell into the ovum) is prevented by chemical and electrical events in the fertilized ovum. Once fertilization has taken place, several rounds of cell division take place until the morula and then the blastula are formed. The blastula is a ball of cells that has a blastocoele cavity within.

**Implantation and formation of embryonic tissues**

The cells outside the inner cell mass of the blastula, the trophoblast, begin to invade the wall of the uterus. The amniotic cavity develops from the inner cell mass so that the layer of cells separating the amniotic cavity from the blastocoele cavity is the embryonic disk, from which the embryo develops. Three cell layers of the embryo then develop: ectoderm, mesoderm, and endoderm

**Fetal nutrition**

The embryo is nourished by blood channels in the uterine wall that wash over the trophoblast cells, but once the fetus has developed, the placenta becomes the site of close apposition of fetal and maternal blood. The placenta receives a rich blood supply. Fetal blood is able to carry more oxygen than maternal blood because of the different quaternary structure of fetal and adult hemoglobin. High circulating levels of glucose in maternal blood (induced by placental hormones) mean that adequate nutrition of the fetus is normally assured.

**Related topics**

Alveolar exchange and gas transport (E4)
Pancreatic hormones (K9)
Overview of the female reproductive system (L2)
Hormonal control of the menstrual cycle (L3)
Pregnancy (L5)

## Fertilization and division

When a **haploid sperm cell** encounters an **ovum** in the female reproductive tract, **fertilization** (the fusion of the two **gametes** to form a **diploid** zygote) may occur. In humans, where fertilization takes place inside the body, contractions of the uterus help to conduct the sperm towards the **oviduct**, and once in the oviduct, the motility of the sperm and attractant chemicals produced by the ovum and the **ovary** help to guide them in their search. Human ova, in cellular terms, are large structures, visible to the naked eye (*Fig. 1*). They are composed of the egg itself, but also a protein coat known as the **zona pellucida**, and outside that a layer of cells derived from the **ovarian follicle** (see Topic L2) called the **corona radiata** layer. Spermatozoa can make their way through the corona radiata, but cannot penetrate the zona pellucida without the use of the enzymes in the **acrosome sac** (see Topic L1). However, the acrosome reaction cannot happen unless the spermatozoa have spent some time in the female reproductive tract, a process known as **capacitation**.

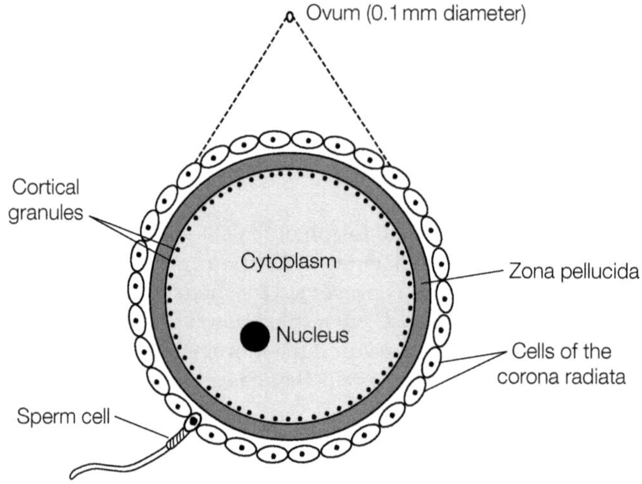


*Fig. 1. Features of the human ovum.*

When a spermatozoon encounters an ovum, the two gametes bind to each other, the acrosome reaction occurs, and the head of the spermatozoon enters the ovum. Once an ovum has been fertilized and has become a **zygote**, the entry of other sperm cells is prevented by alteration of the electrical potential across the membrane of the ovum (depolarization), and by the **cortical granule reaction**. In the cortical granule reaction, granules just inside the membrane of the zygote cross the membrane and cause cross-linking of glycoproteins in the zona pellucida so that entry of further sperm cells is prevented. These reactions prevent **polyspermy** and the creation of a zygote that has an abnormal genetic character. Male and female pronuclei are brought together by a microtubule assembly, and each undergoes a phase of DNA synthesis (S-phase) before mitotic division. It is only after the first division of the zygote to form two cells that diploid nuclei proper are formed. The two cells undergo further cycles of mitotic division until a ball of cells known as the **morula** is formed (*Fig. 2*). The morula is approximately the same size as the original ovum, as little new cytoplasm is made during these initial divisions. Further divisions then take place, and an internal fluid-filled cavity is formed; the sphere of cells is now known as the **blastula**, and the fluid-filled cavity is known as the **blastocoele** (*Fig. 2*).

## Implantation and formation of embryonic tissues

In the blastula, an **inner cell mass (ICM)** forms. The cells immediately outside this ICM are known as the **trophoblast**, and it is this set of cells that begins to implant into the wall of the uterus. Vascular channels form to bathe the embryo in nutrient-rich blood. Implantation is, in truth, an invasion by a (genetically) foreign body; the uterine lining is designed to withstand this invasion, but other areas are not. Ectopic implantation can be very dangerous because hemorrhage can result. The trophoblast becomes the **placenta** and **fetal membranes**, the various membranous bags that surround the fetus in the uterus during gestation; the placenta proper is present at around 10 weeks after fertilization

The ICM then develops a fluid-filled sac, known as the **amniotic cavity** (*Fig. 3*). The cells lining the amniotic cavity are known as **ectoderm**, whereas those lining the original blastocoele cavity are known as **endoderm**. Where endoderm

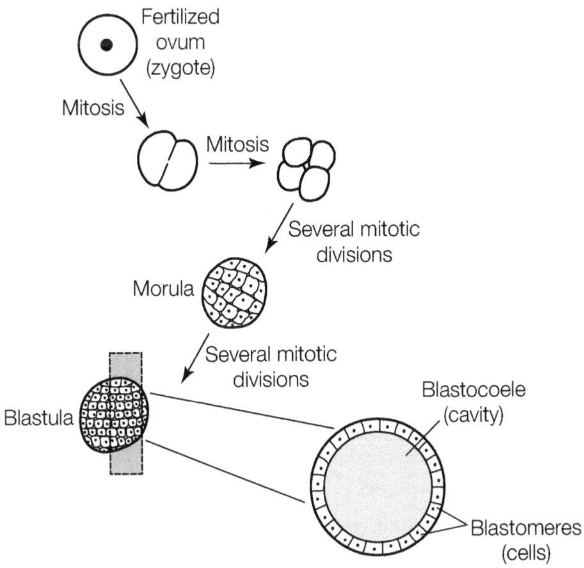


*Fig. 2. Development of the zygote into the stage just before the embryo, the blastula.*

and ectoderm come together is known as the **embryonic disk**: an elliptical disk which, along its longitudinal axis, will become the head–tail axis of the fetus. Some ectodermal cells along the midline migrate down towards the endoderm and spread between the ectoderm and endoderm to form **mesoderm**, the third cell layer of the three-layered embryo (*Fig. 3*). Along the midline, some mesodermal cells are left behind; these become the notochord or primitive backbone.

Further folding and migration of embryonic tissues takes place, so that by 8 weeks after fertilization, the single-celled zygote has become a **fetus**, with the majority of its body organs in place. Over the next 30 weeks, the fetus becomes much larger, with much of the development that takes place involving the brain and nervous system.

**Fetal nutrition**

While this development from a blastocyst to a fetus is taking place, the cells that make up the trophoblast deeply invade the endometrial lining of the uterus, gaining access to the nutrients in the blood supply of the mother (*Fig. 4*). Initially, blood sinuses known as **trophoblast lacunae** form when the trophoblast disrupts maternal blood capillaries. Maternal blood is flowing through these, and this blood supplies the embryo with adequate nutrients and oxygen to allow the early stages of development to take place. Later, finger-like projections (**chorionic villi**) push their way into the trophoblast from embryonic mesoderm, eventually forming blood vessels. The placenta develops from both fetal and maternal tissues, and acts as the interface between the fetus and the mother. Fetal blood and maternal blood come into close apposition in the placenta, but do not mix. The space between the villi is taken up by maternal blood vessels. In the mother, uterine arteries that supply only the uterus outwith pregnancy (100 ml $min^{-1}$), must supply both the uterus and the placenta during pregnancy (around 800 ml $min^{-1}$ at the end of **gestation**).

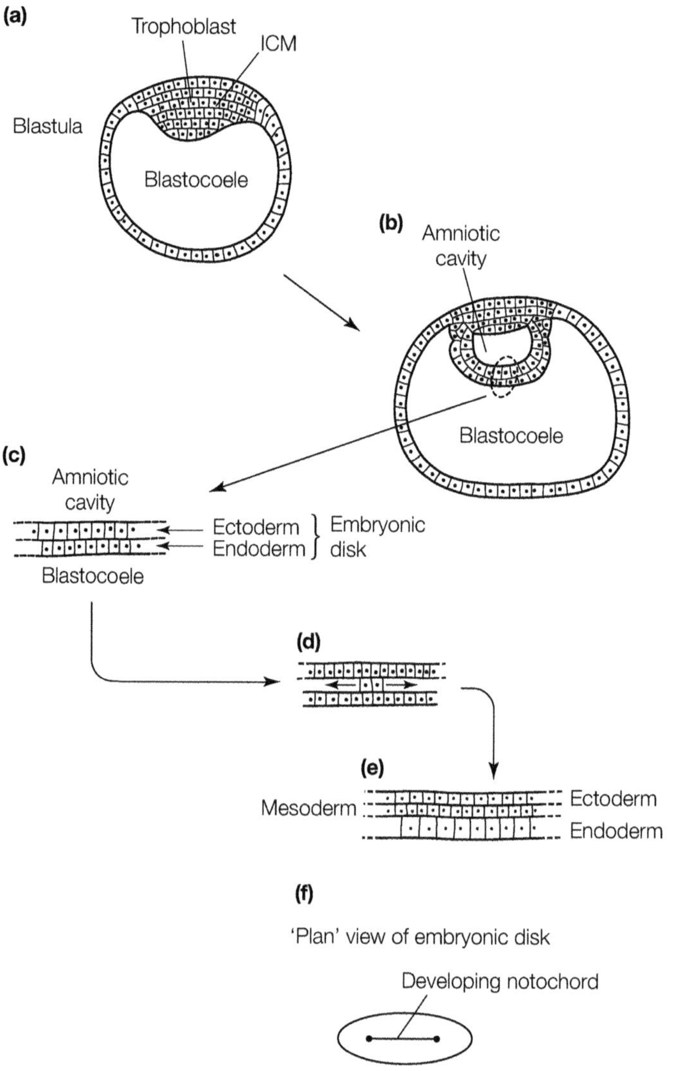


*Fig. 3. Development of the blastula into the embryo: (a) The blastula develops an inner cell mass (ICM) and trophoblast; (b) the ICM develops into two layers of cells, separating the amniotic cavity from the blastocoele; (c) these two layers of cells are known as ectoderm and endoderm. Collectively, these cells are known as the embryonic disk; (d) ectodermal cells in the midline of the embryonic disk migrate towards the endodermal layer; (e) these cells migrate along between the ectoderm and endoderm to form mesoderm; (f) in the midline of the embryonic disk some mesoderm cells will eventually form the notochord or primitive backbone.*

The fetus is supplied entirely by oxygen and nutrients from the bloodstream of the mother, and several mechanisms exist that ensure that the fetus, in a parasitic fashion, gains as much from this relationship as possible. **Fetal hemoglobin (HbF)**, for example, is different from **adult hemoglobin (HbA)**. HbA is composed of two α-globin and two β-globin chains ($\alpha_2\beta_2$), but HbF is composed of two α-globin chains and two γ-globin chains ($\alpha_2\gamma_2$). HbF has a higher affinity for oxygen than does HbA, and fetal blood contains around 20% more hemoglobin

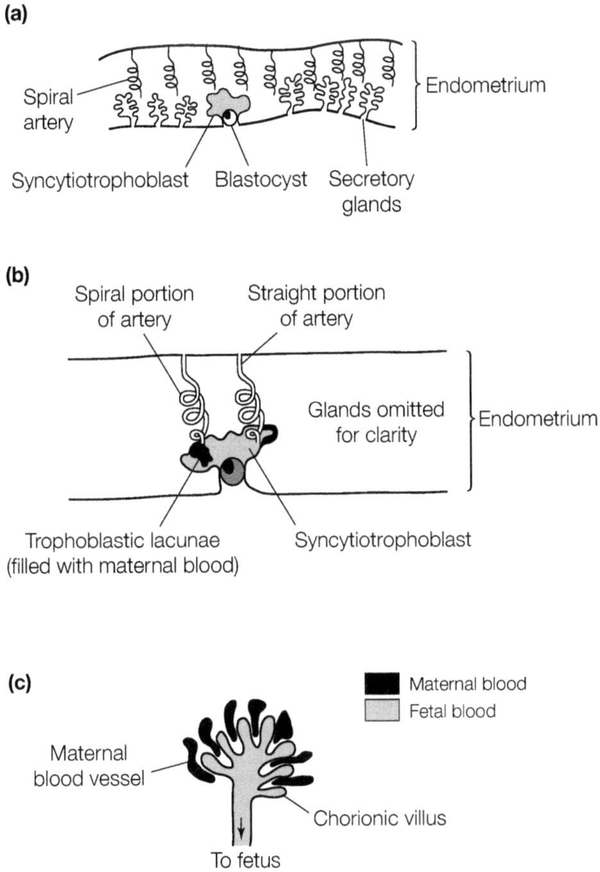


*Fig. 4. Supply of nutrients to the embryo/fetus. (a) The endometrium of the uterus in secretory phase (see Topic L2) is invested with large numbers of glands and spiral arteries; (b) the trophoblastic tissue of the fertilized conceptus forms a large mass (syncytiotrophoblast) and invades the uterine lining, rupturing blood vessels to form trophoblastic lacunae. At this stage, maternal blood washes over the conceptus in the endometrium, supplying some nutrients; (c) at a later stage, the fetus produces mesodermal villi that, within the placenta, come into close apposition with maternal blood vessels, facilitating exchange of nutrients and waste products.*

per liter than adult blood. At a given oxygen tension ($P\text{O}_2$), HbF can carry 2–3 times more oxygen than HbA. Thus fetal blood can carry a lot of oxygen, despite the fact that the $P\text{O}_2$ of blood in the **umbilical vein** (supplying the fetus) may be as low as 30 mmHg. At a $P\text{O}_2$ of 30 mmHg, adult blood would have very little oxygen content (around 100 ml oxygen $l^{-1}$), but in the fetus this $P\text{O}_2$ results in a blood oxygen content of 200 ml $l^{-1}$, close to that of the richly oxygenated blood leaving the lungs in an adult. See Topic E4 for further details of oxygen transport by hemoglobin.

The fetus starts to produce small amounts of HbA from about 13 weeks of pregnancy, and the ratio of HbF:HbA in the third trimester is about 4:1. The genes for both β- and γ-chains of hemoglobin are present on chromosome 11 and, as the baby starts to breathe after birth, gene switching causes a decrease in HbF and an increase in HbA, so that little HbF is present in normal human blood. This necessitates the synthesis of new erythrocytes containing HbA and the destruction of erythrocytes containing HbF. Occasionally this can lead to neonatal **jaundice** when the capacity of the neonatal liver to conjugate **bilirubin**, derived from breakdown of heme groups, is inadequate. Bilirubin is normally excreted into bile conjugated to glucuronic acid (see Topic J5), and the **glucuronyl transferase** enzyme responsible for this reaction is expressed towards the end of the third trimester. Any delay in the expression of this enzyme can lead to neonatal jaundice.

The ability of the fetus to rob the mother of nutrients is shown by the incidence of **gestational diabetes** in pregnant women. Around 3% of pregnant women develop **diabetes mellitus** during pregnancy. This usually manifests itself as tiredness and extreme thirst in the middle of gestation. In all pregnancies, however, hormones produced by the placenta are thought to result in some degree of **insulin resistance**; insulin is less able to induce transport of glucose into maternal tissues (see Topic K9), and plasma glucose levels rise. In some women, gestational diabetes results when they are not able to produce enough insulin to counteract this effect. Inducing insulin resistance in the mother benefits the fetus because more glucose is then available to cross the placenta for use by the fetus. In fact, babies born to mothers with gestational diabetes can be so large that delivery is difficult. In the vast majority of cases, normal control of maternal plasma glucose concentration by insulin returns on delivery of the placenta, although gestational diabetes has been shown to be associated with the development of non-gestational diabetes later in life.

# L5 Pregnancy

## Key Notes

**Defining terms**

It is important that the following terms are understood: zygote, blastomere, blastula, conceptus, embryo, fetus and puerperium.

**Hormonal changes in pregnancy**

The major hormones of pregnancy are human chorionic gonadotropin (HCG), progesterone, estrogen, prolactin and human placental lactogen (HPL). HCG maintains the corpus luteum in the mother's ovary in the early stages of pregnancy. Progesterone and estrogen are produced by the placenta from around week 12 of pregnancy. Prolactin and HPL stimulate breast development in preparation for lactation after birth. Of these hormones, only prolactin levels are maintained after birth; levels of the others decrease because the placenta is delivered after birth of the baby.

**Anatomical changes**

The major anatomical change in pregnancy is the hyperplasia and hypertrophy of the uterine myometrium. This, together with the fetus, placenta and amniotic fluid, results in other organs such as the intestines, bladder and diaphragm becoming displaced.

**Hematological and cardiovascular changes**

A number of hematological and cardiovascular changes take place during pregnancy. Blood volume increases by around 40% and erythrocyte numbers increase by around 20%. Cardiac output increases by around 30%. Smooth muscle relaxation is common, as a result of high circulating levels of progesterone; this can result in lowered blood pressure, particularly in the second trimester of pregnancy.

**Respiratory and renal changes**

Respiratory changes that take place in pregnancy to keep the fetus adequately oxygenated are easily achieved. For example, minute ventilation of the lungs may rise by up to 40%, reflecting an increase in tidal volume. The kidneys become enlarged and renal blood flow increases by as much as 55%, resulting in an increase in glomerular filtration rate. As a result, the levels of creatinine, urea, and uric acid in the blood fall. Some loss of glucose in the urine (glycosuria) can occur.

**Related topics**

Layout and function of the vasculature and lymphatics (D5)
The cardiac cycle, blood pressure, and its maintenance (D6)
Investigating lung function (E3)
Overview of the female reproductive system (L2)
From ovum to fetus (L4)
Glomerular filtration and renal plasma flow (M2)
Reabsorption of electrolytes and glucose (M3)

## Defining terms

For the sake of clarity in the proceeding description of pregnancy it is perhaps worth defining the following:

- **zygote**: a single cell (fertilized ovum), present as such for only a very brief stage
- **blastomeres**: cells at the third to fifth division (30–50 cells) of the original zygote, which form a cell mass termed the **morula**
- **blastula**: the stage of development after the morula enters the uterus. It consists of a peripheral layer of blastomeres forming the **trophoblast** surrounding a mass of cells (**inner cell mass**)
- **conceptus**: the product of conception at any point between conception and birth. This would include the embryo/fetus as well as extra-embryonic membranes such as the placenta. Thus it includes everything that arises from the zygote, at any stage of gestation. It is probably used mostly for the early stages of pregnancy, after which separate names for its different components would be used; for example **trophoblast, embryonic disk, yolk sac, amniotic cavity, chorion, chorionic vesicle** (which becomes the extra-embryonic coelom and yolk sac), **chorionic villi**
- **embryo**: used up to week 8 of pregnancy – that part of the conceptus that gives rise to the fetus (i.e. the embryonic disk). Strictly speaking, in fertility treatment, it is the conceptus which is transplanted, not just the embryo
- **fetus**: this term should be restricted to the embryo from the ninth week onwards, with separate names for other components (e.g. **placenta**)
- **puerperium**: the period after childbirth.

Pregnancy involves a number of physiological changes in the mother that allow growth and development of the conceptus through the period of uterine quiescence, and prepare the mother for nurture of the fetus and delivery. Interestingly, hormone production and secretion by both the conceptus and the mother facilitate these changes in the maternal anatomical, hematological, cardiovascular, respiratory, gastrointestinal, and renal systems.

As outlined in Topic L4, following fertilization of the ovum in the Fallopian tube, the zygote undergoes three or four rounds of cell division to produce blastomeres of the morula. The morula enters the uterus and begins to absorb uterine fluid, eventually forming the blastula. The inner cell mass of the blastula will eventually develop into the embryo and later the fetus, while the trophoblast will surround the embryo/fetus and supply nutrients. Cells of the trophoblast are also an important source of hormones and the outermost layers (chorion) compose the fetal contribution to the placenta, the maternal contribution deriving from the endometrium. The stages of pregnancy are shown in *Table 1*.

*Table 1. Stages of pregnancy*

| Week of pregnancy | Weeks since last menstrual period | Trimester | Stages |
|---|---|---|---|
| 1–13 | | First | Pre-embryonic |
| | 4–10 | | Embryonic development |
| 14–27 | | Second | Fetal development |
| | 23 | | Viability |
| 28–40 | | Third | Maturation |
| 37–42 | | Term | Delivery |
| Delivery + 6 weeks | | | Puerperium |

### Hormonal changes in pregnancy

The changes in the concentrations of **estrogen**, **progesterone**, and **human chorionic gonadotropin (HCG)** during pregnancy are shown in *Fig. 1*. During the first two months of pregnancy virtually all of the estrogen and progesterone is synthesized and secreted by the **corpus luteum**. Estrogen stimulates the growth of uterine muscle and of connective tissue forming the cervix, and promotes the development of ductal and alveolar tissue in the breast (see Topic L6). Progesterone stimulates the formation of the decidua from the endometrium, and has a general relaxant effect on uterine smooth muscle, inhibiting uterine contractility during pregnancy.

The patency of the corpus luteum, which in the absence of fertilization degenerates within a couple of weeks, is maintained by the actions of HCG, a glycoprotein secreted by the trophoblast cells as they merge with and burrow into the endometrium. Like **luteinizing hormone (LH)**, HCG maintains the corpus luteum and promotes the secretion of steroid hormones by the tissue. In this case however, the signal to maintain the corpus luteum is derived from the conceptus, rather than from the mother. The peak secretion of HCG is seen at about week 10 of gestation, and decreases rapidly to a low level at week 12, a level which is then maintained throughout the rest of pregnancy. The fall in HCG concentration is accompanied by degeneration of the corpus luteum, and the loss of this tissue as a source of steroid hormones. From week 12 of pregnancy, the trophoblast cells of the placenta start to produce large amounts of estrogen and progesterone. Interestingly, although the placenta can synthesize progesterone from cholesterol and also has the **aromatase** enzyme required for the conversion of **androgens** to estrogens, it lacks the enzymes required to metabolize progesterone to androgens (see Topic K8). The placenta is able to synthesize estrogens only by being supplied firstly with androgens from the maternal ovaries and adrenals, and/or the fetal adrenals. The raised levels of progesterone and estrogen in the maternal circulation during pregnancy, from the corpus luteum initially and latterly from the placenta, exert negative feedback on the secretion of **gonadotropin-releasing hormone (GnRH)** by the hypothalamus and on the secretion of LH and **follicle-stimulating hormone (FSH)**

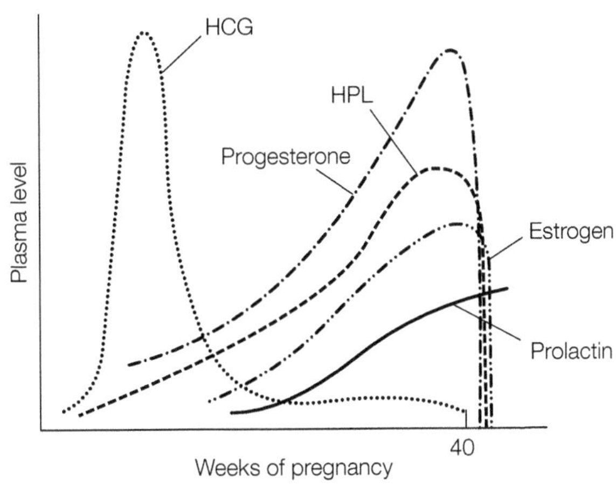


*Fig. 1. Plasma levels of various hormones during pregnancy. HCG, human chorionic gonadotropin; HPL, human placental lactogen.*

by the anterior pituitary (see Topics K4 and K5). This results in the absence of ovarian and menstrual cycles during pregnancy.

Another hormone, which is unique to the placenta, is **human placental lactogen (HPL)**. HPL has some growth hormone-like properties (mobilizes fat, stimulates maternal glucose production, and inhibits maternal glucose utilization). HPL also has prolactin-like properties (stimulates breast development).

## Anatomical changes

The development of the embryo and eventually the fetus takes place in the uterus, surrounded by amniotic fluid, and the balance of hormonal effects allows the uterine muscles to grow both in number (hyperplasia) and size (hypertrophy) without compromising the integrity of the fetus. At term, the uterus is approximately 10 times its pre-pregnant weight of around 100 g. During this time, the uterus becomes a sealed unit as the cervix at its outlet becomes firm and non-compliant due to the laying down of inflexible collagen matrix and the formation of a mucus plug. The major role of the cervix is to retain the conceptus within the uterus. This situation changes markedly as the body prepares for delivery of the fetus (parturition; see Topic L6).

The growing fetus also displaces the maternal diaphragm, heart, and bladder. Coupled with the retention of fluid in all body systems, these anatomical changes result in an increase in maternal weight during pregnancy, with contributions, at parturition, from both the conceptus (placenta, 0.7 kg; amniotic fluid, 0.8 kg; fetus, ~2.8 kg) and the mother (myometrium, 0.9 kg; fat, 4 kg; blood, 1.2 kg). The rate of weight gain during pregnancy is shown in *Table 2*.

*Table 2. Average maternal weight gain during pregnancy*

| Weeks of pregnancy | Rate of weight gain (kg week$^{-1}$) |
|---|---|
| <18 | 0.3 |
| 18–28 | 0.45 |
| 28–term | 0.36–0.41 |

The embryo is almost negligible in weight, but from about 12 weeks onwards, the weight of the fetus is proportional to approximately the cube of its length, growing rapidly during the third trimester of pregnancy (*Table 3*). Pregnancy represents an anabolic state metabolically, and increased body fat (4 kg) and protein are features of the condition.

*Table 3. Length and weight of an average fetus during pregnancy*

| Week of pregnancy | Length of fetus (cm) | Weight of fetus (kg) |
|---|---|---|
| 20 | 28 | 0.35 |
| 28 | 39 | 1.25 |
| 32 | 44 | 2.00 |
| 40 | 48 | 2.80 |

## Hematological and cardiovascular changes

The growing fetus presents an increased demand on the mother in terms of metabolism and blood supply to the placenta. Pregnancy is associated with an approximate 40% increase in blood volume (an extra 1–2 l), and a 20% increase in erythrocyte mass (total volume of red cells in the circulation), in an attempt to satisfy the increased oxygen use (~15%) by the fetus as well as an increase in maternal basal metabolic rate. Interestingly, because plasma volume increases more than the red cell mass, there is a fall in hemoglobin, hematocrit and red cell count during pregnancy. Maternal hemoglobin may be reduced as iron is transferred to the fetus.

The heart becomes slightly enlarged in mid-pregnancy due to wall thickening, and particularly increased venous filling. There is also a 30% rise in **cardiac output** in the first trimester (heart rate increases from 70 to 85 beats $min^{-1}$ and stroke volume from around 65 to 70 ml), and this is increased further during labor. Such increases have major implications for pregnancy in women with pre-existing heart failure or valvular heart disease. Despite these alterations in cardiac output, it is normal for pregnant women to have slightly lower blood pressure than their non-pregnant counterparts, particularly in the first and second trimesters. This is thought to be caused by the general relaxant effect of progesterone on smooth muscle, which lowers **total peripheral resistance** (see Topic D6), to the extent that the influence of increased cardiac output on blood pressure is cancelled out by reduced resistance to blood flow.

Peripheral **edema** is also common in pregnancy, particularly in dependent areas like the ankles. This is caused by decreased oncotic pressure in the blood (see Topic D5); as plasma volume increases to a greater extent than plasma protein production, less fluid is attracted back into the capillaries from the interstitial space.

## Respiratory and renal changes

Although, as mentioned above, there is a 15% rise in oxygen consumption during pregnancy, this does not impose any major problems for the mother, and her respiratory rate does not change. However, tidal volume (see Topic E3) increases by approximately 40%, and thus the **minute ventilation** (breaths per minute × tidal volume) is also increased by the same percentage. Inspiratory volume increases progressively during pregnancy, while various measures of expiratory flow rate remain unchanged.

The kidneys become enlarged during pregnancy due to an increase (by 70% in the third trimester) in renal parenchymal volume. This is accompanied by a 30–55% rise in renal blood flow which is apparent as early as the first trimester. There is a simultaneous increase, by 50% after four months of pregnancy, in **glomerular filtration rate (GFR)**, although this falls somewhat towards term. The resulting increased **creatinine clearance** (see Topic M2) however, is not accompanied by increased creatinine or urea production at this time, and so the plasma concentrations of these metabolic end-products fall. A fall in serum uric acid concentration during pregnancy also reflects increased clearance without an increase in production or reabsorption. The filtered loads of sodium and potassium rise as a consequence of the increase in GFR, and a parallel increase in tubular reabsorption leads to retention of both cations in maternal stores.

Increased glucose excretion, possibly due to the increased GFR and/or decreased tubular reabsorption, is also a feature of pregnancy. This occurs independently of the concentration of blood glucose (but see Topic L4), and may lead to **glycosuria**.

# L6 BIRTH AND LACTATION

## Key Notes

**Birth**

Birth is triggered by a complex series of hormonal and mechanical events that influence contractility of uterine smooth muscle. Oxytocin is released from the posterior pituitary in response to uterine stretch, and helps to coordinate uterine contractions during labor. Ripening of the cervix is also important, and stretch of the cervix, usually by the baby's head, is an important factor in producing more forceful uterine contractions. As labor progresses, uterine smooth muscle converts from multi-unit to single-unit smooth muscle by increasing the number of gap junctions between cells. Uterine contraction persists after birth of the baby, so that the placenta can be delivered and these contractions can continue for several weeks after birth, stimulated by oxytocin released as part of the response to suckling of an infant.

**Breast development and structure**

The breasts are exocrine glands designed to produce milk to nourish a baby after birth. Breast development takes place during puberty, but milk is not produced until after birth, under the control of certain pituitary hormones. Each breast consists of 15–20 separate secretory units.

**Lactation**

During pregnancy, breast tissue develops considerably, under the influence of several hormones including estrogen and prolactin. Later in pregnancy, and directly after birth, colostrum may be produced from the breast. Milk is only produced after birth of the baby when estrogen and progesterone levels fall, and prolactin can then act unopposed. Human milk is composed of primarily water, carbohydrate, lipids and protein.

**Milk ejection**

Milk is ejected by contraction of the myoepithelial cells in the breast lobules in response to oxytocin. Oxytocin release is triggered by suckling at the breast, via a feedback loop in which the afferent arm is neural and the efferent arm is hormonal. Afferent input to the hypothalamus also decreases the secretion of dopamine; this in turn leads to an increase in prolactin (and thereby milk) production.

**Related topics**

Homeostasis and integration of body systems (A2)
Hypothalamic hormones (K4)
Pituitary hormones (K5)
Overview of the female reproductive system (L2)
Pregnancy (L5)

## Birth

The duration of pregnancy is approximately nine months (270 ± 14 days), and throughout this time the uterus undergoes periods of weak and slow rhythmic contractions known as **Braxton Hicks' contractions**. As pregnancy progresses, the contractions become stronger, becoming acute near the time of delivery of the baby. **Parturition**, the process by which the baby is born, relies on the expulsion of the fetus from the uterus, through the concerted action of contractions of abdominal and uterine smooth muscle. A number of factors, both humoral and mechanical, combine to increase the contractility of the uterus such that at parturition the force of contraction is sufficient to expel the baby. As mentioned in Topic L5, the placenta secretes large amounts of both **progesterone** and **estrogens** throughout pregnancy. Progesterone inhibits uterine contractility, and thereby prevents premature expulsion of the fetus. Its action on the cervix causes a softening of the tissue, allowing it to stretch during the first stage of labor. Estrogens, on the other hand, have the opposite effect on uterine contractility, and increased secretion of estrogens relative to progesterone later in pregnancy initiates the stronger contractions occurring at this time. Another hormone, **oxytocin** (from the posterior pituitary), is released just prior to parturition in response to cervical stretching, and stimulates uterine contractions even further. Synthetic analogs of oxytocin are administered in some cases to help strengthen uterine contractions during labor.

The mechanical factors involved in the eventual delivery of the baby arise as a result of the increase in size and movement of the fetus during pregnancy. Thus there is a steady stretching as the fetus grows, and further occasional stretching as the fetus moves in the womb. The effect of such stretching on contraction is illustrated by the observation that when **twins** are present in the uterus they are born on average approximately 19–20 days earlier than when a single baby is present. The gradual stretching of the cervix elicits contraction of uterine muscle, so that as the cervix dilates as the baby's head descends, uterine contractions are enhanced. More than 90% of babies are delivered head first, and this serves to open the birth canal for subsequent delivery of the body. Stretching of the birth canal in turn stimulates neurogenic reflexes which initiate contraction of abdominal muscle. Thus at this point, uterine contractions are accompanied by contractions of abdominal muscle, increasing the force for expulsion of the baby. The combined actions of abdominal and uterine contractions may exert a force in excess of 10 kg at each contraction.

The strong contractions may occur at about every 30 min within two days of delivery, becoming more frequent as labor progresses such that they occur every 2–3 min just prior to parturition. These late contractions can be extremely intense, and allow only a very short period of relaxation between them. Hormonal changes around the time of labor result in increased expression of **gap junction** proteins in the smooth muscle cells of the uterus, so that it is converted from **multi-unit smooth muscle** to **single-unit smooth muscle** (see Topic I6). This allows spread of depolarization from one myometrial cell to another, causing coordinated contraction of the uterus.

As mentioned earlier, softening of the cervix during pregnancy under the action of progesterone allows the tissue to stretch during labor, and in the first stage of labor it dilates such that the cervical opening is as large as the baby's head. Once this occurs, the **fetal membranes** rupture and **amniotic fluid** is lost via the vagina. This initiates the second stage of labor which includes the delivery of the baby, accompanied by acute pain due to stretching of both the cervix and perineum.

Within an hour of delivery, the uterus shrinks markedly and sheds the placenta. This may involve loss of 300–400 ml of blood from placental sinuses, but the act of shrinkage of the uterus post partum constricts the blood vessels originally supplying the placenta and limits this loss. The uterus has normally returned to its pre-pregnant size within six weeks after birth. It is important at this stage that **lactation** begins, so that the baby can be nourished after the first 2–3 days of life. Suckling of the newborn is important in inducing further release of oxytocin (see below), which helps contract the uterus, sometimes resulting in '**after pains**', labor-like pains induced by suckling at the breast.

## Breast development and structure

Breast tissue is present in both males and females, but development begins during puberty in females under the influence of female sex hormones. Breast tissue does not normally develop in males, but can do under certain circumstances, resulting in **gynecomastia**. One possible cause of gynecomastia in males is liver disease; this is because liver disease can affect the ability of the liver to metabolize estrogens.

Adult breast tissue (*Fig. 1*) is composed of 15–20 secretory units composed of epithelial cells and their basement membrane. Some epithelial cells have the capacity to contract in response to certain hormonal influences, and are known as **myoepithelial cells**. These secretory units have ducts that open out onto the surface of the body at the nipple, so this epithelial tissue is continuous with the epidermis of the skin. The secretory units of the breast have ducts, lobes, and lobules. Each lobe is arranged like a bunch of grapes in which each 'grape' would represent a lobule of epithelial cells.

The glandular tissue of the breast is surrounded by connective tissue, a lot of which is adipose tissue. **Suspensory ligaments (of Cooper)** act to tether the breast tissue to the deep connective tissue fascia of the breast, but not to the underlying muscle of the chest wall, the **pectoralis major**.

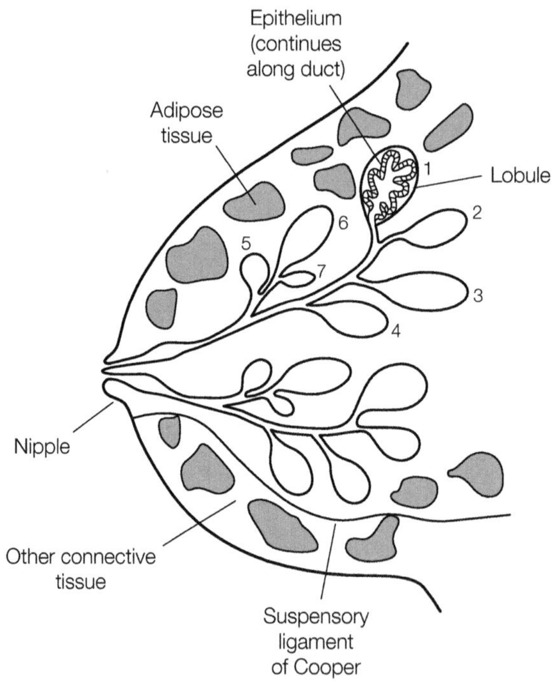


*Fig. 1. Diagram of a cross-sectional view through the human breast. Lobules 1–4 belong to one lobe, and lobules 5–7 belong to another.*

## Lactation

As indicated above, the development of the breast in females begins in puberty in response to the monthly cycle of estrogen release, with the initial growth of the stroma and duct system, and deposition of fat. There is a considerable increase in growth during pregnancy as the tissue prepares for the synthesis and secretion of milk. The hormonal drive for this increase in breast size, which includes growth and branching of the duct system, development of the alveolar system, and deposition of fat in the stroma, comes from the placenta (estrogen, progesterone, and **human chorionic somatomammotropin**), and the anterior pituitary (**prolactin**). There is also a contribution from **thyroid hormone** and **corticosteroids**.

Prolactin is the major stimulus for milk synthesis and secretion, but despite the rise in plasma levels of the hormone during pregnancy to 10 times the non-pregnant level, the high levels of estrogen and progesterone inhibit its action on milk secretion. It is not until after parturition and the loss of the placenta (the source of estrogen and progesterone) that prolactin is able to act unopposed and stimulate milk secretion. The development of the lobulo-alveolar cells during pregnancy prepares the cells for the synthesis and secretion of the components of milk (*Table 1*).

*Table 1. Composition of human milk*

| Component | % Total | Daily output |
|---|---|---|
| Water | 88 | 1.5 l |
| Protein (casein, lactalbumin) | 1.5 | |
| Carbohydrate (lactose) | 7 | 100 g |
| Lipids (triglycerides) | 3.5 | 50 g |
| Calcium and phosphate | | 2–3 g |
| Ions ($K^+$, $Na^+$, $Cl^-$) | Highly variable | |
| Vitamins | Highly variable, except vitamin K which is absent | |

In biochemical terms this equates to a many-fold increase in the capacity of the cells for:

- fatty acid synthesis (pentose phosphate pathway and fatty acid synthase)
- acylation of glycerol phosphate and triglyceride synthesis
- α-lactalbumin and galactosyl transferase for lactose synthesis
- protein synthesis; this includes proliferation of the rough endoplasmic reticulum.

In each of the above, increased synthesis of the individual milk components is due to increased amounts of the synthetic enzymes rather than stimulation of pre-existing enzyme.

## Milk ejection

As mentioned above, milk secretion is under the control of prolactin. Secretion of prolactin, in turn, is controlled by stimulatory (**prolactin-releasing factor**) and inhibitory factors (**dopamine**) released from the **hypothalamus** (see Topic K4), and the balance is such that the level of prolactin is low. During pregnancy, plasma prolactin rises steadily in response to placental estrogen, to a level at parturition that is about 10-fold higher than the non-pregnant value. However, the inhibitory actions of estrogen and progesterone prevent milk secretion, and mean that only **colostrum** (a watery fluid containing protein and lactose but little fat) is released from the gland. The precipitous loss of estrogen and progesterone at parturition allows the high level of prolactin to exert its stimulatory effect on milk secretion by the alveolar cells.

However, milk does not flow easily into the ductal system and expression of milk through the nipple requires both neurogenic and hormonal reflexes set off by suckling as follows:

Suckling → sensory nerves → spinal cord → hypothalamus → oxytocin → blood → breasts → contraction of myoepithelial cells → milk expression by alveoli → ducts → nipple.

Oxytocin from the hypothalamus causes contraction of the myoepithelial cells, leading to ejection of preformed milk in the alveoli. Milk starts to flow within about 30 s of the suckling stimulus. The afferent input into the hypothalamus from the nipple inhibits the release of dopamine.

With the fall in the plasma estrogen level after parturition, the stimulus to prolactin secretion is also lost, and plasma prolactin levels fall, returning to basal levels after a few months. This gradual fall in plasma prolactin is interrupted temporarily by suckling, which initiates a secretory burst of prolactin lasting about an hour, and this brief rise in prolactin is sufficient to drive the breast to make milk for the next nursing period. Cessation of nursing causes the breast to lose the ability to produce milk within a few days, while continued suckling allows milk production for several years, although it is markedly decreased after 7–9 months post partum.

# M1 Layout of the renal system

## Key Notes

**The kidneys**

The kidneys are important in regulating fluid and electrolyte balance, and in the excretion of waste substances from the body. They receive around one-quarter of cardiac output, and up to 20% of renal plasma flow is normally filtered. This amounts to the filtration of 170 l of fluid per day, but only around 1% of this is excreted as urine. The kidneys can reclaim much of the fluid and electrolytes that are filtered. The kidneys also secrete a number of hormones, such as the active form of vitamin D, erythropoietin and thrombopoietin.

**The ureters and bladder**

Urine is conducted from the kidneys, via the ureters, to the urinary bladder. The bladder acts as the site of storage of urine until micturition (the act of urination) can take place at a socially appropriate time. Bladder smooth muscle is activated by stretch, but both branches of the autonomic nervous system have significant influence over the degree of bladder contraction, and of contraction of the internal urinary sphincter. The external urethral sphincter is controlled by voluntary input from the higher centers.

**The urethra**

The urethra conducts urine from the bladder to the outside world. The urethra is significantly longer in men than in women. The male urethra passes through the prostate gland, so prostate disease can affect urine flow.

**Normal values for production of urine**

Only a small proportion of the plasma filtered in the kidneys is lost as urine, amounting to approximately 1.5 l $day^{-1}$. The composition of this urine (e.g. the concentration of solutes therein) can be altered by the kidneys to meet the needs of the body.

**Related topics**

Overview of the male reproductive system (L1)
Glomerular filtration and renal plasma flow (M2)
Reabsorption of electrolytes and glucose (M3)
Hormonal control of fluid balance (M5)

## The kidneys

The **renal system** (*Fig. 1*) is designed to ensure that the body is able to get rid of various waste products from metabolism, in the form of nitrogen-containing compounds like **urea**, acids like phosphoric acid, and potentially toxic metabolites of many prescribed drugs. At the same time, the renal system must ensure that the amount of fluid lost in the **urine** maintains fluid balance. Thus, the **kidneys** are primary homeostatic organs, involved in maintaining the composition of blood and (by extension) extracellular fluid via endocrine and other mechanisms.

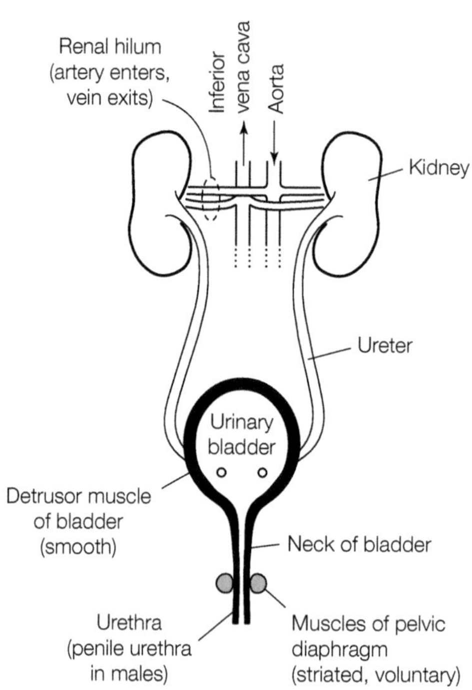


*Fig. 1. Layout of the renal system.*

Between 20% and 25% of cardiac output perfuses the kidneys, so normal **renal blood flow** is approximately 1.1 l $min^{-1}$. With a normal hematocrit of 0.45 (i.e. 45% of blood volume is taken up by blood cells), this corresponds to **renal plasma flow** of 600 ml $min^{-1}$ (1.1 l $min^{-1}$ × 0.55). On passing through the kidneys, 20% of the renal plasma flow (120 ml $min^{-1}$) is filtered, meaning that >170 l of plasma are filtered off by the kidneys in a day. Thankfully, only around 1% of this volume (1.7 l) is actually lost in the urine, the remainder being reabsorbed in the kidneys after filtration, otherwise we would have to spend our entire lives drinking salty water in the vicinity of a toilet. In addition to adjusting the volume of urine produced, the kidneys can make dilute or concentrated urine to suit the needs of the body. For example, if an excess of fluid has been ingested, then dilute urine will be produced, but if only a small amount of fluid has been ingested more concentrated urine can be made, with an osmolality of up to 1200 mOsmol (kg $H_2O)^{-1}$ (see Topic M5). As humans, we need to excrete 900 mOsmol of solute each day to balance intake and metabolism. Imagine that we could only produce urine with a similar osmolality to plasma (around 300 mOsmol $kg^{-1}$). This would mean that we would have to excrete 3 kg (3 l) of $H_2O$ with the solute. By being able to produce urine that is significantly hypertonic to plasma (up to 1200 mOsmol $kg^{-1}$), we need only produce 0.75 l urine, in order to excrete the 900 mOsmol of solute, avoiding an additional loss of 2.25 l of $H_2O$.

In addition to their role in the excretion of waste products in urine, the kidneys are also important in a number of other bodily processes. For example, they play an important role in middle- to long-term maintenance of **blood pressure** (see Topics D6 and M5). The kidneys are also important in the regulation of

**calcium homeostasis** (see Topic K7) by producing the active form of vitamin D (**1,25-dihydroxy vitamin D**, or calcitriol), which facilitates calcium absorption from the gut. The kidneys also respond to both **calcitonin** (see Topic K6), and **parathyroid hormone** (see Topic K7) as part of the mechanisms of calcium balance. In response to low levels of oxygen in the blood, the kidneys also produce **erythropoietin**, which increases the rate of formation of erythrocytes. Another hormone produced by the kidneys (and liver) is **thrombopoietin**, which stimulates the proliferation of **megakaryocytes**, the precursor cells from which platelets are formed in the bone marrow.

## The ureters and bladder

The urine that is made by the kidneys flows to the **bladder** via the **ureters** and leaves the bladder via the **urethra** (*Fig. 1*). Urine is conveyed to the bladder by peristaltic contractions of the smooth muscle in the wall of the ureters, and the ureters join the bladder inferiorly. This means that as the bladder fills with urine, there is no need for the ureters to move. Once in the bladder, urine is

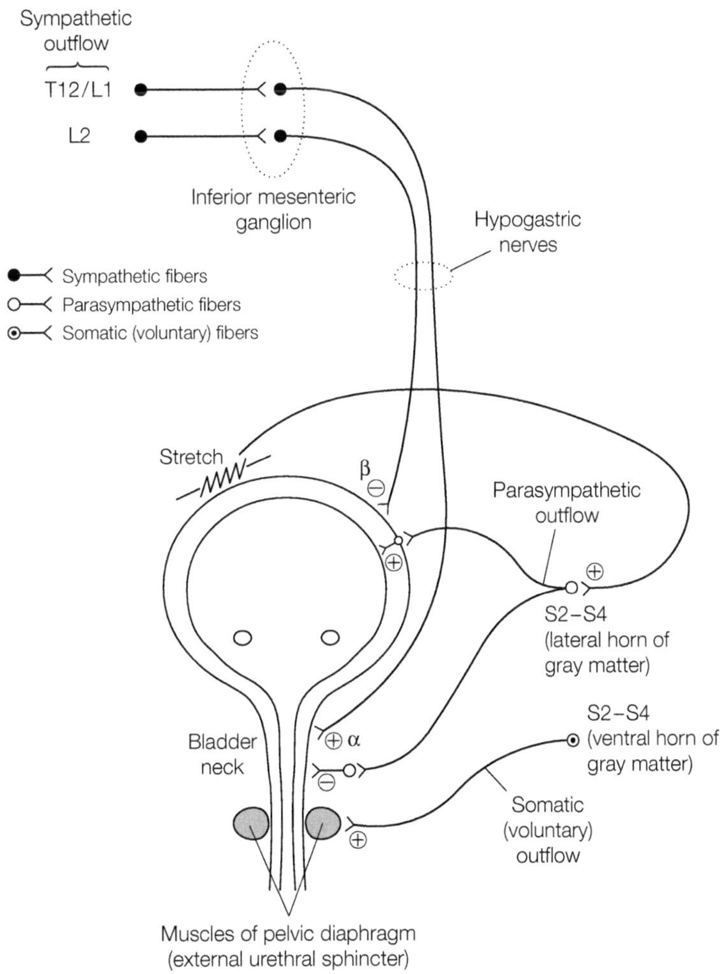


*Fig. 2. Neural control of urinary continence and reflexes involved in micturition. See text for details. α, α-adrenoceptors; β, β-adrenoceptors.*

stored until it is socially appropriate to void (**micturition**), or until the pressure in the bladder is so great that micturition is inevitable. The wall of the bladder is composed primarily of smooth muscle known as **detrusor muscle**, and the internal surface is lined with a specialized epithelium that is known as stratified **transitional epithelium** (the shape of the cells changes as the volume of the organ increases, to maintain patency of the epithelium). The detrusor muscle and the **bladder neck** are innervated by both branches of the autonomic nervous system. **Parasympathetic** input promotes micturition by contracting the detrusor muscle and relaxing the bladder neck, whereas **sympathetic** input inhibits micturition by relaxing the detrusor muscle and contracting the bladder neck. The detrusor muscle is stretch activated – when bladder volume reaches around 250 ml, the stretch of the muscle fibers activates a parasympathetic reflex that induces contraction of the detrusor muscle and relaxation of the internal urinary sphincter. The act of micturition can be aided by increasing intra-abdominal pressure, effectively pushing down on the bladder. It is only by voluntary contraction of the skeletal muscle of the **external urethral sphincter** (the **pelvic diaphragm**) that we are able to remain continent, a skill that is normally only mastered towards the end of the second year, or in the third year, of life. Painful sensations result when bladder volume reaches around 500 ml, and loss of control is normally unavoidable, even in adults, at bladder volumes of around 700 ml.

## The urethra

Urine flows from the bladder out of the body via the urethra (*Fig. 1*), which is significantly longer in males (20 cm) than in females (4 cm). The first parts of the urethra are also lined with transitional epithelium, and **urethral stretch** serves to reinforce the parasympathetic input to the bladder, encouraging rapid emptying. The male urethra is composed of three segments: the **prostatic urethra**, the **membranous urethra**, and the **penile urethra** (see Topic L1, *Fig. 1*). The prostatic component of the urethra runs through the body of the **prostate gland**, the membranous component through the pelvic diaphragm, and the penile urethra through the **penis**. In men with prostatic disease (either carcinoma or hypertrophy), flow of urine through the urethra may be impeded by the swollen prostatic tissue, leading to hesitancy in initiating micturition, slow flow, and inadequate emptying of the bladder.

## Normal values for production of urine

In a normal, healthy 70 kg male, daily water gain is matched to daily water loss and is normally around 2.5 l. In terms of intake, around 2.2 l is ingested in food and drink, and the remaining 0.3 l comes from oxidation of energy substrates by cellular metabolism; fats and carbohydrates are burned in oxygen to produce carbon dioxide and water (consider $C_6H_{12}O_6$ (glucose) + $6O_2 \rightarrow 6CO_2 + 6H_2O$ + energy). Around 0.2 l of water is lost in the feces daily, and a surprisingly large volume (0.8 l) is lost via skin secretions (including sweat) and via the lungs (**insensible water loss**). This leaves 1.5 l water in the urine as the balance to make up the total of 2.5 l; 170 l of plasma are filtered by the kidneys each day (see above and Topic M2), so more than 99% of filtered fluid is reabsorbed before it reaches the urine, and only small alterations (say 1%) in reabsorption result in large changes (100%) in the amount of urine produced daily. Normal values for urinary composition in a healthy 70 kg male are given in *Table 1*.

Kidney function is complex and exquisitely controlled, as described in the subsequent topics in this section.

*Table 1. Normal urine composition in humans*

| Parameter | Average measurement (or range) |
|---|---|
| Urinary volume produced | 1.5 l $day^{-1}$ |
| Urinary osmolality | 80–1200 mOsmol $kg^{-1}$ |
| Urinary urea | 11 g $day^{-1}$ |
| Urinary creatinine | 1.8 g $day^{-1}$ |
| Urinary urobilinogen | < 4 mg $day^{-1}$ |
| Urinary protein | < 0.15 g $day^{-1}$ |
| Urinary amino acids | 0.7 g $day^{-1}$ |
| Urinary glucose | 0.15 g $day^{-1}$ |
| pH | 6.0 |

# M2 Glomerular filtration and renal plasma flow

## Key Notes

**Layout of the nephron**

Each kidney is composed of around one million coiled tubes of epithelial tissue known as nephrons. Each nephron is composed of a renal corpuscle, where the plasma is filtered, and a renal tubule, where substances are reabsorbed from or added to the filtrate. The various parts of the nephron have specific names. The renal corpuscle is made up of the glomerular capillaries and Bowman's capsule, whereas the renal tubule is made up of the proximal tubule, the loop of Henle, the distal tubule, the collecting tubule, and the collecting duct.

**The glomerular filter**

Plasma is filtered at the glomerulus, and around 20% of renal plasma flow is normally filtered into Bowman's space here, amounting to 170 l day$^{-1}$ under normal circumstances. The glomerular filter is composed of endothelial cells of the glomerular capillaries (which are fenestrated), the epithelial cells of Bowman's capsule (which have foot processes, hence the reason why they are called podocytes), and between these two layers, the fused basement membranes of the endothelial cells and the epithelial cells. The glomerular filter selects substances for filtration on the basis of size and charge. Smaller, positively charged molecules are filtered, whereas larger negatively charged molecules (e.g. proteins) are repelled. The net filtration pressure at the glomerulus is determined by the same Starling's forces that govern fluid exchange between the capillaries and the intersititial fluid in other tissues.

**Glomerular filtration rate**

The normal quoted figure for glomerular filtration rate (GFR) is 120 ml min$^{-1}$, in a healthy 70 kg male. The clearance of a substance is defined as the theoretical volume of plasma cleared of the substance in a given period of time. GFR can be estimated by measuring the clearance of substances that are freely filtered at the glomerulus, but which are neither reabsorbed from nor secreted into the renal tubule, or metabolized by the kidney. Two such substances are creatinine and inulin, although in the case of creatinine, the clearance of the molecule is only an estimate of GFR, albeit a relatively accurate one.

**Renal plasma flow measurement**

Around 20% of renal plasma flow is filtered off as glomerular filtrate, and the remaining 80% flows around the renal tubules in peritubular capillaries. If a molecule is freely filtered at the glomerulus and secreted but not reabsorbed in the renal tubule, and its concentration in the plasma is low enough that the transport processes for its secretion into the urine remove all the remaining material, such that the mass excreted per unit time is equal to the mass presented to the kidney in the inflowing plasma, then the clearance of that molecule will be a relatively accurate estimate of renal plasma flow (RPF). One such molecule is para-aminohippuric acid (PAH). PAH clearance is approximately

600 ml $min^{-1}$, indicating that the normal filtration fraction (GFR/RPF) is 0.2. Renal blood flow would be commensurately higher than RPF, since only around 55% of blood volume is plasma (the rest is cells).

**Related topics**

Epithelia and connective tissue (C2)
Layout and function of the vasculature and lymphatics (D5)
Reabsorption of electrolytes and glucose (M3)

## Layout of the nephron

The majority of the tissue in the kidneys is epithelial; each kidney is composed of up to one million extensively coiled tubes, each in potential contact with the outside world. These coiled epithelial tubes are known as **nephrons**, each of which is the smallest functional unit of the kidney (*Fig. 1*). The nephron is the site of blood filtration, reabsorption of substances from this filtrate, and addition of substances to this filtrate to make urine. Each nephron is composed of a **glomerulus**, a proximal tubule (convoluted and straight parts), a **loop of Henle**, a distal tubule (convoluted and straight parts), a collecting tubule, and a **collecting duct**. Each of these parts of the nephron has a particular role in the formation of an appropriate volume of urine of the appropriate composition, as outlined in *Table 1*. At all points along the nephron, there exists the tubular space (containing filtered fluid that will become urine), epithelial cells (of which

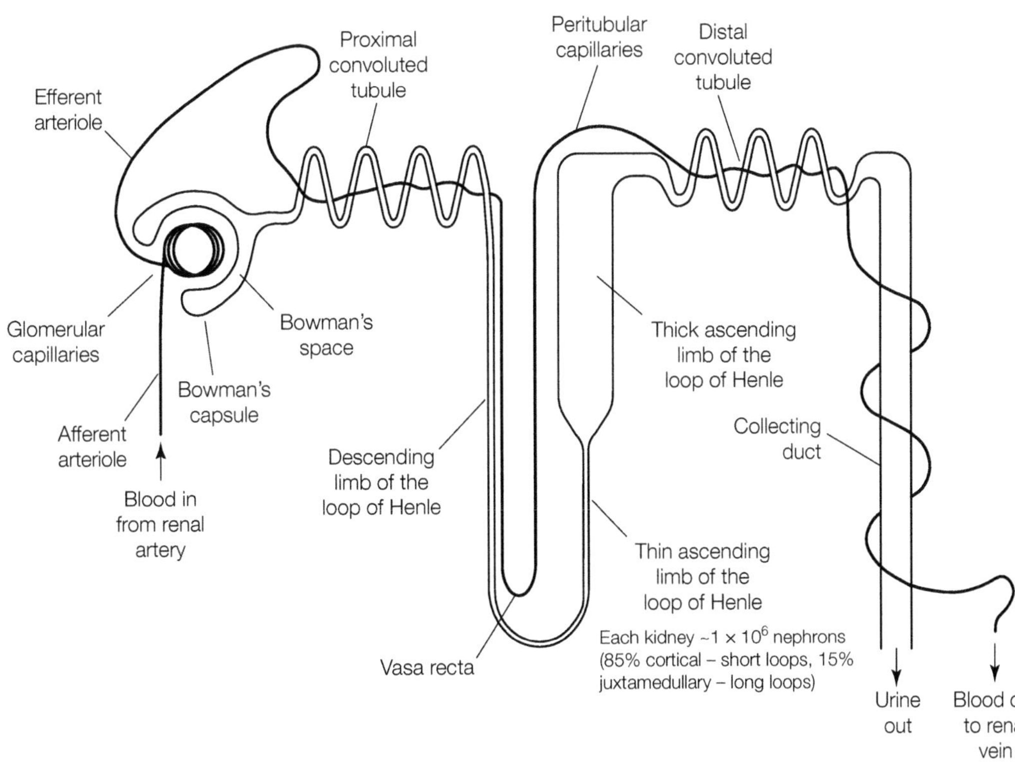


*Fig. 1. Schematic diagram of the nephron, showing its constituent parts.*

*Table 1. Major functions of the parts of the nephron*

| Part of the nephron | Major role |
|---|---|
| Glomerulus | Filtration of blood to form glomerular filtrate |
| Proximal tubule | Reabsorption of the majority of filtered fluid, electrolytes and nutrients (e.g. glucose); secretion of organic ions; acidification of the urine |
| Loop of Henle | Formation of concentrated urine; acidification of the urine |
| Distal tubule and collecting tubule | Final modifications to urine in terms of reabsorption of certain electrolytes (e.g. $Ca^{2+}$ ions) |
| Collecting duct | Formation of variably concentrated urine; important in $K^+$ balance |

the tubule is composed), intersititial space (between epithelial and other cells in the kidney), and blood vessels (containing blood). More detail on the function of these various parts of the nephron is given in subsequent topics in this section.

If one were small enough to be introduced into the urinary system, one could crawl up inside the **bladder**, into one of the two **ureters** and from there take as many as one million different routes inside either of the two kidneys to arrive at the site of filtration of the blood in the renal cortex. The site of filtration of the blood is known as the **renal corpuscle**, and the remainder is known as the **renal tubule.** *Figure 2* shows what the nephron would look like were it unraveled.

At the renal corpuscle, bundles of leaky capillaries allow some components of the blood plasma to enter **Bowman's space**. Bowman's space is the intraluminal compartment of the blind-end of each of the one million potential nephron routes that one could take into the kidney. As the filtered fluid passes along the renal tubule, many substances are reabsorbed into the blood, and some substances are added so that by the end of the renal tubule (collecting duct), the filtered and modified fluid has become urine. The first stage in urine production is filtration of the plasma at the glomerulus.

## The glomerular filter

Under normal circumstances, **glomerular filtrate** has a composition much like plasma, but without the plasma proteins; it has the same concentration of ions,

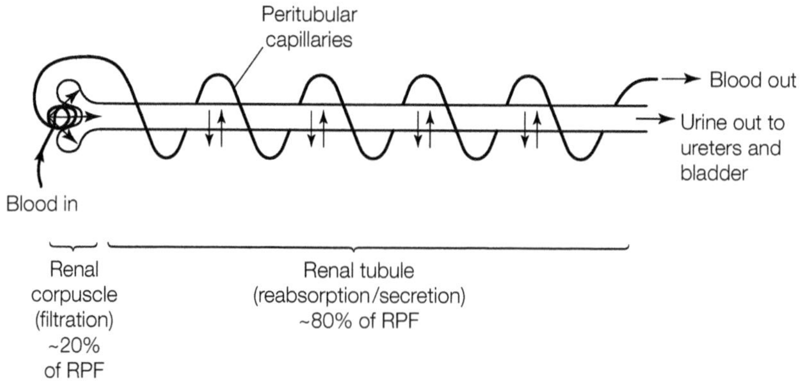


*Fig. 2. Events in the renal corpuscle (glomerular capillaries and Bowman's capsule), and the renal tubule (remainder of the nephron). RPF, renal plasma flow.*

glucose and amino acids, and has the same pH. Proteins and blood cells are normally excluded from the glomerular filtrate because they are too large to pass through the glomerular filtration barrier (*Fig. 3*).

The glomerular filtration barrier is composed of three layers:

1. the endothelial cells of **glomerular capillaries**
2. the foot processes of **podocytes**, the epithelial cells lining Bowman's space
3. *between* these two layers, the fused basement membrane of both these cellular layers.

The endothelial cells lining the glomerular capillaries are **fenestrated**. Not only are there small gaps between the cells as in most capillaries in the body, but the cells have pores through them, which measure 50–100 nm in diameter. Thus, glomerular capillaries are extremely leaky; plasma proteins, with a much smaller molecular size, might be able to pass through these pores, but in general they do not. The epithelial cells lining Bowman's space (podocytes) interdigitate with one another through their foot processes. Between these foot processes there exists a **filtration slit diaphragm**, made up of specific protein elements that act as a molecular filter, allowing molecules of no more than 20 nm diameter through into Bowman's space. The **glomerular basement membrane** (between the endothelial cells and the podocytes) is composed of layers of interwoven proteins, with small spaces in between.

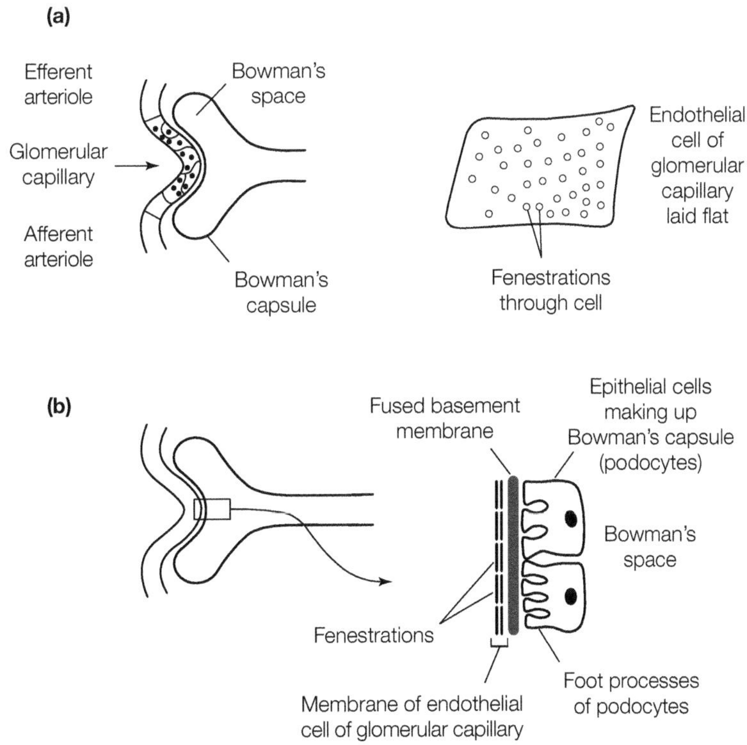


*Fig. 3. Diagram showing (a) fenestrations in the endothelial cells making up glomerular capillaries; and (b) the layers of the glomerular filtration barrier.*

The likelihood of a molecule passing from the plasma into the glomerular filtrate is determined by two factors: molecular size and molecular electrical charge. Molecules with a molecular weight less than 10 000 Da (1 Da (Dalton) = weight of a hydrogen atom) can pass through the glomerular filter, provided that they are not bound to plasma proteins. Above this molecular weight, molecules find it more difficult to pass through the barrier, until at around 100 000 Da, none can pass through. The various studies that have been conducted into glomerular filtration have estimated that if the glomerular filter functioned on the basis of molecular size alone, the pores should have a size of between 7 and 10 nm in diameter, but no such small-diameter pores have ever been seen. Thus, much of the sieving properties of the glomerular filter must be conferred in other ways.

All elements of the glomerular filter contain fixed negative charges (**glycocalyx** on the cellular elements, and **sialic acid** on the basement membrane) such that, above a given molecular weight, negatively charged (anionic) molecules are less likely to be filtered than neutral or cationic molecules. This is the major reason why serum albumin, despite its molecular diameter of approximately 7.5 nm, is not present in glomerular filtrate under normal circumstances; at physiological pH most serum proteins are negatively charged and are repelled by the fixed negative charges on the glomerular filtration barrier. Small anions such as $Cl^-$ and $HCO_3^-$ are completely filterable because of their extremely small size (35 Da and 61 Da, respectively). Glomerular disease often presents with **proteinuria** (protein in the urine) caused by disruption of the glomerular filtration barrier.

As with movement of fluid out of capillaries in other tissues (see Topic D5), Starling's forces are important in determining the **ultrafiltration pressure (UP)** at the glomerulus:

$$\text{UP} = (P_{\text{cap}} + \pi_{\text{BS}}) - (P_{\text{BS}} + \pi_{\text{cap}}) \quad (1)$$

**Glomerular capillary hydrostatic pressure ($P_{cap}$)** and **Bowman's space hydrostatic pressure ($P_{BS}$)** oppose each other. **Blood colloid osmotic pressure ($\pi_{cap}$)** and **Bowman's space colloid osmotic pressure ($\pi_{BS}$)** also oppose each other, but because blood contains far more protein than the glomerular filtrate in Bowman's space (for reasons see above), $\pi_{BS}$ can be discounted under normal circumstances. Thus the equation that describes glomerular UP normally takes the form:

$$\text{UP} = P_{\text{cap}} - P_{\text{BS}} - \pi_{\text{cap}} \quad (2)$$

In normal function, $P_{cap}$ is around 60 mmHg, $P_{BS}$ is around 15 mmHg, and $\pi_{cap}$ is around 25 mmHg, meaning that UP is normally around 20 mmHg. Filtration is a passive process. None of the components of the filtrate is actively transported from the plasma into the fluid in Bowman's space, and as UP increases or decreases, so does **glomerular filtration rate (GFR)**.

## Glomerular filtration rate

In a normal healthy 70 kg male, approximately 20–25% of cardiac output perfuses the kidneys. Thus, renal blood flow amounts to 1.1 l $min^{-1}$, and with a normal **hematocrit** of 0.45, this means that renal plasma flow is 600 ml $min^{-1}$. Normally, around 20% of this renal plasma flow is filtered at the glomerulus, resulting in a GFR of 120 ml $min^{-1}$. One of the ways in which a substance is removed (cleared) from the plasma is via filtration. The equation for calculating the **clearance** of substance X (the theoretical volume of plasma completely cleared of the substance in a given period of time, $C_X$) can be derived in the following manner:

1. measure the concentration of substance X in the plasma ($P_X$)
2. measure the concentration of substance X in the urine ($U_X$)
3. measure urine output over a period of 24 h ($V$).

Since substances are cleared into the urine, the amount cleared should equal the amount that is found in the urine. Thus the relationship between $C_X$, $P_X$, $U_X$, and $V$ takes the form:

$$C_X\, P_X = U_X\; V \tag{3}$$

Equation 3 can be re-arranged to form an equation for calculating $C_X$:

$$C_X = (U_X\, V)/P_X \tag{4}$$

**Creatinine** is one of the by-products of protein catabolism. Creatinine is freely filtered at the glomerulus but *on the whole* is not altered as it moves along the tubule, i.e. it is neither secreted by the renal tubule nor reabsorbed there. This means that the theoretical volume of plasma completely cleared of creatinine in a given period of time (the **creatinine clearance**) is an estimate of GFR. In actual fact, to calculate GFR accurately, one should infuse a subject with the plant polysaccharide, **inulin,** until plasma inulin concentrations reach equilibrium, then calculate **inulin clearance**. Inulin is freely filtered at the glomerulus, and as a plant polysaccharide there are no cellular transport proteins for it in human cells, so it is *never* secreted or reabsorbed by the renal tubule.

Note that if a given substance (e.g. $H^+$) is secreted by the renal tubule in addition to being freely filtered, then the numerator on the right of Equation 4 will increase, and clearance will be greater than for substances like inulin or creatinine. If the substance (e.g. glucose) is reabsorbed by the renal tubule in addition to being freely filtered, then the numerator on the right of Equation 4 will be lower, and clearance will also be lower than for inulin or creatinine.

As an example, using Equation 4, one can calculate $C_{\text{creatinine}}$ to obtain an estimate of GFR if one knows $P_{\text{creatinine}}$ (e.g. 80 μmol $l^{-1}$ ), $U_{\text{creatinine}}$ (e.g. 7.1 mmol $l^{-1}$), and $V$ (e.g. 2 l $\text{day}^{-1}$):

$$C_{\text{creatinine}} = (U_{\text{creatinine}}\; V)/\; P_{\text{creatinine}} \tag{5}$$

$$C_{\text{creatinine}} = (7100\ \mu\text{mol}\ l^{-1}\ 2\ l\ \ \text{day}^{-1})/80\ \mu\text{mol}\ \ l^{-1}$$

$$C_{\text{creatinine}} = 177.5\ l\ \ \text{day}^{-1} = 0.123\ l\ \ \text{min}^{-1} = 123\ \text{ml}\ \ \text{min}^{-1}$$

Thus a reasonable estimate of GFR from the data given above is 123 ml $\text{min}^{-1}$. It is important to realize that one is only estimating GFR by calculating $C_{\text{creatinine}}$. This is because some creatinine is secreted by the renal tubule. Therefore, $U_{\text{creatinine}}$ can be as much as 20% higher than if no creatinine were secreted by the renal tubule. However, the normal methods for measuring $P_{\text{creatinine}}$ are not entirely accurate either, so that the errors in both the numerator and the denominator on the right side of Equation 5 cancel each other out over a wide range of GFR. In modern clinical practice, $P_{\text{creatinine}}$, age, sex, and the race of the patient can be used to estimate GFR (via the MDRD (Modification of Diet in Renal Disease) algorithm) just as accurately as calculating $C_{\text{creatinine}}$, so this is now being done more often.

## Renal plasma flow measurement

**Renal plasma flow (RPF)** is normally around 600 ml $\text{min}^{-1}$, of which 120 ml $\text{min}^{-1}$ is filtered at the glomerulus. The remaining 80% of renal plasma flow then makes up the blood that perfuses the rest of the kidney in the **peritubular capillaries** (*Fig.* 2). As outlined above, if a substance is secreted into the urine from the plasma as it passes along the peritubular capillaries, its clearance will be greater than GFR, and if it is reabsorbed from the urine into the peritubular capillaries, its clearance will be lower than GFR. This principle can be applied to calculating renal plasma flow. Low GFR might indicate blocked glomeruli or, since UP (Equation 2) is largely dependent upon $P_{cap}$, lowered kidney perfusion. Thus it is worthwhile considering whether low GFR is symptomatic of glomerular damage or of low kidney perfusion, and estimating RPF is an important diagnostic tool in some circumstances.

Many molecules absorbed from the lumen of the nephron into the blood, or secreted into the lumen from the blood are transported across cell membranes via specific transport proteins (see Topic M3). However, the rate of transport of substances into or out of the lumen of the nephron is limited by the **transport maximum ($T_m$)** for the molecule in question (usually quoted in mmol $min^{-1}$). $T_m$ is determined partly by the kinetics of the transporter and the number of transporter molecules available. Two molecules that, in addition to being freely filtered at the glomerulus, are subject to transport across the epithelium of the nephron are glucose (which is reabsorbed) and **para-aminohippuric acid (PAH)** (which is secreted). If the delivery of glucose to the renal tubules rises above the $T_m$ (2 mmol $min^{-1}$), then some glucose will find its way into the urine. Under normal circumstances, plasma glucose levels are low enough that the amount of glucose filtered into renal tubules is below the $T_m$, and all of the glucose is reabsorbed into the blood. This means that $C_{glucose}$ is normally 0 ml $min^{-1}$, and any increase in $C_{glucose}$ is either indicative of a high plasma glucose concentration (>15 mmol $l^{-1}$, e.g. in **diabetes mellitus**), or poor reabsorption of glucose in the renal tubules (e.g. in **acute renal failure**) or both. The plasma concentration of PAH is normally zero but if it is infused until plasma concentrations equilibrate at levels lower than 800 μmol $l^{-1}$, all of the PAH that has not been filtered at the glomerulus is secreted into the renal tubule; at these plasma concentrations the rate of delivery of PAH to the renal tubule via the peritubular capillaries is lower than the $T_m$ for PAH. Thus, because all of the blood entering the glomerular capillaries under these circumstances is cleared of PAH by a combination of filtration and secretion, **PAH clearance ($C_{PAH}$)** is an accurate measure of RPF:

$$\text{RPF} = C_{PAH} = (U_{PAH}\, V)/P_{PAH} \qquad (6)$$

$$\text{RPF} = (67\,000\ \mu\text{mol l}^{-1}\ 2\ \text{l day}^{-1})/150\ \mu\text{mol l}^{-1}$$

$$\text{RPF} = 893\ \text{l day}^{-1} = 0.62\ \text{l min}^{-1} = 620\ \text{ml min}^{-1}$$

Using the estimates of GFR (123 ml $min^{-1}$) and RPF (620 ml $min^{-1}$) that we have shown above, using Equations 5 and 6, we can also calculate a **filtration fraction** (GFR/RPF) of 0.20, which indicates that 20% of RPF is filtered at the glomerulus. The filtration fraction can be relevant to the function of other parts of the nephron because if it falls, the colloid osmotic pressure of the blood flowing through the peritubular capillaries also falls (because less plasma leaves at the glomerulus), and this can lead to problems with reabsorption of a number of other substances in the renal tubule (see Topic M3).

# M3 REABSORPTION OF ELECTROLYTES AND GLUCOSE

## Key Notes

**Epithelial transport mechanisms**

Renal tubule cells carry out many absorptive and secretory functions that depend on movement of ions between the two sides of the renal tubule epithelial cells. These cells have luminal membranes in contact with the tubular fluid, and basolateral membranes in contact with interstitial fluid. Substances secreted into the renal tubule move from interstitial fluid to the luminal side of the cell, and substances absorbed move from the luminal side of the cell to the interstitial fluid. Much of the movement of ions and, in some cases other substances, is dependent upon active pumping of ions against their concentration gradients at one side of the renal tubule cell. Mechanisms are different in the different sections of the nephron.

**$Na^+$ and $K^+$ balance**

Large amounts of $Na^+$ and $K^+$ ions are filtered at the glomeruli each day, and the renal tubule must reabsorb around 96% and 85% of the filtered amount of these ions, respectively. Most $Na^+$ and $K^+$ ions are reabsorbed in the renal tubule by active transport processes, which are affected by hormonal signals. Problems with $Na^+$ and $K^+$ balance can have serious consequences.

**$Ca^{2+}$ balance**

$Ca^{2+}$ ions are essential for the function of nerve and muscle, and in the formation of bone. More than 99% of the filtered $Ca^{2+}$ ions must be reabsorbed in the renal tubule. $Ca^{2+}$ balance is maintained by a number of hormones, including 1,24-dihydroxy vitamin D, parathyroid hormone, and calcitonin.

**Carrier-mediated transport of glucose**

Glucose is freely filtered at the glomerulus, and specific transport mechanisms exist for its reabsorption in the nephron. If plasma glucose is low, then all of the filtered glucose will be reabsorbed by sodium-dependent glucose transport molecules in the proximal parts of the renal tubule, and glucose clearance will be 0 ml $min^{-1}$. If plasma glucose concentrations are elevated, e.g. in diabetes mellitus, these transport processes can be overwhelmed and result in some glucose entering the urine. In this case, glucose clearance will be greater than 0 ml $min^{-1}$. Glucose transport in the nephron is a form of secondary active transport, using the energy stored in $Na^+$ ion gradients to reabsorb glucose. If energy-producing processes in the kidneys are impaired, e.g. by ischemia, then glucose clearance may also rise.

**Related topics**

Bone and muscle (C5)
The cardiac cycle, blood pressure, and its maintenance (D6)
Thyroid hormones (K6)
Parathyroid hormone (K7)
Glomerular filtration and renal plasma flow (M2)
Acid–base balance (M4)
Hormonal control of fluid balance (M5)

## Epithelial transport mechanisms

As in all cells in the body, the function of a renal tubule cell is completely dependent upon the cell's ability to regulate the concentration of various ions on either side of the cell membrane. For example, neurons and muscle cells maintain high concentrations of $Na^+$ ions in the extracellular fluid, and high concentrations of $K^+$ ions in the intracellular fluid (see Topic B4). However, because **renal tubule** cells are epithelial, and form a surface that is in potential contact with the outside world, this arrangement of ions is made more complicated; one part of the cell (the luminal surface) is in contact with the tubular fluid, whereas other parts of the cell (the basolateral surfaces) are in contact with the extracellular fluid between the cell and the bloodstream (the **interstitial fluid**). This cellular arrangement allows the cell to preferentially send various **ion channels, ATPase pumps** and **exchanger proteins** to different cell surfaces in order to effect absorption of substances from or secretion of substances into the tubular fluid. Renal tubule cells do this in a manner similar to cells performing other forms of **exocrine secretion** or absorption, such as those producing gastric acid (see Topic J3) or enterocytes that absorb glucose, lipids and amino acids in the small intestine (see Topic J6).

An example of the ability of renal tubule cells to reabsorb substances that have been filtered at the **glomerulus** is the reabsorption of NaCl in the thick part of the ascending limb of the **loop of Henle** (*Fig. 1*). The epithelial cells of the thick part of the ascending limb of the loop of Henle are important in reabsorption of around 20% of the $Na^+$ ions filtered at the glomerulus. These cells send more **$Na^+/K^+$ ATPase** pumps and **$K^+/Cl^-$ cotransporters** to the basolateral surface

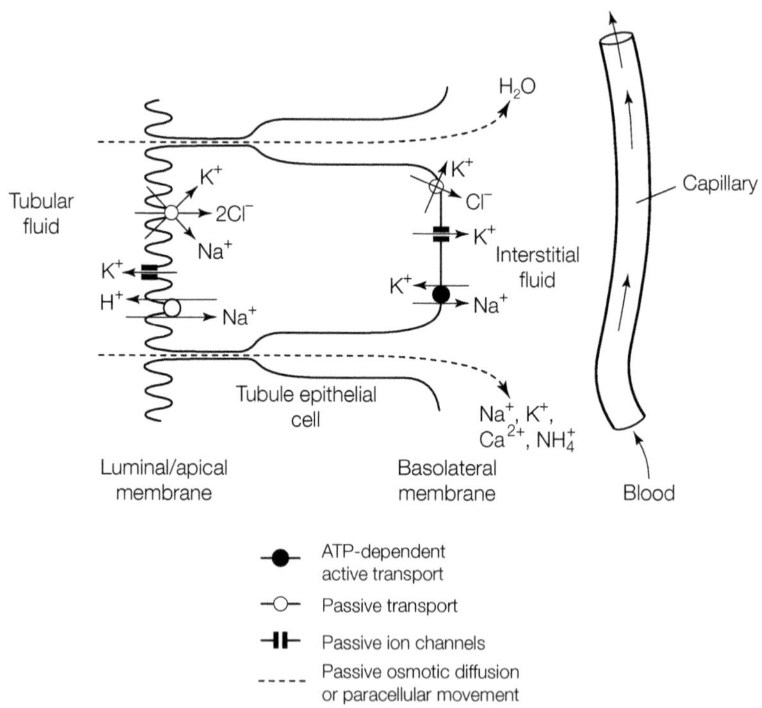


*Fig. 1. Cell model for reabsorption of solute in the thick ascending portion of the loop of Henle.*

(blood side) of the cell than to other parts, and send more **$Na^+/K^+/2Cl^-$ cotransporters** and **$Na^+/H^+$ exchanger** proteins to the luminal surface (urine side) of the cell. $K^+$ and $Cl^-$ ion channels are also placed on specific cell surfaces. Active pumping of $Na^+$ and $K^+$ ions against their concentration gradients (out of and into the cell, respectively) at the basolateral surfaces results in a high concentration of $K^+$ ions inside the cell, and a high concentration of $Na^+$ ions in the interstitial fluid. Some $K^+$ ions are able to leak back out of the cell through $K^+$ ion channels on the basolateral membrane, and others 'drag' $Cl^-$ ions with them through the $K^+/Cl^-$ cotransporters; this process maintains function of the $Na^+/K^+$ ATPase, which would stop working if $K^+$ ion concentrations fell to too low a level. The function of all tubular epithelial cells requires the formation of large amounts of **adenosine triphosphate (ATP)**, and so most such cells have a high number of mitochondria. The pumping of $Na^+$ ions out of the cell at the basolateral side also sets up a low concentration of $Na^+$ ions intracellularly, creating a concentration gradient for $Na^+$ ions to move into the cell. $Na^+$ ions enter the cell at the luminal surface, bringing with them a single $K^+$ ion and two $Cl^-$ ions through the $Na^+/K^+/2Cl^-$ cotransporters. Other $Na^+$ ions enter the cells in exchange for $H^+$ ions, using the $Na^+/H^+$ exchanger on the luminal membrane. Some $K^+$ ions also leak out through $K^+$ ion channels on the luminal surface, helping to maintain the function of the $Na^+/K^+/2Cl^-$ cotransporter.

These various ionic movements lead to a slightly more electropositive environment in the tubular fluid than in the interstitial fluid, resulting in some passive movement of small cations such as $Na^+$, $K^+$, $Ca^{2+}$ and $NH_4^+$ from the tubular fluid into the interstitial fluid, via tiny gaps between the epithelial cells. Water also moves from the luminal fluid to the interstitial fluid, following the net movement of solute. In fact, the major portion of fluid reabsorption from the tubular fluid goes via this mechanism in all parts of the nephron. The luminal $Na^+/K^+/2Cl^-$ cotransporter proteins are blocked by 'loop' diuretic drugs, such as **furosemide**, and if reabsorption of $Na^+$ ions falls, so does the net movement of water, resulting in production of greater volumes of urine (**diuresis**).

These epithelial cells utilize a wide range of membrane transport processes in a unique way, in order to effect reabsorption of $Na^+$ and other ions, and excretion of acid. The movement of substances between the tubular fluid and the interstitial fluid is controlled by similar, yet subtly different, mechanisms in different parts of the nephron, in order to bring about reabsorption or secretion of different substances at different parts of the tubule. See Topic M4 for an example of secretion, in that case of $H^+$ ions.

## $Na^+$ and $K^+$ balance

$Na^+$ and $K^+$ ions exist in the plasma at concentrations of approximately 140 mM and 4.2 mM, respectively, meaning that approximately 25 000 mmol of $Na^+$ ions, and 760 mmol of $K^+$ ions are filtered at the glomerulus each day. In order for our bodies to function effectively, almost all of these filtered ions have to be reabsorbed by the kidney following filtration; this is done by active transport processes in the renal tubular epithelial cells, as shown above. In general, almost all of the $Na^+$ and $K^+$ ions that we ingest via our diet (around 200 mmol and 100 mmol, respectively) is balanced by $Na^+$ and $K^+$ excretion in the urine. In the case of $Na^+$, this is normally done as part of our body's normal mechanism for maintaining **plasma osmolality**.

**Plasma** is a solution of various electrolytes with equal concentrations of negatively and positively charged ionized species. In plasma, $Na^+$ is the major cation, with $Cl^-$ and $HCO_3^-$ the major anions. Plasma osmolality has contributions from the ions and other dissolved substances such as glucose and urea; the clinical equation for calculating plasma osmolality is as follows:

$$\text{plasma osmolality} = (2 \times [Na^+]) + [\text{glucose}] + [\text{blood urea nitrogen}]$$

Under normal circumstances, this means that:

$$\text{Plasma osmolality} = (280 + 5.6 + 3.6)\ \text{mM} = 289\ \text{mOsmol kg}^{-1}$$

It is clear that plasma $Na^+$ concentrations play a large part in determining plasma osmolality. Body $Na^+$ balance and plasma osmolality are tightly controlled. Alterations in plasma $Na^+$ concentration have large effects on the movement of water into or out of the plasma, to maintain plasma osmolality. Increases in plasma osmolality (caused by increased $Na^+$ intake, $Na^+$ retention or a lack of water intake) also results in activation of osmoreceptors in the hypothalamus, and secretion of **antidiuretic hormone (ADH, vasopressin)** from the posterior part of the pituitary gland (see Topics D6, K4, and K5). ADH acts on the epithelial cells of the **collecting duct** in the kidney to promote water reabsorption, helping to reduce plasma osmolality. More details of this process are given in Topic M5.

$K^+$ balance is also important for body function; although plasma $K^+$ ion concentrations are low in comparison to $Na^+$ ions, maintaining extracellular $K^+$ within a very tight range is extremely important for the function of nerve, skeletal muscle, and the heart. The renal system is normally very good at balancing $K^+$ intake and excretion for this reason. If plasma $K^+$ rises (**hyperkalemia**), for example in **Addison's disease** (in which $K^+$ excretion by the kidneys is deficient due to low circulating levels of the hormone **aldosterone**), this can make repolarization of heart muscle more difficult, and result in cardiac arrhythmias. In people with end-stage renal failure who are undergoing hemodialysis, a change in diet can be just as dangerous (e.g. cheese and bananas are particularly rich in $K^+$).

If plasma $K^+$ falls (**hypokalemia**), for example as a side-effect of diuretic drug treatment, this can also lead to muscle weakness and **arrhythmias**. Ninety per cent of the $K^+$ ions in the body are inside cells, so the intracellular compartment (67% of body water) acts as a reservoir of $K^+$ ions for the rest of body fluid (extracellular fluid, lymph and plasma). However, in order to maintain electroneutrality of the extracellular fluid, entry of $K^+$ into cells is associated with release of $H^+$ ions into the extracellular fluid, and release of $K^+$ from cells is associated with a rise in extracellular fluid pH as $H^+$ ions enter cells. Additionally, $K^+$ ions are exchanged for $H^+$ ions in cells of the distal parts of the nephron via a **$H^+$/$K^+$ ATPase** that functions like that in parietal cells of the stomach (see Topic J3). Thus $K^+$ balance is intimately related to acid–base balance (see Topic M4). $K^+$ depletion can lead to **metabolic alkalosis** as the body attempts to shift $K^+$ from inside cells, whereas $K^+$ excess can result from **metabolic acidosis**, as the body attempts to shift $H^+$ ions into cells from the extracellular space, in exchange for $K^+$ ions.

## $Ca^{2+}$ balance

$Ca^{2+}$ ions, like all small inorganic ions, are freely filtered at the glomerulus. $Ca^{2+}$ ions are essential for the function of all cells in the body, but are especially important in the mineralization of bone (see Topic C5), neurotransmitter release (see Topic F3), and muscle cell contraction (see Topics I3 and I6). Only around 3 mmol $Ca^{2+}$ is absorbed from the diet each day, so it is important that the majority of the $Ca^{2+}$ ions that are filtered at the glomerulus (360 mmol) are reabsorbed in the renal tubule. In most parts of the renal tubule, $Ca^{2+}$ ions move by diffusion, either paracellularly or through $Ca^{2+}$ channels on renal tubule cells, via mechanisms that are similar to those for other ions (see above). However, in the late distal tubule, the epithelial cells utilize a luminal $Ca^{2+}$ channel and **$Na^+/Ca^{2+}$ exchangers**, and a **$Ca^{2+}$ ATPase** on the basolateral membrane. Within the cytoplasm, $Ca^{2+}$ is transported by a $Ca^{2+}$-binding protein known as **calbindin**. The action of calbindin is dependent on **1,24-dihydroxy vitamin D**, the formation of which is controlled by **parathyroid hormone (PTH**, see Topic K7). When serum free $Ca^{2+}$ levels fall below the normal physiological range, and PTH is secreted in increased amounts, some of the effect of PTH to increase serum free $Ca^{2+}$ levels is mediated via this action in the kidney. PTH also inhibits the reabsorption of phosphate ions in the proximal convoluted tubule, resulting in increased phosphate excretion in the urine, in an effort to limit the effect on serum phosphate concentrations of the bone resorption that is increased by PTH (see Topic C5).

## Carrier-mediated transport of glucose

Glucose is an uncharged monosaccharide sugar with a molecular weight of 180 Da, and a molecular radius of <0.4 nm. These three characteristics (charge, weight and size) mean that glucose is freely filtered at the glomerulus. Plasma glucose concentrations are normally around 6 mM, and so with a **glomerular filtration rate (GFR)** of 120 ml $min^{-1}$, the filtration rate for glucose is 0.72 mmol $min^{-1}$. With normal renal function, all of this glucose would normally be absorbed well before reaching the urine (usually by the end of the proximal tubule), because the **transport maximum ($T_m$)** for glucose is normally around 1.8 mmol $min^{-1}$. Thus, with normal renal function and a GFR of 120 ml $min^{-1}$, plasma glucose levels would have to rise above 15 mM before any glucose would find its way into the urine. This is the reason why people with **diabetes mellitus** (either due to lack of insulin secretion or **insulin resistance**, see Topic K9) usually have trace amounts of glucose in their urine.

Over the course of a day, our kidneys must reabsorb around 1 mol of filtered glucose; this is equivalent to an amazing 180 g of glucose. How are the kidneys able to do this, especially if the plasma glucose concentration in the peritubular capillaries is roughly the same as that of the glomerular filtrate? Glucose transport in the proximal part of the nephron is linked to the reabsorption of $Na^+$ ions (*Fig. 2*). The epithelial cells of the proximal tubule have large numbers of $Na^+/K^+$ ATPase pumps on their basolateral membranes, actively pumping out $Na^+$ ions and pumping in $K^+$ ions. The $K^+$ ions are able to leave the basolateral surface of the cell and keep the $Na^+/K^+$ ATPases supplied with $K^+$ via 'leak' $K^+$ channels. This creates a low $Na^+$ ion concentration in the proximal tubule epithelial cell. The proximal tubule cells also have large numbers of **glucose/$Na^+$ cotransporters** on their luminal membrane, and these transporters use the large concentration gradient for $Na^+$ ions to 'drag' glucose into the proximal tubule cell (the glomerular filtrate will have a $Na^+$ concentration like that of plasma, around 140 mM, see Topic M2). This movement of glucose into the proximal tubule cell is a form of **secondary active transport**, using the 'energy' stored in the $Na^+$ concentration gradient to move glucose against its concentration gradient. Glucose is then able to move from the intracellular space to the interstitial space between the proximal tubule cells and the blood, via another glucose transporter protein on the basolateral membrane. Reabsorption of amino acids and phosphate occurs by a similar process, via other specific $Na^+$-cotransporters and basolateral transporter proteins.

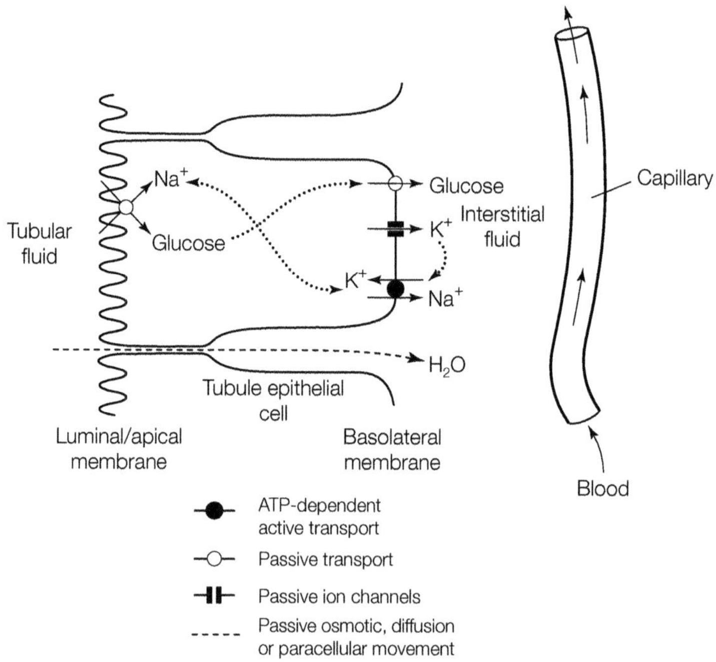


*Fig. 2. Cell model for glucose reabsorption in the proximal parts of the nephron.*

The $T_m$ for glucose is a measure of how efficiently glucose can be removed from the tubular fluid; if cells have more luminal surface glucose/$Na^+$ cotransporters, then $T_m$ will be high, whereas if the number of cotransporters is low or some cells are starved of oxygen or ATP, $T_m$ will be lower.

# M4 ACID–BASE BALANCE

## Key Notes

**Renal compensation for acid–base imbalance**

Acids are produced via cellular metabolism. These can be carbonic acid, produced from carbon dioxide, or non-volatile acids such as lactic acid. Acid–base balance in the body is tightly regulated, and the respiratory system can 'blow off' carbon dioxide to help with this. However, in some circumstances the kidneys must take over a compensatory role by altering the excretion of $H^+$ ions or $HCO_3^-$ ions.

**Acidification of the urine**

Marked acidification takes place between the formation of glomerular filtrate and the formation of urine. Many parts of the nephron contribute to this acidification of the tubular fluid by either excreting $H^+$ ions combined with $HPO_4^{2-}$ or $NH_3$, or by excreting fewer $HCO_3^-$ ions. There are a number of different cellular mechanisms for excretion of acid, such as $Na^+/H^+$ exchange or active pumping of $H^+$ ions into the luminal fluid.

**Modulation of $HCO_3^-$ excretion**

The majority of the $HCO_3^-$ ions that are filtered at the glomerulus are reabsorbed in the renal tubule; this can take the form of absorption of $HCO_3^-$ ions or production of new $HCO_3^-$ ions in renal tubule cells. Different cellular mechanisms exist for these two processes.

**Modulation of $H^+$ excretion**

$H^+$ ions can be excreted into the urine combined with $HPO_4^{2-}$ as titratable acid, or combined with $NH_3$ (ammonia) as $NH_4^+$ (ammonium ions). The excretion of a proton into the urine normally results in a $HCO_3^-$ ion being returned to the blood. When large amounts of acid are excreted, the majority of this takes the form of $NH_4^+$.

**Related topics**

Metabolic processes (B3)
Chemoreceptors and respiratory acid–base balance (E6)

## Renal compensation for acid–base imbalance

For a description of **acids**, **bases**, and the major **buffers** in the body, see Topic E6. Cellular metabolic processes result in the production of **carbon dioxide ($CO_2$)** and, in some cases, non-volatile acids such as **lactic acid** (see Topics B3 and E6). $CO_2$ can be 'blown off' in the lungs by increasing the rate of ventilation (see Topic E6), but in order to maintain plasma pH between 7.45 and 7.35 ($H^+$ concentration of 35–45 nM), other mechanisms must exist to rid the body of these other acids. The **acid–base balance** of blood plasma is monitored by **chemoreceptors** in the carotid and aortic bodies, which sense pH and $P_{CO_2}$ (they also monitor $P_{O_2}$, but that is largely irrelevant in discussion of acid–base balance). Central chemoreceptors monitor the $P_{CO_2}$ and $P_{O_2}$ of the blood. When $P_{CO_2}$ rises or pH falls, respiratory areas in the brainstem produce an increased drive to ventilation, helping to 'blow off' $CO_2$ and bring pH and $P_{CO_2}$ back within the normal range (**negative feedback**). However, for a number of acid–base imbalances, $CO_2$ cannot be formed from the non-volatile acids, or

**respiratory compensation** mechanisms are insufficiently powerful. In such cases, alteration of the rate of excretion of $H^+$ and/or $HCO_3^-$ by the kidneys comes into play (**renal compensation**).

## Acidification of the urine

The **glomerular filtrate** produced in the renal cortex is more or less iso-osmotic with the plasma, and has the same pH of 7.4. Urine produced and stored in the urinary bladder has an average pH of 6.0. Under normal conditions, the kidneys excrete the same amount of $H^+$ ions as are produced by cellular respiration. This is normally around 70 mmol day$^{-1}$. If the kidneys produce 1.5 l urine day$^{-1}$ and the urinary pH is 6.0, then only 1.5 μmol are free in the urine (since the pH scale indicates free $H^+$ ions). The majority of $H^+$ ions excreted by the body in the urine (70 mEq) are combined with anions in the form of **titratable acid ($H_2PO_4^-$)** and **ammonium ($NH_4^+$)** ions.

The majority of $H^+$ ions that are secreted into the urine enter the tubular fluid in the proximal parts of the renal tubule (up to the end of the loop of Henle). In the **proximal convoluted tubule (PCT)**, $H^+$ ions are added to the tubular fluid by luminal **$Na^+/H^+$ exchanger** and **$H^+$-ATPases**. By the time the tubular fluid has reached the end of the PCT, pH has fallen to around 6.7 (*Fig. 1*). As the fluid flows into the descending limb of the loop of Henle, water reabsorption increases (see Topic M6), until at the bottom of the loop, $HCO_3^-$ concentrations are greater than at the top; this results in an increase in the luminal pH to around 7.4. As fluid flows up the ascending limb, it is acidified by the action of $Na^+/H^+$ exchangers on the luminal surface of the epithelial cells of the tubule, so that by the beginning of the **distal convoluted tubule (DCT)**, the pH of the

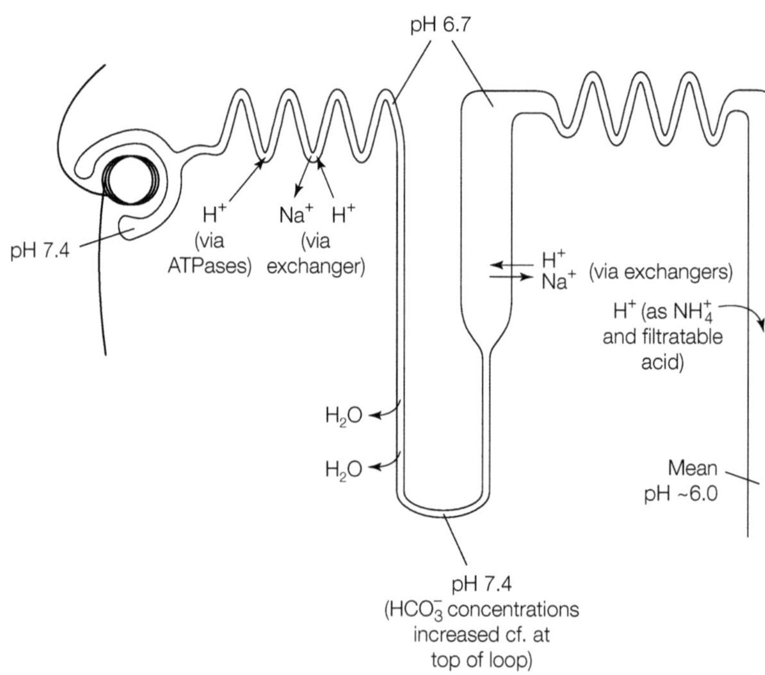


*Fig. 1. Schematic diagram of the nephron, showing sites of acidification and alkalinization of the tubular fluid.*

tubular fluid has fallen again to around 6.7. Additionally, there are a number of mechanisms for reabsorption of $NH_4^+$ ions in the ascending limb of the loop of Henle; dissociation of $NH_4^+$ provides $H^+$ ions to serve as a substrate for the $Na^+/H^+$ exchangers, and the remaining $NH_3$ is secreted into the collecting duct from the interstitium.

Further acidification of the urine takes place in the distal parts of the nephron. Two types of 'intercalated' cells of the collecting duct exist; **α-intercalated cells**, and **β-intercalated cells**. The α-intercalated cells secrete $H^+$, and the β-intercalated cells secrete $HCO_3^-$. In the collecting duct, $H^+$ ions secreted by α-intercalated cells can combine with the $NH_3$ that diffuses into the ductal fluid from the interstitium, leading to the formation of $NH_4^+$, in addition to titratable acid.

## Modulation of $HCO_3^-$ excretion

Virtually all of the $HCO_3^-$ ions filtered at the glomerulus (>4000 mmol $day^{-1}$) must be reabsorbed by tubular epithelial cells if blood pH is to maintain much of its buffering capacity, because $HCO_3^-$ is the major buffer in the body (see Topic E6). Some processes in the nephron allow filtered $HCO_3^-$ ions to be reclaimed, but the formation of *new* $HCO_3^-$ ions is a central part of the process for producing $H^+$ ions and acidification of the urine when **acidemia** (blood pH <7.35) cannot be corrected by respiratory mechanisms. The model for the reclamation of $HCO_3^-$ ions from the tubular fluid in the proximal tubule is shown in *Fig. 2*.

Within the cells of the proximal tubule, the enzyme **carbonic anhydrase** converts water and carbon dioxide to $H^+$ ions and $HCO_3^-$ ions. Let us leave the $HCO_3^-$ ions for the moment; the $H^+$ ions are exchanged for $Na^+$ ions that have been filtered at the glomerulus, via a $Na^+/H^+$ exchanger on the luminal surface of the epithelial cell. The $H^+$ ions (now in the lumen of the tubule) can then combine with filtered $HCO_3^-$ ions to form water and carbon dioxide (this reaction is catalyzed by carbonic anhydrase associated with the **brush border** of the

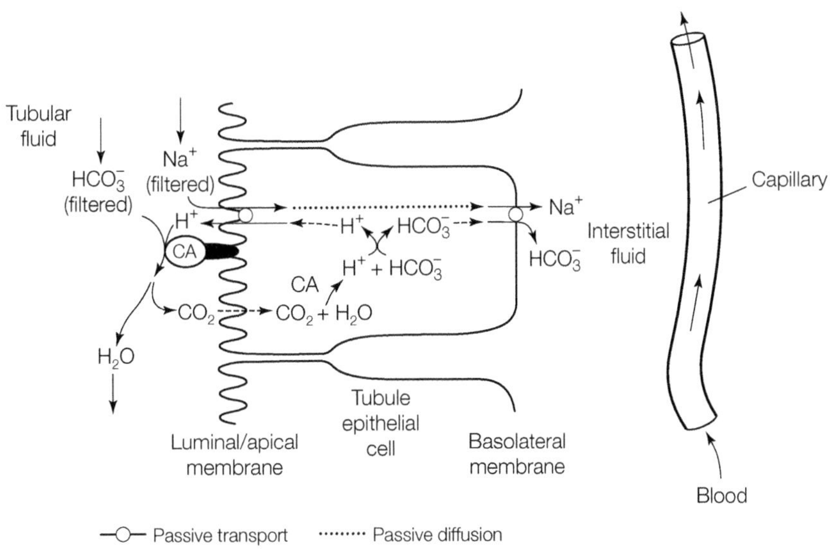


*Fig. 2. Cell model for reclamation of $HCO_3^-$ ions in the proximal parts of the nephron. See text for details. CA, carbonic anhydrase.*

epithelial cell). The water remains in the tubular fluid, but the carbon dioxide can diffuse into the epithelial cell, where it becomes one of the substrates (along with water) for formation of $H^+$ ions and $HCO_3^-$ ions via cellular carbonic anhydrase (see above). The $HCO_3^-$ ions formed leave the cell via a **$Na^+/HCO_3^-$ cotransporter** on the basolateral surface of the cell, to add a *reclaimed* $HCO_3^-$ ion to the interstitial fluid. It is important to note that no new $HCO_3^-$ ions are formed by this process in the proximal tubule; filtered $HCO_3^-$ ions are converted to carbon dioxide, which is then used by the cell to form intracellular $HCO_3^-$ ions that can only leave via the basolateral membrane.

## Modulation of $H^+$ excretion

An example of the ability of renal tubule cells to secrete substances into the tubular fluid, the secretion of $H^+$ ions as titratable acid ($H_2PO_4^-$), is shown in *Fig. 3*. Acid secretion is a property of many of the epithelial cells forming the walls of the nephron, but is most apparent in the proximal tubule, the ascending limb of the loop of Henle and the collecting duct. Epithelial cells lining the tubule use a number of different methods for secreting acid into the lumen, of which *Fig. 3* shows only one.

$Na_2HPO_4$ filtered at the glomerulus flows down the nephron. Within the epithelial cells, the enzyme carbonic anhydrase converts water and carbon dioxide to $H^+$ ions and $HCO_3^-$ ions. The $H^+$ ions are exchanged for one of the two $Na^+$ ions associated with the $HPO_4^{2-}$ ions flowing in the tubular fluid, via a $Na^+/H^+$ exchanger on the luminal surface of the epithelial cell. The $H^+$ ion is then excreted in the urine associated with the $HPO_4^{2-}$ ions as $NaH_2PO_4$ (a form of titratable acid). The $HCO_3^-$ ions formed by the action of carbonic anhydrase in the cell leave the cell on the basolateral side, via a $Na^+/HCO_3^-$ cotransporter, to add a new $HCO_3^-$ ion to the interstitial fluid. Thus, on average, for each $H^+$ ion that is secreted into the urine by this process, a new $HCO_3^-$ ion is added to the interstitial fluid. It is worth mentioning that, for the kidneys to excrete large amounts of acid, induction of ammonia production in renal tubule cells must occur, and $NH_4^+$ ions must be formed, because the supply of $HPO_4^{2-}$ ions is limited.

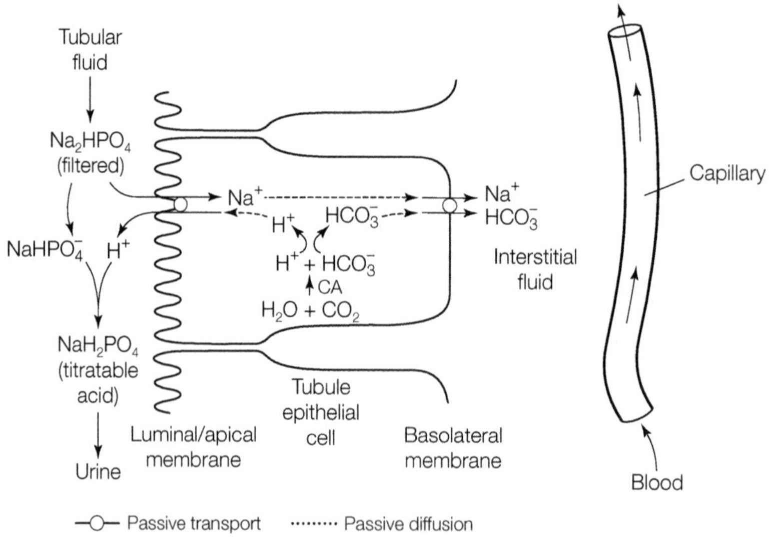


*Fig. 3. Cell model for secretion of $H^+$ ions as titratable acid ($NaH_2PO_4$). See text for details. CA, carbonic anhydrase.*

# M5 Hormonal control of fluid balance

## Key Notes

**Overview of fluid reabsorption**

Approximately 99% of the plasma filtered at the glomerulus is reabsorbed in the renal tubule. If this did not happen, we would rapidly become dehydrated and short of electrolytes. Plasma water and plasma electrolytes are in equilibrium with interstitial fluid water and electrolytes, upon which body cells are reliant. One of the ways in which fluid balance of the body can be altered to meet demand is by regulation of renal tubular reabsorption of electrolytes and water. In general, the primary mechanisms for solute and water reabsorption draw solute out of the tubular fluid, and water moves paracellularly or via water channels in cell membranes, drawn by osmotic forces.

**The role of the loop of Henle**

The loop of Henle is important in setting up an osmolar gradient between the renal cortex and the renal medulla. This is done by active pumping of electrolytes from the tubular fluid in the ascending limb, without concomitant movement of water in that section because the ascending limb of the loop of Henle is impermeable to water. The increasing osmolality of interstitial fluid as one passes from the renal cortex to the renal medulla has increasing influence on water movement from the tubular fluid in the collecting duct. Variations in the permeability of the collecting duct to water also play an important role in altering the osmolality of the urine, between 160 and 1200 mOsmol $kg^{-1}$.

**The role of urea**

Urea is recycled in the kidney nephron so that around 50% of the urea in the kidney participates in setting up the osmolar gradient that exists between the renal cortex and the renal medulla. In people who have a low-protein diet (the major metabolic source of urea), the ability to maximally concentrate the urine is adversely affected.

**Fluid reabsorption in the distal convoluted tubule**

The distal convoluted tubule (DCT) is a site of reabsorption of solute (principally NaCl) and water. The mechanisms here are subject to alterations by a number of factors, one of which is the hormone aldosterone. Slight variations in solute and water reabsorption in the DCT can account for large changes in the amount of fluid reabsorbed or excreted in the kidneys.

**Hormonal influences on fluid balance**

A number of different hormone systems exert major influence on solute and water reabsorption in the renal tubule. These include antidiuretic hormone (ADH, vasopressin), aldosterone and atrial natriuretic peptide (ANP). ADH regulates the production of water channels (aquaporin-2) in the renal tubule epithelial cells, increasing fluid reabsorption, and thereby acts to maintain blood volume. Aldosterone, the final hormonal product of the renin–angiotensin–aldosterone system, increases the reabsorption of $Na^+$ and water, and the excretion of $K^+$ and $H^+$ by the principal cells of the collecting duct. ANP reduces the reabsorption of $Na^+$ and water, thereby lowering circulating blood volume.

**Diuretic drugs**

A number of therapeutic agents that are used to lower blood pressure and reduce edema in peripheral tissues act as diuretic agents. These include 'loop' diuretics such as furosemide, $K^+$-sparing diuretics such as spironolactone (which antagonizes the action of aldosterone), carbonic anhydrase inhibitors such as acetazolamide, and osmotic diuretics such as isosorbide.

**Related topics**

The cardiac cycle, blood pressure, and its maintenance (D6)
Hypothalamic hormones (K4)
Pituitary hormones (K5)
Adrenal hormones (K8)
Reabsorption of electrolytes and glucose (M3)

## Overview of fluid reabsorption

In an average 70 kg male, 60% of body weight is taken up by water, amounting to 42 kg. This **body water** is distributed in three major compartments: intracellular space, interstitial and lymphatic fluid, and blood plasma. **Plasma water** (3.5 l) makes up only around 5% of body weight; bear in mind that the fluid inside blood cells is not included in this measure. Additionally, a very small amount of fluid is found in synovial joints and in the eyes, but this is rarely, if ever, important in fluid balance.

The **interstitial fluid**, which bathes the majority of the cells of the body, can be considered a filtered form of plasma. As outlined in Topic D5, **Starling's forces** determine the amount of exchange of fluid and electrolytes that takes place between the plasma and interstitial fluid, and the electrolyte composition of the interstitial fluid normally mirrors that of the plasma. Free exchange of proteins between the plasma and the interstitial fluid does not usually occur, so the interstitial fluid has a very low protein content compared to plasma. For these reasons, plasma electrolyte composition is normally considered to be a good marker of the composition of interstitial fluid, and clinical analysis of blood samples relies on this fact. Plasma $Na^+$ ion concentrations are held to be indicative of the concentration of $Na^+$ ions in the larger, but inaccessible, interstitial fluid compartment. This free exchange of fluid and electrolytes between the plasma and interstitial fluid is also important in reabsorption of substances into the blood from the **renal tubule**; in fact substances are reabsorbed into the interstitial space on the blood side of the epithelial cells rather than directly into the blood. Movement of fluid and electrolytes from the interstitial fluid into the blood is governed by the same Starling forces. Plasma (and therefore blood) volume is a major determinant of **blood pressure** (see Topic D5). If blood volume rises, so does **venous return** to the heart, and this must be matched by greater **cardiac output**. Increased cardiac output leads to an increase in **mean arterial blood pressure** if **peripheral resistance** remains the same.

Reabsorption of fluid in the renal tubule is obviously very important in maintaining water balance. If only 98% of filtered fluid were reabsorbed, compared with the normal 99%, this would mean that we would need to ingest 1.8 l more fluid than normal every day. The converse is also true; by drinking 0.9 l less in a day, we would only have to reabsorb 98% of the filtered fluid, rather than the normal 99%. One of the major ways in which fluid balance is maintained is by monitoring the osmolality of the plasma; this is done by specialized **osmoreceptor** cells in the anterior parts of the **hypothalamus**, which signal to other hypothalamic neurons to promote drinking and also to release or retain various hormones (particularly **antidiuretic hormone, ADH**). Another method for maintaining fluid balance is by monitoring blood pressure; activation of **baroreceptors** in the heart and the arteries also influences the secretion of hormones that act on the kidney. Understanding how these various hormones have their effect on plasma volume is dependent upon understanding the normal mechanisms by which the kidneys reabsorb fluid.

### Role of the loop of Henle

Around 70% of the fluid and electrolytes filtered at the glomerulus is reabsorbed in the **proximal convoluted tubule (PCT)**, by mechanisms outlined in Topic M3. **Urinary osmolality** can be as high as 1200 mOsmol $kg^{-1}$ if very little fluid is available through lack of intake or excessive losses in the sweat. The kidneys are able to produce such concentrated urine because of the arrangement of the **loop of Henle** and **vasa recta** in the **renal medulla**, and the differing permeability of various parts of the loop of Henle to water.

NaCl reabsorption in the ascending limb of the loop of Henle follows the model shown in *Fig. 1* of Topic M3. In the ascending limb, water is unable to follow solute into the interstitial space because this part of the loop is impermeable to water. Imagine beginning from a starting point (*Fig. 1a*) where fluid entering the loop of Henle is iso-osmotic with plasma (since water reabsorption in the PCT follows solute), and imagine fluid being added in stages to the descending limb. As NaCl is reabsorbed from the ascending limb into the interstitium (*Fig. 1b*), and because water cannot follow the solute, fluid in the ascending limb becomes hypo-osmolar. The intersititium would then become hyperosmolar and the fluid in the descending limb would then also become hyperosmolar over time as water left the descending limb to equilibrate with the interstitial fluid. On the next introduction of fluid and solute to the descending limb (*Fig. 1c*), more solute would be pumped out of the tubule in the ascending limb, water would not follow, and the tubular fluid in the ascending limb would become more hypo-osmolar still (*Fig. 1d*). The fluid in the descending limb and the interstitium would then become more hyperosmolar to balance the solute concentration in the interstitium.

If this chain of events were continued many times over, the situation could arise where interstitial fluid osmolality would become progressively greater as one moved deeper into the medulla (*Fig. 1e–f*), although the degree of hyperosmolality of the interstitial fluid would depend on how much NaCl were being actively reabsorbed from the ascending limb.

Variations in solute reabsorption in the loop of Henle, and variations in the permeability of the collecting duct epithelium to water (see below) mean that we can make hyperosmolar or hypo-osmolar urine to suit our requirements for fluid and electrolyte balance. The **$Na^+/K^+/2Cl^-$ cotransporter** on the luminal side of the cells in the thick portion of the ascending limb of the loop of Henle provides the major source of the solute that is pumped out into the interstitium (see Topic M3). Hence, factors that change the expression of the genes encoding this cotransporter or the activity of the transporter could have a major effect on this mechanism in the loop of Henle.

The hairpin loop organization of the vasa recta ensures that these blood vessels help to maintain this osmolar gradient, rather than flushing solute away, as would happen if the blood vessels simply flowed past one particular region

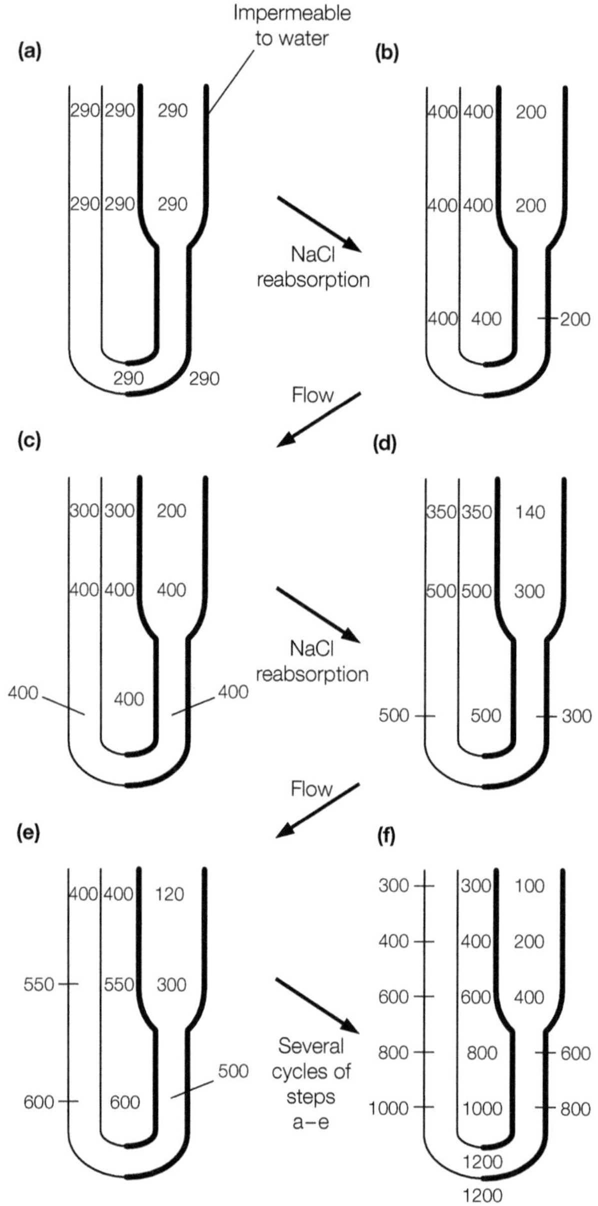


*Fig. 1. Schematic diagram showing the method by which hyperosmolar conditions are set up in the renal medullary interstitium. Figures shown are osmolality of tubular fluid or interstitial fluid, in mOsmol kg$^{-1}$.*

of the loop. Thus, with the hairpin arrangement, solute picked up by the inflowing plasma in the descending vasa recta leaves the ascending vasa recta plasma as this flows through regions with a lower solute concentration.

The purpose of setting up this osmotic gradient (from 100 to 1200 mOsmol kg$^{-1}$) as one moves deeper into the medulla from the cortex is to make it possible to make more concentrated urine as the tubular fluid passes down the **collecting**

**duct** towards the **ureter**. Movement of water from the tubular fluid in the collecting duct takes place via specific water pores formed by the protein **aquaporin-2**. If the cells of the collecting duct have a large number of aquaporin-2 molecules on their cell membrane, due to the action of circulating ADH, then it is easier for water to leave the tubular fluid under the osmotic drive of the hyperosmolar medullary interstitial fluid. The converse is true if fewer aquaporin-2 molecules are present on the luminal side of these cells. Hence, by altering the number of aquaporin-2 molecules available to the cells of the collecting duct, and by actively inserting more or fewer aquaporin molecules in the cell membrane, one could alter the osmolality of the urine being produced, as a direct consequence of varying circulating ADH levels in response to plasma osmolality and/or volume (see above and below).

**The role of urea**

**Urea** is the major nitrogenous compound in human urine and is formed from metabolism of amino acids and purines. The pathway for urea production is shown in Fig. 2. **Uric acid** can also be formed from purine metabolism, but uric acid, **ammonia** (as $NH_3$ and $NH_4^+$) and **creatinine** normally make up <20% of the nitrogen-containing compounds in the urine. Significant amounts of uric acid are only found in the urine when there are problems with metabolism. Under those circumstances, uric acid crystals may form in the body, and this can lead to the painful condition known as **gout**.

Urea is also very important in the urinary concentrating mechanisms of the kidney. This was first hinted at when it was noticed that people who had diets

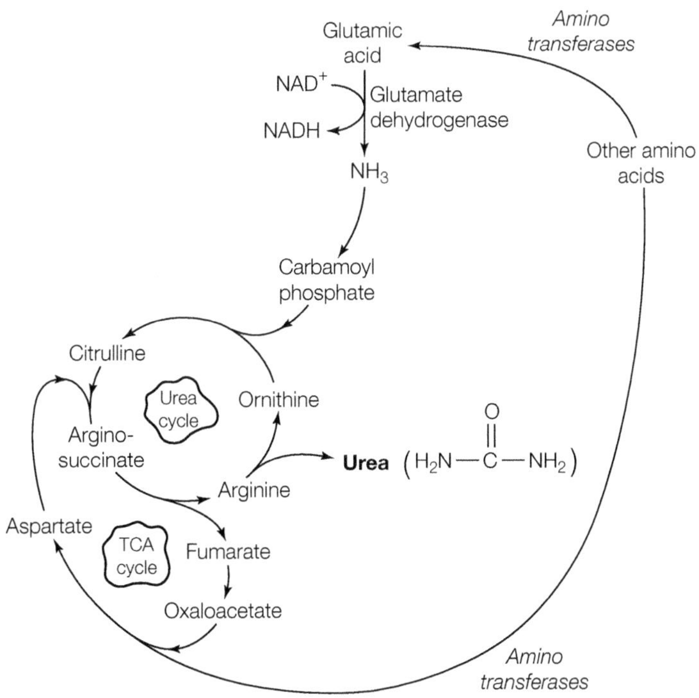


*Fig. 2. Biochemical pathways for the formation of urea. $NH_3$, ammonia; TCA, tricarboxylic acid.*

low in protein, leading to decreased urea production, could not produce maximally concentrated urine. Urea is recycled from the collecting duct to the loop of Henle by a mechanism demonstrated in *Fig. 3*.

Urea is freely filtered at the glomerulus, but because the PCT is permeable to urea, by the time the filtrate has reached the top of the loop of Henle, around 50% of this urea has been reabsorbed. The cells of the loop of Henle secrete urea so that by the time the tubular fluid has reached the **distal convoluted tubule (DCT)**, the amount of urea present in the tubular fluid is similar to that in the glomerular filtrate. The thick portion of the loop of Henle, and the rest of the renal tubule up to and including the outer medullary parts of the collecting duct are impermeable to urea, so little urea is present in the outer medulla. However, the inner medullary parts of the collecting duct are highly permeable to urea via a **urea transporter** (activated by ADH), so urea leaves the tubular fluid in the inner medulla where it contributes to the osmotic potential of the medullary interstitial fluid, playing an important role in the kidney's major urinary concentrating mechanism. Once this mechanism reaches equilibrium, the urea being recycled between the collecting duct and the loop of Henle amounts to about 50% in excess of the filtered load, and of the filtered load, 50% is reabsorbed in the PCT, 30% finds its way into the vasa recta leaving the medulla, and 20% is excreted in the urine.

## Fluid reabsorption in the distal convoluted tubule

The DCT is a further site of fluid reabsorption in the nephron. In the epithelial cells of the DCT, basolateral **$Na^+$/$K^+$ ATPases** pump $Na^+$ ions out of the cell and $K^+$ ions into the cell. $K^+$ ions are able to 'leak' back out of the basolateral side of the cell via $K^+$ channels, to maintain function of the $Na^+$/$K^+$ ATPase. The consequent low $Na^+$ concentration inside the cells of the DCT creates a concentration

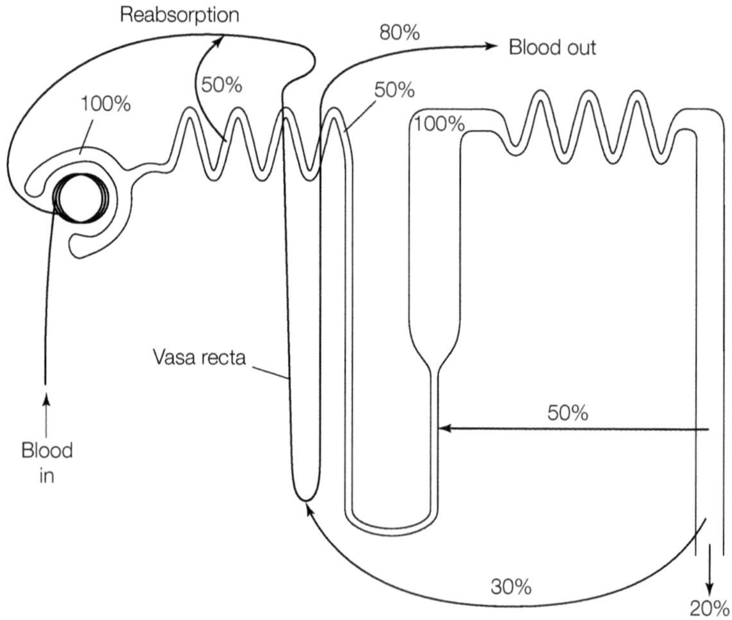


*Fig. 3. Urea recycling in the nephron. Of the urea filtered, only around 20% finds its way into the urine. Around 50% is utilized in the osmolar gradient in the inner medulla, helping to concentrate the urine. Figures presented are % of filtered load.*

gradient for $Na^+$ ions to enter the DCT cells from the luminal fluid, a process facilitated by a **$Na^+/Cl^-$ cotransporter** on the luminal membrane. $Cl^-$ ions are able to leave the cell on the basolateral side via **$Cl^-$ channels**. This DCT transport of NaCl facilitates the movement of water from the luminal fluid into the interstitial space. The permeability of the DCT to water is low, and so the osmolality of the fluid in the DCT can remain low. Despite this, any factor that affects transport of NaCl in the DCT could have some limited effect on fluid reabsorption.

## Hormonal influences on fluid balance

Three major hormones are involved in the regulation of fluid balance in the kidney. These are **antidiuretic hormone (ADH)**, **aldosterone**, and **atrial natriuretic peptide (ANP)**. All three hormones act on various parts of the renal tubule to alter fluid absorption. ADH is also known as **vasopressin**, in view of its effect of increasing the tone of blood vessels.

ADH, as its name suggests, promotes absorption of fluid by the kidney. ADH is a hypothalamic peptide hormone released from neuronal processes that extend down from the **supraoptic nuclei** and the **paraventricular nuclei** via the **infundibulum** to the **posterior pituitary** (see Topics K4 and K5). Circulating levels of ADH are normally low, but are increased by raised **plasma osmolality** when osmoreceptor cells in the hypothalamus are activated. Above the normal plasma osmolality of around 280 mOsmol $kg^{-1}$, there is a direct linear relationship between plasma osmolality and ADH secretion. The major effect of ADH is on the epithelial cells of the kidney collecting duct; ADH acts on these cells to increase the trafficking of aquaporin-2 to the luminal membrane, and to increase the expression of the gene for aquaporin-2. As outlined above, increasing the permeability of the collecting duct to water in this way promotes the concentrating effect of the high medullary osmolality set up by the loop of Henle, resulting in increased water reabsorption and concentrated urine. Under normal circumstances, this increased water reabsorption should correct the initial increase in osmolality of the plasma that stimulated release of ADH. ADH also activates the urea transporter molecules in the cells of the inner medullary collecting duct (see above), thereby increasing the osmotic potential in the medulla. Circulating levels of ADH are extremely low in the disease **diabetes insipidus**, in which either the infundibulum or the posterior pituitary (see Topic K4) have been damaged. As one might expect, this leads to production of copious amounts of dilute urine (polyuria, 5–20 l $day^{-1}$), accompanied by excessive drinking (polydipsia) in order to offset this level of fluid loss.

Aldosterone is the major **mineralocorticoid** in the body, released from the **adrenal cortex** (see Topic K8). Its secretion is controlled in part by **angiotensin II**, one of the hormones of the **renin–angiotensin–aldosterone system** (*Fig. 4*). **Renin** is an enzyme released from cells in the wall of the glomerular afferent arteriole in response to two stimuli: reduced kidney perfusion pressure, and decreased **glomerular filtration rate** (sensed as reduced delivery of $Na^+$ ions to the DCT). Such changes could be brought about by a reduction in blood pressure caused by either a fall in peripheral resistance or a fall in circulatory volume. The interaction between the glomerulus and the DCT has led to this area of the nephron being named the **juxtaglomerular apparatus**. Renin is a small peptide enzyme that acts in the bloodstream to convert **angiotensinogen** (produced by the liver) to **angiotensin I**. Angiotensin I is inactive, but can be converted to an active form, angiotensin II, by the action of **angiotensin-converting enzyme (ACE)**. ACE is found in multiple sites throughout the body, but is present at high levels in the vascular endothelial cells of the lungs.

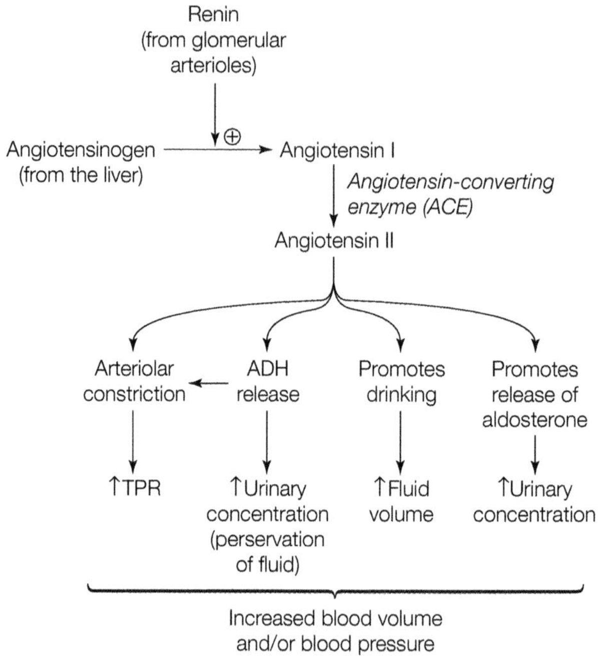


*Fig. 4. The renin–angiotensin–aldosterone system in fluid balance and long-term regulation of blood pressure. ADH, antidiuretic hormone; TPR, total peripheral resistance.*

Angiotensin II has four major effects, mediated by specific angiotensin II receptors: it constricts arterioles, promotes the release of ADH, stimulates release of aldosterone from the adrenal cortex, and promotes drinking through an action on the hypothalamus. All four of these effects are consistent with an attempt to raise blood pressure and/or blood volume. Aldosterone increases the activity of basolateral $Na^+/K^+$ ATPases in the principal cells of the collecting duct, and also increases the likelihood that luminal $Na^+$ and $K^+$ channels will be open in these cells. This results in increased reabsorption of $Na^+$ from, and increased secretion of $K^+$ into, the tubular fluid. The increase in the amount of solute transported into the interstitium, coupled with an increase in ADH secretion, can result in the production of more concentrated urine, and an increase in plasma volume.

Another hormone, atrial natriuretic peptide (ANP), is released from the atria of the heart in response to stretch. This hormone acts on the kidney to reduce $Na^+$ reabsorption, hence its name. As we have seen, as more $Na^+$ is excreted in the urine, water follows. Thus ANP can cause both **natriuresis** and **diuresis** and reduce or limit the expansion of blood volume. ANP also has negative effects on renin, aldosterone, and ADH secretion.

## Diuretic drugs

In view of the close relationship between fluid balance, blood volume, and blood pressure, a number of different drugs that have diuretic effects are used to treat **hypertension** and **edema**. These include 'loop' diuretics such as **furosemide**, thiazide diuretics such as **bendroflumethiazide**, aldosterone receptor antagonists such as **spironolactone**, carbonic anhydrase inhibitors such as **acetazolamide**, and osmotic diuretics such as **isosorbide** and **mannitol**.

Furosemide blocks the action of the $Na^+/K^+/2Cl^-$ cotransporters on the cells of the thick ascending limb of the loop of Henle that are so important in concentration of the urine, leading to increased fluid loss in the urine. Bendroflumethiazide blocks the $Na^+/Cl^-$ cotransporter on the epithelial cells of the DCT, and hence is a less potent diuretic than furosemide, since only a small amount of filtered fluid is reabsorbed in the DCT. One potential adverse effect of these diuretics is that they promote $K^+$ loss in the urine, which can lead to **hypokalemia** in some cases. This effect is due to increased presentation of $Na^+$ to the aldosterone-sensitive site at which exchange for $K^+$ occurs, and also to the increased tubular flow rate causing $K^+$ washout.

Spironolactone prevents the urinary concentrating effects of aldosterone by blocking the action of aldosterone on its (mineralocorticoid) receptors. Spironolactone and other mineralocorticoid receptor antagonists are known as $K^+$-sparing diuretics (see above) and, in some circumstances, are preferred to 'loop' diuretics like furosemide.

Acetazolamide blocks the action of **carbonic anhydrase**, the major role of which is to allow reabsorption of $HCO_3^-$ ions in the PCT (see Topic M4). This increases losses of $HCO_3^-$ (water follows), leading to **diuresis** and, in some cases, **metabolic acidosis**. They are rarely used clinically because, unlike the other drugs mentioned here, they are not effective single-agent diuretics.

Another class of substances that have no specific molecular action are osmotic diuretics like isosorbide and mannitol. These substances are freely filtered at the glomerulus, but are not markedly reabsorbed by the renal tubule; thus they help to keep water in the tubular fluid by osmotic drive, and can induce diuresis.

# FURTHER READING


**Section B**

Baynes, J.W. and Dominiczak M.H. (eds) (2005) *Medical Biochemistry,* 2nd Edn. Elsevier Mosby, Philadelphia, Chapters 8, 11–15, pp. 99–111, 143–208.

Pukkila, P.J. (2001) Molecular biology: the central dogma. In: *Encyclopedia of Life Sciences.* Wiley Interscience. www.els.net (accessed 19 July 2006).

Matthews, G.G. (2003) *Cellular Physiology of Nerve and Muscle,* 4th Edn. Blackwell, Oxford.

Murray, R.K, Granner, D.K., Mayes, P.A. and Rodwell, V.W. (2000) *Harper's Biochemistry,* 25th Edn. Appleton and Lange, Stamford, Chapters 12–24, pp. 123–249.

**Section C**

Baynes, J.W. and Dominiczak M.H. (eds) (2005) *Medical Biochemistry,* 2nd Edn. Elsevier Mosby, Philadelphia, Chapters 3, 24, pp. 25–34, 345–358.

Freeman, W.H. and Bracegirdle, B. (1976) *An Advanced Atlas of Histology.* Heinemann Educational, London.

Kierszenbaum, A.L. (2002) *Histology and Cell Biology. An introduction to pathology.* Mosby, St Louis.

Monkhouse, S. (2001) *Clinical Anatomy.* Churchill Livingstone, Edinburgh, Chapters 1–9, pp. 7–59.

Sambrook, P., Schrieber, F., Taylor, T. and Ellis, A. (2001) *The Musculoskeletal System.* Churchill Livingstone, Edinburgh.

**Section D**

Berne R.M. and Levy, M.N. (2001) *Cardiovascular Physiology,* 8th Edn. Mosby, St Louis.

Douglas, S.A., Ohlstein, E.H. and Johns, D.G. (2004) Cardiovascular pharmacology and drug discovery in the 21st century. *Trends Pharmacol Sci.* **25**, 225–233.

Hampton, J.R. (2003) *The ECG Made Easy,* 6th Edn. Churchill Livingstone, Edinburgh.

Joyner, M.J. and Halliwill, J.R. (2000) Neurogenic vasodilation in human skeletal muscle: possible role in contraction-induced hyperaemia. *Acta Physiol Scand.* **168**, 481–488.

Katz, A.M. (2006) *Physiology of the Heart,* 4th Edn. Lippincott Williams Wilkins, Philadelphia, Chapter 2, pp. 40–81.

Klabunde, R.E. (2005) *Cardiovascular Physiology Concepts.* Lippincott Williams and Wilkins, Philadelphia.

Levick, J.R. (2003) *An Introduction to Cardiovascular Physiology,* 4th Edn. Arnold, London, Chapter 6, pp. 70–95.

Wilmore JH, Costill DL. (2005) *Physiology of Sport and Exercise,* 3rd Edn. Human Kinetics, Champaign, Chapter 4, pp. 118–157.

**Section E**

Cotes, J.E., Chinn, D.J. and Miller, M.R. (2006) *Lung Function: physiology, measurement and application in medicine,* 6th Edn. Blackwell Publishing, Malden.

Davies, A.S. and Moores, C. (2003) *The Respiratory System.* Churchill Livingstone, Edinburgh, Chapter 8, pp. 109–130.

Levitzky, M.G. (2003) *Pulmonary Physiology,* 6th Edn. McGraw-Hill, New York, Chapter 8, pp. 163–187.

Smith, C., Marks, A.D. and Lieberman, M. (2005) *Marks' Basic Medical Biochemistry,* 2nd Edn. Lippincott Williams & Wilkins, Philadelphia, Section 4, pp. 337–476.

West, J.B. (2005) *Respiratory Physiology: the essentials*, 7th Edn. Lippincott Williams & Wilkins, Philadelphia.

**Section F**

Aidley, D.J. (1998) *The Physiology of Excitable Cells*, 4th Edn. Cambridge University Press, Cambridge.

Barker, R., Barasi, S. and Neal, M.J. (2003) *Neuroscience at a Glance*, 2nd Edn. Blackwell, Malden/Oxford.

Kandel, E.R., Schwartz, J.H. and Jessell, T.M.(2000) *Principles of Neural Science*, 4th Edn. McGraw-Hill Publishing Co, New York/London.

Rapport, R. (2005) *Nerve Endings, the Discovery of the Synapse: the quest to find how brain cells communicate*. WW Norton & Co Ltd, New York.

Siegel, G.J., Albers, R.W., Brady, S. and Price, D.L. (eds) (2006). *Basic Neurochemistry: molecular, cellular and medical aspects*, 7th Edn. Elsevier Academic Press, Burlington/London.

Squire, L.R., Bloom, F.E., McConnell, S.K., Roberts, J.L, Spitzer, N.C. and Zigmond, M.J. (eds) (2003) *Fundamental Neuroscience*, 2nd Edn. Academic Press, Amsterdam/London.

Webster, R. (ed.) (2001) *Neurotransmitters, Drugs and Brain Function*. John Wiley & Sons Ltd, Chichester.

**Section G**

Barker, R., Barasi, S. and Neal, M.J. (2003) *Neuroscience at a Glance*, 2nd Edn. Blackwell, Malden/Oxford.

Kandel, E.R., Schwartz, J.H. and Jessell, T.M.(2000) *Principles of Neural Science*, 4th Edn. McGraw-Hill Publishing Co, New York/London.

Maguire, E.A., Gadian, D.G., Johnsrude, I.S., Good, C.D., Ashburner, J., Frackowiak, R.S. and Frith, C.D. (2000) Navigation-related structural change in the hippocampi of taxi drivers. *Proc Natl Acad Sci U S A.* **97**, 4398–4403.

Schieber, M.H. (2001) Constraints on somatotopic organization in the primary motor cortex. *J Neurophysiol.* **86**, 125–143.

Snell, R.S. (2005) *Clinical Neuroanatomy*, 6th Edn. Lippincott Williams & Wilkins, Philadelphia.

Squire, L.R., Bloom, F.E., McConnell, S.K., Roberts, J.L., Spitzer, N.C. and Zigmond, M.J. (eds) (2003). *Fundamental Neuroscience*, 2nd Edn. Academic Press, Amsterdam/London.

Zeki, S. (1993) *A Vision of the Brain: the visible world and the cortex*. Blackwell Scientific, Oxford.

**Section H**

Clarac, F., Cattaert, D. and Le Ray, D. (2000) Central control components of a 'simple' stretch reflex. *Trends Neurosci.* **23**, 199–208.

Melzack, R. and Wall, P.D. (1996) *The Challenge of Pain*, 2nd Edn. Penguin, London.

**Section I**

Bagshaw, C.R. (1993) *Muscle Contraction*, 2nd Edn. Chapman and Hall, London.

Matthews GG. (2003) *Cellular Physiology of Nerve and Muscle*, 4th Edn. Blackwell, Oxford.

**Section J**

Arias, I.M. (ed.) (1994) *The Liver: biology and pathobiology*, 3rd Edn. Raven, New York.

Johnson, L.R. (2001) *Gastrointestinal Physiology*, 6th Edn. Mosby, St Louis, Chapter 11, pp. 119–141.

Smith, M.E. and Morton, D.G. (2001) *The Digestive System*, Churchill Livingstone, Edinburgh, Chapter 8, pp. 133–156.

**Section K**

Baynes, J.W. and Dominiczak, M.H. (eds) (2005) *Medical Biochemistry*, 2nd Edn. Elsevier Mosby, Philadelphia, Chapters 12, 20 and 37, pp. 157–173, 273–297, 523–540.

Laycock, J. and Wise, P. (1996) *Essential Endocrinology*, 3rd Edn. Oxford University Press, Oxford.

Padgett, D.A. and Glaser, R. (2003) How stress influences the immune response. *Trends Immunol.* **24**, 444–448.

Porterfield, S.P. (2001) *Endocrine Physiology*, 2nd Edn. Mosby, St Louis.

Wilson, J.D., Foster, D.W., Kronenberg, H.M. and Larsen, P.R. (eds) (1998) *Williams Textbook of Endocrinology*, 10th Edn. W.B. Saunders, Philadelphia.

**Section L**

Baynes, J.W. and Dominiczak, M.H. (eds) (2005) *Medical Biochemistry*, 2nd Edn. Elsevier Mosby, Philadelphia, pp. 531–536 and 539–540.

Holstege, G., Georgiadis, J.R., Paans, A.M.J., Meiners, L.C., van der Graaf, F.C.H.E. and Reinders, A.A.T.S. (2003) Brain activation during human male ejaculation. *J Neurosci.* **23**, 9185–9193.

Sadler, T.W. (2004) *Langman's Medical Embryology*, 9th Edn. Lippincott Williams & Wilkins, Philadelphia.

Symonds, E.M. and Symonds, I. (2004) *Essentials of Obstetrics and Gynaecology*, 4th Edn. Churchill Livingstone, Edinburgh, Chapters 3 and 4, pp. 21–45.

Yen, S.S.C., Jaffe, R.B. and Barbieri, R.L. (1991) *Reproductive Endocrinology*, 3rd Edn. WB Saunders, Philadelphia.

**Section M**

Abelow, B. (1998) *Understanding Acid–Base*. Williams & Wilkins, Baltimore.

Field, M., Pollock, C. and Harris D. (2001) *The Renal System*. Churchill Livingstone, Edinburgh.

O'Callaghan, C. and Brenner, B.M. (2000) *The Kidney at A Glance*. Blackwell Science, Oxford.

# INDEX

For Product Safety Concerns and Information please contact our EU representative GPSR@taylorandfrancis.com Taylor & Francis Verlag GmbH, Kaufingerstraße 24, 80331 München, Germany

**Batch number: 10726410**

Printed by Printforce, the Netherlands